D1807142

Applications of Deep Learning in Electromagnetics

The ACES Series on Computational and Numerical Modelling in Electrical Engineering

Andrew F. Peterson, PhD – Series Editor

The volumes in this series will encompass the development and application of numerical techniques to electrical and electronic systems, including the modelling of electromagnetic phenomena over all frequency ranges and closely related techniques for acoustic and optical analysis. The scope includes the use of computation for engineering design and optimization, as well as the application of commercial modelling tools to practical problems. The series will include titles for senior undergraduate and postgraduate education, research monographs for reference, and practitioner guides and handbooks.

Titles in the Series

K. Warnick, **"Numerical Methods for Engineering,"** 2010.

W. Yu, X. Yang and W. Li, **"VALU, AVX and GPU Acceleration Techniques for Parallel FDTD Methods,"** 2014.

A.Z. Elsherbeni, P. Nayeri and C.J. Reddy, **"Antenna Analysis and Design Using FEKO Electromagnetic Simulation Software,"** 2014.

A.Z. Elsherbeni and V. Demir, **"The Finite-Difference Time-Domain Method in Electromagnetics with MATLAB Simulations, 2nd Edition,"** 2015.

M. Bakr, A.Z. Elsherbeni and V. Demir, **"Adjoint Sensitivity Analysis of High Frequency Structures with MATLAB,"** 2017.

O. Ergul, **"New Trends in Computational Electromagnetics,"** 2019.

D. Werner, **"Nanoantennas and Plasmonics: Modelling, design and fabrication,"** 2020.

K. Kobayashi and P.D. Smith, **"Advances in Mathematical Methods for Electromagnetics,"** 2020.

V. Lancellotti, **"Advanced Theoretical and Numerical Electromagnetics, Volume 1: Static, stationary and time-varying fields,"** 2021.

V. Lancellotti, **"Advanced Theoretical and Numerical Electromagnetics, Volume 2: Field representations and the method of moments,"** 2021.

S. Roy, **"Uncertainty Quantification of Electromagnetic Devices, Circuits, and Systems,"** 2021.

Applications of Deep Learning in Electromagnetics

Teaching Maxwell's equations to machines

Edited by
Maokun Li and Marco Salucci

The Institution of Engineering and Technology

Published by SciTech Publishing, an imprint of The Institution of Engineering and Technology, London, United Kingdom

The Institution of Engineering and Technology is registered as a Charity in England & Wales (no. 211014) and Scotland (no. SC038698).

© The Institution of Engineering and Technology 2022

First published 2022

This publication is copyright under the Berne Convention and the Universal Copyright Convention. All rights reserved. Apart from any fair dealing for the purposes of research or private study, or criticism or review, as permitted under the Copyright, Designs and Patents Act 1988, this publication may be reproduced, stored or transmitted, in any form or by any means, only with the prior permission in writing of the publishers, or in the case of reprographic reproduction in accordance with the terms of licences issued by the Copyright Licensing Agency. Enquiries concerning reproduction outside those terms should be sent to the publisher at the undermentioned address:

The Institution of Engineering and Technology
Futures Place
Kings Way, Stevenage
Herts, SG1 2UA, United Kingdom

www.theiet.org

While the author and publisher believe that the information and guidance given in this work are correct, all parties must rely upon their own skill and judgement when making use of them. Neither the author nor publisher assumes any liability to anyone for any loss or damage caused by any error or omission in the work, whether such an error or omission is the result of negligence or any other cause. Any and all such liability is disclaimed.

The moral rights of the author to be identified as author of this work have been asserted by him in accordance with the Copyright, Designs and Patents Act 1988.

British Library Cataloguing in Publication Data
A catalogue record for this product is available from the British Library

ISBN 978-1-83953-589-5 (hardback)
ISBN 978-1-83953-590-1 (PDF)

Typeset in India by MPS Ltd
Printed in the UK by CPI Group (UK) Ltd, Croydon

Cover Image: ThinkNeo / DigitalVision Vectors via Getty Images

Contents

About the editors

Maokun Li (senior member, IEEE) is an associate professor in the Department of Electronic Engineering at Tsinghua University, Beijing, China. He received his BS degree in electronic engineering from Tsinghua University, Beijing, China, in 2002, and his MS and PhD degrees in electrical engineering from University of Illinois at Urbana-Champaign in 2004 and 2007, respectively. After graduation, he worked in Schlumberger-Doll Research as a research scientist before he joined Tsinghua University in 2014.

Marco Salucci (senior member, IEEE) received an MS degree in Telecommunication Engineering from the University of Trento, Italy, in 2011, and his PhD degree from the International Doctoral School in Information and Communication Technology of Trento in 2014. He was a postdoctoral researcher at CentraleSupélec, in Paris, France, and then at the Commissariat à l'Énergie Atomique et aux Énergies Alternatives (CEA), in France. He is currently an assistant professor at the Department of Civil, Environmental, and Mechanical Engineering (DICAM) at the University of Trento, and a Research Fellow of the ELEDIA Research Center. Dr Salucci is a member of the IEEE Antennas and Propagation Society and he was a member of the COST Action TU1208 "Civil Engineering Applications of Ground Penetrating Radar." He is the associate editor for communications and memberships of the *IEEE Transactions on Antennas and Propagation*. Moreover, he serves as an associate editor of the *IEEE Transactions on Antennas and Propagation* and of the *IEEE Open Journal of Antennas and Propagation*, and as a reviewer for different international journals including *IEEE Transactions on Antennas and Propagation, IEEE Transactions on Microwave Theory and Techniques, IEEE Journal on Multiscale and Multiphysics Computational Techniques*, and *IET Microwaves, Antennas & Propagation*. His research activities are mainly concerned with inverse scattering, biomedical and GPR microwave imaging techniques, antenna synthesis, and computational electromagnetics with focus on system-by-design methodologies integrating optimization techniques and artificial intelligence for real-world applications.

Foreword

With the help of big data, massive parallelization, and computational algorithms, deep learning (*DL*) techniques have been developed rapidly during the recent years. Complex artificial neural networks, trained by large amounts of data, have demonstrated unprecedented performance in many tasks in artificial intelligence, such as image and speech recognition. This success also leads *DL* into many other fields of engineering. And electromagnetics (*EM*) is one of them.

This book is intended to overview the recent research progresses in applying *DL* techniques in *EM* engineering. Traditionally, research and development in this field have been always based on *EM* theory. The *EM* field distribution in engineering problems is modeled and solved by means of Maxwell's equations. The results can be very accurate, especially with the help of modern computational tools. However, when the system gets more complex, it is tough to solve because the increase in the degree-of-freedom exceeds the modeling and computational capabilities. Meanwhile, the demand for real-time computing also poses a significant challenge in the current *EM* modeling procedure.

DL can be used to alleviate some of the above challenges. First, it can "learn" from measured data and master some information about the complex scenarios for the solution procedure, which can improve the accuracy of modeling and data processing. Second, it can reduce the computational complexity in *EM* modeling by building fast surrogate models. Third, it can discover new designs and accelerate the design process while combining with other design tools. More engineering applications are being investigated with deep learning techniques, such as antenna design, circuit modeling, *EM* sensing and imaging, etc. The contents of the book are as follows.

In Chapter 1, a brief introduction to machine learning with a focus on *DL* is discussed. Basic concepts and taxonomy are presented. The classification of *DL* techniques is summarized, including supervised learning, unsupervised learning, and reinforcement learning. Moreover, popular *DL* architectures such as convolutional neural networks, recurrent neural networks, generative adversarial networks, and auto-encoders are presented.

Chapter 2 reviews the recent advances in *DL* techniques as applied to *EM* forward modeling. Traditional *EM* modeling uses numerical algorithms to solve Maxwell's equations, such as the method of moments, the finite element method, and the finite difference time domain method. In this context, *DL* can establish the mapping between a physical model and the corresponding field distribution. In other words, it is possible to predict the field distribution without solving partial differential equations, resulting in a much faster computation speed. In addition to fully data-driven approaches,

DL can be incorporated into traditional forward modeling algorithms to improve efficiency. Physics inspired forward modeling includes physics into neural networks to enhance their interpretability and generalization ability.

Chapter 3 discusses the application of *DL* to free-space inverse scattering. The surveyed techniques can be classified into three broad categories. The "black-box" approaches directly map measured far-field data to parametric values. Differently, learning-augmented iterative methods apply a learning-based solution to an iterative procedure, either by approximating the solution of the forward problem with a learning-based surrogate model or by integrating deep artificial neural networks into the entire iterative process. Finally, non-iterative learning approaches obtain the solution for the inverse problem directly. Still, they combine the prior knowledge of the problem, and for such a reason they cannot be regarded as simple black-box solutions.

Chapter 4 describes the use of *DL*-based methods for non-destructive testing and evaluation. The discussion is categorized based on the domain of application, including energy, transportation and civil infrastructures, manufacturing and agrifood sectors. The application to higher frequency methods is also reviewed, such as infrared thermography testing, terahertz wave testing, and radiographic testing. Moreover, the challenges and future trends of *DL* in non-destructive testing and evaluation are carefully discussed.

Chapter 5 reviews recent *DL* research as applied to subsurface imaging with a focus on *EM* methods. The state-of-the-art techniques, including purely data-driven approaches, physics-embedded data-driven approaches, and learning-assisted physics-driven approaches, are discussed. Several *DL*-based methods for seismic data inversion are also included in the Chapter. Furthermore, different techniques for constructing training datasets are discussed, which is essential for learning-based procedures.

Chapter 6 focuses on the current state-of-the-art *DL* methods used in medical imaging approaches. The physics of electromagnetic medical imaging techniques and their related physical imaging methods are first reviewed. Then, the commonly used deep neural networks with their applications in medical imaging are discussed. Recent studies on synergizing learning-assisted and physics-based imaging methods are presented, as well.

Chapter 7 presents an overview of how *DL* can be exploited for direction-of-arrival (*DoA*) estimation. After introducing the mathematical formulation of this problem under different conditions, the most common *DL* frameworks that have been applied to *DoA* estimation are reviewed, including their neural network configurations and the most widely used algorithmic implementations. Finally, a hierarchical deep neural network framework is presented to solve the *DoA* estimation problem.

Chapter 8 reviews the application of *DL* to remote sensing. With the accumulation of years of vast data, *DL* can effectively use these data in an automatic manner to serve many practical applications. The fields of target recognition, land cover and land use, weather forecasting, and forest monitoring are discussed, with a focus on how various *DL* models are employed and fitted into these specific tasks.

Chapter 9 discusses *DL*-based methods to improve digital satellite communications. *DL* can be employed to automate resource allocation, noise characterization, and nonlinear distortion in digital satellite communication links. Both are essential tasks in efficient digital satellite communications. Moreover, these strategies can be extended to other domains, such as *EM* compatibility or signal integrity.

Chapter 10 focuses on applying *DL* in task-oriented sensing, such as imaging and gesture recognition, based on metasurfaces. Three recent research progresses are discussed: intelligent metasurface imagers, variational-auto-encoder-based intelligent integrated metasurface sensors, and free-energy-based intelligent integrated metasurface sensors.

Chapter 11 reviews the application of *DL* to the design of metamaterials and metasurfaces. The design strategies are categorized into four groups: discriminative learning approach, generative learning approach, reinforcement learning approach, and optimization hybrid approach. With the help of *DL*, the inverse design of metamaterials and metasurfaces can become more flexible, efficient, and feasible.

Chapter 12 describes *DL* as applied to microwave circuit modeling, an important area of computer-aided design for fast and accurate microwave design and optimization. The feed-forward deep neural network and the vanishing gradient problem during its training process are introduced. Various recurrent neural networks for nonlinear circuit modeling are presented. Several application examples are presented to demonstrate the capabilities of deep neural network modeling techniques. As widely demonstrated through the Chapter, the powerful learning ability of *DL* makes it a suitable choice for modeling the complex input-output relationship of microwave circuits.

Based on these discussions, Chapter 13 summarizes the pros and cons of *DL* when applied to *EM* engineering, envisaging challenges and future trends in this area, and drawing some concluding remarks.

Despite the recent rapid progress in research, *DL* is still in its early stage in solving *EM* problems. Compared with imaging and speech processing, its application to *EM* engineering is more challenging considering the available data, the complexity of the scenarios, the requirement of learning and generalization ability, etc. A hybridization of physics and data may provide a way to address some long-term challenges.

Maokun Li
Marco Salucci

Acknowledgment

We sincerely thank all the contributors to this book. Special thanks to Ms Olivia Wilkins for her help in editing it. We sincerely hope this manuscript can help more researchers to work in this promising field, and we look forward to your feedback.

Chapter 1

An introduction to deep learning for electromagnetics

Marco Salucci[1,2], Maokun Li[3,4], Xudong Chen[5] and Andrea Massa[1,2,4,6]

1.1 Introduction

Nowadays, deep learning (*DL*) is arguably one of the hottest research topics in the scientific community [1–4]. According to the Google Trends analytics [5], there has been worldwide an exponential growth of interest in this topic during the last ten years [Figure 1.1(a)], South Korea, Singapore, Ethiopia, and China being the five regions with the highest occurrences of *DL*-related searches [Figure 1.1(b)]. Moreover, it comes as no surprise that Machine Learning (*ML*), Artificial Intelligence (*AI*), and Convolutional Neural Networks (*CNN*s) are currently ranked at the 2nd, 3rd, and 21st positions of the "top search terms" list of the *IEEEXplore* database [6], respectively. In such a framework, we have witnessed an unprecedented development of *DL* methods for solving complex problems in electromagnetics (*EM*) with very high computational efficiency [7–9]. For instance, *DL* is an extremely powerful paradigm (whose capabilities have been yet widely unexplored) for addressing fully non-linear imaging and inverse scattering problems on a pixel-wise basis with almost real-time performance [10–16]. Otherwise, impressive and highly encouraging results have been recently documented in fields including (but not limited to) remote sensing [17], ground penetrating radar [18], wireless communications [19,20], *EM* forward modeling [21,22], antenna synthesis [23–25], and metamaterial design [26–29].

Despite such a wide success, according to the well-known "*no-free lunch*" (*NFL*) theorems [30,31], there is no *ML/DL* algorithm universally performing better than others in any type of prediction problem. For instance, in a classification task, "*the*

[1]ELEDIA Research Center (ELEDIA@UniTN – University of Trento), DICAM – Department of Civil, Environmental, and Mechanical Engineering, Italy
[2]CNIT – "University of Trento" ELEDIA Research Unit, Italy
[3]Institute of Microwave and Antenna, Department of Electronic Engineering, China
[4]ELEDIA Research Center (ELEDIA@TSINGHUA – Tsinghua University), China
[5]Department of Electrical and Computer Engineering, National University of Singapore, Singapore
[6]ELEDIA Research Center (ELEDIA@UESTC – UESTC), School of Electronic Science and Engineering, University of Electronic Science and Technology of China, China

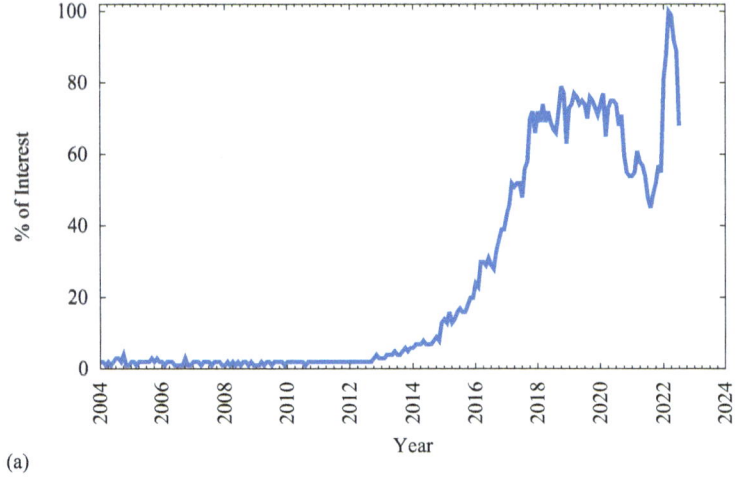

(a)

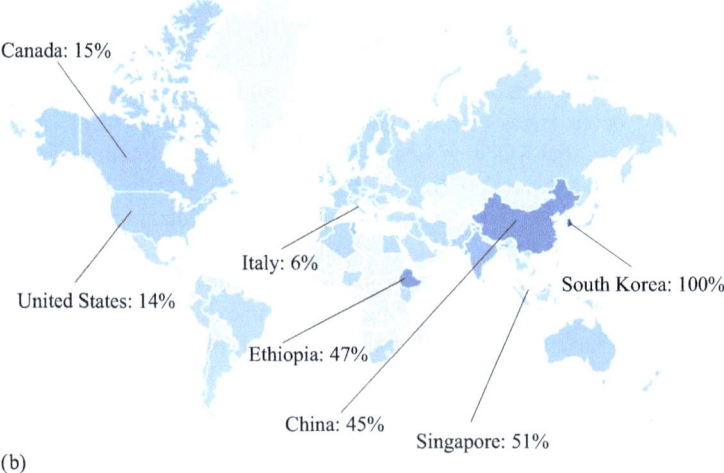

(b)

Figure 1.1   Worldwide search interest on DL *according to the Google Trends website [5]: (a) popularity over time ("a value of 100 is the peak popularity for the term" [5]) and (b) corresponding World map ("100 is the location with the most popularity as a fraction of total searches in that location" [5]).*

most sophisticated algorithm we can conceive of has the same average performance (over all possible tasks) as merely predicting that every point belongs to the same class" [1]. Therefore, due to impossibility to define a "holy grail" technique, it is paramount to understand the concepts, theory, and main features of each algorithm to select (and properly customize) the most suitable one for the problem at hand.

Following this line of reasoning, the goal of this chapter is to provide the reader with a general overview of *DL*, describing its pillar concepts and taxonomy, as well as providing a survey of the most widespread architectures found in the recent literature. The chapter refers to *EM* applications of these architectures where necessary.

1.2 Basic concepts and taxonomy

Before entering the details of the most common *DL* techniques and architectures in the recent *EM* literature, this section briefly recalls some basic concepts and provides a taxonomy of the most used terms in this field that will be helpful to the reader through this book.

1.2.1 What is deep learning?

With reference to Figure 1.2, *DL* techniques are commonly regarded as a small subset of *AI*, which in turn comprises all "smart" algorithms (i.e., automated instructions) enabling computers to perform tasks mimicking human intelligence by exploiting logic, decision trees, if-then rules, optimization, and *ML*. As for this latter, *ML* identifies a class of *AI* techniques allowing machines to *learn* from data and make

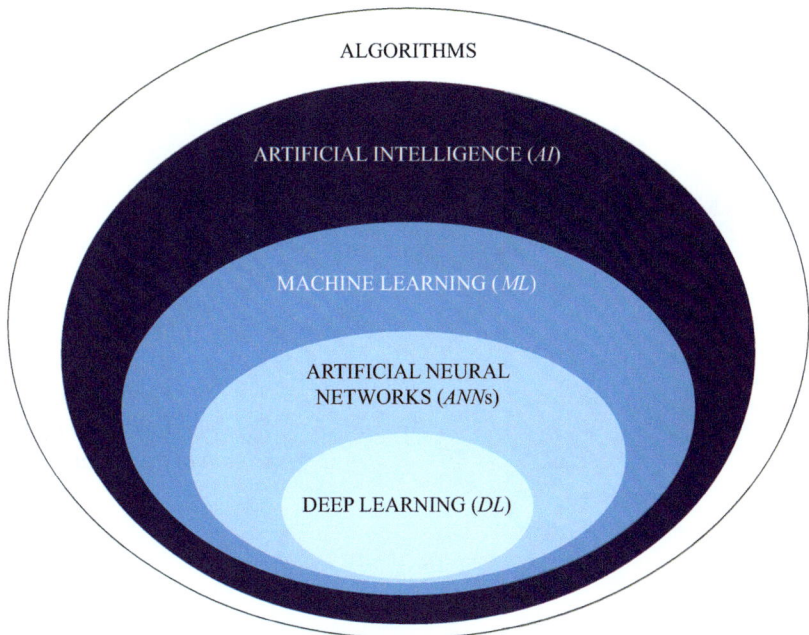

Figure 1.2　Classification of algorithms showing the hierarchical structure of AI, ML, ANNs, and DL techniques

decisions based on the observed patterns/relationships/associations within it, without being explicitly programmed and possibly improving at performed tasks with experience. Finally, *DL* is a subset of *ML* exploiting brain-inspired Artificial Neural Networks (*ANNs*) composed by a large amount of hidden layers and neurons (or units) adapting and learning from vast amounts of data. In other words, the following hierarchical structure exists within the large family of computer algorithms (Figure 1.2)

$$DL \subset ANNs \subset ML \subset AI. \tag{1.1}$$

Clearly, the following question immediately arises: *how deep is "deep"?* It is worth pointing out that there is not a standardized definition of the term "deep." However, it is often said that *ANNs* with at least three hidden layers (i.e., layers that do not coincide neither with the input nor the output of the network itself) are considered deep, while *ANNs* with a lower number of hidden layers are generally referred to as "shallow" architectures [7].

1.2.2 Classification of deep learning techniques

Although many nuances exist making impossible a unique classification, *ML* techniques (and accordingly, *DL* ones) are generally categorized into three distinct macro areas as follows (Figure 1.3):

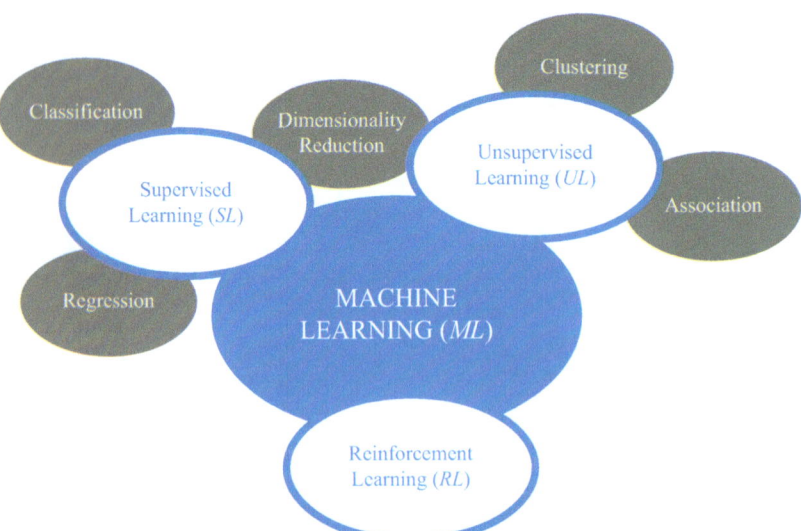

Figure 1.3   Subdivision of ML techniques into SL, UL, and RL branches

1. **Supervised learning (*SL*)** – *ML* techniques belonging to this group are also referred to as "Learning-by-Examples" (*LBE*) [32]. They rely on the off-line generation of a training dataset $\mathbb{D}$ of N input/output (*I/O*) pairs

$$\mathbb{D} = \left\{ \left(\underline{\mathscr{X}}^{(n)}, \underline{\mathscr{Y}}^{(n)} \right); n = 1, ..., N \right\} \tag{1.2}$$

where $\underline{\mathscr{X}}^{(n)} \in \mathbb{R}^K$ and $\underline{\mathscr{Y}}^{(n)} \in \mathbb{R}^Q$ are the nth ($n = 1, ..., N$) sample of the K-dimensional input space,

$$\underline{\mathscr{X}}^{(n)} = \left\{ \mathscr{X}_k^{(n)}; k = 1, ..., K \right\}, \tag{1.3}$$

$\mathscr{X}_k^{(n)}$ being the kth input variable or "feature," and the corresponding Q-dimensional output,

$$\underline{\mathscr{Y}}^{(n)} = \left\{ \mathscr{Y}_q^{(n)}; q = 1, ..., Q \right\} \tag{1.4}$$

being

$$\underline{\mathscr{Y}}^{(n)} = \mathscr{F}\left\{ \underline{\mathscr{X}}^{(n)} \right\}; \quad n = 1, ..., N \tag{1.5}$$

where $\mathscr{F}\{.\}$ is the *I/O* relationship that must be learned from the examples/observations stored within $\mathbb{D}$. The nature of $\underline{\mathscr{Y}}^{(n)}$ determines the specific learning task at hand, i.e.,

   (a)   ***Classification task*** – $\underline{\mathscr{Y}}^{(n)}$ is generally an integer scalar (i.e., $Q = 1$ and $\mathscr{Y}^{(n)} \in \mathbb{Z}$) representing the class/label associated to $\underline{\mathscr{X}}^{(n)}$ [Figure 1.4(*a*)];

   (b)   ***Regression task*** – $\underline{\mathscr{Y}}^{(n)}$ is typically a vector of $Q \geq 1$ continuous variables ($\underline{\mathscr{Y}}^{(n)} \in \mathbb{R}^Q$) dependent on $\underline{\mathscr{X}}^{(n)}$ [Figure 1.4(*b*)].

Starting from the information collected within $\mathbb{D}$, during the training phase *SL* techniques build a computationally-fast *surrogate model* (*SM*) of $\mathscr{F}\{.\}$ capable of making predictions of the (unknown) output associated to a new input sample $\underline{\mathscr{X}}$,

$$\underline{\tilde{\mathscr{Y}}} = \tilde{\mathscr{F}}\left\{ \underline{\mathscr{X}} \mid \underline{\mathscr{H}}^{opt} \right\}. \tag{1.6}$$

Such a task is accomplished by determining the optimal setting of A *hyperparameters* $\underline{\mathscr{H}}^{opt}$ ($\underline{\mathscr{H}}^{opt} = \left\{ \mathscr{H}_a^{opt}; a = 1, ..., A \right\}$) best predicting the N examples inside $\mathbb{D}$. Towards this end, a proper *loss function* $\mathscr{L}\{.\}$ quantifying the mismatch between actual and estimated outputs is minimized by means of a properly chosen optimization algorithm, yielding

$$\underline{\mathscr{H}}^{opt} = \arg\left[\min_{\underline{\mathscr{H}}} \mathscr{L}\left\{ \left(\underline{\mathscr{Y}}^{(n)}, \underline{\tilde{\mathscr{Y}}}^{(n)} \big|_{\underline{\mathscr{H}}} \right); n = 1, ..., N \right\} \right] \tag{1.7}$$

where

$$\underline{\tilde{\mathscr{Y}}}^{(n)} \big|_{\underline{\mathscr{H}}} = \tilde{\mathscr{F}}\left\{ \underline{\mathscr{X}}^{(n)} \mid \underline{\mathscr{H}} \right\} \tag{1.8}$$

is the *SM* prediction for the nth ($n = 1, ..., N$) training sample $\underline{\mathscr{X}}^{(n)}$. Clearly, the goal of a *SL* model is to go beyond the training data and correctly predict the output associated to previously unseen inputs. For such a reason, *SL*

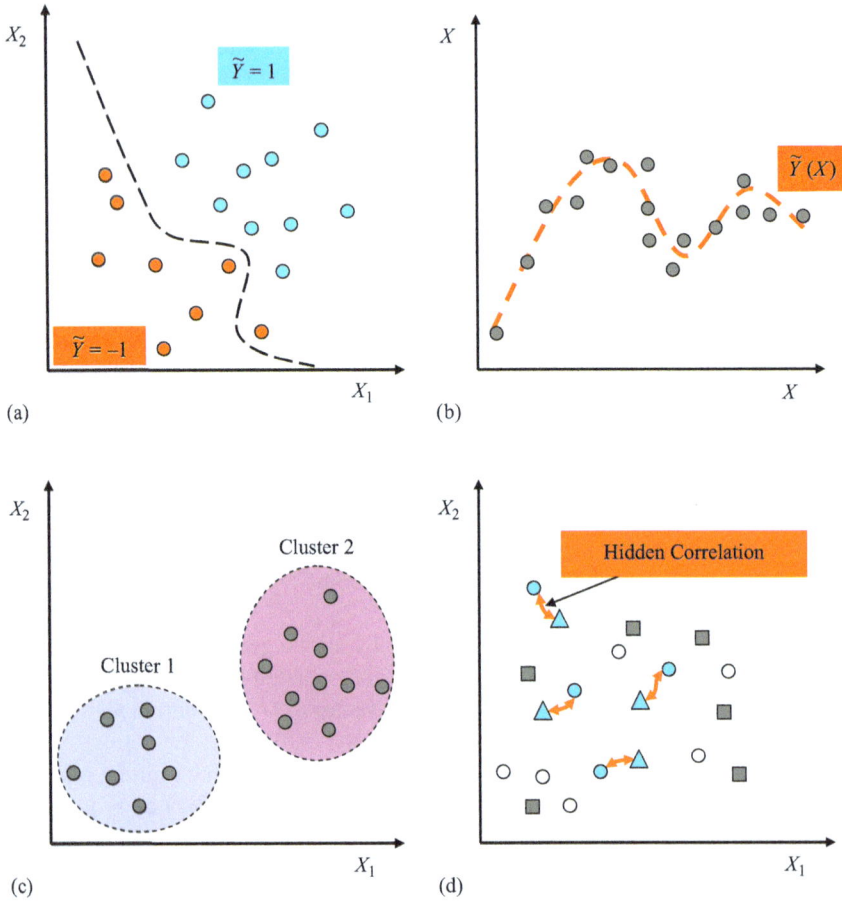

Figure 1.4   *Pictorial representation of the most common (a) and (b)* SL *tasks [(a) classification and (b) regression] and (c) and (d)* UL *tasks [(c) clustering and (d) association]*

methods are generally considered more powerful than traditional interpolation techniques (e.g., linear interpolation, nearest neighbor interpolation) since they exhibit remarkably higher *generalization* capabilities.

2.   **Unsupervised learning (UL)** – Differently from *SL*, *UL* algorithms process *unlabeled* data (i.e., without the corresponding output in the training data) and they are often trained to find *patterns* within the K-dimensional input space [33]. Their training can rely on different definitions of the loss function, which is often in the form of

$$\mathscr{L}\left\{\left(\underline{\mathscr{X}}^{(n)}, \underline{\widetilde{\mathscr{X}}}^{(n)}\right); n = 1, ..., N\right\} \tag{1.9}$$

where $\widetilde{\mathscr{X}}^{.(n)}$ is the reconstructed version of the input sample $\mathscr{X}^{.(n)}$ starting from a lower-dimensionality and/or noisy representation of it (see Section 1.3.4). The methods belonging to this group are generally classified into [33]

(a) **Clustering techniques** – Discover inherent groupings in the data, dividing them by similarity [Figure 1.4(c)].
(b) **Association techniques** – Determine rules that describe large portions of data, finding hidden dependencies/correlations within it [Figure 1.4(d)].

3. **Reinforcement learning (RL)** – This term refers to *ML* models (often called "agents") capable of performing autonomous decisions depending on their inter-actions with the external environment, progressively learning how to achieve a given goal [34–36] (Figure 1.5). Toward this end, a "game-like" situation is faced by the agent, which employs a *trial-and-error* approach to find a solution to the assigned problem, getting rewards or penalties for the performed actions. Although the designer sets the reward policy (i.e., the "rules of the game"), no hints/suggestions are given to the agent, which has to figure out how to perform the task by maximizing the total reward. Often, such a process starts from totally random trials and ends with sophisticated tactics and super-human skills. As a matter of fact, differently from human beings, *AI* can gather experience from thousands of parallel gameplays if a *RL* algorithm is run on a sufficiently powerful computing infrastructure, and the recent advances in computational technologies are opening the way to completely new applications in this field.

Autonomous driving is a common representative applicative scenario of deep *RL*, where the programmer cannot predict everything that could happen on the road [37]. Instead of building lengthy "if–then" instructions, he "prepares" the agent to learn from rewards and penalties, with the goal of guaranteeing safety, obeying the rules of law, minimizing costs and ride time, maximizing passengers comfort, and minimizing pollution. Starting from the interactions with the sur-rounding environment, the collected observations from it, the performed actions, and the gained rewards, the training algorithm defines the internal decision policy of the agent (Figure 1.5). After this training phase, the agent can autonomously drive exploiting the readings from the on-board sensors and the learnt policy.

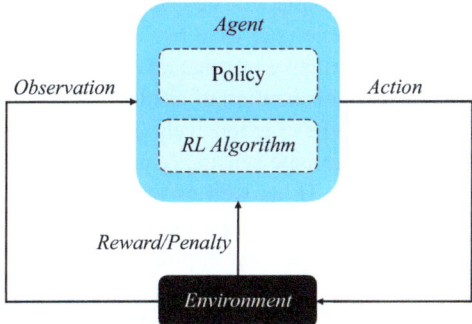

Figure 1.5   Block scheme of a RL technique

The term **Dimensionality Reduction** embraces both *SL* or *UL* methods [38,39] (Figure 1.3) aimed at finding a mapping rule from the original K-dimensional variables to a smaller subset of $J \ll K$ features carrying the same amount of information. In this way, it is possible to reduce the complexity of a *ML* learning task at hand, counteracting the so-called *curse of dimensionality* [40]. As for this latter, it refers to the exponential growth of the required amount of training data when the input space dimensionality increases. In other words, when K grows, the volume of the input space increases much faster, making the available data samples sparse (i.e., very far from each other) since the number of possible configurations of $\mathcal{X}$ is much larger than N.

Finally, **Transfer Learning** refers to recently developed approaches that allow to exploit a previously trained *ANN* for accelerating the training of a new *ANN* performing a new (but similar) task with a lower computational (time/CPU) cost and a smaller amount of training data [7,41].

1.3   Popular *DL* architectures

This section overviews the basic theory and fundamental concepts of some of the most popular *DL* architectures, whose applications in several *EM*-related fields will be revised in the next chapters.

1.3.1   *Convolutional neural networks*

*CNN*s are arguably one of the most popular *DL* architectures for prediction problems dealing with (or referable to) images [1–4,7,9–11].

   CNNs have recently attracted a lot of attention in the *EM* community, since they proved to be very effective and computationally efficient in the solution of many complex problems. For instance, they have been successfully exploited in *EM* forward modeling, where they have been trained to solve ordinary/partial differential equations (e.g., *2D* and *3D* Poisson's equations – see Chapter 2). In inverse scattering, subsurface imaging, and biomedical imaging *CNN*s implemented as "*U-Nets*" are probably the most popular *DNN* architectures capable of performing reconstructions on a pixel basis almost in real-time (see Chapters 3, 5, and 6). *CNN*s have been also widely employed in non-destructive testing and evaluation (*NDT/NDE*) applications including, for instance, the localization of casting defects in X-ray images (see Chapter 4). Moreover, they have been extensively exploited in direction-of-arrival (*DoA*) problems (see Chapter 7), remote sensing (e.g., to classify the Sentinel-1 *SAR* time-series images – see Chapter 8), as well as in satellite communications (e.g., to estimate the characteristics of the channel and the transmitter directly from the received signals – see Chapter 9). Successful applications of *CNN*s can be also found in the field of gesture recognition (Chapter 10) and metamaterials analysis and design (see Chapter 11).

   Their main advantage is that they automatically detect and extract informative/meaningful features from the input without any supervision. As a matter of fact,

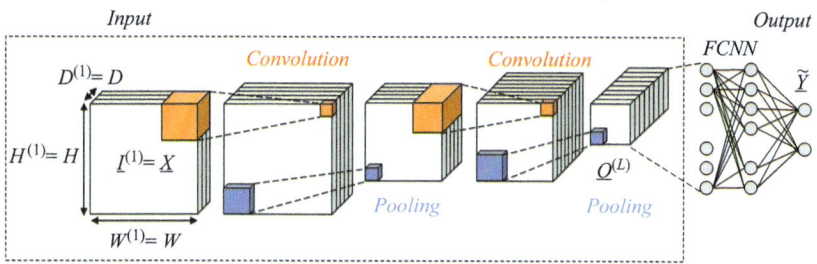

Figure 1.6   Block scheme of a CNN composed by L convolutional/pooling layers

such a "feature learning" from training data with translational/rotational invariance is the main difference of *CNN*s over standard *ANN*s [1,2]. Therefore, the focus of the following discussion will be on how such a powerful capability is enabled.

In the most general case, the input of a *CNN* is arranged as a 3D tensor $\mathscr{X} \in \mathbb{R}^{H \times W \times D}$ [i.e., the input space dimensionality is $K = (H \times W \times D)$], where H, W, and D denote the height, width, and depth of $\mathscr{X}$, respectively (Figure 1.6). For instance, in image classification $(H \times W)$ is the number of pixels in the input picture, while D is the number of "channels" (i.e., $D = 3$ for *RGB* images, where each dth channel includes the values of the pixels associated to the red, green, or blue components, respectively). Otherwise, in *EM* imaging and inverse scattering the depth is often set to $D = 2$, the two input channels being respectively associated to the real and imaginary parts of a complex-valued image such as, for instance, a coarse guess of the unknown contrast function or the incident field distribution emitted by the source [9–11,16].

A *CNN* architecture typically consists of a cascaded sequence of L layers alternating convolution and pooling operations to progressively transform the input of the first convolution layer, $\mathscr{I}^{(1)} = \mathscr{X}$, to the final "feature map" outputted by the Lth layer, $\mathscr{O}^{(L)}$, which is then inputted to a standard fully connected *NN* (*FCNN*) to predict $\widetilde{\mathscr{Y}}$ (Figure 1.6).* At each ℓth convolution layer, a set of $F^{(\ell)} \geq 1$ convolutional filters (or "kernels")

$$\underline{\Phi}^{(\ell,f)} \in \mathbb{R}^{C^{(\ell)} \times C^{(\ell)} \times D^{(\ell)}}; \quad f = 1, ..., F^{(\ell)} \tag{1.10}$$

being

$$\underline{\Phi}^{(\ell,f)} = \left\{ \Phi_{i,j,d}^{(\ell,f)}; i = 1, ..., C^{(\ell)}; j = 1, ..., C^{(\ell)}; d = 1, ..., D^{(\ell)} \right\} \tag{1.11}$$

is applied to the input, $\mathscr{I}^{(\ell)} \in \mathbb{R}^{H^{(\ell)} \times W^{(\ell)} \times D^{(\ell)}}$, to compute the output, $\underline{\mathscr{O}}^{(\ell)} \in \mathbb{R}^{U^{(\ell)} \times V^{(\ell)} \times F^{(\ell)}}$, as the combination of all performed convolutions, i.e.,

$$\underline{\mathscr{O}}^{(\ell)} = \left\{ \underline{\mathscr{O}}^{(\ell,f)}; f = 1, ..., F^{(\ell)} \right\}. \tag{1.12}$$

---

*Remembering that the output of each convolution and pooling layer is a 3D tensor, $\mathscr{O}^{(L)}$ is first flattened to a 1D vector in order to become the input layer of the *FCNN*.

More in detail, the result of each fth ($f = 1, ..., F^{(\ell)}$) convolution,

$$\underline{\mathcal{O}}^{(\ell,f)} = \left\{ O_{u,v}^{(\ell,f)}; u = 1, ..., U^{(\ell)}; v = 1, ..., V^{(\ell)} \right\}$$

is computed as (Figure 1.7)

$$\underline{\mathcal{O}}^{(\ell,f)} = \Psi \left\{ \underline{\mathcal{I}}^{(\ell)} \circledast \underline{\Phi}^{(\ell,f)} \right\} \tag{1.13}$$

where $\circledast$ denotes the convolution operator and $\Psi\{\,.\,\}$ is a local non-linear "activation" function.

More specifically, each (u, v)th entry ($u = 1, ..., U^{(\ell)}$, $v = 1, ..., V^{(\ell)}$) of $\underline{\mathcal{O}}^{(\ell,f)}$ is computed as

$$O_{u,v}^{(\ell,f)} = \Psi \left\{ \sum_{i=1}^{C^{(\ell)}-1} \sum_{j=1}^{C^{(\ell)}-1} \sum_{d=1}^{D^{(\ell)}} \Phi_{i,j,d}^{(\ell,f)} \times \mathcal{I}_{(u-1+i),(v-1+j),d}^{(\ell)} \right\} \tag{1.14}$$

where $U^{(\ell)} = \left(H^{(\ell)} - C^{(\ell)} + 1 \right)$ and $V^{(\ell)} = \left(W^{(\ell)} - C^{(\ell)} + 1 \right)$. In case $\underline{\mathcal{O}}^{(\ell,f)}$ must preserve the width and height of the input (i.e., $U^{(\ell)} = W^{(\ell)}$ and $V^{(\ell)} = H^{(\ell)}$), zero-padding can be applied to $\underline{\mathcal{I}}^{(\ell)}$ before entering the convolution (i.e., surrounding $\underline{\mathcal{I}}^{(\ell)}$ with zeros) [1].

As for the activation function, its purpose is to increase the degree of non-linearity of the *CNN* in order to model complex non-linear *I/O* relationships. One of the most adopted definitions of $\Psi\{\,.\,\}$ is the rectified linear unit (*ReLU*) function, which is defined for a generic scalar input $\xi \in \mathbb{R}$ as (Figure 1.8)

$$\Psi\{\xi\} = \max\{0, \xi\}. \tag{1.15}$$

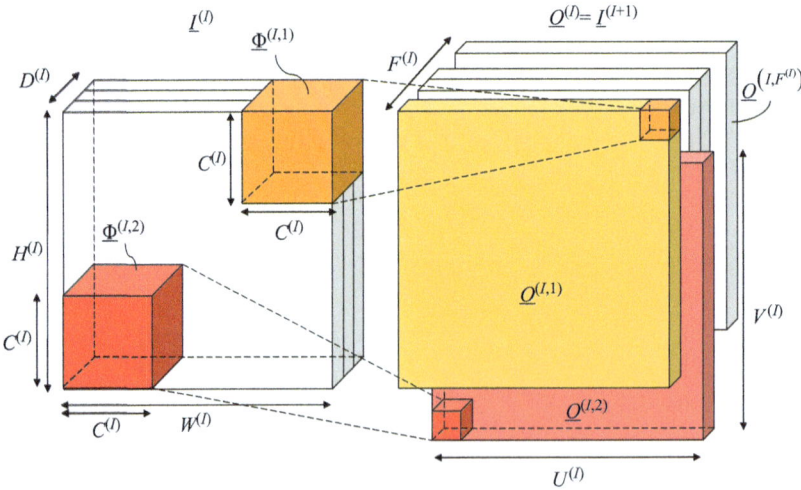

Figure 1.7 *Pictorial description of the convolution operation performed at the ℓth hidden layer of a CNN*

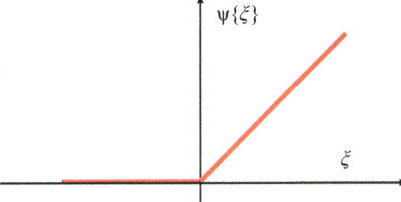

Figure 1.8   The ReLU non-linear activation function

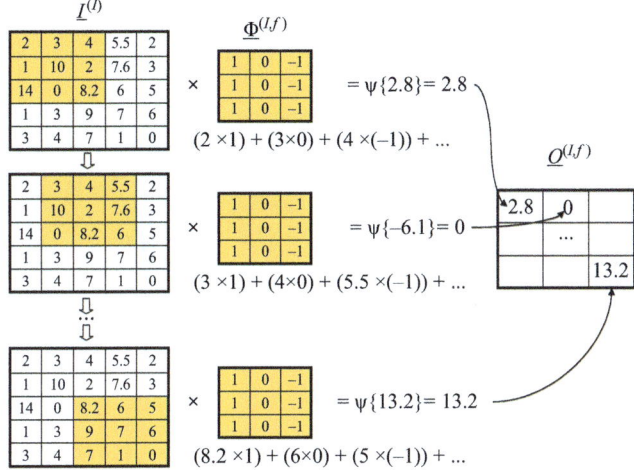

Figure 1.9   Numerical example of the convolution between an input feature map $\underline{\mathscr{I}}^{(\ell)} \in \mathbb{R}^{5 \times 5 \times 1}$ and a kernel $\underline{\Phi}^{(\ell,f)} \in \mathbb{R}^{3 \times 3 \times 1}$ without zero-padding, yielding $\underline{\mathscr{O}}^{(\ell,f)} \in \mathbb{R}^{3 \times 3 \times 1}$

To let the reader better understand the meaning of the previous expressions, Figure 1.9 shows a numerical example concerned with the convolution between an input feature map $\underline{\mathscr{I}}^{(\ell)} \in \mathbb{R}^{5 \times 5 \times 1}$ and a kernel $\underline{\Phi}^{(\ell,f)} \in \mathbb{R}^{3 \times 3 \times 1}$, yielding (without zero-padding) an output feature map $\underline{\mathscr{O}}^{(\ell,f)} \in \mathbb{R}^{3 \times 3 \times 1}$. According to (1.14), at every location of the kernel an element-wise multiplication between a portion of the input $\underline{\mathscr{I}}^{(\ell)}$ (also called the "receptive field") and the filter entries is performed, then the result is summed and inputted to the *ReLU* function to derive the entries of the output feature map (Figure 1.9).

The entries of the convolution filters constitute the set of *CNN* hyper-parameters, i.e.,

$$\underline{\mathscr{H}}_{CNN} = \left\{ \underline{\Phi}^{(\ell,f)}; f = 1, ..., F^{(\ell)}; \ell = 1, ..., L_{conv} \right\} \tag{1.16}$$

$L_{conv} < L$ being the number of convolution layers, and they are automatically derived from the training data during the training phase. In this regard, one evident benefit of *CNN*s over other *ANN* architectures is the fact that all spatial locations of the input to a given layer share the same convolution kernel, which is slided to compute (1.14), greatly reducing the number of parameters that must be learnt for each layer of the network, and making it independent on the dimensions of the input image [1].

Concerning the *CNN* pooling layers (Figure 1.6), their purpose is to reduce the width and height of the input. More in detail, in a "max pooling" layer, the operator maps a sub-region of $(P^{(\ell)} \times P^{(\ell)})$ neighboring entries of $\mathscr{I}^{(\ell)}$ belonging to the same dth channel ($d = 1, ..., D^{(\ell)}$) to its maximum value. Accordingly, indicating with $\mathscr{O}^{(\ell)} \in \mathbb{R}^{U^{(\ell)} \times V^{(\ell)} \times D^{(\ell)}}$ the output of the ℓth pooling layer, its (u, v, d)th ($u = 1, ..., U^{(\ell)}$, $v = 1, ..., V^{(\ell)}$, $d = 1, ..., D^{(\ell)}$) entry is computed as

$$\mathscr{O}^{(\ell)}_{u,v,d} = \max_{i,j=0,...,P^{(\ell)}-1} \left\{ \mathscr{I}^{(\ell)}_{[(u-1)\times s+i],[(v-1)\times s+j],d} \right\} \tag{1.17}$$

where $s \geq 1$ is the so-called "stride" determining the interval between two consecutive pooling windows. Otherwise, in a "average pooling" layer it turns out that

$$\mathscr{O}^{(\ell)}_{u,v,d} = \frac{1}{P^{(\ell)} \times P^{(\ell)}} \sum_{i=1}^{P^{(\ell)}} \sum_{j=1}^{P^{(\ell)}} \mathscr{I}^{(\ell)}_{[(u-1)\times P^{(\ell)}+i],[(v-1)\times P^{(\ell)}+j],d} \tag{1.18}$$

Figure 1.10 reports an example of max-pooling operation performed on an input feature map $\mathscr{I}^{(\ell)} \in \mathbb{R}^{3\times3\times1}$ with $P^{(\ell)} = 2$ and $s = 1$. It is worth highlighting that, differently from convolutional layers, pooling ones do not have any weighting parameter that must be learnt during the training phase.

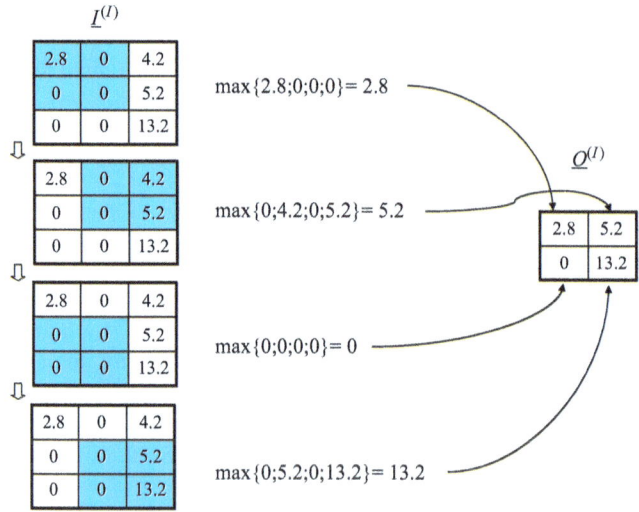

Figure 1.10   Numerical example of the max-pooling operation performed on an input feature map $\mathscr{I}^{(\ell)} \in \mathbb{R}^{3\times3\times1}$ with $P^{(\ell)} = 2$ and $s = 1$

1.3.2   Recurrent neural networks

Recurrent Neural Networks (*RNNs*) are a widespread *ANN* architecture that has recently attracted significant research in the field of sequential data processing and forecasting, being mainly used to detect patterns in temporal sequences [42–47]. As a matter of fact, *RNNs* and their variants (such as, for instance, Long Short-Term Memory, *LSTM* networks [45]) are particularly suitable in those contexts where the temporal dependency in the data is an important implicit feature. On the one hand, they process sequential information, performing the same operations on every element (time step) of the input sequence. On the other hand, their output at each time step also depends on the previous inputs and past computations, allowing the network to develop a sort of *memory* (encoded within in its "hidden state") of the previous events. Therefore, the main difference between *RNNs* and traditional Feed-forward *NNs* (*FFNN*, or Multi-Layer Perceptrons, *MLPs*) relies in how the information gets passed through the layers. More in detail, while *FFNNs* pass information through the network from input to output layers without backward connections or cycles, *RNNs* include cycles that allow the backward transmission of information into itself [44–47].

RNNs are quite popular in the recent *EM* literature. For instance, they have been exploited in forward modeling to process sequential *EM* data or model time-domain *EM* phenomena (see Chapter 2), such as those arising in forward scattering problems (see Chapter 3). In *NDT/NDE*, *RNNs* have been exploited – for instance – to analyze temperature time series to classify different defects that typically affect carbon fiber reinforced plastic material in assembled structures (see Chapter 4). Alternatively, they proved to be effective in estimating source *DoAs* (e.g., by processing a sequence of covariance matrices – see Chapter 7). Finally, *RNNs* have been successfully exploited for non-linear microwave circuit modeling (see Chapter 12).

Figure 1.11(a) gives a schematic representation of a simple *RNN* architecture, where $z^{-1}\{.\}$ indicates the unit delay operator, while Figure 1.11(b) reports the same network after being unfolded (or unrolled) to explicitly show the input time series and the corresponding output predictions. Let us denote the input sample and the hidden state at time step t ($t = 1, ..., T$, T being the number of time instants) as $\mathscr{I}^{(t)} = \mathscr{X}^{(t)} \in \mathbb{R}^K$ and $\mathscr{L}^{(t)} \in \mathbb{R}^G$, respectively, where G is the number of hidden units/neurons. Accordingly, $\mathscr{L}^{(t)}$ is computed as follows (Figure 1.11) [53]

$$\mathscr{L}^{(t)} = \Psi_{\mathscr{L}}\left\{ \mathscr{I}^{(t)} \times \underline{\underline{\mathscr{W}}}_{\mathscr{I}\mathscr{L}} + \mathscr{L}^{(t-1)} \times \underline{\underline{\mathscr{W}}}_{\mathscr{L}\mathscr{L}} + B_{\mathscr{L}} \right\} \tag{1.19}$$

where $\underline{\underline{\mathscr{W}}}_{\mathscr{I}\mathscr{L}} \in \mathbb{R}^{K \times G}$ and $\underline{\underline{\mathscr{W}}}_{\mathscr{L}\mathscr{L}} \in \mathbb{R}^{G \times G}$ are the input and hidden layers weight matrices, respectively, $B_{\mathscr{L}} \in \mathbb{R}^G$ is a bias parameter, and $\Psi_{\mathscr{L}}\{.\}$ is a non-linear activation operator typically set to the logistic sigmoid or tanh function [44,45]. As for the tth output $\mathscr{O}^{(t)} = \mathscr{Y}^{(t)} \in \mathbb{R}^Q$, it is computed as (Figure 1.11)

$$\mathscr{O}^{(t)} = \Psi_{\mathscr{O}}\left\{ \mathscr{L}^{(t)} \times \underline{\underline{\mathscr{W}}}_{\mathscr{L}\mathscr{O}} + B_{\mathscr{O}} \right\} \tag{1.20}$$

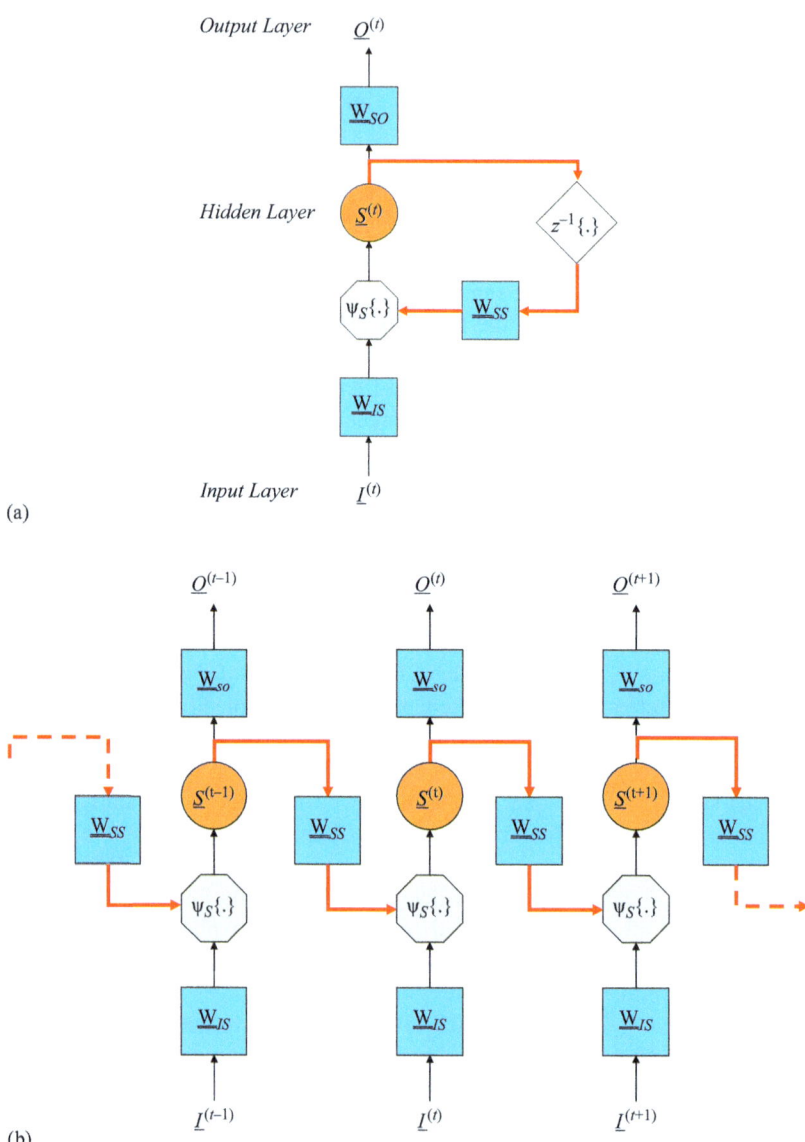

Figure 1.11   Block scheme of a RNN (a) before and (b) after unfolding

where $\Psi_{\mathscr{O}}\{\,.\,\}$ and $\underline{B}_{\mathscr{O}} \in \mathbb{R}^{Q}$ are the output activation function and bias, respectively, while $\underline{\underline{\mathscr{W}}}_{\mathscr{S}\mathscr{O}} \in \mathbb{R}^{G \times Q}$ is the output weighting matrix. The entries of $\underline{\underline{\mathscr{W}}}_{\mathscr{I}\mathscr{S}}$, $\underline{\underline{\mathscr{W}}}_{\mathscr{S}\mathscr{S}}$, $\underline{B}_{\mathscr{S}}$, $\underline{\underline{\mathscr{W}}}_{\mathscr{S}\mathscr{O}}$, and $\underline{B}_{\mathscr{O}}$ are the hyper-parameters of the *RNN* and they are determined during the training phase.

To better highlight the differences between *RNN*s and traditional *FFNN*s, let us recall that, using a similar notation but neglecting the time index t since it is not taken into account, the output of this latter would be simply equal to

$$\underline{\mathcal{O}} = \Psi_{\mathcal{O}} \left\{ \underline{\mathcal{S}} \times \underline{\underline{\mathcal{W}}}_{\mathcal{S}\mathcal{O}} + \underline{B}_{\mathcal{O}} \right\} \tag{1.21}$$

where the hidden variable is not dependent on previous computations, i.e.,

$$\underline{\mathcal{S}} = \Psi_{\mathcal{S}} \left\{ \underline{\mathcal{I}} \times \underline{\underline{\mathcal{W}}}_{\mathcal{I}\mathcal{S}} + \underline{B}_{\mathcal{S}} \right\}. \tag{1.22}$$

As for deep *RNN*s (*DRNN*s) made of $L > 1$ hidden layers to address highly non-linear forecasting problems, they are simply obtained by stacking shallow *RNN*s (Figure 1.12). Mathematically, it means that the tth ($t = 1, ..., T$) hidden state at the ℓth ($\ell = 1, ..., L$) layer, $\underline{\mathcal{S}}^{(t,\ell)}$, is computed as [46]

$$\underline{\mathcal{S}}^{(t,\ell)} = \begin{cases} \Psi_{\mathcal{S}}^{(\ell)} \left\{ \underline{\mathcal{I}}^{(t)}, \underline{\mathcal{S}}^{(t-1,\ell)} \right\} & \text{if} \quad \ell = 1 \\ \Psi_{\mathcal{S}}^{(\ell)} \left\{ \underline{\mathcal{S}}^{(t,\ell-1)}, \underline{\mathcal{S}}^{(t-1,\ell)} \right\} & \text{otherwise} \end{cases} \tag{1.23}$$

or, equivalently, as

$$\underline{\mathcal{S}}^{(t,\ell)} = \Psi_{\mathcal{S}}^{(\ell)} \left\{ \underline{\mathcal{S}}^{(t,\ell-1)} \times \underline{\underline{\mathcal{W}}}_{\mathcal{I}\mathcal{S}}^{(\ell)} + \underline{\mathcal{S}}^{(t-1,\ell)} \times \underline{\underline{\mathcal{W}}}_{\mathcal{S}\mathcal{S}}^{(\ell)} + \underline{B}_{\mathcal{S}}^{(\ell)} \right\}; \ \ell = 0, ..., L \tag{1.24}$$

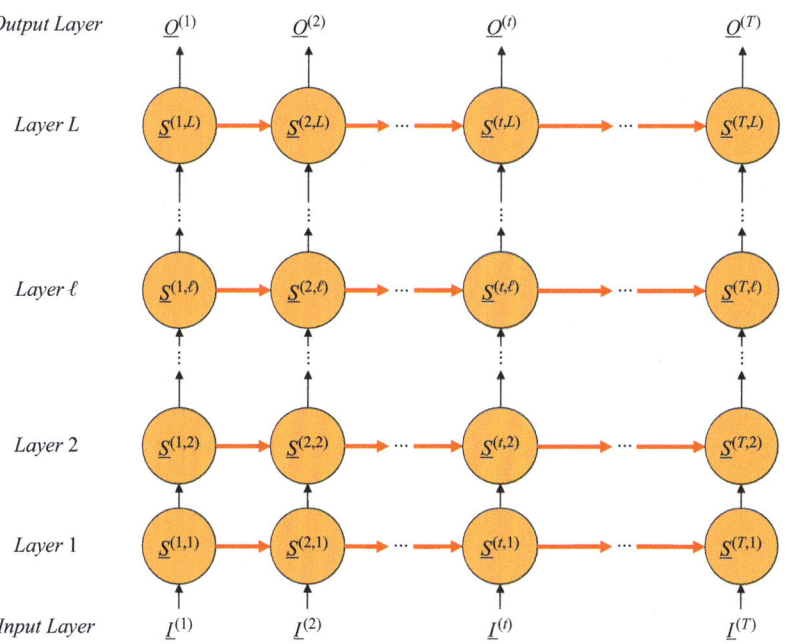

Figure 1.12   Block scheme of a DRNN composed by L hidden layers

where $\left.\underline{\underline{\mathscr{L}}}^{(t,\ell)}\right|_{\ell=0} = \underline{\underline{\mathscr{I}}}^{(t)}, \underline{\underline{\mathscr{W}}}^{(\ell)}_{\mathscr{I}\mathscr{S}}, \underline{\underline{\mathscr{W}}}^{(\ell)}_{\mathscr{S}\mathscr{S}}$, and $\underline{B}^{(\ell)}_{\mathscr{S}}$ being the hyper-parameters of the ℓth layer. Similarly, the tth *DRNN* output is derived from the hidden state of the last ($\ell = L$) layer (Figure 1.12)

$$\underline{\mathcal{O}}^{(t)} = \Psi_{\mathcal{O}} \left\{ \underline{\underline{\mathscr{L}}}^{(t,L)} \times \underline{\underline{\mathscr{W}}}_{\mathscr{S}\mathcal{O}} + \underline{B}_{\mathcal{O}} \right\} \tag{1.25}$$

where $\underline{\underline{\mathscr{W}}}_{\mathscr{S}\mathcal{O}}$ and $\underline{B}_{\mathcal{O}}$ are the learnt parameters of the output layer.

1.3.3   Generative adversarial networks

Generative Adversarial Networks (GANs) recently gained particular attention from the scientific community starting from their introduction back in 2014 by Goodfellow *et al.* [48].

They have recently attracted particular attention in the EM community. For instance, *GAN*s have been exploited to model the induced currents in *EM* scattering problems (see Chapter 2). Moreover, they have been widely employed in inverse scattering (see Chapter 3), subsurface imaging (see Chapter 5), and biomedical imaging (e.g., for segmentation of magnetic resonance imaging and computed tomography images – see Chapter 6). Moreover, *GAN*s have been exploited in remote sensing where, for instance, they have been exploited to to generate *SAR*-alike images to substitute part of real images for training (see Chapter 8). *GAN*s are inspired by the Game Theory and they exploit a pair of *ANN*s generally referred to as the *Generator*, $\mathscr{G}$, and the *Discriminator*, $\mathscr{D}$, networks (Figure 1.13) [49,50]. During the training process, such networks *compete* with each other in order to achieve the Nash equilibrium [51,52]. More specifically, the generator aims at generating "fake" data mimicking as much as possible the real data distribution of a given dataset, while the discriminator is a binary classifier that must correctly distinguish real samples from fake ones outputted by the generator. A commonly used analogy in the real world is

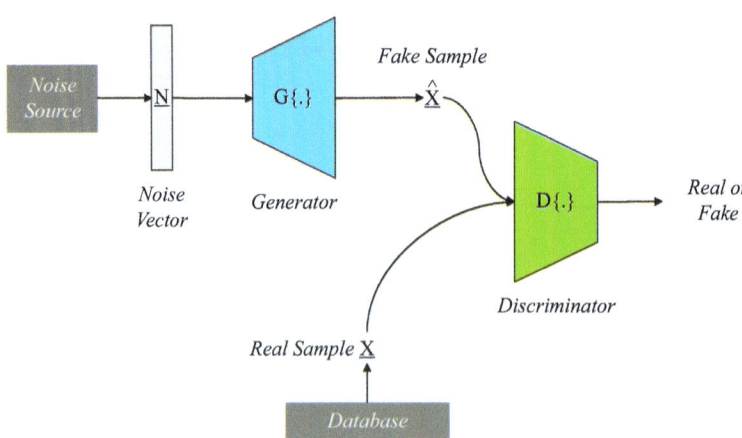

Figure 1.13   Block scheme of a GAN

to think to the generator as an art forger and to the discriminator as an art expert [49]. On the one hand, the forger ($\mathscr{G}$) generates forgeries as much as possible similar to real artworks. On the other hand, the expert ($\mathscr{D}$) aims at detecting whether the painting at hand is authentic or not. Starting from the errors made by the discriminator, the generator can progressively improve himself, by generating forgeries of increasing quality.

In practice, differently from the discriminator, the generator has no direct access to the real data and the unique way to learn how to create faithful fakes is through its interactions with $\mathscr{D}$, generally transforming an input random normally distributed noise vector $\mathscr{N}$ into a fake sample,

$$\widehat{\mathscr{X}} = \mathscr{G}\{\mathscr{N}\}, \tag{1.26}$$

and trying to progressively improve itself by maximizing the detection error made by $\mathscr{D}$ (Figure 1.13). When the discriminator cannot determine whether the input is real or fake, the optimal state is reached and the generator has correctly learned the (unknown) distribution of the real data.

Mathematically, the output of the discriminator for a given input sample $\mathscr{X}$ (either real or fake) can be expressed as the sigmoid probability function [25]

$$\mathscr{D}\{\mathscr{X}|\mathscr{H}_{\mathscr{D}}\} = \frac{1}{1 + \exp\left[-\alpha\left(\mathscr{X}|\mathscr{H}_{\mathscr{D}}\right)\right]} \tag{1.27}$$

where $\alpha\left(\mathscr{X}|\mathscr{H}_{\mathscr{D}}\right) \in \mathbb{R}$ is the discriminator prediction for a given setting of its hyper-parameters $\mathscr{H}_{\mathscr{D}}$. In the previous expression, $\mathscr{D}\{\mathscr{X}|\mathscr{H}_{\mathscr{D}}\} = 1$ and $\mathscr{D}\{\mathscr{X}|\mathscr{H}_{\mathscr{D}}\} = 0$ indicate that $\mathscr{X}$ is considered by the discriminator as real or fake with maximum probability, respectively, while $\mathscr{D}\{\mathscr{X}|\mathscr{H}_{\mathscr{D}}\} = 0.5$ indicates the impossibility to determine the correct class for $\mathscr{X}$. Accordingly, the discriminator hyper-parameters are trained to minimize the *cross-entropy loss*

$$\mathscr{H}_{\mathscr{D}} = \arg\left\{\min_{\mathscr{H}_{\mathscr{D}}}\left[-\mathscr{Y}\log\mathscr{D}\{\mathscr{X}|\mathscr{H}_{\mathscr{D}}\} - (1-\mathscr{Y})\log\left(1 - \mathscr{D}\{\mathscr{X}|\mathscr{H}_{\mathscr{D}}\}\right)\right]\right\} \tag{1.28}$$

where $\mathscr{Y} \in \mathbb{Z}$ is the actual label corresponding to $\mathscr{X}$ (i.e., $\mathscr{Y} = 1$ if $\mathscr{X}$ is real, $\mathscr{Y} = 0$ otherwise).

On the other hand, the goal of the generator is to "fool" the discriminator to classify $\widehat{\mathscr{X}}$ as real, i.e., by letting it predict $\mathscr{D}\{\mathscr{G}\{\mathscr{N}\}\} \simeq 1$. Toward this end, for a fixed setting of the discriminator hyper-parameters $\mathscr{H}_{\mathscr{D}}$, those of the generator, $\mathscr{H}_{\mathscr{G}}$, are updated such that

$$\mathscr{H}_{\mathscr{G}} = \arg\left\{\min_{\mathscr{H}_{\mathscr{G}}}\left[-\log\mathscr{D}\{\mathscr{G}\{\mathscr{N}\}|\mathscr{H}_{\mathscr{D}}\}\right]\right\}. \tag{1.29}$$

The generator is optimal when $\mathscr{D}$ is maximally confused and cannot distinguish real data from fake one, and the training phase is often performed by alternating the training of the two networks until reaching the Nash equilibrium [49].

One of the most straightforward applications of *GAN*s is *image synthesis*, where the generator is exploited to generate new images with specific attributes in order to – for instance – increase the number of samples within a given training set (i.e., data augmentation). Otherwise, *GANs* proved excellent *image-to-image* translation capabilities, that is, the task of transforming one image into another one [49,50]. *GANs* applications include also *super-resolution*, allowing a high resolution image to be generated from a low-resolution one, by inferring photo-realistic details while performing the up-sampling [49]. As concerns *EM* applications, *GANs* have been recently exploited for solving both inverse scattering and antenna design problems. For instance, *GANs* have been exploited to enhance the resolution of coarse guesses of unknown dielectric targets generated by means of fast inversion tools (e.g., back-propagation) [54,55]. Differently, *GANs* successfully learned to design log-periodic folded dipole antennas (*LPFDAs*) with specific (user-defined) Q-factors in [56].

1.3.4   Autoencoders

Autoencoders (*AEs*) are a specific class of unsupervised *FFNNs* that are trained to make an approximated copy, $\widetilde{\underline{\mathscr{X}}}$, of the input, $\underline{\mathscr{X}}$ [57,58].

In electromagnetics, *AE*-based architectures have been studied in *NDT/NDE* (e.g., to improve visibility of rear surface cracks during inductive thermography of metal plates – see Chapter 4). Moreover, they have been successfully applied to *DoA* estimation (see Chapter 7) and to signal-to-noise (*SNR*) estimation in satellite communications (see Chapter 9), just to mention a few relevant examples.

They generally consist in (i) an Encoder $\Gamma\{.\}$, (ii) a Decoder $\Gamma^{-1}\{.\}$, and (iii) a code, $\underline{\mathscr{C}}$ (Figure 1.14). The goal of the encoding function $\Gamma\{.\}$ is to represent the input in terms of a compressed sequence/code (also called "latent space representation"), i.e.,

$$\underline{\mathscr{C}} = \Gamma\left\{\underline{\mathscr{X}}\right\} \tag{1.30}$$

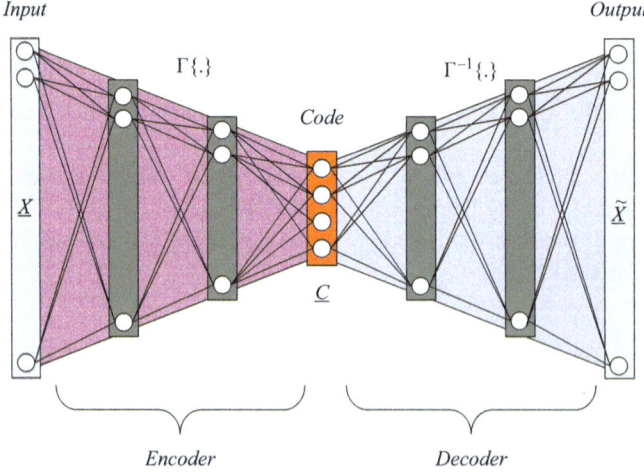

Figure 1.14   Block scheme of an AE

whereas the encoder aims at faithfully reconstructing $\underline{\mathscr{X}}$ from $\underline{\mathscr{C}}$, i.e.,

$$\underline{\widetilde{\mathscr{X}}} = \Gamma^{-1} \left\{ \underline{\mathscr{X}} \right\}. \tag{1.31}$$

Both encoder and decoder are typically implemented as *FCNN*s. Traditionally, *AE*s have been used for dimensionality reduction thanks to their capability to encode into $\underline{\mathscr{C}}$ a lower-dimension set of features carrying the largest possible amount of information to recover $\underline{\mathscr{X}}$ (i.e., $\underline{\mathscr{X}} \in \mathbb{R}^K$ and $\underline{\mathscr{C}} \in \mathbb{R}^J$, with $J \ll K$ – Figure 1.14). Such *AE* implementations are often called "under-complete", and their training process is aimed to instruct the decoder to capture the most salient features of the training data by minimizing the loss function [1]

$$\mathscr{L}\left(\underline{\mathscr{X}}, \Gamma^{-1} \left\{ \Gamma \left\{ \underline{\mathscr{X}} \right\} \right\} \right) \tag{1.32}$$

which penalizes the mismatch between $\underline{\widetilde{\mathscr{X}}}$ and $\underline{\mathscr{X}}$ (e.g., $\mathscr{L}\{.\}$ is the L2-norm of their difference). Clearly, the final interest is not in replicating the input, but rather in yielding at the output of the decoder a code sequence $\underline{\mathscr{C}}$ containing all relevant information on the input in a remarkably lower number of features.

More recently, de-noising *AE*s (*DAE*s) gained particular attention given their capability to subtract the noise corrupting blurred inputs. In this case, the loss function is defined as [1]

$$\mathscr{L}\left(\underline{\mathscr{X}}, \Gamma^{-1} \left\{ \Gamma \left\{ \underline{\mathscr{X}}_{\mathscr{N}} \right\} \right\} \right) \tag{1.33}$$

where

$$\underline{\mathscr{X}}_{\mathscr{N}} = \underline{\mathscr{X}} + \underline{\mathscr{N}} \tag{1.34}$$

is a noise-corrupted version of $\underline{\mathscr{X}}$, $\underline{\mathscr{N}}$ being an additive noise vector (Figure 1.15). Therefore, *DAE*s are not meant to exactly replicate the input, but rather to learn only useful features that are necessary to recover a noise-free version of the input.

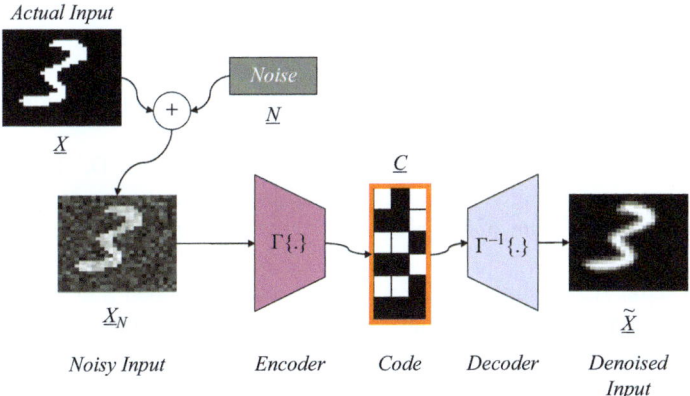

Figure 1.15 Pictorial example of a DAE performing the de-noising of a noisy-corrupted input image

Recent applications of *AE*s in *EM* include the design of metasurfaces [28] and direction of arrival (*DoA*) estimation [59]. The reader is referred to the following chapters for a detailed overview of such applicative scenarios.

1.4 Conclusions

This chapter provided a gentle introduction to *DL*, recalling the most used terms in this field as well as describing some of the most widespread architectures in the recent literature and their applications in *EM*.

Acknowledgments

This work benefited from the networking activities carried out within the Project "EMvisioning – Cyber-Physical Electromagnetic Vision: Context-Aware Electromagnetic Sensing and Smart Reaction" (Grant no. 2017HZJXSZ) funded by the Italian Ministry of Education, University, and Research within the PRIN2017 Program (CUP: E64I19002530001).

References

[1]   Goodfellow I., Bengio Y., and Courville A. *Deep Learning*. Boston, MA: MIT Press, 2016.

[2]   LeCun Y., Bengio Y., and Hinton G. 'Deep learning'. *Nature*. 2015;**521**(7553), pp. 436–444.

[3]   Schmidhuber J. 'Deep learning in neural networks: an overview'. *Neural Netw.* 2015;**61**, pp. 85–117.

[4]   Dong S., Wang P., and Abbas K. 'A survey on deep learning and its applications'. *Comput. Sci. Rev.* 2021;**40**, p. 100379.

[5]   Google. *Google Trends* [online]. 2022. Available from https://trends.google.com [accessed 19 July 2022].

[6]   IEEE. *IEEEXplore Database* [online]. 2022. Available from https://ieeexplore.ieee.org [accessed 19 July 2022].

[7]   Campbell S.D., Jenkins R.P., O'Connor P.J., and Werner D. 'The explosion of Artificial Intelligence in antennas and propagation: how deep learning is advancing our state of the art'. *IEEE Antennas Propag. Mag.* 2021;**63**(3), pp. 16–27.

[8]   Bayraktar A., Anagnostou D.E., Goudos S.K., Campbell S.D., Werner D.H., and Massa A. 'Guest Editorial: Special cluster on machine learning applications in electromagnetics, antennas, and propagation'. *IEEE Antennas Wireless Propag. Lett.* 2019;**18**(11), pp. 2220–2224.

[9]   Massa A., Marcantonio D., Chen X., Li M., and Salucci M. 'DNNs as applied to electromagnetics, antennas, and propagation – a review'. *IEEE Antennas Wireless Propag. Lett.* 2019;**18**(11), pp. 2225–2229.

[10]   Salucci M., Arrebola M., Shan T., and Li M. 'Artificial Intelligence: new frontiers in real-time inverse scattering and electromagnetic imaging'. *IEEE Trans. Antennas Propag.*, 2022;**70**(8), pp. 6349–6364, doi: 10.1109/TAP.2022.317755.

[11]   Li M., Guo R., Zhang K., *et al.* 'Machine learning in electromagnetics with applications to biomedical imaging: a review'. *IEEE Antennas Propag. Mag.* 2021;**63**(3), pp. 39–51.

[12]   Li L., Wang L.G., Teixeira F.L., Liu C., Nehorai A., and Cui T.J. 'DeepNIS: deep neural network for nonlinear electromagnetic inverse scattering'. *IEEE Trans. Antennas Propag.* 2019;**67**(3), pp. 1819–1825.

[13]   Jin K.H., McCann M.T., Froustey E., and Unser M. 'Deep convolutional neural network for inverse problems in imaging,' *IEEE Trans. Image Process.* 2017;**26**(9), pp. 4509–4522.

[14]   McCann M.T., Jin K.H., and Unser M. 'Convolutional neural networks for inverse problems in imaging: a review'. *IEEE Signal Process. Mag.*, 2017; **34**(6), pp. 85–95.

[15]   Hamilton S.J. and Hauptmann A. 'Deep D-Bar: real-time electrical impedance tomography imaging with deep neural networks'. *IEEE Trans. Med. Imag.* 2018;**37**(10), pp. 2367–2377.

[16]   Wei Z. and Chen X. 'Physics-inspired convolutional neural network for solving full-wave inverse scattering problems'. *IEEE Trans. Antennas Propag.* 2019;**67**(9), pp. 6138–6148.

[17]   Zhu X.X., Tuia D., Mou L., *et al.* 'Deep learning in remote sensing: a comprehensive review and list of resources'. *IEEE Geosci. Remote Sens. Mag.* 2017;**5**(4), pp. 8–36.

[18]   Travassos X., Avila S., and Ida N. 'Artificial neural networks and machine learning techniques applied to ground penetrating radar: a review'. *Appl. Comput. Inform.*, 2020;**17**(2), pp. 296–308.

[19]   Jiang C., Zhang H., Ren Y., Han Z., Chen K.-C., and Hanzo L. 'Machine learning paradigms for next-generation wireless networks'. *IEEE Wireless Commun.* 2017;**24**(2), pp. 98–105.

[20]   Zappone A., Di Renzo M., and Debbah M. 'Wireless networks design in the era of deep learning: model-based, AI-based, or both?'. *IEEE Trans Commun.* 2019;**67**(10), pp. 7331–7376.

[21]   Ma Z., Xu K., Song R., Wang C.-F., and Chen X. 'Learning-based fast electromagnetic scattering solver through generative adversarial network'. *IEEE Trans. Antennas Propag.*, 2021;**69**(4), pp. 2194–2208.

[22]   Yan L., Jin Y., Qi C., *et al.* 'Deep learning-assisted real-time forward modeling of electromagnetic logging in complex formations'. *IEEE Geosci. Remote Sens. Lett.* 2022;**19**, pp. 1–5.

[23]   Khan M.M., Hossain S., Mozumdar P., Akter S., and Ashique R.H. 'A review on machine learning and deep learning for various antenna design applications'. *Helyon.* 2022;**8**(4), pp. 1–22.

[24]   Kim J.H. and Bang J. 'Antenna impedance matching using deep learning'. *Sensors*, 2021;**2021**(21), p. 6766.

[25]   Zhang B., Jin C., Cao K., Lv Q., and Mittra R. 'Cognitive conformal antenna array exploiting deep reinforcement learning method'. *IEEE Trans. Antennas Propag.*, 2022;**70**(7), pp. 5094–5104, doi: 10.1109/TAP.2021.309699.

[26]   Liu P., Chen L., and Chen Z.N. 'Prior-knowledge-guided deep-learning-enabled synthesis for broadband and large phase-shift range metacells in metalens antenna'. *IEEE Trans. Antennas Propag.*, 2022;**70**(7), pp. 5024–5034, doi: 10.1109/TAP.2021.3138517.

[27]   Hu J., Zhang H., Bian K., Di Renzo M., Han Z., and Song L. 'MetaSensing: intelligent metasurface assisted RF 3D sensing by deep reinforcement learning'. *IEEE J. Sel. Areas Commun.* 2021;**39**(7), pp. 2182–2197.

[28]   Wei Z., Zhou Z., Wang P., *et al.* 'Equivalent circuit theory-assisted deep learning for accelerated generative design of metasurfaces'. *IEEE Trans. Antennas Propag.*, 2022;**70**(7), pp. 5120–5129, doi: 10.1109/TAP.2022.3152592.

[29]   Naseri P., Pearson S., Wang Z., and Hum S.V. 'A combined machine-learning/optimization-based approach for inverse design of nonuniform bianisotropic metasurfaces'. *IEEE Trans. Antennas Propag.*, 2022;**70**(7), pp. 5105–5119, doi: 10.1109/TAP.2021.3137496.

[30]   Wolpert D.H. and Macready W.G. 'No free lunch theorems for optimization'. *IEEE Trans. Evol. Comput.* 1997;**1**(1), pp. 67–82.

[31]   Wolpert D.H. 'The supervised learning No-Free-Lunch theorems'. in Roy R., Koppen M., Ovaska S., Furuhashi T., and Hoffmann F. (eds.), *Soft Computing and Industry*. London: Springer; 2022, pp. 25–42.

[32]   Massa A., Oliveri G., Salucci M., Anselmi N., and Rocca P. 'Learning-by-examples techniques as applied to electromagnetics'. *J. Electromagn. Waves Appl.* 2018;**32**(4), pp. 516–541.

[33]   Hastie T., Tibshirani R., and Friedman J. *The Elements of Statistical Learning: Data mining, Inference, and Prediction*. New York, NY: Springer; 2008.

[34]   Kaelbling L., Littman M., and Moore A. 'Reinforcement learning: a survey'. *J. Artif. Intell. Res.* 1996;**4**, pp. 237–285.

[35]   Francois-Lavet V., Henderson P., Islam R., Bellemare M., and Pineau J. 'An introduction to deep reinforcement learning'. *Found. Trends Mach. Learn.* 2018;**11**(3–4), pp. 219–354.

[36]   Arulkumaran K., Deisenroth M.P., Brundage M., and Bharath A.A., 'Deep reinforcement learning: a brief survey'. *IEEE Signal Proc. Mag.* 2017;**34**(6), pp. 26–38.

[37]   Kiran B.R., Sobh I., Talpaert V., *et al.* 'Deep reinforcement learning for autonomous driving: a survey'. *IEEE Trans. Intell. Transp. Syst.* 2022;**23**(6), pp. 4909–4926.

[38]   Vogelstein J.T., Bridgeford E.W., Tang M., *et al.* 'Supervised dimensionality reduction for big data'. *Nature Commun.* 2021;**12**, p. 2872.

[39]   Jia W., Sun M., Lian J., and Hou S. 'Feature dimensionality reduction: a review'. *Complex Intell. Syst.* 2022;**8**(3), pp. 2663–2693.

[40]   Bellman R.E. *Adaptive Control Processes*. Princeton, NJ: Princeton University Press, 1961.

[41] Torrey L. and Shavlik J. 'Transfer learning', in Soria E., Martin J., Magdalena R., Martinez M., and Serrano A. (eds.), *Handbook of Research on Machine Learning Applications and Trends: Algorithms, Methods, and Techniques*. Hershey, PA: IGI Global; 2009.

[42] Noakoasteen O., Wang S., Peng Z., and Christodoulou C. 'Physics-informed deep neural networks for transient electromagnetic analysis'. *IEEE Open J. Antennas Propag.* 2020;**1**, pp. 404–412.

[43] Antczak K. 'Deep recurrent neural networks for ECG signal denoising'. 2018; arXiv:1807.11551.

[44] Lipton Z.C., Berkowitz J., and Elkan C. 'A critical review of recurrent neural networks for sequence learning'. 2015; arXiv:1506.00019.

[45] Bianchi F.M., Maiorino E., Kampffmeyer M.C., Rizzi A., and Jenssen R. 'An overview and comparative analysis of Recurrent Neural Networks for short term load forecasting'. 2018; arXiv:1705.04378.

[46] Schmidt R.M. 'Recurrent Neural Networks (RNNs): a gentle Introduction and overview'. 2019; arXiv:1912.05911.

[47] Chen G. 'A gentle tutorial of Recurrent Neural Network with error backprop-agation'. 2018; arXiv:1610.02583.

[48] Goodfellow I., Pouget-Abadie J., Mirza M., *et al.* 'Generative adversarial nets', in *Proc. Advances Neural Information Processing Systems Conf.*, 2014, pp. 2672–2680.

[49] Creswell A., White T., Dumoulin V., Arulkumaran K., Sengupta B., and Bharath A.A. 'Generative adversarial networks: an overview'. *IEEE Signal Proc. Mag.* 2018;**35**(1), pp. 53–65.

[50] Pan Z., Yu W., Yi X., Khan A., Yuan F., and Zheng Y. 'Recent progress on Generative Adversarial Networks (GANs): a survey'. *IEEE Access.* 2019;**7**, pp. 36322–36333.

[51] Holt C. and Roth A. 'The Nash equilibrium: a perspective'. *Proc. Natl. Acad. Sci.* 2004;**101**(12), pp. 3999–4002.

[52] Rao R.C. 'Game theory, overview', in *Encyclopedia of Social Measurement;* 2005; pp. 85–97.

[53] Zhang A., Lipton Z.C., Li M., and Smola A.J. *Dive Into Deep Learning*. [Online]. 2022. Available from Available: http://www.d2l.ai.2019 [Accessed 19 July 2022].

[54] Ye X., Bai Y., Song R., Xu K., and An J. 'An inhomogeneous background imaging method based on Generative Adversarial Network'. *IEEE Trans. Microw. Theory Techn.* 2020;**68**(11), pp. 4684–4693.

[55] Song R., Huang Y., Xu K., Ye X., Li C., and Chen X. 'Electromagnetic inverse scattering with perceptual Generative Adversarial Networks'. *IEEE Trans. Computat. Imag.* 2021;**7**, pp. 689–699.

[56] Noakoasteen O., Vijayamohanan J., Gupta. A., and Christodoulou C. 'Antenna design using a GAN-based synthetic data generation approach'. *IEEE Open J. Antennas Propag.* 2022;**3**, pp. 488–494.

[57]   Dong G., Liao G., Liu H., and Kuang G. 'A review of the Autoencoder and its variants: a comparative perspective from target recognition in synthetic-aperture radar images'. *IEEE Geosci. Remote Sens. Mag.* 2018;**6**(3), pp. 44–68.

[58]   Tschannen M., Banchem O., and Lucic M. 'Recent advances in autoencoder-based representation learning'. 2018; arXiv:1812.05069.

[59]   Liu Z., Zhang C., and Yu P.S. 'Direction-of-arrival estimation based on deep neural networks with robustness to array imperfections'. *IEEE Trans Antennas Propag.* 2018;**66**(12), pp. 7315–7327.

Chapter 2

Deep learning techniques for electromagnetic forward modeling

Tao Shan[1] and Maokun Li[1]

2.1 Introduction

Electromagnetic forward modeling (EMFD) is a ubiquitous tool for theoretical research and engineering applications in the field of electromagnetics [1–4], such as scientific simulation, engineering design, information processing, etc. The commonly applied computational algorithms for EMFD include finite element method (FEM) [5], finite difference method (FDM) [6], method of moments (MoM) [7,8]. All of them perform forward modeling by solving the governing equations formulated in the form of differential and/or integral equations. The solving process usually involves discretizing and converting the formulated governing equations into a linear system of matrix equations, which leads to a large number of unknowns, heavy computational cost, and immense memory load. This poses a long-standing challenge for EMFD, especially in the context of increasingly complicated electromagnetic systems and diverse electrical scales. Computational efficiency has long been a core issue in electromagnetic forward modeling. Many efforts have been devoted to improving the computational efficiency of EMFD. Reducing redundant calculations based on the physical laws is one important acceleration approach, for example, conjugate gradient-fast Fourier transform [9], adaptive integral method [10], fast multipole method [11], domain decomposition method [12], etc. Another approach is to divide the whole computation into online and offline parts where offline computation can reduce the online computational burden and further accelerate online computing, such as reduced basis method [13] and machine learning-based method [14]. With the increase in computing performance of the central process unit (CPU), parallel computing is widely applied to improve the computational efficiency of FDTD, MoM, FMM, etc. However, the speed of parallel computing is further limited by the communication speed between CPU nodes and between the computing cores and memory in a single CPU node, which is far from real-time computing. Therefore, it is still a big challenge to perform EMFD in real time.

[1]Beijing National Research Center for Information Science and Technology (BNRist), Department of Electronic Engineering, Tsinghua University, China

The rapid developments in the graphics processing unit (GPU) and high-performance computing have led to an explosion of deep learning (DL) technologies by enabling the optimization of large-scale deep neural networks (DNNs) [15]. Boasting powerful learning capacity and approximation ability, DNNs can independently extract, abstract, and nest hierarchical features from high dimensional data. DL has achieved ground-breaking contributions to image, speech, and video processing and led to dramatic improvements in the corresponding algorithms [15]. The great success of DL also ignites the flare of applying DL into the fields of mathematics, physics, and engineering [16–18]. In applied mathematics, the relationship between DL and ordinary/partial differential equations (ODEs/PDEs) is widely investigated. DL can help reduce the curse of dimensionality when solving ODEs/PDEs, and conversely, the theory of ODEs/PDEs can further help improve the robustness and interpretability of DL [19]. In intuitive physics, DL can establish the physical relationship between different objects and further predict the future changes of objects' physical states [20]. Besides, DL can learn and capture the exact physical laws to model the corresponding physical phenomena, such as the Hamiltonian neural network motivated by Hamiltonian mechanics [21], the Symplectic recurrent neural networks incorporated with Symplectic integration [22], the Lagrangian neural networks emulating Lagrangians [23], etc. Computational fluid dynamics usually need to solve high-dimensional, nonlinear, nonconvex, and multiscale problems, which is where DL excels. DL is devoted to solving problems hard to solve by traditional CFD methods which can be categorized into three types [24]. The first is to apply DL to solve closure terms for improved precision of CFD by exploiting the approximation ability of DL [25]. Second, DL can directly solve the governing equations of fluid dynamics, including Navier-Stokes equations or Euler equations, by avoiding the iterative process of traditional CFD methods [26]. Third, DL can be incorporated with traditional CFD methods to enhance computing efficiency and precision [27].

The successful applications of DL in image, video, and speech processing, especially in applied mathematics, physics, engineering, make DL a promising candidate for fast and efficient EMFD. Before the emergence of DL, machine learning (ML) techniques have been applied to accelerate EMFD by training offline and accelerating online [14,28,29], for example, microwave circuit design, antenna design, microwave detection, etc. Due to the small parameter sets, ML possesses limited learning capacity and approximation ability which further restricts the applications and performance of ML. DL boasts strong learning capacity and approximation ability compared to traditional machine learning techniques, as DL usually has a massive parameter set that can be fine-tuned with the help of stochastic optimization algorithms and massive parallel computing platforms (GPUs). Recently, many works have been reported to incorporate DL into traditional computational electromagnetic algorithms for acceleration, such as MoM [30], and FDTD [31]. Compared to traditional EMFD, which builds and solves mathematical or physical models with Maxwell's equation as a starting point, DL learns to abstract the intrinsic physical laws from a large amount of electromagnetic data. In fact, the ultimate goal of DL and EMFD is the same and it can be summarized as computing precision, computing efficiency, and generalization (universality), as shown in Figure 2.1. Therefore, it poses the great potential

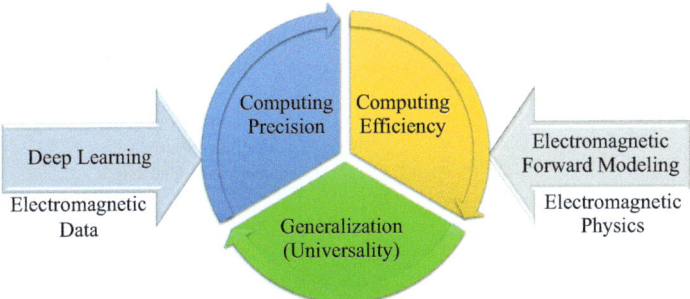

Figure 2.1   Schematic of relationship between deep learning and electromagnetic forward modeling

to incorporate DL into EMFD for better computational precision, computational efficiency, and universality by leveraging the power of both EM data and physics.

2.2   DL and ordinary/partial differential equations

Differential equations are important tools to model physical phenomena across vast areas of physics and engineering, including ordinary and partial differential equations. The link between DL and ODEs/PDEs has been widely investigated in many mature fields founding on nonlinear ODEs/PDEs, which can be concluded in Figure 2.2. On the one hand, DL can help overcome the dimensional curse and improve computational efficiency when solving ODEs/PDEs; on another hand, many deep neural networks (DNNs) can be interpreted within the theory of ODEs/PDEs which further guide the design of effective DNN architectures. Maxwell's equations are the basics of electromagnetic engineering and they are also an important set of PDEs. Therefore, it is thought-provoking to review the works contributing to the relationship between DL and ODEs/PDEs, which further provides new insights into applications of DL in EMFD.

Artificial neural networks (ANNs) have already been employed to solve ODEs/PDEs numerically before the explosive development of DL [32,33]. The common approach of ANN-based methods is to learn the direct mappings between the solutions and corresponding variables of ODEs/PDEs but the performance is limited by the scale of ANN parameters and computing platforms. Recently, with the development of DL, many works are devoted to applying DL to solve various ODEs/PDEs in the fields of physics and engineering. Convolutional neural networks (CNNs) are trained to solve Schrödinger equation to predict the ground-state energy of an electron [34]. Variational quantum Monte Carlo is applied to train the DNN-based wavefunction ansatz as the solution of electronic Schrödinger equation [35]. Navier–Stokes equations are solved by deep learning techniques to model flow dynamics

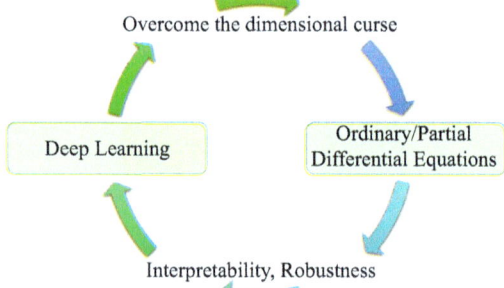

Figure 2.2   Schematic of relationship between deep learning and ordinary/partial differential equations

[36]. High-dimensional Burgers' equations are solved by DNNs with the differential operator, initial condition, and boundary conditions as constraints [37]. The heat equation is solved to predict the steady states by employing a weakly supervised training scheme [38]. The fast solver of Poisson's equation is built based on CNNs to provide reliable predictions of electric potentials [39]. Space–time fractional advection–diffusion equations are solved by encoding PDEs into the loss function of DNNs [40]. DL is also applied to solve Hamilton–Jacobi–Bellman equations to calculate the optimal feedback control of nonlinear systems [41]. The empirical risk is modeled by applying deep artificial neural networks to solve Black–Scholes equations [42].

The approaches of solving ODEs/PDEs based on DL techniques can be tentatively divided into two categories: fully data-driven and physics-driven approaches. The fully data-driven approach directly trains DNNs to learn the inner laws between input and output from massive data by leveraging the learning capacity of DNNs [36,43,44]. It is noted that there is no physical prior information to constrain the training of DNNs in the fully data-driven approach. It is usually time-consuming to generate sufficient data samples for training DNNs. The performance of fully data-driven approaches depends on the quality of training data and the generalization ability is also limited. Physics-driven approaches incorporate the physics priori or mathematical models into the DNNs that demonstrate better performance and improved generalization ability after sufficient training. Traditional numerical methods of ODEs/PDEs inspire the way DL techniques are applied. The Galerkin method motivates DNNs to introduce the PDEs and the corresponding initial, boundary conditions into their objective functions [37]. PDEs can also be converted into the variational form and solved by the DNN in the context of the Ritz method [45]. The trial solution or weak formulation can be combined with DNNs to learn the parametric solutions of PDEs [46,47]. The DNN can be trained to learn the update rules of the fixed-point iteration method for solving PDEs [48]. Physics-informed neural networks (PINNs) are another important approach to solve ODEs/PDEs based on DL techniques. PINNs are first proposed in [49] as a deep learning framework of forward and inverse problems governed by PDEs. The PDE parameterized by θ can be expressed as:

$$\Phi\left(x, t, \theta, u, \nabla u, \nabla^2 u, \ldots\right) = 0, \quad x \in \Omega, \tag{2.1}$$

where Φ denotes the PDE, x is defined in Ω, t is the temporal domain and u denotes the solution of Φ. u also satisfies the BCs:

$$\Psi(u, x, t) = 0, \quad x \in \partial\Omega. \tag{2.2}$$

PINNs regard the DNN as the surrogate of the ODE/PDE solutions:

$$u^* = \mathscr{F}(\Theta, x, t), \tag{2.3}$$

where $\mathscr{F}$ and Θ denote the DNN and the corresponding parameter set. Due to that the surrogate solution u^* should satisfy the PDE and BCs, it is natural to incorporate the weighted sum of the PDE and the corresponding BCs into the loss function:

$$\mathscr{L}(\mathscr{F}, \Theta) = w_\Phi \|\Phi(\mathscr{F}(\Theta, x, t))\|^2 + w_\Psi \|\Psi(\mathscr{F}(\Theta, x, t))\|^2, \tag{2.4}$$

Automatic differentiation of DL applies the chain rule to calculate the differentiations of Φ to constrain the training of network $\mathscr{F}$, which further enables PINNs mesh-free. The gradient descent-based optimization method is employed to determine the optimal parameter set Θ of the DNNs by minimizing the loss function $\mathscr{L}(\mathscr{F}, \Theta)$ [50]. PINNs are also extended to solve fractional PDEs [40] and stochastic PDEs [51], solve PDEs based on multi-fidelity data [52], etc. The extreme learning machine is applied to improve the computing efficiency instead of taking DNNs as surrogate solutions of PDEs [53]. The library of PINNs is published for solving forward and inverse PDEs numerically and it is named as DeepXDE [50].

ODEs, PDEs, and their corresponding numerical methods provide new insights into the reasoning and design of effective DNN architectures for better robustness and interpretability. The numerical stability of the residual neural network (ResNet) [54] is analyzed by relating the exploding and vanishing gradient phenomenon within the theory of ODEs [55,56], which further guide the design of stable DNN architectures. A theoretical framework is built to analyze the reversibility of ResNet by interpreting ResNet as ODEs [57]. The lesioning properties of ResNet are analyzed from the view of dynamical systems to allow the acceleration of training [58]. The ResNet is also related with the dynamical systems, such as recurrent neural network (RNN) [59], characteristic lines of the transport equations [60], etc. The gap between DNN structures and numerical methods of ODEs/PDEs is bridged in [61] and DNNs are also interpreted as the numerical schemes of ODEs in [61]. In [61], several effective DNN models including ResNet, PolyNet [62], FractalNet [63], and RevNet [64] are interpreted as the forward Euler scheme, backward Euler scheme, Runge–Kutta scheme, and forward Euler scheme of ODEs. Based on these observations, the linear multi-step ResNet is designed by combining ResNet with the linear multi-step scheme of ODEs. The similarity between the finite difference operator and the convolution operator is investigated and the PDE-Net is built to uncover the underlying PDEs of dynamic systems [65]. Motivated by ODEs, the neural ODEs are proposed as the continuous-depth DNN models and the adjoint sensitivity method is applied to compute gradients [66].

The ResNet is interpreted as a discretization of the space–time differential equation in [19], which further motivates new CNN models by incorporating parabolic and hyperbolic PDEs. MgNet is designed based on the connections between the CNN and multigrid methods for numerically solving PDEs [67]. The DNN is trained to learn the mappings between the discretized diffusion PDEs and the prolongation operators of the multi-grid method [68]. By regarding the CNN as a multi-period dynamical system, RKNet is proposed by combining Runge–Kutta methods with the CNN for better accuracy [69]. The multiscale ANN is designed to solve the nonlinear PDEs derived from the Schrödinger equation and the Kohn–Sham density functional theory by introducing the ANN in each spatial scale of hierarchical matrices [70].

2.3   Fully data-driven forward modeling

Fully data-driven forward modeling (FD) designs effective DNN architectures for specific scenarios of EMFD and trains the designed DNN with a large amount of data to learn the inner physical mappings by leveraging its powerful learning capacity. The trained DNN can provide a real-time response with fair precisions in the online computing of EMFD. The main disadvantage of a fully data-driven FD is that its performance depends on the quality of the training data and its generalization ability is limited. That means that the performance of a fully data-driven FD may deteriorate sharply when the data has a different distribution from the training one.

The first approach of the fully data-driven FD is to consider the solution of a matrix equation as an optimization problem of DNN parameters. In the EMFD, Maxwell's equations are solved by converting them into a system of matrix equations:

$$\mathbb{A} \cdot u = b, \tag{2.5}$$

where $\mathbb{A}$, u and b denote coefficient matrix, solution and right hand side. The multiplication between u and the ith row of $\mathbb{A}$ can be expressed as:

$$\sum_j \mathbb{A}_{i,j} \cdot u_j = b_i. \tag{2.6}$$

Similarly, the calculation of a single layer perceptron in the fully connected network (FCN) can be written as:

$$y = \sum_{j=1}^{J} w_j \cdot x_j + \xi, \tag{2.7}$$

where w_j and ξ denote the weight and bias of the perceptron. If ξ is neglected, the matrix multiplication is the same as the calculation of a single layer perceptron. The training of the FCN is to employ optimization algorithms to tune the weight parameters w_j. By regarding w_j as the jth element of the solution u_j, then solving a matrix equation is transformed as the optimization of the FCN, as shown in Figure 2.3. The input of the FCN is the ith row of the coefficient matrix $\mathbb{A}$ and the output of the FCN is the i-th element of the right-hand side. It is noted that the bias item in the calculation of

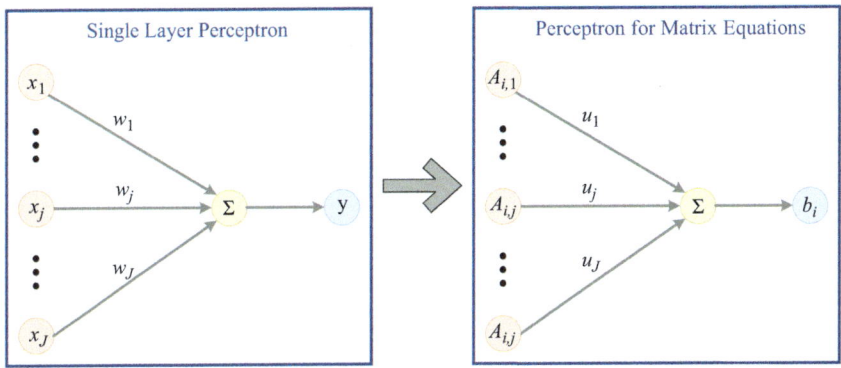

Figure 2.3 *Schematic of solving a matrix equation based on the fully connected layer*

FCN must be avoided. The objective function of the FCN can be expressed as the Euclidean distance for simplicity here:

$$obj = \|\mathbf{b}' - \mathbf{b}\|^2, \tag{2.8}$$

where $\mathbf{b}'$ and $\mathbf{b}$ denote the output of FCN and the ground truth. The optimization algorithms for training DNNs are usually gradient-based methods, including stochastic gradient descent (SGD) method [71], AdaGrad method [72], RMSProp method [73], and adaptive moment estimation (Adam) [74]. In [75], the Adam optimizer of DL is adopted to solve the matrix equations in the MoM. The rows of the coefficient matrix in the matrix equations are randomly selected at a certain ratio to train the FCN. Then the trained FCN can provide reliable solutions of the matrix equations with the same level of precisions as the traditional methods, including conjugate gradient, generalized minimal residual algorithm. As only a portion of the rows in the coefficient matrix are involved in the computation, the computational complexity is reduced compared with the traditional methods. The computational efficiency is further improved by parallel computation of the FCN based on the GPUs. The feasibility of solving matrix equations in low-frequency EM problems based on the FCN is further verified in [76]. The matrix equations of low-frequency EM problems usually have worse condition numbers especially in the cases of dielectric materials, and it is difficult for MoM to solve such matrix equations [77]. The FCN-based approach demonstrates good computational precisions and efficiency even for the cases of complicated structures [77]. A similar approach is adopted in [78] to accelerate the computation of MoM by applying FCNs to solve matrix equations for parasitic capacitance extraction.

The second fully data-driven approach is to design and train a DNN to learn the mappings between the desired physical quantities and the corresponding variables. This approach regards the DNN as a "black-box" approximator to learn the parametric

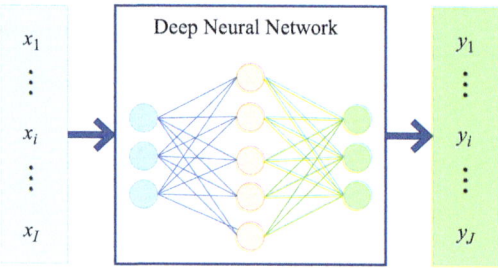

Figure 2.4   Schematic of fully data-driven electromagnetic forward modeling

function between different physical quantities. As shown in Figure 2.4, it can be expressed as:

$$y_1, \cdots, y_j, \cdots, y_J = \mathscr{F}(\Theta, x_1, \cdots, x_i, \cdots, x_I), \tag{2.9}$$

where $\mathscr{F}$, Θ denote the DNN and its parameter set, $y_1, \cdots, y_j, \cdots, y_J$ denote the J physical quantities in the DNN output, and $x_1, \cdots, x_i, \cdots, x_I$ denote the I physical quantities in the DNN input. The output and the input of the DNN should be chosen according to their physical relationship and the output is usually related to or dependent on the input. In the case where the DNN output is a single physical quantity, the objective function can adopt a metric function to evaluate the discrepancy between the DNN output and ground truth:

$$obj_{single} = \mathscr{M}(y_1', y_1^*), \tag{2.10}$$

where $\mathscr{M}$, y_1', and y_1^* denote the metric function, DNN output, and ground truth, respectively.

If the DNN is applied to model multiple physical quantities, its objective function needs to measure the discrepancy of all physical quantities:

$$obj_{multiple} = \alpha_1 \mathscr{M}_1(y_1', y_1^*) + \cdots + \alpha_j \mathscr{M}_j(y_j', y_j^*) + \cdots \alpha_J \mathscr{M}_J(y_J', y_J^*), \tag{2.11}$$

where y_j' and y_j^* denote the jth physical quantities output by the DNN and the jth ground truth, $\mathscr{M}_j$ denotes the jth metric functions, and α_j is the weight of the jth discrepancy. The multi-task learning scheme can be adopted to train the DNN to model multiple physical quantities by regarding each quantity as a task. It can improve the learning efficiency and reduce the computational load compared to building different models for different quantities. As the multiple physical quantities are usually related to each other, the generalization ability and computational precisions of the DNN model can be further improved by applying multiple objective functions to constrain the training. Despite the great advantages, it is still a big challenge to balance different tasks in the objective function (2.11). The performance of multi-task learning depends on the choice of the weights in (2.11). The multiple physical quantities may be measured on independent scales and their magnitude may have a huge difference. If the weights are simply uniform or manually tuned, it is usually far from optimal. It is prohibitively computationally expensive to search for the optimal weights of different tasks. Here, we introduce two multi-task learning schemes including weighing based on the uncertainty [79] and weighing based on the Pareto optimal solution [80].

The first multi-task learning scheme weighs different tasks based on homoscedastic uncertainty. In Bayesian modeling, the homoscedastic uncertainty is dependent on the tasks regardless of their input and it can capture the relative confidence of different tasks [79,81]. The Gaussian probability model can describe the homoscedastic uncertainty of forward modeling and it can be expressed as [79,81]:

$$\rho(y_j^*, x, \Theta) = \frac{1}{\sqrt{2\pi\sigma_j^2}} \exp\left(-\frac{\left\|y_j^* - \mathscr{F}_j(\Theta, x)\right\|^2}{2\sigma_j^2}\right), \tag{2.12}$$

where σ_i denotes the uncertainty of jth task, $x = [x_1, \cdots, x_i, \cdots, x_I]^T$ denotes the DNN input, y_j^* and $\mathscr{F}(\Theta, x)$ are the ground truth and the jth DNN output. Then, (2.12) can be re-written in the form of maximum likelihood [79,81]:

$$\log \rho(y_j^*, x, \Theta) \propto -\frac{\left\|y_j^* - \mathscr{F}_j(\Theta, x)\right\|^2}{2\sigma_j^2} - \log \sigma_j. \tag{2.13}$$

The Gaussian probability model of J output can be formulated as [79,81]:

$$\rho(y_1^*, \cdots, y_J^*, x, \Theta) = \rho(y_1^*, x, \Theta) \cdots \rho(y_j^*, x, \Theta) \cdots \rho(y_J^*, x, \Theta). \tag{2.14}$$

The maximum likelihood of (2.14) can be expressed as [79,81]:

$$\log \rho(y_1^*, \cdots, y_J^*, x, \Theta) \propto \sum_j -\frac{\left\|y_j^* - \mathscr{F}_j(\Theta, x)\right\|^2}{2\sigma_j^2} - \log \sigma_j. \tag{2.15}$$

It can be observed in (2.15) that the uncertainty σ_j is the coefficient of the discrepancy between the ground truth and the jth DNN output. By regarding the uncertainty σ_j as the weight of jth task, then (2.15) can be employed as the objective function of the multi-task learning scheme.

The second scheme links the multi-task learning to the multi-objective optimization problem [80]. In this scheme, the parameter set of Θ is divided into the shared one Θ_s and task-specific one Θ_j. Then (2.11) can be re-written as [80]:

$$obj_{multiple} = \sum_{j=1}^{J} \alpha_j \mathscr{M}_j(\Theta_s, \Theta_j). \tag{2.16}$$

The goal of multi-task learning is to find the optimal Θ_s, Θ_j that minimize the (2.16), which can be described as a multi-objective optimization problem [80]:

$$\min_{\Theta_s, \Theta_1, \cdots, \Theta_J} \mathbf{M}(\Theta_s, \Theta_1, \cdots, \Theta_J) = \min_{\Theta_s, \Theta_1, \cdots, \Theta_J} \sum_{j=1}^{J} \alpha_j \mathscr{M}_j(\Theta_s, \Theta_j), \tag{2.17}$$

where $\mathbf{M}(\Theta_s, \Theta_1, \cdots, \Theta_J) = [\mathscr{M}_1(\Theta_s, \Theta_1), \cdots, \mathscr{M}_j(\Theta_s, \Theta_j), \cdots, \mathscr{M}_J(\Theta_s, \Theta_J)]^T$ is a vector of the calculated discrepancies. The Pareto optimality is the optimization goal of (2.17) and it can be defined as [80]:

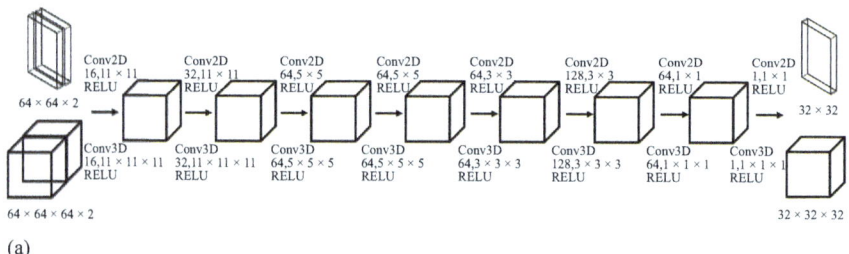

(a)

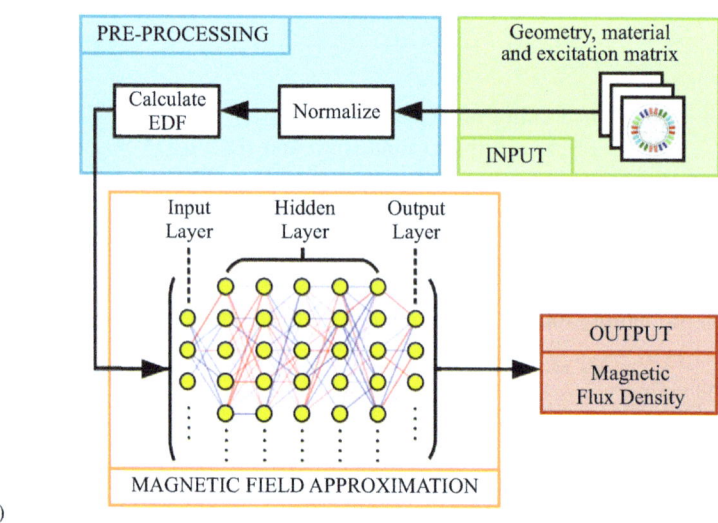

(b)

Figure 2.5   Fully data-driven approaches for electrostatic and magnetostatic field modeling, (a) the fully CNN for 2D and 3D Poisson's equations (source: [39]) and (b) modeling magnetic fields based on the DNN (source: [83])

Definition 1. *1. A solution Θ dominates a solution $\bar{\Theta}$ if $\mathcal{M}_j(\Theta_s, \Theta_t) \leq \mathcal{M}_j(\bar{\Theta}_s, \bar{\Theta}_t)$ for all tasks j and $\mathbf{M}_{\Theta_s, \Theta_1, \cdots, \Theta_J} \neq \mathbf{M}_{\bar{\Theta}_s, \bar{\Theta}_1, \cdots, \bar{\Theta}_J}$; 2. A solution Θ^* is called Pareto optimal if there exists no solution Θ that dominates Θ^*.*

The Pareto solutions can be obtained by the multiple gradient descent algorithm [82] that founds on the Karush–Kuhn–Tucker (KKT) conditions of (2.17).

The fully data-driven approach has been widely applied to model the forward scattering by solving different Maxwell's equations, including the electrostatic and magnetostatic field modeling, wave physics modeling, etc.

First, the full data-driven models are built based on DNNs to model the electrostatic and magnetostatic fields. The 1D Poisson's equation of homogeneous media with Dirichlet BCs is solved by building a multiple-input DNN model [84]. The activation function of the multiple-input DNN model is formulated based on the sinc- and cosine-type trial functions to emulate the governing physical equations [84]. A fast solver based on the fully CNN model is built to solve 2D and 3D Poisson's equation in [39], as shown in Figure 2.5(a). With the varying permittivity distributions and excitation source locations, both 2D and 3D Poisson's equations are solved by FDM to generate training and testing data sets. The fully CNN models for 2D and 3D Poisson's equations share the same architecture and reduce the computational time significantly compared to FDM. The magnetic fields at low frequencies are solved by separately building two different DNN models, including the ANN-based model, the modified U-Net model [83], as shown in Figure 2.5(b). In [83], a new pre-processing method of the DNN input is derived based on the analytical formula of magnetic potential for a more condensed and compact representation of the input matrix of geometry, excitations, and boundaries. With the pre-processed data, the computational precisions and training time of the DNN models are improved. The data-driven CNN is applied to model low-frequency electromagnetic devices by learning the mapping between the magnetic field distributions and the topologies of devices [85]. The training data of the proposed CNN model is generated by the FEM, and the Monte Carlo dropout is applied to improve the computational precisions and evaluate the confidence of predictions. In [86], a reduced-order model based on the CNN is presented to model the magnetization dynamics of magnetic thin film elements that play an important role in magnetic sensors. With the training data generated by solving the Landau–Lifshitz–Gilbert equation, the CNN encodes the magnetic states into the low-dimensional latent space to reduce the dimensionality for acceleration.

Second, wave physics can be modeled by the DNN models that are trained with massive physical data. The EM scattering problem is solved to predict the magnetic fields by building a CNN model in [87], as shown in Figure 2.6(a). Based on an encoder-decoder structure, the proposed CNN model combines the structures of U-Net and ResNet with the introduction of the skip connections. The training data is generated by applying 2D FDTD to solve the magnetic fields of random scatterers. The input of the CNN model consists of two matrices representing the scatterers and excitation sources and the output includes the real and imaginary parts of magnetic fields. The EMFDs also need to process sequential data, such as time-domain modeling, and wave propagation modeling. The recurrent neural networks (RNNs) are deep learning models for processing sequential data, including vanilla RNNs, long short-term memory network, and their variants. Based on such observation, RNNs are modified and trained to process sequential EM data or model time-domain EM phenomena. The LSTM model is trained to model the acoustic-magnetic radiation caused by underwater pressure wave [89]. FEM is applied to solve magneto-hydrodynamics equations to generate training and testing data samples. The proposed LSTM model demonstrates good precisions and significantly improved computational efficiency compared to the conventional numerical methods. An LSTM model is built to solve time-domain electric fields of various scatterers in 2D and 3D cases [88], as shown in Figure 2.6(b). Compared with the commercial software, the proposed LSTM model has good computational precision and efficiency. The LSTM model can learn the short-range and long-range dependencies between the temporal electrical fields and further perform reliable simulations of time-harmonic propagation. In [90], a forward model is built based on the ANN to accelerate the 3D EM simulations for the ground-penetrating

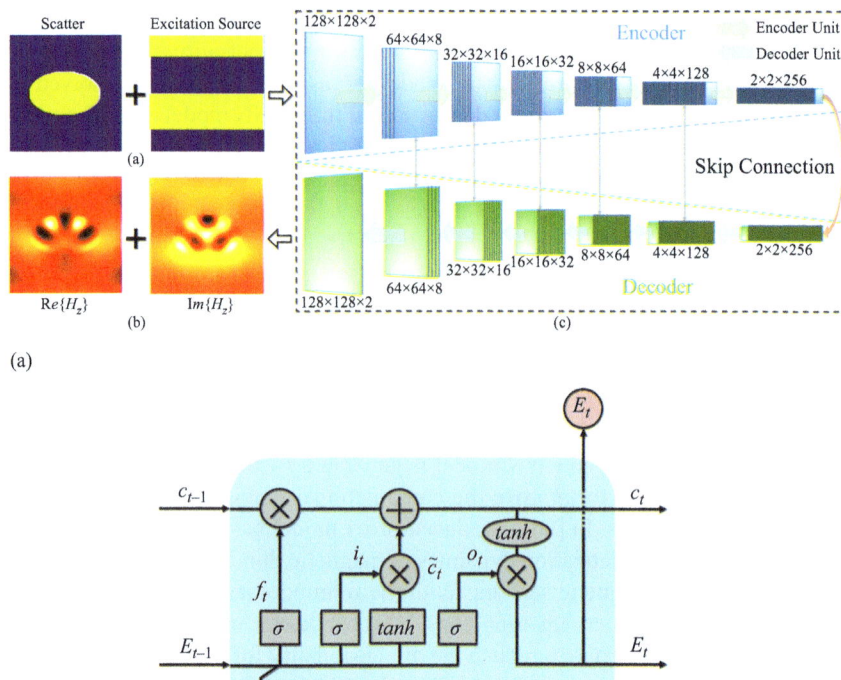

Figure 2.6 *Fully data-driven approach for modeling wave physics, (a) CNN model for electromagnetic scattering (source: [87]) and (b) LSTM cell for time domain electromagnetic scattering (source: [88])*

radar. The ANN-based solver can provide the near-real-time response compared to the traditional methods that have a high computational load, which can further improve the full-waveform inversion. The synthetic and real data are both used to verify the effectiveness of the ANN-based solver. The calculation of radar cross-section (RCS) is accelerated by the ANN model in [91]. The effectiveness of the ANN model is verified in the cases of both single target and multiple targets.

Third, the fully-data driven models can be trained to model the EM properties of microwave devices which lays the foundation for the design and optimization process. The parameter extraction of microwave filters is modeled in a high dimension by building an ANN model [92]. As the commonly used rectified linear unit (ReLU) is a switch function and not smooth for learning the continuous mapping, a modified and smooth ReLU is presented based on the quadratic function. With the S parameters of microwave filers as input, the ANN model can produce the corresponding coupling parameters. The effectiveness of the ANN model is verified by modeling a fourth-order cavity filter and a sixth-order multicoupled cavity filter. In [93], an ANN model consisting of two parts is presented for parametric modeling of passive microwave filters. The first part encodes the geometrical properties of filters and the encoded vectors along with the working frequency are taken as the input of the second part. The numerical results of the three-pole H-plane filter and fifth-order waveguide bandpass filter are demonstrated to validate the proposed ANN model. The modeling and design workflows of the inductor are proposed based on the artificial neural network (ANN) [94]. The training data of the ANN is generated by 3D FEM and all relevant factors affecting the modeling and design of inductors are taken into account. The channel simulation of Fin Field-Effect transistors is accelerated by building the ANN models to learn the mapping between the FinFET transistors and the corresponding EM properties [95]. The graph neural network (GNN) model is presented to compute the S parameters of the resonator filters in the design of distributed circuits [96]. The nodes and edges in the GNN model denote the resonators and their EM coupling effects in the circuit respectively. The node attributes include the parameters of resonators and the edge attributes contain the gap, shift, and the relative position between two resonators. The GNN model encodes the whole circuit layout into a global high-level representation and then a neural network performs predictions based on the encoded representations.

Fourth, the forward modeling of nano-structures EM properties also benefits from the fully data-driven approaches [98,99], including modeling photonic devices [100, 101], modeling scattering-spectra [102,103], predicting electric polarization [97], etc. The recurrent U-Net with residual and shortcut connections is proposed to model long-distance coupling effects by predicting the near field in a large neighborhood of nanopillars [99]. The optical properties of the photonic crystal fibers can be predicted by the ANN model, including effective index, effective mode area, dispersion loss, etc. [100]. The CNN and FCN are combined as the response predictor to model the responses of the target nanostructures and the response predictor is integrated into the inverse design for acceleration [101]. The DL framework based on U-Net is proposed to model the EM wave scattering of nano-structures by avoiding solving Maxwell's equations [102]. The proposed DL framework takes as input the incident fields and the scatterer images, then outputs the corresponding real and imaginary parts of magnetic fields. The light scattering of multilayer nanoparticles is modeled based on the ANN with orders of magnitude speeding up of computation [103]. The 3D fully convolutional neural network is built to model the nano-optical effects of nanostructures [97], and the workflow is shown in Figure 2.7. The proposed 3D fully CNN predicts the real and imaginary parts of the electric fields ($x-, y-, z-$ components) with the volume discretization of nanostructures. Then multiple related physical quantities can be derived based on the predicted electric fields, i.e. cross-sections, far-field scattering pattern, near fields, etc.

Fifth, multi-task learning can help model different but related physical quantities in the same physical or mathematical model. It also provides a new way to weigh different parts in the objective function of DNN models. In [104], the multi-task learning is applied in the 2D forward modeling of magnetotellurics (MT) to predict

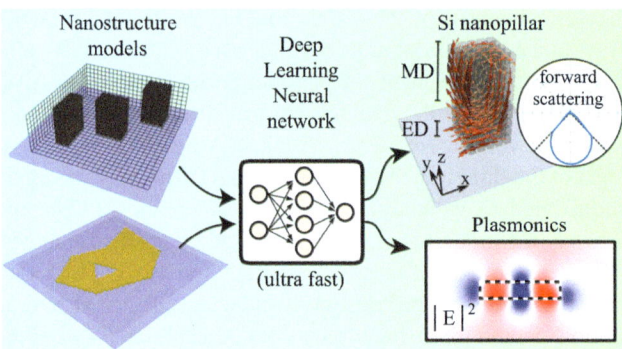

*Figure 2.7 Workflow of modeling nanophotonics based on deep learning
(Reprinted (adapted) with permission from [97]. Copyright 2022
American Chemical Society.)*

the apparent resistivity and impedance phase. The losses of sub-tasks in the objective function are weighed based on the homoscedastic uncertainty and the structural similarity regularization is incorporated to improve the precisions of predictions. The proposed CNN model has an identical encoder and two independent decoders for two physical quantities, as shown in Figure 2.8. The dilated convolutions and atrous spatial pyramid pooling are adopted due to the inner multi-scale characteristics of MT problems. An end-to-end DNN model is presented to optimize and design the 3D chiral metamaterials under the constraints of circular dichroism, right and left circularly polarized spectra [105]. The multi-task joint learning is implemented to realize the forward modeling and inverse design of chiral metamaterials in a bidirectional DNN model by avoiding the auxiliary networks for different physical quantities. Although the applications of multi-task learning have not yet been widely launched in EMFDs, it poses a great potential for hybrid modeling in many fields, such as residual value forward modeling in the business applications [106], spatial–temporal hydrological modeling [107], etc.

2.4   DL-assisted forward modeling

The DL-assisted forward modeling aims to design DNN models to replace the parts with high computational complexity in traditional EM modeling methods. With the trained DNN models, the online computation of traditional EM modeling methods can be speeded up. As shown in Figure 2.9, the computation of a traditional EM modeling method can be divided into several parts, and the parts with high computational complexity can be replaced by the trained DNN. Although the DNN is integrated into the traditional EM modeling method, it is still regarded as a "black-box" approximator to learn the inner mappings from the massive data.

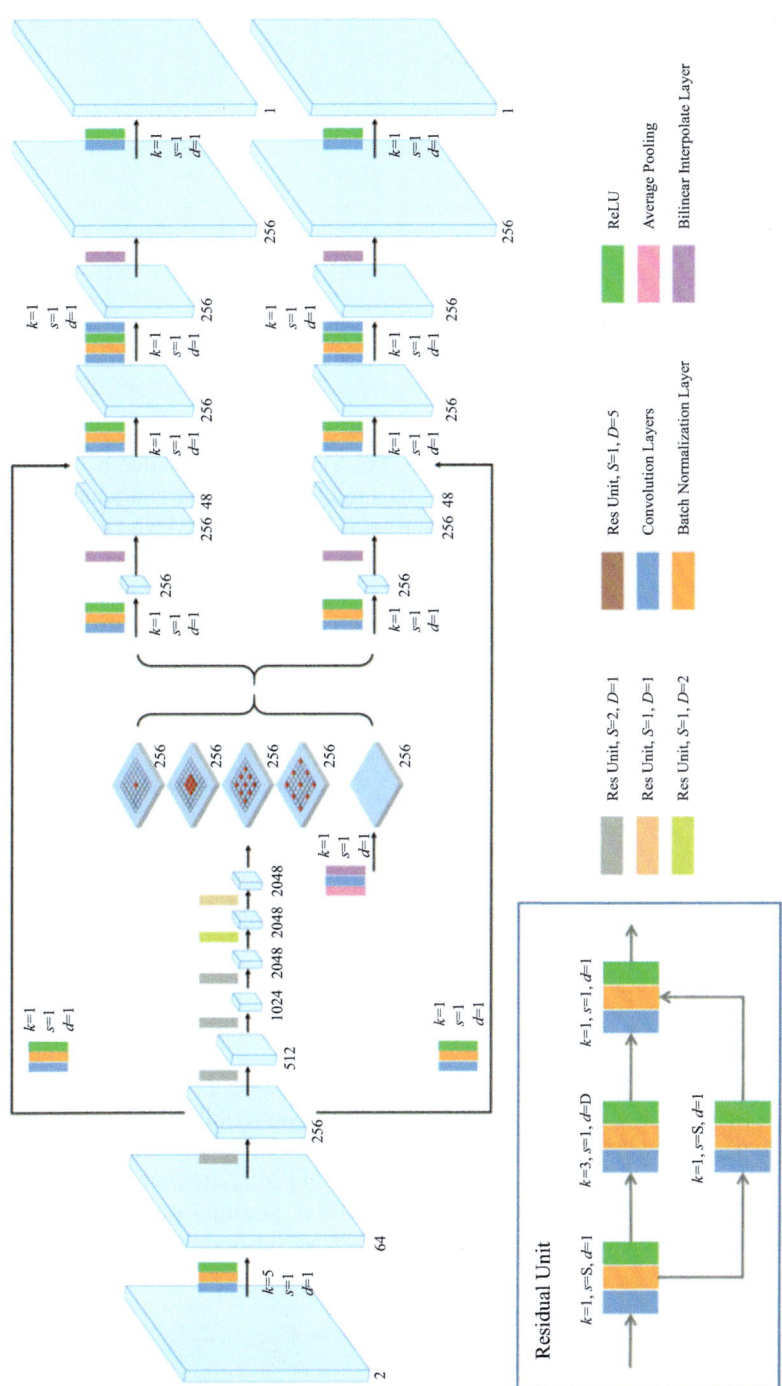

Figure 2.8 Multi-task learning convolutional neural network (source: [104])

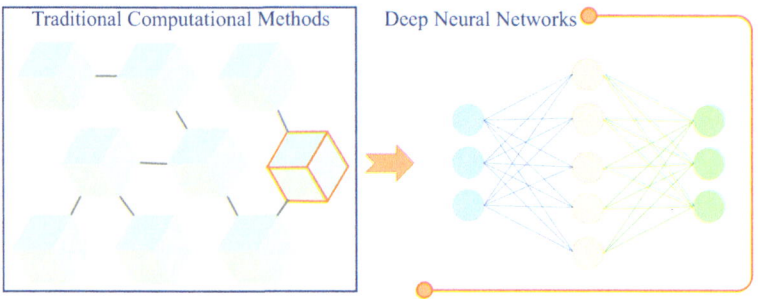

Figure 2.9   Schematic of DL-assisted forward modeling

First, the absorbing boundary conditions (ABCs) of the finite difference time domain method can be modeled based on the DNN models to improve computational efficiency. The hyperbolic tangent basis function neural network (HTBFNN) is built to approximate the perfectly matched layer (PML) that is one of the important absorbing boundary conditions (ABCs) for FDTD [31]. There are four components in the PML with Yee cell, including E_x, E_y, H_{zx}, and H_{zy}, and their relationship can be expressed as [31]:

$$E_{x,i,j+\frac{1}{2}}^{n+\frac{1}{2}} = e^{-\sigma_{y,j+\frac{1}{2}}\frac{\Delta t}{\varepsilon_0}} E_{x,i,j+\frac{1}{2}}^{n-\frac{1}{2}} - \frac{1-e^{-\sigma_{y,j+\frac{1}{2}}\frac{\Delta t}{\varepsilon_0}}}{\Delta y \sigma_{y,j+\frac{1}{2}}} \left[H_{z,i,j}^n - H_{z,i,j+1}^n \right]$$

$$E_{y,i+\frac{1}{2},j}^{n+\frac{1}{2}} = e^{-\sigma_{x,i+\frac{1}{2}}\frac{\Delta t}{\varepsilon_0}} E_{y,i+\frac{1}{2},j}^{n-\frac{1}{2}} - \frac{1-e^{-\sigma_{x,i+\frac{1}{2}}\frac{\Delta t}{\varepsilon_0}}}{\Delta x \sigma_{y,i+\frac{1}{2}}} \left[H_{z,i+1,j}^n - H_{z,i,j}^n \right]$$

$$H_{zx,i,j}^{n+1} = e^{-\sigma_{mx,i}\frac{\Delta t}{\mu_0}} H_{zx,i,j}^n - \frac{1-e^{-\sigma_{mx,i}\frac{\Delta t}{\mu_0}}}{\Delta x \sigma_{mx,i}} \left[E_{y,i+\frac{1}{2},j}^{n+\frac{1}{2}} - E_{y,i-\frac{1}{2},j}^{n+\frac{1}{2}} \right]$$

$$H_{zy,i,j}^{n+1} = e^{-\sigma_{my,j}\frac{\Delta t}{\mu_0}} H_{zy,i,j}^n - \frac{1-e^{-\sigma_{my,j}\frac{\Delta t}{\mu_0}}}{\Delta x \sigma_{mx,j}} \left[E_{x,i,j+\frac{1}{2}}^{n+\frac{1}{2}} - E_{x,i,j-\frac{1}{2}}^{n+\frac{1}{2}} \right]$$

$$(2.18)$$

where δ is the thickness of PML, $\sigma(\rho)$ is the conductivity at the distance ρ from the interface and σ_{max} denotes the conductivity at the outmost layer of PML. Trained with the field data of the first cell layer in PML, the HTBFNN can replace PML in FDTD and significantly reduce the computational complexity of PML. The configuration of the PML based on HTBFNN is illustrated in Figure 2.10(a). The input of the HTBFNN is chosen based on the configuration of the PML [31], as shown in Figure 2.10(a):

$$\mathbf{x}_{ij}^t = \left[H_{z,i,j+1}^t, H_{z,i,j}^t, H_{z,i,j-1}^t, H_{z,i+1,j}^t, E_{x,i,j+\frac{1}{2}}^t, E_{x,i,j-\frac{1}{2}}^t, \right.$$

$$(2.19)$$

$$\left. E_{x,i,j-\frac{3}{2}}^t, E_{x,i+1,j-\frac{1}{2}}^t, E_{y,i+\frac{1}{2},j+1}^t, E_{y,i+\frac{1}{2},j}^t, E_{y,i+\frac{1}{2},j-1}^t, E_{y,i+\frac{3}{2},j}^t \right]$$

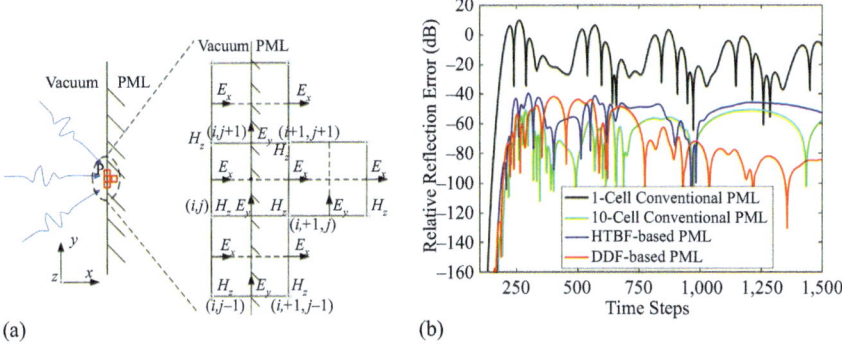

(a) (b)

Figure 2.10 *Deep learning based perfectly matched layer in FDTD, (a) source:*
[31] and (b) source: [109]

the output is the field on the interface between the target domain and the PML [31], as shown in Figure 2.10(a):

$$\mathbf{y}_{ij}^t = \left[H_{z,i+1,j}^{t+1}, \quad E_{x,i+1,j-\frac{1}{2}}^{t+1}, \quad E_{y,i+\frac{3}{2},j}^{t+1} \right] \tag{2.20}$$

The HTBFNN-based PML demonstrates a similar error level to the conventional 5-cell PML. Furthermore, the enhanced LSTM-based PML is proposed to accelerate the computation of FDTD in [108]. The LSTM-based PML is also trained by the field data in the first cell layer of PML and demonstrates better computational precisions than the HTBFNN-based PML. The learned perfectly matched monolayer is trained based on the deep differentiable forest (DDF) to replace the conventional multilayer PML of FDTD [109]. Deep differentiable forest combines the strengths of the classification trees and the learning functionality based on the DNNs [110]. The unsplit field scheme is adopted to enable DDF-based PML good absorption performance over a large time span, as shown in Figure 2.10(b).

Second, the evolution of the FDTD method in the time domain can be modeled by the DNN models, especially the recurrent neural networks (RNNs). The temporal evolution of fields is modeled by building a DNN model in transient electrodynamics [111]. Adopting an encoder-decoder structure, the proposed DNN model consists of an encoder, an LSTM, and a decoder. They are all built based on the convolutional operations and residual blocks. The convolutional encoder extracts the features of input data into the latent vectors, the convolutional LSTM stimulates the temporal dynamics of wave physics with the encoded latent vectors as input, the decoder outputs the desired electric and magnetic fields. The computational domain size of the DNN model is extended based on the non-overlapping domain decomposition scheme. The initial value problem in the time-domain simulation can be described as [111]:

$$u^* = \mathbb{D}u, \quad u(t) = u_t, \tag{2.21}$$

where $\mathbb{D}$ denotes the spatial discretization matrix. Then the solution of (2.21) at t time step can be expressed as [111]:

$$u_{t+1} = e^{\mathbb{D}t}u_t = (\mathbb{I} + \mathbb{D}t + \frac{1}{2!}\mathbb{A}^2 t^2 + \cdots + \frac{1}{n!}\mathbb{A}^n t^n + \cdots)u_t \tag{2.22}$$

The matrix exponential operator $e^{\mathbb{D}t}$ can be evaluated by the Runge–Kutta 4 (RK4) integration scheme. When the size of the computational domain grows larger, the computational efficiency of RK4 could deteriorate due to the matrix–vector multiplications in the RK4 integration scheme. Therefore the preconditioning technique is utilized to decompose the matrix $\mathbb{D}$ into a block-diagonal matrix $\mathbb{B}$ and a remainder matrix $\mathbb{R} = \mathbb{D} - \mathbb{B}$. Then the matrix exponential operator can be approximated in block-wise fashion [111]:

$$e^{\mathbb{B}} = \text{diag}\left(e^{\mathbb{B}_1}, \ldots, e^{\mathbb{B}_n}\right) \tag{2.23}$$

As the remainder matrix, $\mathbb{R}$ is highly sparse with the information of the coupling effect, the computational load of matrix multiplications with $\mathbb{R}$ is acceptable. Motivated by this integration scheme, the matrix exponential is replaced by the DNN model as the basic building block in the RK4 scheme, and then the DNN model can extend its size of the computational domain. In [111], the proposed DNN model is trained on a 128×128 domain and its effectiveness is verified on the 256×256 and 512×512 domains.

Third, the DNN models are trained to replace the parts with high complexity to speed up the computation of MoM. A forward-induced current learning method (FICLM) is presented based on the pix2pix generative adversarial network (GAN) to model the induced currents in the electromagnetic scattering problems, and then the scattered fields can be calculated based on the predicted induced currents [112], as shown in Figure 2.11. The FICLM regards the induced current predictions as the image-to-image translations and the input and output of the GAN are selected based on the physical relationship [112]:

$$\mathbf{J} = \left(\mathbb{I} - \chi_{\text{diag}} \cdot \mathbb{G}_D\right)^{-1} \cdot \left(\chi_{\text{diag}} \cdot \mathbf{E}^{inc}\right) \tag{2.24}$$

where $\mathbb{I}$, $\mathbf{J}$, $\mathbb{G}_D$, $\mathbf{E}^{inc}$, and χ_{diag} denote the identity matrix, induced current, Green's function, incident field, and the diagonal contrast matrix. Three input scheme are demonstrated in [112]: χ_{diag}, $\chi_{\text{diag}} \cdot \mathbf{E}^{inc}$, and their concatenations $\chi_{\text{diag}} \oplus \chi_{\text{diag}} \cdot \mathbf{E}^{inc}$, as shown in Figure 2.11. The ANN is employed to accelerate the computation of numerical Green's function in the EM scattering of multiple dielectric cylinders [113]. The training data of the ANN is generated by extracting the numerical Green's function (NGF) based on the electric field integral equation with the free-space Green's function. With the extracted NGF, the ANN is trained to learn the mappings between the NGF matrix elements and their Cartesian coordinates. Then, the NGF can be replaced by the trained ANN that has fewer parameters than the NGF in the online computation. In [30], the multilevel fast multipole algorithm (MLFMA) is accelerated by eliminating the EM interactions progressively and iteratively. During the eliminations, the ANN is trained to determine which EM interactions

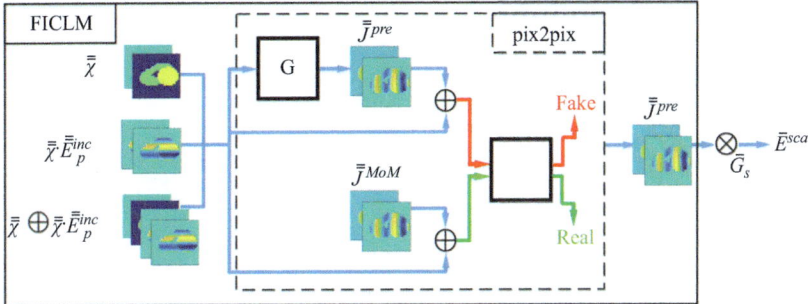

*Figure 2.11   Workflow of the forward-induced current learning method
(source: [112])*

can be omitted in the far zone by predicting the error levels of the equivalent surface currents.

In [114], the translator function of the MLFMA is learned based on the generalized regression neural network (GRNN) to improve the computational efficiency. The translator function of the MLFMA can be written as [114]:

$$T_p \left(\hat{k}_p, \hat{r} \right) = \sum_{l=0}^{L} i^l (2l + 1) h_l^{(2)}(kr) P_l \left(\hat{k}_p \cdot \hat{r} \right) . \tag{2.25}$$

According to (2.25), the GRNN takes as input the $\hat{k}_p$, $\hat{r}$, and the level label, and produces the real and imaginary parts of $T_p \left(\hat{k}_p, \hat{r} \right)$. The effectiveness of the GRNN is verified in applying the two-level 2D MLFMA to solve the RCS of a perfect electrically conductor. Furthermore, a hybrid method is presented to combine the ANN and GRNN to model the translation phase of MLFMA to improve the computational efficiency [115]. In the 2D MLFMA, the translator can be simplified from (2.25) [114]:

$$\widetilde{\alpha}_{m'm}(\alpha) = \sum_{p=-P}^{P} H_p^{(1)} \left(k\rho_{m'm} \right) e^{-ip(\phi_{m'm} - \alpha + \pi/2)} \tag{2.26}$$

The ANN and GRNN approximate the translation function at the coarse levels and fine levels respectively. Both of them take $\rho_{m'm}$ and $\phi_{m'm}$ as input, then produce the real and imaginary parts of $\widetilde{\alpha}_{m'm}(\alpha)$. The effectiveness of the proposed hybrid method is validated by calculating the bistatic RCS of different scatterers.

Fourth, the forward modeling and design of microwave devices can be assisted and speeded up by deep learning techniques. The model-order reduction based on the neuro-transfer function models is presented to calculate the frequency responses of the microwave passive components [116,117]. With the FEM and model-order

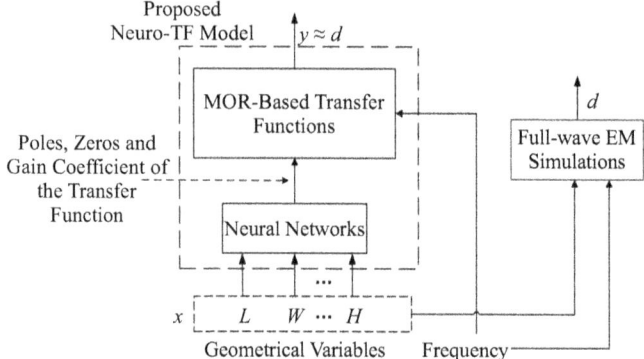

Figure 2.12   The neuro-transfer function model (source: [116])

reduction, the transfer function of a microwave passive structure can be described as [116,117]:

$$H(x, \omega) = K(x)\gamma(\omega)\frac{\prod_{i=1}^{q-1}(\sigma(\omega) - z_i(x))}{\prod_{i=1}^{q}(\sigma(\omega) - p_i(x))} + c \qquad (2.27)$$

where $K(x)$, $p_i(x)$, and $z_i(x)$ are the gain coefficient, poles, and zeros, respectively, q denote the order of $H(x, \omega)$, ω is the frequency, c is a constant. In the neuro-transfer function model, the DNNs are applied to approximate the gain coefficient, poles and zeros as $K_{NN}(x, \Theta_K)$, $P_{NN}(x, \Theta_P)$, $Z_{NN}(x, \Theta_Z)$ respectively, as shown in Figure 2.12. Then the DNN approximations are input into the transfer function to calculate the frequency responses. In [118], two CNNs are combined with the binary particle swarm optimization to accelerate the topology optimization of a synchronous reluctance motor. The first CNN is built to evaluate the torque properties of the motor, and the second CNN helps accelerate the computation of the finite element method by learning the mapping between the BCs and the magnetic fields.

2.5   Physics-inspired forward modeling

Physics-inspired forward modeling designs and trains the DNN models by drawing on the physical or mathematical models and their numerical algorithms. By incorporating physics or mathematics, the DNN models boast improved robustness and interpretability, which further enables the better inner reasonings of DNN models.

First, the physics-informed neural networks (PINNs) are applied to various EMFD scenarios. The PINNs are constrained and guided by introducing the governing equations of EMFD scenarios into the objective functions. The Wave Y-Net of an encoder-decoder structure is proposed based on the U-Net as a fast EM simulator of the periodic dielectric nanoridge array [120]. Similar to the PINNs, the objective function of Wave Y-Net not only includes the data loss between the ground truth and predictions but also incorporates the governing Maxwell's equation with proper BCs, which can be expressed as [120]:

$$\mathcal{M} = \mathcal{M}_{\text{data}} + \omega\mathcal{M}_{\text{physics}}$$

$$\mathcal{M}_{\text{data}} = \frac{1}{N} \sum_{n=1}^{N} \|H^n - H^{n*}\|_1$$

$$\mathcal{M}_{\text{physics}} = \frac{1}{N} \sum_{n=1}^{N} \left\| \nabla \times \left(\frac{1}{\varepsilon^n} \nabla \times H^{n*} \right) - \omega^2 \mu_0 H^{n*} \right\|_1$$

(2.28)

where ω balances the contributions of the governing Maxwell's equation and it is determined by trial and error, $\| \cdot \|_1$ denotes the mean absolute error, H^n and H^{n*} denote the true and predicted fields. With the training data generated by FDTD, the WaveY-Net takes the nanoridge array structure as input and outputs the corresponding magnetic fields. Although the introduction of $\mathcal{M}_{\text{physics}}$ demonstrates a modest improvement in computational precisions, it enables the magnetic fields predicted by the WaveY-Net self-consistent with the magnetic field wave equation. Furthermore, the electric fields derived from the predicted magnetic fields are more accurate and also consistent with the electric field wave equation. The beam dynamics of a particle accelerator is studied by building a PINN to model the electromagnetic coupling of a particle beam [121]. The PINN is trained to calculate the space charge fields of particle beams in the accelerator vacuum chamber. Different from the one in [120], the employed objective function is directly defined as the sum of mean squared error (MSE) of the governing wave equation and the corresponding BCs, which can be written as [121]:

$$\mathcal{M} = \mathcal{M}_{\text{PDE}} + \mathcal{M}_{\text{BC}}$$

$$\mathcal{M}_{\text{PDE}} = \frac{1}{N_{PDE}} \sum_{p=1}^{N_{PDE}} \left| \left(\frac{\partial^2}{\partial x^2} + \frac{\partial^2}{\partial y^2} \right) E_z^* - \frac{k^2}{\gamma^2} E_z^* + \frac{jk}{\varepsilon_0 \gamma^2} \rho_\perp \right|^2$$

$$\mathcal{M}_{\text{BC}} = \frac{1}{N_{BC}} \sum_{p=1}^{N_{BC}} |E_z^*|^2$$

(2.29)

where $E_z^* = E_z^*(x_p, y_p; \Theta)$ is the predicted fields of DNNs parameterized by Θ, N_{PDE} and N_{BC} are numbers of sampling points. Equation (2.29) separately enforces the predicted fields to satisfy the governing wave equations and the corresponding BCs. In (2.29), $\mathcal{M}_{\text{PDE}}$ and $\mathcal{M}_{\text{BC}}$ calculate MSE of the E_z^* sampled inside the target domain and on the boundary respectively. The effectiveness of the proposed PINN is validated by calculating space charge fields in accelerator vacuum chambers and the PINN predictions demonstrate good computational accuracy compared with the analytical solutions. In [119], the PINN is trained by an unsupervised training scheme to solve Maxwell's equations in the time domain, with no need of label data. The PINN takes the temporal and spatial variables as input and predicts the corresponding electric and magnetic fields. According to the uniqueness theorem, the solutions of Maxwell's

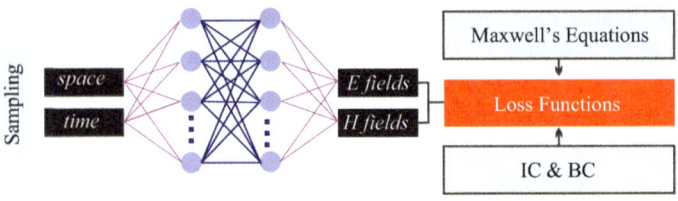

Figure 2.13   Schematic of physics-informed objective functions (source: [119])

equations can be determined by the BCs and ICs. Motivated by this, the objective function of the PINN combines the governing wave equations with the corresponding initial conditions (ICs) and BCs to constrain the training process, as shown in Figure 2.13. The objective function can be described as [119]:

$$\mathcal{M} = \mathcal{M}_{PDE} + \mathcal{M}_{BC} + \mathcal{M}_{IC}$$

$$\mathcal{M}_{PDE} = \frac{1}{N_{PDE}} \sum_{i=1}^{N_{PDE}} \left(\frac{1}{\mu} \left| \nabla \times E_i^* + \mu \frac{\partial H_i^*}{\partial t} \right|^2 \right.$$

$$\left. + \frac{1}{\varepsilon} \left| \nabla \times H_i^* - \varepsilon \frac{\partial E_i^*}{\partial t} - \sigma E_i^* - J_i \right|^2 \right) \tag{2.30}$$

$$\mathcal{M}_{BC} = \frac{1}{N_{BC}} \sum_{i=1}^{N_{BC}} \left(|E_i^*|^2 + |(\nabla \times H_i^*)|^2 \right)$$

$$\mathcal{M}_{IC} = \frac{1}{N_{IC}} \sum_{i=1}^{N_{IC}} \left(|E_i^* - E_i| + |H_i^* - H_i| \right)$$

where E_i^* and E_i, H_i^* and H_i denote the predicted and true electric and magnetic fields, N_{PDE}, N_{BC}, and N_{IC} are numbers of sampling points. It is noted that $\mathcal{M}_{WavePDE}$, $\mathcal{M}_{BC}$, and $\mathcal{M}_{IC}$ are defined on different temporal or spatial domains. The limited-memory BFGS algorithm is used to tune the parameters of the PINN to minimize (2.30). The homogeneous and inhomogeneous media are taken as numerical examples to verify the efficacy of the proposed PINN.

The objective function of the PINNs usually comprises multiple components and the multi-task learning scheme can be employed to weigh them for better performance. In [122], the PDEs with point source are solved based on the PINNs, including 2D time-domain wave equations, Poisson's equation, the governing PDEs of the Barry and Mercer's source problem. The multi-task learning scheme applies lower bound-constrained uncertainty to weigh different loss components in the objective function of the PINN. The objective function in [122] adopts the weighted sum of the wave equation, BCs, and ICs:

$$\mathcal{M} = \mathcal{M}_{PDE} + \omega_{BC} \mathcal{M}_{BC} + \omega_{IC} \mathcal{M}_{IC} \tag{2.31}$$

As the point source is approximated by the Dirac delta function, the $\mathcal{M}_{\text{PDE}}$ tends to be dominated by the sampling points around the origin of the Dirac delta function, which further makes it hard to train and optimize the PINN. Therefore, the target domain is divided into two subdomains, one containing the origin and one covering the complementary regions [122]:

$$\Omega = \Omega_0 \cup \Omega_1 \quad \Omega_0 \cap \Omega_1 = \emptyset \tag{2.32}$$

Then $\mathcal{M}_{\text{PDE}}$ can be evaluated separately on Ω_0 and Ω_1 [122]:

$$\mathcal{M}_{\text{PDE}} = \omega_{\text{PDE},0}\mathcal{M}_{\text{PDE},0} + \omega_{\text{PDE},1}\mathcal{M}_{\text{PDE},1} \tag{2.33}$$

With such decomposition, (2.31) can be written as [122]:

$$\mathcal{M} = \omega_{\text{PDE},0}\mathcal{M}_{\text{PDE},0} + \omega_{\text{PDE},1}\mathcal{M}_{\text{PDE},1} + \omega_{\text{BC}}\mathcal{M}_{\text{BC}} + \omega_{\text{IC}}\mathcal{M}_{\text{IC}} \tag{2.34}$$

The weights of different components in (2.34) play an important role in the performance of the PINN. Based on the multi-task learning scheme in [79], the uncertainty with lower bound is used to weigh the components of (2.34) [122] :

$$\mathcal{M} = \sum_{i=1}^{m} \frac{1}{2\left(\omega_i'\right)} \mathcal{M}_i + \log\left(\omega_i'\right) , \tag{2.35}$$

where $\omega_i' = w_i^2 + \varepsilon^2$ denotes the uncertainty, w_i is trainable and the lower bound is controlled by ε^2.

Second, the connections between RNNs and dynamics of wave physics in the time domain are investigated. The time-varying dynamics of wave physics is first linked to the computation of RNN to build analog signal processors in [123]. Built by operating the system of wave physics as an RNN, the wave-based RNN can be employed to process signals in an analogue way by the standard RNN training techniques. The standard RNN can be expressed as [123]:

$$\begin{aligned} h_t &= \sigma_h\left(W_h \cdot h_{t-1} + W_x \cdot x_t\right) \\ y_t &= \sigma_y\left(W_y \cdot h_t\right) \end{aligned} \tag{2.36}$$

where h_t denotes the hidden states, $\sigma_h(\cdot)$ and $\sigma_y(\cdot)$ are nonlinear activation functions for the hidden states and output, W_h, W_x, W_y are the weights of RNN. The scalar wave dynamics governed by a second-order PDE is taken into account [123]:

$$\frac{\partial^2 u(x,y,z,t)}{\partial t^2} - c^2 \nabla^2 u(x,y,z,t) = f(x,y,z,t) , \tag{2.37}$$

where $\nabla^2 = \frac{\partial^2}{\partial x^2} + \frac{\partial^2}{\partial y^2} + \frac{\partial^2}{\partial z^2}$, $c = c(x,y,z)$ is the wave speed, and $f(x,y,z,t)$ is the excitation source. With Δt denoting the time step, then (2.37) can be discretized as [123]:

$$\frac{u(t+1) - 2u(t) + u(t-1)}{\Delta t^2} - c^2 \nabla^2 u(t) = f(t) \tag{2.38}$$

Then (2.38) can be written in the form of a matrix equation [123]:

$$\begin{bmatrix} u(t+1) \\ u(t) \end{bmatrix} = \begin{bmatrix} 2 + \Delta t^2 \cdot c^2 \cdot \nabla^2 & -1 \\ 1 & 0 \end{bmatrix} \cdot \begin{bmatrix} u(t) \\ u(t-1) \end{bmatrix} + \Delta t^2 \cdot \begin{bmatrix} f(t) \\ 0 \end{bmatrix} \tag{2.39}$$

Defining $h_t = \begin{bmatrix} u(t) \\ u(t-1) \end{bmatrix}$ as the hidden state, (2.39) can operate as the update equation of a standard RNN and it can be re-written as [123]:

$$h_t = \mathbb{A}\,(h_{t-1}) \cdot h_{t-1} + \mathbb{P}^{(i)} \cdot x_t$$
$$y_t = |\mathbb{P}^{(o)} \cdot h_t|^2$$

(2.40)

where $\mathbb{P}^{(o)}$ and $\mathbb{P}^{(i)}$ are two linear projectors of input and output respectively. The similarity between the Laplacian operator ∇^2 and the convolution operation is also applied in the update of h_t [123]:

$$\nabla^2 u(t) = \frac{1}{\Delta s^2} \begin{bmatrix} 0 & 1 & 0 \\ 1 & -4 & 1 \\ 0 & 1 & 0 \end{bmatrix} * u(t)$$

(2.41)

where Δs is the spacial step. The effectiveness of the wave-based RNN is verified by classifying vowels based on the continuous speech recordings.

The theory-guided recurrent neural network (RNN) is built to emulate the computation of the FDTD method for modeling the electromagnetic wave propagation in [124], as shown in Figure 2.14. The 2D FDTD method for the transverse magnetic mode is taken into account [124]:

$$H_{x,i,j+\frac{1}{2}}^{n+\frac{1}{2}} = \alpha_1(m)H_{x,i,j+\frac{1}{2}}^{n-\frac{1}{2}} - \alpha_2(m)\frac{E_{z,i,j+1}^n - E_{z,i,j}^n}{\Delta y}$$

$$H_{y,i+\frac{1}{2},j}^{n+\frac{1}{2}} = \alpha_1(m)H_{y,i+\frac{1}{2},j}^{n-\frac{1}{2}} - \alpha_2(m)\frac{E_{z,i+1,j}^n - E_{z,i,j}^n}{\Delta x}$$

$$E_{z,(i,j)}^{n+1} = \alpha_3(m)E_{z,i,j}^n + \alpha_4(m)\left[\frac{H_{y,i+\frac{1}{2},j}^{n+\frac{1}{2}} - H_{y,i-\frac{1}{2},j}^{n+\frac{1}{2}}}{\Delta x} - \frac{H_{x,i,j+\frac{1}{2}}^{n+\frac{1}{2}} - H_{x,i,j-\frac{1}{2}}^{n+\frac{1}{2}}}{\Delta y} - J_z^{n+\frac{1}{2}}\right]$$

$$\alpha_1(m) = \frac{\frac{\mu(m)}{\Delta t} - \frac{\sigma_m(m)}{2}}{\frac{\mu(m)}{\Delta t} + \frac{\sigma_m(m)}{2}}, \quad \alpha_2(m) = \frac{1}{\frac{\mu(m)}{\Delta t} + \frac{\sigma_m(m)}{2}}$$

$$\alpha_3(m) = \frac{\frac{\varepsilon(m)}{\Delta t} - \frac{\sigma(m)}{2}}{\frac{\varepsilon(m)}{\Delta t} + \frac{\sigma(m)}{2}}, \quad \alpha_4(m) = \frac{1}{\frac{\varepsilon(m)}{\Delta t} + \frac{\sigma(m)}{2}},$$

(2.42)

where n denotes the time step, and the Yee space lattice is used. It can be observed from (2.5) that the $H_x^{n+1/2}(i, j+\frac{1}{2})$, $H_y^{n+1/2}(i+\frac{1}{2},j)$, and $E_z^{n+1}(i, j)$ of the present time step can be calculated by the differentiating the ones of the previous time step. The $\alpha_1(m)$, $\alpha_2(m)$, $\alpha_3(m)$, and $\alpha_4(m)$ are material-related and maintain unchanged over time and they can be regarded as constants. This update scheme in the time domain is consistent with the conventional RNNs but the specific update rule in each time step is different. Therefore, the FDTD can be described in the form of RNN by implementing (2.5) as the RNN update rule. In this way, the modified RNN can emulate the computing process of FDTD and achieve improved computational efficiency compared with the conventional FDTD. If the magnetic and electric fields are known but the $\alpha_1(m)$, $\alpha_2(m)$, $\alpha_3(m)$, and $\alpha_4(m)$ are unknown, this RNN can be easily extended to solve inverse modeling problem to retrieve the material-related

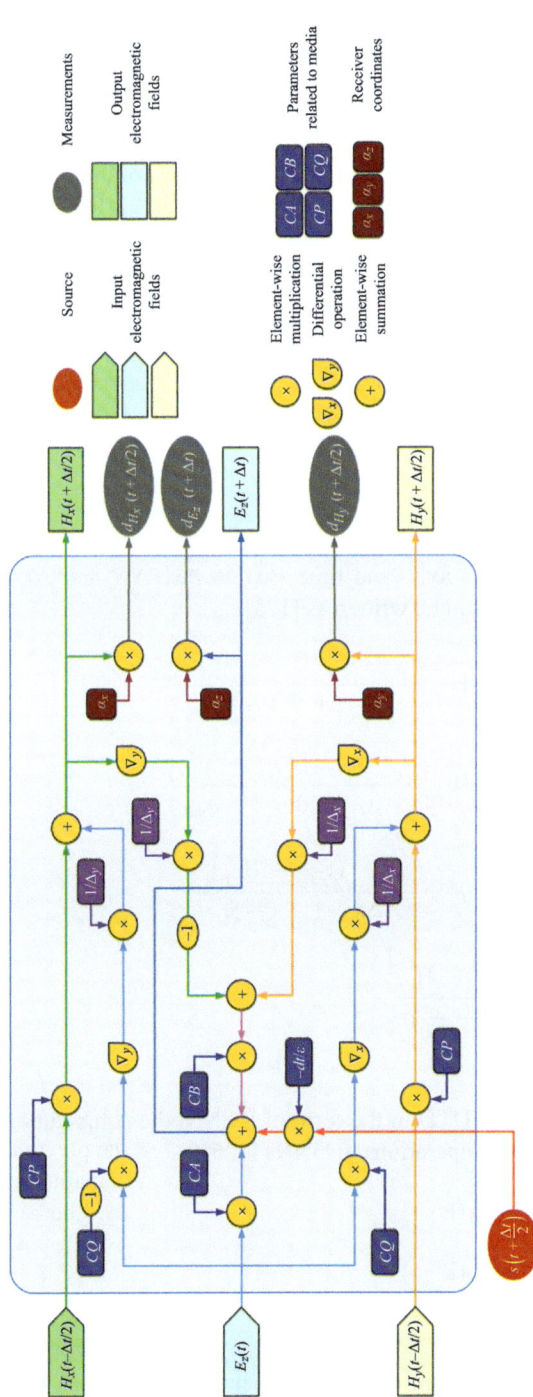

Figure 2.14 The theory-guided RNN for emulating the FDTD method (source: [124])

parameters. The automatic differentiation in the field of DL can directly determine the material-related parameters by avoiding the computation of forward modeling in the conventional inverse modeling problems.

The recurrent convolution neural network (RCNN) is presented as an equivalent of the traditional 2D FDTD method by investigating the mathematical similarities between the RNN, CNN, and 2D FDTD [125]. The weights of the FDTD-RCNN do not need to be optimized and they are determined based on the 2D FDTD formulations. Therefore, the FDTD-RCNN avoids the standard training process of RNN and it can perform full-wave EM modeling once its weights are formulated. The EM wave of TM mode in an isotropic domain is considered in [125]:

$$\frac{\partial E_z}{\partial y} = -\mu \frac{\partial H_x}{\partial t} - \sigma_m H_x$$

$$\frac{\partial E_z}{\partial x} = \mu \frac{\partial H_y}{\partial t} + \sigma_m H_y \qquad (2.43)$$

$$\frac{\partial H_y}{\partial x} - \frac{\partial H_x}{\partial y} = \varepsilon \frac{\partial E_z}{\partial t} + \sigma E_z$$

By discretizing the x-axis, y-axis, and time step as Δx, Δy, and Δt, the update equations of the 2D FDTD can be written as [125]:

$$H_{x,i,j}^{t+\frac{1}{2}} = \frac{1 - \frac{\Delta t}{2\mu_{i,j}}\sigma_{m,i,j}}{1 + \frac{\Delta t}{2\mu_{i,j}}\sigma_{m,i,j}} H_{x,i,j}^{t-\frac{1}{2}} - \frac{\Delta t}{\mu_{i,j}\Delta y} \frac{E_{z,i,j+1}^t - E_{z,i,j}^t}{1 + \frac{\Delta t}{2\mu_{i,j}}\sigma_{m,i,j}}$$

$$H_{y,i,j}^{t+\frac{1}{2}} = \frac{1 - \frac{\Delta t}{2\mu_{i,j}}\sigma_{m,i,j}}{1 + \frac{\Delta t}{2\mu_{i,j}}\sigma_{m,i,j}} H_{y,i,j}^{t-\frac{1}{2}} - \frac{\Delta t}{\mu_{i,j}\Delta x} \frac{E_{z,i+1,j}^t - E_{z,i,j}^t}{1 + \frac{\Delta t}{2\mu_{i,j}}\sigma_{m,i,j}}$$

$$E_{z,i,j}^{t+1} = \frac{1 - \frac{\Delta t}{2\varepsilon_{i,j}}\sigma_{i,j}}{1 + \frac{\Delta t}{2\varepsilon_{i,j}}\sigma_{i,j}} E_{z,i,j}^t + \frac{\Delta t}{\varepsilon_{i,j}\Delta x} \frac{H_{y,i,j}^{t+\frac{1}{2}} - H_{y,i-1,j}^{t+\frac{1}{2}}}{1 + \frac{\Delta t}{2\varepsilon_{i,j}}\sigma_{i,j}} \qquad (2.44)$$

$$- \frac{\Delta t}{\varepsilon_{i,j}\Delta y} \frac{H_{x,i,j}^{t+\frac{1}{2}} - H_{x,i,j-1}^{t+\frac{1}{2}}}{1 + \frac{\Delta t}{2\varepsilon_{i,j}}\sigma_{i,j}}$$

In order to operate the 2D FDTD in the form of RCNN, the computations in (2.44) can be replaced by standard operations in DNNs [125]:

$$H_{x,i,j}^{t+\frac{1}{2}} = c_1(H_{x,i,j}^{t-\frac{1}{2}}) + d_1(E_{z,i,j}^t)$$

$$H_{y,i,j}^{t+\frac{1}{2}} = c_2(H_{y,i,j}^{t-\frac{1}{2}}) + d_2(E_{z,i,j}^t) \qquad (2.45)$$

$$E_{z,i,j}^{t+1} = c_3(E_{z,i,j}^t) + d_{3y}(H_{y,i,j}^{t+\frac{1}{2}}) - d_{3x}(H_{x,i,j}^{t+\frac{1}{2}})$$

In (2.45), d_1, d_2, d_{3x}, and d_{3y} are differential operations and it can be replaced as 1D convolutions in CNNs, as shown in Figure 2.15(a). The coefficient matrices

$W_1, \cdots, W_7$ of c_1, c_2, c_3 are invariant and can be regarded as the weight matrices of the RNN, then the temporal update of (2.45) can operate as the calculation of a RNN, as shown in Figure 2.15(b). Adopting PML as its ABC, the 2D-FDTD can be expressed as [125]:

$$
\left.
\begin{aligned}
\frac{\partial E_z}{\partial y} &= -\kappa_y \frac{\partial B_x}{\partial t} - \frac{\sigma_y}{\varepsilon_0} B_x \\
\frac{\partial E_z}{\partial x} &= \kappa_x \frac{\partial B_y}{\partial t} + \frac{\sigma_x}{\varepsilon_0} B_y
\end{aligned}
\right\} \quad \text{update } E_z \rightarrow B_x, B_y
$$

$$
\left.
\begin{aligned}
\kappa_x \frac{\partial B_x}{\partial t} + \frac{\sigma_x}{\varepsilon_0} B_x &= \mu_1 \frac{\partial H_x}{\partial t} \\
\kappa_y \frac{\partial B_y}{\partial t} + \frac{\sigma_y}{\varepsilon_0} B_y &= \mu_1 \frac{\partial H_y}{\partial t}
\end{aligned}
\right\} \quad \text{update } B_x, B_y \rightarrow H_x, H_y
$$

$$
\left.
\begin{aligned}
\frac{\partial H_y}{\partial x} - \frac{\partial H_x}{\partial y} &= \varepsilon_1 \frac{\partial P_z}{\partial t} + \sigma_1 P_z \\
\frac{\partial P_z}{\partial t} &= \kappa_x \frac{\partial D_z}{\partial t} + \frac{\sigma_x}{\varepsilon_0} D_z
\end{aligned}
\right\} \quad \text{update } H_x, H_y \rightarrow P_z \rightarrow D_z
$$

$$
\left.
\frac{\partial D_z}{\partial t} = \kappa_y \frac{\partial E_z}{\partial t} + \frac{\sigma_y}{\varepsilon_0} E_z
\right\} \quad \text{update } D_z \rightarrow E_z
$$

$$(2.46)$$

where B_x, B_y, P_z, D_z are the auxiliary variables in the PML, $\sigma_x, \sigma_y, \kappa_x, \kappa_y, B_x, B_y$ are the conductivity, relative permittivity, magnetic flux density along $x-, y-$ axes in the PML, respectively, ε_1, μ_1 denote the permittivity and permeability inside the PML. The update of 2D-FDTD with PML is depicted in Figure 2.15(c). By implementing the FDTD in the form of the RCNN, the FDTD can be accelerated significantly on the parallel computing platform, i.e. GPUs, by fully leveraging the parallel computing power. The 1D and 2D FDTD-RCNN are implemented in [125] and they both demonstrate a significant reduction in the computing time compared to the conventional FDTD, especially in the cases that have a large amount number of unknowns.

Third, the iterative solvers of the matrix equations motivate the design of effective DNN architectures of EMFDs, including the conjugate gradient (CG) and fixed-point iteration method. The physics embedded DNN is presented to solve the 2D volume integral equation (VIE) by combining the conjugate gradient method with the DNN in [126]. The VIE is usually discretized as a linear system of matrix equations $\mathbb{A} \cdot x = b$ to solve in MoM. The CG method is a commonly-applied iterative solver of the matrix equations and it is concluded in Algorithm 1. The proposed physics-embedded DNN applies the DNNs to predict the p_{k+1} and $\alpha_{k+1} p_{k+1}$ [126]:

$$
\begin{aligned}
p_{k+1} &= \mathscr{F}_p^k \left(p_k, r_k, r_{k-1}, \Theta_p^k \right) \\
\alpha_{k+1} p_{k+1} &= \mathscr{F}_{dx}^k \left(p_{k+1}, A p_{k+1}, r_k, \Theta_{dx}^k \right)
\end{aligned}
$$

$$(2.47)$$

where $\mathscr{F}_p^k$ and $\mathscr{F}_{dx}^k$ are two independent DNNs parameterized by Θ_p^k and Θ_{dx}^k respectively. It can be observed in (2.47) that the input of the DNN $\mathscr{F}_p^k$ and $\mathscr{F}_{dx}^k$ are determined based on the relationship in the CG method. The matrix-vector multiplications in $\mathscr{F}_p^k$ and $\mathscr{F}_{dx}^k$ are computed numerically. The update learning of the physics-embedded DNN is concluded in Algorithm 2. The architecture of an iterative block in the physics-embedded DNN is depicted in Figure 2.16. It is designed to solve volume integral equations that can be written as a matrix equation $\overline{\overline{A}} \cdot \overline{E}^{tot} = \overline{E}^{inc}$. The similar physics-embedded DNN is also applied to solve the combined field integral

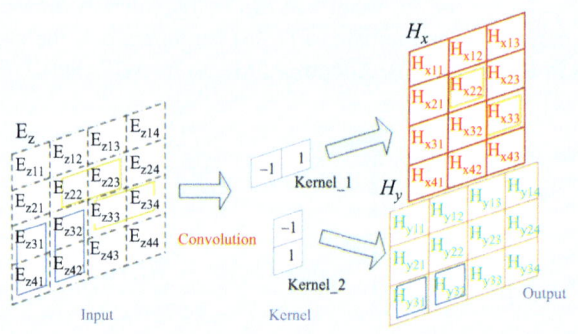

(a)

(b)

(c)

Figure 2.15   Schematics of RCNN-FDTD (source: [125]). (a) The convolutions for
differential operations, (b) the update of 2D FDTD as the RNN, and
(c) the update of 2D FDTD with PML.

Algorithm 1: Conjugate gradient method

1: Input x_0
2: $r_0 = \mathbf{b} - \mathbb{A}x_0$, $p_1 = r_0$, $\alpha_1 = \left(r_0^T r_0\right) / p_1^T \left(\mathbb{A}p_1\right)$
3: $x_1 = x_0 + \alpha_1 p_1$
4: **for** $k = 1, 2, \ldots$, until $\|r_k\| \leq \varepsilon$ **do**
5: $\quad r_k = r_{k-1} - \alpha_k \left(\mathbb{A}p_k\right)$
6: $\quad \beta_{k+1} = \left(r_k^T r_k\right) / \left(r_{k-1}^T r_{k-1}\right)$
7: $\quad p_{k+1} = r_k + \beta_{k+1} p_k$
8: $\quad \alpha_{k+1} = \left(r_k^T r_k\right) / p_{k+1}^T \left(\mathbb{A}p_{k+1}\right)$
9: $\quad x_{k+1} = x_k + \alpha_{k+1} p_{k+1}$
10: **end for**

---

equation in [127], which further verify the efficacy of the proposed approach. The physics-embedded DNN is further extended by replacing the ANN as the U-Net to model EM scattering by solving the induced currents of the dielectric objects in [128].

Algorithm 2: Update learning of physics-embedded DNN

1: Input x_0
2: $r_0 = b - \mathbb{A}x_0$, $p_1 = r_0$, $x_1 = x_0$
3: **for** $k = 1, 2, \ldots, N$ **do**
4: $\quad r_k = b - \mathbb{A}x_k$
5: $\quad p_{k+1} = \mathscr{F}_p^k \left(p_k, r_k, r_{k-1}, \Theta_p^k\right)$
6: $\quad x_{k+1} = x_k + \mathscr{F}_{dx}^k \left(p_{k+1}, \mathbb{A}p_{k+1}, r_k, \Theta_{dx}^k\right)$
7: **end for**

The physics-informed supervised residual learning (PhiSRL) is designed as an effective deep learning framework for EM forward modeling. It is based on the mathematical connection between the ResNet and the fixed-point iteration method [129,130]. The fixed-point iteration method is an iterative solver for the linear matrix equations $\mathbb{A}x = b$ and it can be described as:

$$x_{k+1}^a = x_k^a + \mathbb{A}^{a-1}(b - \mathbb{A}x_k^a), \tag{2.48}$$

where $\mathbb{A}^a$ is an approximation of the $\mathbb{A}$ to reduce the computational complexity. With the $\mathscr{L}$ denoting the matrix multiplication, the update equation of the stationary iterative scheme can be written as:

$$x_{k+1}^a = x_k^a + \mathscr{L}(b - \mathbb{A}x_k^a, \mathbb{A}^{a-1}), \tag{2.49}$$

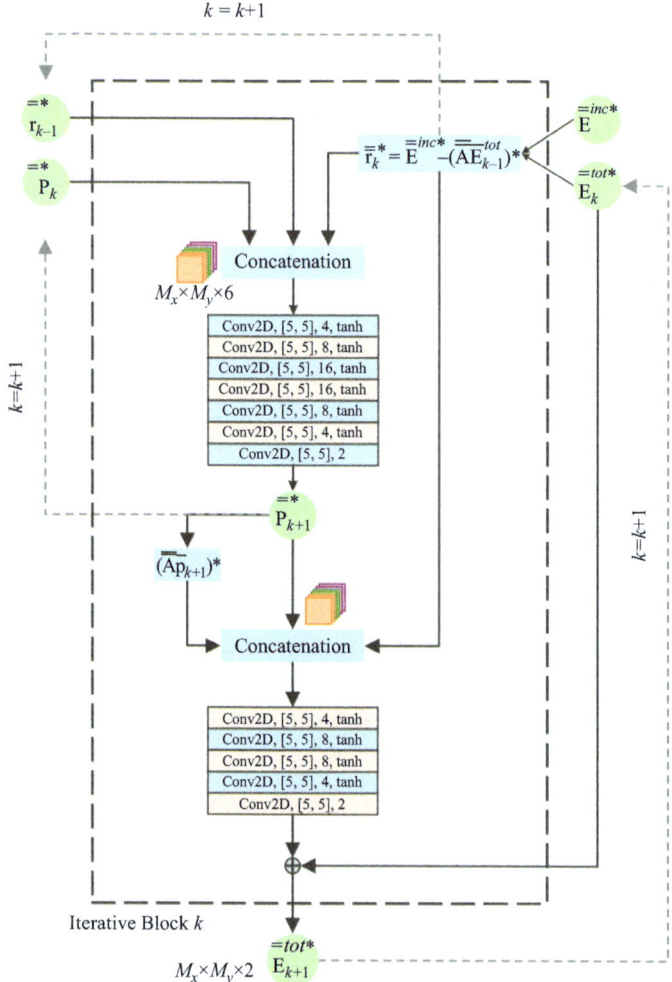

Figure 2.16   The architecture of an iterative block in the physics-embedded DNN
(source: [126])

If the approximated coefficient matrix is different at each iteration, then (2.49) can be transformed into the nonstationary iterative scheme:

$$x_{k+1}^a = x_k^a + \mathcal{L}(b - \mathbb{A}x_k^a, \mathbb{A}_k^{a-1}).   \tag{2.50}$$

As shown in Figure 2.17(a), the update equation of the typical ResNet can be written as [129,130]:

$$
\begin{aligned}
y_k &= \widetilde{h}(x_k) + \mathscr{F}(x_k, \Theta_k), \\
x_{k+1} &= \widetilde{N}(y_k),
\end{aligned}
\tag{2.51}
$$

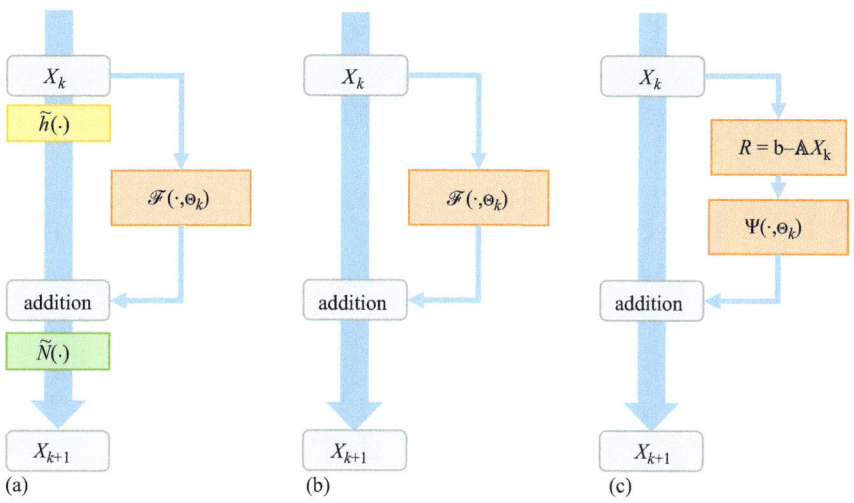

Figure 2.17   Schematics of (a) the general ResNet block, (b) the ResNet block with identity mapping, and (c) the proposed PhiSRL block

where $\widetilde{h}$, $\widetilde{N}$ denote the linear projection and the nonlinear activation, $\mathscr{F}$, Θ_k denote the CNN and the corresponding parameter set. The ResNet proves to achieve better performance by introducing the identity mappings to replace the linear projection $\widetilde{h}$ and the nonlinear activation $\widetilde{N}$. As shown in Figure 2.17(b), the update equation of the ResNet with identity mappings can be formulated based on the (2.51) [129,130]:

$$x_{k+1} = x_k + \mathscr{F}(x_k, \Theta_k). \tag{2.52}$$

Equations (2.49) and (2.50) have the similar update schemes with the (2.52). Based on this connection, the physics-informed supervised residual learning is designed by incorporating the fixed-point iteration method into the ResNet, as shown in Figure 2.17(c). The stationary scheme of the physics-informed supervised residual learning can be formulated as [129,130]:

$$x_{k+1} = x_k + \Psi^{Si}(b - \mathbb{A}x_k, \Theta), \tag{2.53}$$

and the nonstationary one is [129,130]:

$$x_{k+1} = x_k + \Psi^{Ni}(b - \mathbb{A}x_k, \Theta_k), \tag{2.54}$$

where Ψ and Θ denote the CNNs and the corresponding parameter set. Two DNNs are presented based on (2.53) and (2.54) including the stationary and non-stationary iterative physics-informed residual neural network (SiPhiResNet and NiPhiResNet), as shown in Figure 2.18. In both of them, the CNN is employed to learning mappings between the residual $R_k = b - \mathbb{A}x_k$ and the modification of the candidate solution $\Delta x_k = \Psi^{Si}(b - \mathbb{A}x_k, \Theta)$ or $\Psi^{Ni}(b - \mathbb{A}x_k, \Theta_k)$. In the SiPhiResNet, the CNN Ψ^{Si} adopts the architecture of the U-Net because the U-Net has a strong learning capacity

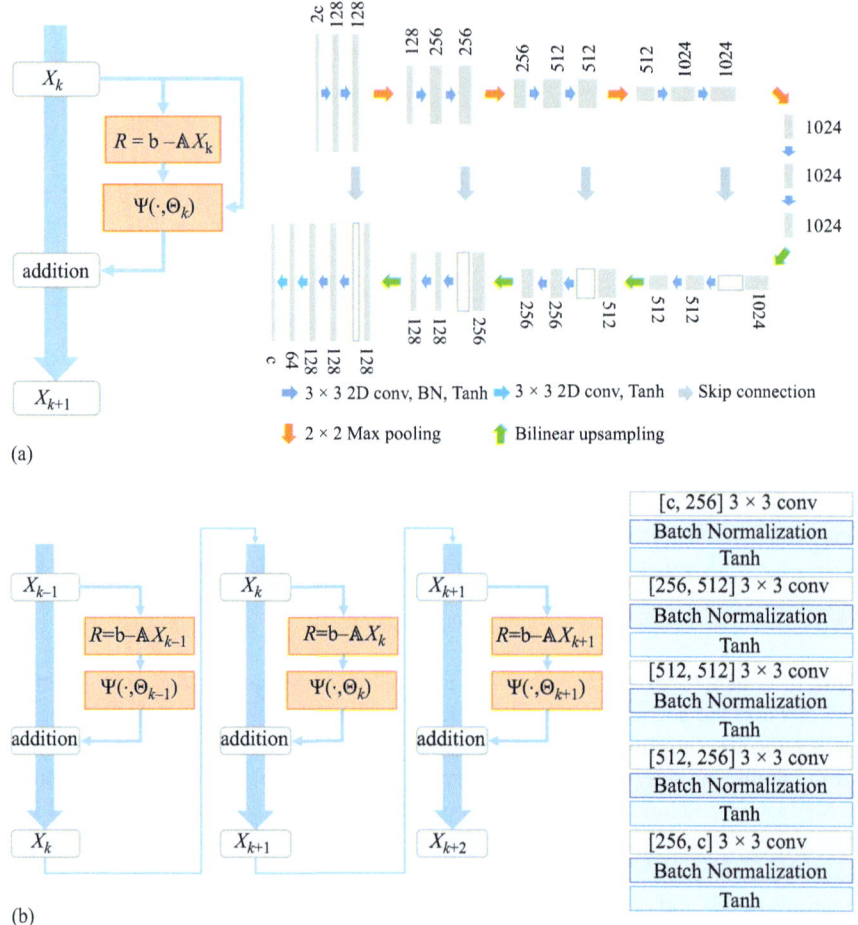

(a)

(b)

Figure 2.18   Schematics of non-stationary iterative physics-informed residual neural network

for various mappings between Δx_k and R_k at all iterative steps. In the NiPhiResNet, the CNN is independent at each iterative step and it can adopt a simple CNN that consists of five stacked layers of convolutional layer, batch normalization, and tanh nonlinear activation functions, as shown in Figure 2.18.

2.6   Summary and outlook

In this chapter, we introduce the approaches of applying deep learning techniques to electromagnetic forward modeling. These approaches are divided into three types: fully data-driven forward modeling, deep learning-assisted forward modeling, and physics inspired forward modeling. In fully data-driven forward modeling, the deep neural networks demonstrate powerful learning capacity and approximating ability to learn and abstract the inner physical laws from the massive training data. In deep learning-assisted forward modeling, the learning tasks of the deep neural networks are simplified and made explicit based on the physical or mathematical models. Thus, the performance of the deep neural networks is further improved and the forward modeling assisted by deep learning boasts better computational efficiency and precision. The physics-inspired forward modeling focuses on the inner reasoning of the deep neural networks to enable effective deep learning models for electromagnetics. The design and training of the deep neural networks are guided, motivated, or inspired by the mathematical or physical models for better robustness and interpretability.

References

[1] Knott EF, Schaeffer JF, Tulley MT. *Radar Cross Section*. SciTech Publishing; 2004.

[2] Nikolova NK. Microwave biomedical imaging. *Wiley Encyclopedia of Electrical and Electronics Engineering*. 1999;49:1–22.

[3] Jin JM, Riley DJ. *Finite Element Analysis of Antennas and Arrays*. Wiley Online Library; 2009.

[4] Poljak D. *Advanced Modeling in Computational Electromagnetic Compatibility*. New York, NY: John Wiley & Sons; 2007.

[5] Jin JM. *The Finite Element Method in Electromagnetics*. New York, NY: John Wiley & Sons; 2015.

[6] Jin JM. *Theory and Computation of Electromagnetic Fields*. New York, NY: John Wiley & Sons; 2011.

[7] Harrington RF. *Field Computation by Moment Methods*. New York, NY: Wiley-IEEE Press; 1993.

[8] Chew WC, Tong MS, Hu B. *Integral Equation Methods for Electromagnetic and Elastic Waves*. Williston, VT: Morgan and Claypool Publishers; 2007.

[9] Peters TJ, Volakis JL. Application of a conjugate gradient FFT method to scattering from thin planar material plates. *IEEE Transactions on Antennas and Propagation*. 1988;36(4):518–526.

[10] Bleszynski E, Bleszynski M, Jaroszewicz T. AIM: adaptive integral method for solving large-scale electromagnetic scattering and radiation problems. *Radio Science*. 1996;31(5):1225–1251.

[11] Rokhlin V. Rapid solution of integral equations of classical potential theory. *Journal of Computational Physics*. 1985;60(2):187–207.

[12] Lee SC, Vouvakis MN, Lee JF. A non-overlapping domain decomposition method with non-matching grids for modeling large finite antenna arrays. *Journal of Computational Physics*. 2005;203(1):1–21.

[13] Dang X, Li M, Yang F, *et al.* Quasi-periodic array modeling using reduced basis method. *IEEE Antennas and Wireless Propagation Letters*. 2017;16:825–828.

[14]   Burrascano P, Fiori S, Mongiardo M. A review of artificial neural networks applications in microwave computer-aided design (invited article). *International Journal of RF and Microwave Computer-Aided Engineering.* 1999;9(3):158–174.

[15]   Goodfellow I, Bengio Y, Courville A, *et al. Deep Learning*, Vol. 1. Cambridge: MIT Press; 2016.

[16]   Girdhar R, Gustafson L, Adcock A, van der Maaten, L. Forward prediction for physical reasoning. arXiv preprint arXiv:2006.10734. 2020.

[17]   Brunton SL, Noack BR, Koumoutsakos P. Machine learning for fluid mechanics. *Annual Review of Fluid Mechanics.* 2020;52:477–508.

[18]   Han J, Jentzen A, Weinan E. Solving high-dimensional partial differential equations using deep learning. *Proceedings of the National Academy of Sciences.* 2018;115(34):8505–8510.

[19]   Ruthotto L, Haber E. Deep neural networks motivated by partial differential equations. *Journal of Mathematical Imaging and Vision.* 2019; pp. 1–13.

[20]   Lerer A, Gross S, Fergus R. Learning physical intuition of block towers by example. arXiv preprint arXiv:160301312. 2016.

[21]   Greydanus S, Dzamba M, Yosinski J. Hamiltonian neural networks. arXiv preprint arXiv:190601563. 2019.

[22]   Chen Z, Zhang J, Arjovsky M, *et al.* Symplectic recurrent neural networks. arXiv preprint arXiv:190913334. 2019.

[23]   Cranmer M, Greydanus S, Hoyer S, *et al.* Lagrangian neural networks. arXiv preprint arXiv:200304630. 2020.

[24]   Kochkov D, Smith JA, Alieva A, *et al.* Machine learning accelerated computational fluid dynamics. arXiv preprint arXiv:210201010. 2021.

[25]   Maulik R, San O, Rasheed A, *et al.* Sub-grid modelling for two-dimensional turbulence using neural networks. arXiv preprint arXiv:180802983. 2018.

[26]   Erichson NB, Muehlebach M, Mahoney MW. Physics-informed autoencoders for Lyapunov-stable fluid flow prediction. arXiv preprint arXiv:190510866. 2019.

[27]   Sirignano J, MacArt JF, Freund JB. DPM: a deep learning PDE augmentation method with application to large-eddy simulation. *Journal of Computational Physics.* 2020;423:109811.

[28]   Wang F, Zhang QJ. Knowledge-based neural models for microwave design. *IEEE Transactions on Microwave Theory and Techniques.* 1997;45(12): 2333–2343.

[29]   Salucci M, Anselmi N, Oliveri G, *et al.* Real-time NDT-NDE through an innovative adaptive partial least squares SVR inversion approach. *IEEE Transactions on Geoscience and Remote Sensing.* 2016;54(11): 6818–6832.

[30]   Karaosmanoğlu B, Ergül Ö. Acceleration of MLFMA simulations using trimmed tree structures. *IEEE Transactions on Antennas and Propagation.* 2020;69(1):356–365.

[31]   Yao HM, Jiang L. Machine-learning-based PML for the FDTD method. *IEEE Antennas and Wireless Propagation Letters*. 2018;18(1):192–196.

[32]   Lagaris IE, Likas A, Fotiadis DI. Artificial neural networks for solving ordinary and partial differential equations. *IEEE Transactions on Neural Networks*. 1998;9(5):987–1000.

[33]   Lee H, Kang IS. Neural algorithm for solving differential equations. *Journal of Computational Physics*. 1990;91(1):110–131.

[34]   Mills K, Spanner M, Tamblyn I. Deep learning and the Schrödinger equation. *Physical Review A*. 2017;96(4):042113.

[35]   Hermann J, Schätzle Z, Noé F. Deep-neural-network solution of the electronic Schrödinger equation. *Nature Chemistry*. 2020;12(10):891–897.

[36]   Miyanawala TP, Jaiman RK. An efficient deep learning technique for the Navier-Stokes equations: application to unsteady wake flow dynamics. arXiv preprint arXiv:171009099. 2017.

[37]   Sirignano J, Spiliopoulos K. DGM: a deep learning algorithm for solving partial differential equations. *Journal of Computational Physics*. 2018;375: 1339–1364.

[38]   Sharma R, Farimani AB, Gomes J, *et al.* Weakly-supervised deep learning of heat transport via physics informed loss. arXiv preprint arXiv:180711374. 2018.

[39]   Shan T, Tang W, Dang X, *et al.* Study on a fast solver for Poisson's equation based on deep learning technique. *IEEE Transactions on Antennas and Propagation*. 2020;68(9):6725–6733.

[40]   Pang G, Lu L, Karniadakis GE. fPINNs: fractional physics-informed neural networks. *SIAM Journal on Scientific Computing*. 2019;41(4): A2603–A2626.

[41]   Nakamura-Zimmerer T, Gong Q, Kang W. Adaptive deep learning for high-dimensional Hamilton–Jacobi–Bellman equations. *SIAM Journal on Scientific Computing*. 2021;43(2):A1221–A1247.

[42]   Berner J, Grohs P, Jentzen A. Analysis of the generalization error: empirical risk minimization over deep artificial neural networks overcomes the curse of dimensionality in the numerical approximation of Black–Scholes partial differential equations. *SIAM Journal on Mathematics of Data Science*. 2020;2(3):631–657.

[43]   Tompson J, Schlachter K, Sprechmann P, *et al.* Accelerating eulerian fluid simulation with convolutional networks. In: *International Conference on Machine Learning*. PMLR; 2017. pp. 3424–3433.

[44]   Zhang Z, Zhang L, Sun Z, *et al.* Solving Poisson's Equation using Deep Learning in Particle Simulation of PN Junction. In: *2019 Joint International Symposium on Electromagnetic Compatibility, Sapporo and Asia-Pacific International Symposium on Electromagnetic Compatibility (EMC Sapporo/ APEMC)*. IEEE; 2019. pp. 305–308.

[45]   Weinan E, Yu B. The deep Ritz method: a deep learning-based numerical algorithm for solving variational problems. arXiv preprint arXiv:171000211. 2017.

[46]   Beidokhti RS, Malek A. Solving initial-boundary value problems for systems of partial differential equations using neural networks and optimization techniques. *Journal of the Franklin Institute*. 2009;346(9):898–913.

[47]   Zang Y, Bao G, Ye X, *et al.* Weak adversarial networks for high-dimensional partial differential equations. *Journal of Computational Physics*. 2020;411:109409.

[48]   Hsieh JT, Zhao S, Eismann S, *et al.* Learning neural PDE solvers with convergence guarantees. arXiv preprint arXiv:190601200. 2019.

[49]   Raissi M, Perdikaris P, Karniadakis GE. Physics-informed neural networks: a deep learning framework for solving forward and inverse problems involving nonlinear partial differential equations. *Journal of Computational Physics*. 2019;378:686–707.

[50]   Lu L, Meng X, Mao Z, *et al.* DeepXDE: a deep learning library for solving differential equations. *SIAM Review*. 2021;63(1):208–228.

[51]   Zhang D, Lu L, Guo L, *et al.* Quantifying total uncertainty in physics-informed neural networks for solving forward and inverse stochastic problems. *Journal of Computational Physics*. 2019;397:108850.

[52]   Meng X, Karniadakis GE. A composite neural network that learns from multi-fidelity data: application to function approximation and inverse PDE problems. *Journal of Computational Physics*. 2020;401:109020.

[53]   Dwivedi V, Srinivasan B. Physics informed extreme learning machine (PIELM)—a rapid method for the numerical solution of partial differential equations. *Neurocomputing*. 2020;391:96–118.

[54]   He K, Zhang X, Ren S, *et al.* Deep residual learning for image recognition. In: *Proceedings of the IEEE Conference on Computer Vision and Pattern Recognition*; 2016. pp. 770–778.

[55]   Weinan E. A proposal on machine learning via dynamical systems. *Communications in Mathematics and Statistics*. 2017;5(1):1–11.

[56]   Haber E, Ruthotto L. Stable architectures for deep neural networks. *Inverse Problems*. 2017;34(1):014004.

[57]   Chang B, Meng L, Haber E, *et al.* Reversible architectures for arbitrarily deep residual neural networks. In: *Proceedings of the AAAI Conference on Artificial Intelligence*. vol. 32; 2018.

[58]   Chang B, Meng L, Haber E, *et al.* Multi-level residual networks from dynamical systems view. arXiv preprint arXiv:171010348. 2017.

[59]   Liao Q, Poggio T. Bridging the gaps between residual learning, recurrent neural networks and visual cortex. arXiv preprint arXiv:160403640. 2016.

[60]   Li Z, Shi Z. Deep residual learning and pdes on manifold. arXiv preprint arXiv:170805115. 2017.

[61]   Lu Y, Zhong A, Li Q, *et al.* Beyond finite layer neural networks: bridging deep architectures and numerical differential equations. In: *International Conference on Machine Learning*. PMLR; 2018. pp. 3276–3285.

[62]   Zhang X, Li Z, Change Loy C, *et al.* Polynet: a pursuit of structural diversity in very deep networks. In: *Proceedings of the IEEE Conference on Computer Vision and Pattern Recognition*; 2017. pp. 718–726.

[63] Larsson G, Maire M, Shakhnarovich G. Fractalnet: ultra-deep neural networks without residuals. arXiv preprint arXiv:160507648. 2016.

[64] Gomez AN, Ren M, Urtasun R, *et al.* The reversible residual network: backpropagation without storing activations. In: *Proceedings of the 31st International Conference on Neural Information Processing Systems*; 2017. pp. 2211–2221.

[65] Long Z, Lu Y, Ma X, *et al.* Pde-net: learning pdes from data. In: *International Conference on Machine Learning*. PMLR; 2018. pp. 3208–3216.

[66] Chen RT, Rubanova Y, Bettencourt J, *et al.* Neural ordinary differential equations. arXiv preprint arXiv:180607366. 2018.

[67] He J, Xu J. MgNet: a unified framework of multigrid and convolutional neural network. *Science China Mathematics*. 2019;62(7):1331–1354.

[68] Greenfeld D, Galun M, Basri R, *et al.* Learning to optimize multigrid PDE solvers. In: *International Conference on Machine Learning*. PMLR; 2019. pp. 2415–2423.

[69] Zhu M, Chang B, Fu C. Convolutional neural networks combined with Runge–Kutta methods. arXiv preprint arXiv:180208831. 2018.

[70] Fan Y, Lin L, Ying L, *et al.* A multiscale neural network based on hierarchical matrices. *Multiscale Modeling & Simulation*. 2019;17(4):1189–1213.

[71] LeCun Y, Bottou L, Bengio Y, *et al.* Gradient-based learning applied to document recognition. *Proceedings of the IEEE*. 1998;86(11):2278–2324.

[72] Duchi J, Hazan E, Singer Y. Adaptive subgradient methods for online learning and stochastic optimization. *Journal of Machine Learning Research*. 2011;12(Jul):2121–2159.

[73] Tieleman T, Hinton G. Lecture 6.5-rmsprop: Divide the gradient by a running average of its recent magnitude. *COURSERA: Neural Networks for Machine Learning*. 2012;4(2):26–31.

[74] Kingma DP, Ba J. *Adam: A Method for Stochastic Optimization*. CoRR. 2014;abs/1412.6980.

[75] Guo L, Li M, Xu S, *et al.* Application of stochastic gradient descent technique for method of moments. In: *2020 IEEE International Conference on Computational Electromagnetics (ICCEM)*. New York, NY: IEEE; 2020. pp. 97–98.

[76] Guo L, Li M, Xu S, *et al.* Investigation of Adam for low-frequency electromagnetic problems. In: *2020 IEEE MTT-S International Conference on Numerical Electromagnetic and Multiphysics Modeling and Optimization (NEMO)*. New York, NY: IEEE; 2020. pp. 1–2.

[77] Casey KF. Low-frequency electromagnetic penetration of loaded apertures. *IEEE Transactions on Electromagnetic Compatibility*. 1981;(4):367–377.

[78] Yao HM, Jiang LJ, Qin YW. Machine learning based method of moments (ML-MoM). In: *2017 IEEE International Symposium on Antennas and Propagation & USNC/URSI National Radio Science Meeting*. New York, NY: IEEE; 2017. pp. 973–974.

[79] Kendall A, Gal Y, Cipolla R. Multi-task learning using uncertainty to weigh losses for scene geometry and semantics. In: *Proceedings of the IEEE Conference on Computer Vision and Pattern Recognition*; 2018. pp. 7482–7491.

[80] Sener O, Koltun V. Multi-task learning as multi-objective optimization. arXiv preprint arXiv:181004650. 2018.

[81] Kendall A, Gal Y. What uncertainties do we need in Bayesian deep learning for computer vision? arXiv preprint arXiv:170304977. 2017.

[82] Désidéri JA. Multiple-gradient descent algorithm (MGDA) for multiobjective optimization. *Comptes Rendus Mathematique*. 2012;350(5–6): 313–318.

[83] Wu H, Zhang Y, Fu W, *et al*. A novel pre-processing method for neural network-based magnetic field approximation. *IEEE Transactions on Magnetics*. 2021;57(10):1–9.

[84] Bhardwaj S, Gohel H, Namuduri S. A multiple-input deep neural network architecture for solution of one-dimensional Poisson equation. *IEEE Antennas and Wireless Propagation Letters*. 2019;18(11):2244–2248.

[85] Khan A, Ghorbanian V, Lowther D. Deep learning for magnetic field estimation. *IEEE Transactions on Magnetics*. 2019;55(6):1–4.

[86] Kovacs A, Fischbacher J, Oezelt H, *et al*. Learning magnetization dynamics. *Journal of Magnetism and Magnetic Materials*. 2019;491:165548.

[87] Qi S, Wang Y, Li Y, *et al*. Two-dimensional electromagnetic solver based on deep learning technique. *IEEE Journal on Multiscale and Multiphysics Computational Techniques*. 2020;5:83–88.

[88] Wu F, Fan M, Liu W, *et al*. An efficient time-domain electromagnetic algorithm based on LSTM neural network. *IEEE Antennas and Wireless Propagation Letters*. 2021;20(7):1322–1326.

[89] Zhu Y, Zhou Y, Javaid F, *et al*. Fast multi-physics simulation approach in underwater exploration via deep learning technique. *Electronics Letters*. 2022;58(5):200–202.

[90] Giannakis I, Giannopoulos A, Warren C. A machine learning-based fast-forward solver for ground penetrating radar with application to full-waveform inversion. *IEEE Transactions on Geoscience and Remote Sensing*. 2019;57(7):4417–4426.

[91] Guo J, Li Y, Cai S, *et al*. Fast prediction of electromagnetic scattering characteristics of targets based on deep learning. In: *2021 International Applied Computational Electromagnetics Society (ACES-China) Symposium*. New York, NY: IEEE; 2021. pp. 1–2.

[92] Jin J, Zhang C, Feng F, *et al*. Deep neural network technique for high-dimensional microwave modeling and applications to parameter extraction of microwave filters. *IEEE Transactions on Microwave Theory and Techniques*. 2019;67(10):4140–4155.

[93] Jin J, Feng F, Zhang J, *et al*. A novel deep neural network topology for parametric modeling of passive microwave components. *IEEE Access*. 2020;8:82273–82285.

[94] Guillod T, Papamanolis P, Kolar JW. Artificial neural network (ANN) based fast and accurate inductor modeling and design. *IEEE Open Journal of Power Electronics*. 2020;1:284–299.

[95] Kim I, Park SJ, Jeong C, *et al*. Simulator acceleration and inverse design of fin field-effect transistors using machine learning. *Scientific Reports*. 2022;12:1140.

[96] Zhang G, He H, Katabi D. Circuit-GNN: graph neural networks for dis-tributed circuit design. In: *International Conference on Machine Learning*. PMLR; 2019. pp. 7364–7373.

[97] Wiecha PR, Muskens OL. Deep learning meets nanophotonics: a gen-eralized accurate predictor for near fields and far fields of arbitrary 3D nanostructures. *Nano Letters*. 2019;20(1):329–338.

[98] Jiang J, Chen M, Fan JA. Deep neural networks for the evaluation and design of photonic devices. *Nature Reviews Materials*. 2021;6(8): 679–700.

[99] Li X, Kojima K, Brand M. Predicting long-and variable-distance coupling effects in metasurface optics. In: *2021 IEEE Photonics Conference (IPC)*. New York, NY: IEEE; 2021. pp. 1–2.

[100] Chugh S, Gulistan A, Ghosh S, *et al.* Machine learning approach for computing optical properties of a photonic crystal fiber. *Optics Express*. 2019;27(25):36414–36425.

[101] Song Y, Wang D, Qin J, *et al.* Physical information-embedded deep learning for forward prediction and inverse design of nanophotonic devices. *Journal of Lightwave Technology*. 2021;39(20):6498–6508.

[102] Li Y, Wang Y, Qi S, *et al.* Predicting scattering from complex nano-structures via deep learning. *IEEE Access*. 2020;8:139983–139993.

[103] Peurifoy J, Shen Y, Jing L, *et al.* Nanophotonic particle simulation and inverse design using artificial neural networks. *Science Advances*. 2018;4(6):eaar4206.

[104] Shan T, Guo R, Li M, *et al.* Application of multitask learning for 2-D modeling of magnetotelluric surveys: TE case. *IEEE Transactions on Geoscience and Remote Sensing*. 2021;60:1–9.

[105] Ashalley E, Acheampong K, Besteiro LV, *et al.* Multitask deep-learning-based design of chiral plasmonic metamaterials. *Photonics Research*. 2020;8(7):1213–1225.

[106] Rashed A, Jawed S, Rehberg J, *et al.* A deep multi-task approach for residual value forecasting. In: *ECML/PKDD*, Vol. 3; 2019. pp. 467–482.

[107] Kraft B, Jung M, Körner M, *et al.* Hybrid modeling: fusion of a deep approach and physics-based model for global hydrological modeling. *The International Archives of Photogrammetry, Remote Sensing and Spatial Information Sciences*. 2020;43:1537–1544.

[108] Yao HM, Jiang L. Enhanced PML based on the long short term memory network for the FDTD method. *IEEE Access*. 2020;8:21028–21035.

[109] Feng N, Chen Y, Zhang Y, *et al.* An expedient DDF-based implementation of perfectly matched monolayer. *IEEE Microwave and Wireless Components Letters*. 2021;31(6):541–544.

[110] Kontschieder P, Fiterau M, Criminisi A, *et al.* Deep neural decision forests. In: *Proceedings of the IEEE International Conference on Computer Vision*; 2015. pp. 1467–1475.

[111] Noakoasteen O, Wang S, Peng Z, *et al.* Physics-informed deep neural networks for transient electromagnetic analysis. *IEEE Open Journal of Antennas and Propagation*. 2020;1:404–412.

[112]   Ma Z, Xu K, Song R, *et al.* Learning-based fast electromagnetic scattering solver through generative adversarial network. *IEEE Transactions on Antennas and Propagation.* 2020;69(4):2194–2208.

[113]   Hao W, Chen YP, Chen PY, *et al.* Solving two-dimensional scattering from multiple dielectric cylinders by artificial neural network accelerated numerical Green's function. *IEEE Antennas and Wireless Propagation Letters.* 2021;20(5):783–787.

[114]   Sun JJ, Sun S, Chen Y, *et al.* Machine learning based multilevel fast multipole algorithm. In: *2018 IEEE International Symposium on Antennas and Propagation & USNC/URSI National Radio Science Meeting.* New York, NY: IEEE; 2018. pp. 2311–2312.

[115]   Sun JJ, Sun S, Chen YP, *et al.* Machine-learning-based hybrid method for the multilevel fast multipole algorithm. *IEEE Antennas and Wireless Propagation Letters.* 2020;19(12):2177–2181.

[116]   Zhang J, Feng F, Zhang QJ. Rapid yield estimation of microwave passive components using model-order reduction based neuro-transfer function models. *IEEE Microwave and Wireless Components Letters.* 2021;31(4): 333–336.

[117]   Feng F, Na W, Jin J, *et al.* ANNs for fast parameterized EM modeling: the state of the art in machine learning for design automation of passive microwave structures. *IEEE Microwave Magazine.* 2021;22(10):37–50.

[118]   Barmada S, Fontana N, Sani L, *et al.* Deep learning and reduced models for fast optimization in electromagnetics. *IEEE Transactions on Magnetics.* 2020;56(3):1–4.

[119]   Zhang P, Hu Y, Jin Y, *et al.* A Maxwell's equations based deep learning method for time domain electromagnetic simulations. *IEEE Journal on Multiscale and Multiphysics Computational Techniques.* 2021;6:35–40.

[120]   Chen M, Lupoiu R, Mao C, *et al.* WaveY-Net: physics-augmented deep-learning for high-speed electromagnetic simulation and optimization. In: *High Contrast Metastructures XI.* 2022;12011:63–66.

[121]   Fujita K. Physics-informed neural network method for space charge effect in particle accelerators. *IEEE Access.* 2021;9:164017–164025.

[122]   Huang X, Liu H, Shi B, *et al.* Solving partial differential equations with point source based on physics-informed neural networks. arXiv preprint arXiv:211101394. 2021.

[123]   Hughes TW, Williamson IA, Minkov M, *et al.* Wave physics as an analog recurrent neural network. *Science Advances.* 2019;5(12):eaay6946.

[124]   Hu Y, Jin Y, Wu X, Chen J. A theory-guided deep neural network for time domain electromagnetic simulation and inversion using a differentiable programming platform. *IEEE Transactions on Antennas and Propagation.* 2021;70:767–772.

[125]   Guo L, Li M, Xu S, *et al.* Electromagnetic modeling using an FDTD-equivalent recurrent convolution neural network: accurate computing on a deep learning framework. *IEEE Antennas and Propagation Magazine.* 2023;65:93–102.

[126]   Guo R, Shan T, Song X, *et al*. Physics embedded deep neural network for solving volume integral equation: 2D case. *IEEE Transactions on Antennas and Propagation*. 2021;70:6135–6147.

[127]   Guo R, Lin Z, Shan T, *et al*. Solving combined field integral equation with deep neural network for 2-D conducting object. *IEEE Antennas and Wireless Propagation Letters*. 2021;20(4):538–542.

[128]   Xue BW, Wu D, Song BY, *et al*. U-Net conjugate gradient solution of electromagnetic scattering from dielectric objects. In: *2021 International Applied Computational Electromagnetics Society (ACES-China) Symposium*. New York, NY: IEEE; 2021. pp. 1–2.

[129]   Shan T, Song X, Guo R, *et al*. Physics-informed supervised residual learning for 2D electromagnetic forward modeling. arXiv:2104.13231; 2021.

[130]   Shan T, Song X, Guo R, *et al*. Physics-informed supervised residual learning for electromagnetic modeling. In: *2021 International Applied Computational Electromagnetics Society Symposium (ACES)*. New York, NY: IEEE; 2021. pp. 1–4.

Chapter 3
Deep learning techniques for free-space inverse scattering
Julio L. Nicolini[1] and Fernando L. Teixeira[1]

The problem of electromagnetic inverse scattering consists of detecting the unknown properties of an object, the scatterer, from the information provided by the scattered electromagnetic fields measured after interaction with said object [1–3]. The sought-after properties can be material (unknown permittivity, permeability, conductivity), geometrical (unknown dimensions, shape), or a combination thereof. This is the *inverse* problem of *forward* scattering, which consists of calculating the resulting (unknown) electromagnetic fields scattered by an object with known material and geometrical properties. Electromagnetic inverse scattering is a subset of inverse scattering problems, which also comprise problems pertaining to the scattering of mechanical waves found in seismology and acoustics, the scattering of electrons in scanning electron microscopes, the scattering of elementary particles in high energy physics and quantum field theory, among others. In turn, these comprise a subset of mathematical inverse problems in general, which consist of obtaining the set of parameters in a model that lead to a given set of "outputs" (set of observations or measurements) for a given set of inputs.

As expected, inverse problems in general, and inverse scattering in particular, have myriad applications in the fields of radiology, archeology, biology, atmospheric science, geophysics, oceanography, plasma physics, materials science, astrophysics, quantum information, and other areas of science and engineering; whenever "direct" observations are impossible or unfeasible and the nature of a physical phenomenon must be indirectly determined, an inverse problem must be solved. While there are many similarities between different types of inverse problems, there are also peculiarities that arise from the specific equations that model the phenomenon of interest. In the context of this work, we will refer to "scattering" and "inverse scattering" as the forward and inverse problems of the electromagnetic case, unless explicitly stated otherwise.

The focus of this chapter is mostly restricted to discussing, as the title suggests, applications of deep learning techniques to *free-space* inverse scattering. The

[1]ElectroScience Laboratory, Department of Electrical and Computer Engineering, The Ohio State University, USA

qualifier free-space used here means that we focus on inverse scattering problems where all transmitters and receivers are located in free space and the scattering object has an unknown shape and/or inhomogeneous spatial distribution for the permittivity, permeability, and/or conductivity material properties but it is otherwise embedded in free-space. A few of the examples considered in this chapter deviate from these scenarios but are also included because they serve to illustrate ideas that can be promptly translated or adapted to the free-space inverse scattering scenario as well.

This chapter is organized as follows: first, we introduce the general statements of the forward and inverse scattering problems in Section 3.1. In Section 3.2, we briefly describe traditional methods used to solve these problems, which include approximate methods that simplify the forward problem to make it feasible to obtain a direct solution of the inverse problem, and iterative procedures that obtain the solution to the inverse problem through optimization-based techniques. A brief general description of artificial neural networks and their applicability to inverse problems is given in Section 3.3, followed by a description of the first forays into using (shallow) artificial neural networks in Section 3.4, which due to the limitations of shallow networks consist mostly of "black-box" approaches to obtaining parametric values of simple problems instead of full-resolution inversion. Then, we classify the current deep learning solutions to inverse scattering problems into three broad categories. This classification is not sharp and is employed simply to facilitate the discussion. In Section 3.5, we describe the black-box type of solutions that share similarities with the initial shallow artificial neural network approaches, but which due to the increased power of deep networks can actually work with the high-resolution "pixel base," that is, obtain the reconstruction of the material distribution directly; in Section 3.6, we describe the methodologies that apply a learning-based solution to an otherwise iterative procedure, either by approximating the solution of the forward problem with a learning-based surrogate model or by integrating deep artificial neural networks into the entire iterative process; and in Section 3.7 we describe non-blackbox, non-iterative learning solutions, that is, algorithms that obtain the solution for the inverse problem directly but that integrate knowledge of the structure of the problem such that the surrogate learning model is not a direct black-box solution.

We note that, thorough this chapter, we have attempted to make our notation and definitions self-consistent. This means that the notation used in the text might differ from that found in the various references. Because of this, care must be exercised when trying to cross-reference the notation employed here with the different notations adopted in the cited references.

The application of deep learning techniques for free-space inverse scattering is a quickly evolving area of research, with a large body of very recent work and with many research groups actively working on the topic. Because of this, we do not claim that our coverage of the topic in this chapter is in any way complete. Nevertheless, we hope this chapter can serve as a good pointer for some important works of interest and as a way of (non-exhaustively) capturing some of the evolution and key trends of the topic.

3.1 Inverse scattering challenges

The fundamental equation for the *forward scattering* problem is given by

$$\mathbf{E}(\mathbf{r}) = \mathbf{E}_{inc}(\mathbf{r}) + \mathbf{E}_{scat}(\mathbf{r}), \tag{3.1}$$

where $\mathbf{E}_{inc}$ is the incident field illuminating the (known) scattering object and $\mathbf{E}_{scat}(\mathbf{r})$ is the scattered field given by the volume integral

$$\mathbf{E}_{scat}(\mathbf{r}) = \int_V \overline{\mathbf{G}}(\mathbf{r}, \mathbf{r}') \cdot O(\mathbf{r}')\mathbf{E}(\mathbf{r}')d\mathbf{r}', \tag{3.2}$$

where $\overline{\mathbf{G}}$ is the dyadic (tensor) Green's function of the problem and $O(\mathbf{r}) = k^2(\mathbf{r}) - k_0^2 = \omega^2\mu_0(\varepsilon(\mathbf{r}) - \varepsilon_0)$ is the contrast function describing the varying material properties of the scatterer [1]. Another common description for the distribution of the material properties is given by the dielectric contrast defined as $\chi(\mathbf{r}) = \varepsilon_r(\mathbf{r}) - 1$; the two descriptions can be linked by the relation $O(\mathbf{r}) = \omega^2\mu_0\varepsilon_0\chi(\mathbf{r})$. For simplicity, we assume a linear medium. We also assume a non-magnetic medium such that μ_0 is constant thorough the domain and an isotropic medium such that the permittivity $\varepsilon_r(\mathbf{r})$ is a scalar number. The above problem statement can be easily generalized to include magnetic and/or anisotropic media as well. A typical set-up for a two-dimensional scattering problem is shown in Figure 3.1, where the incident field is generated by a set of transmitters set around the region of interest that contains the scatterer, and the scattered field is measured by a set of receivers. This particular setup corresponds to a "full-aspect angle" acquisition. Partial aspect angle acquisitions are also possible, where the transmitters and receivers are located over a limited angular sector around the scatterer.

While Maxwell's equations are linear, the scattered field is a nonlinear functional of $O(\mathbf{r})$ since the total field $\mathbf{E}$ is also a function of $O(\mathbf{r})$. This nonlinear dependence can be understood as the effect from multiple scatterings inside the object, and it can be clearly noticed in the following simple example: assume a fixed transmitter that generates the incident field, a fixed receiver that measures the scattered field, and two scattering cylinders S_1 and S_2, as shown in Figure 3.2. If the problem were linear, the scattered field measured from the system with the two cylinders would be the sum of the scattered field from each cylinder by itself, but that is clearly not the case due to the mutual scattering that occurs when both scatterers are present.

This non-linearity compounds the challenges for solving an inverse problem [2]. The (interrelated) challenges for solving inverse scattering problems can be summarized briefly as:

(i) *Non-uniqueness*: The inverse problem is non-unique (or ill-posed), that is, different scattering objects can produce the same observed scattered field, especially in the far-field region where the information contained in evanescent spectrum is lost.

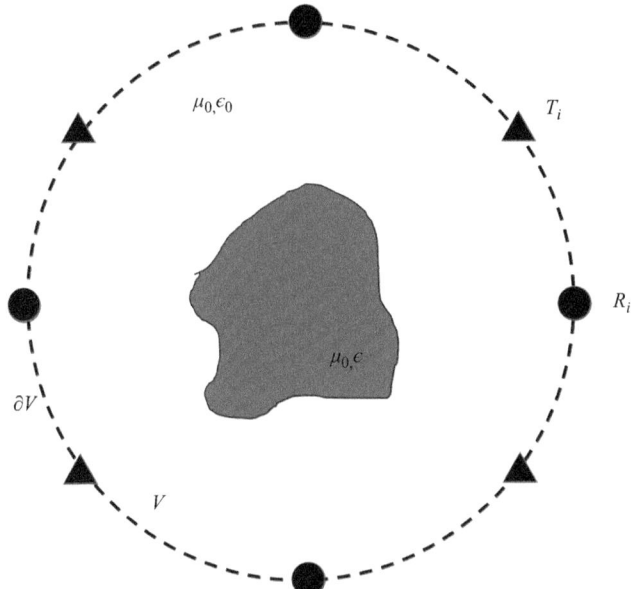

Figure 3.1   *Schematic representation of a two-dimensional scattering problem. The incident field $\mathbf{E}_{inc}$ is generated by a set of transmitters T_i, denoted by triangles, around the region of interest V. The scattered field is measured at a set of receivers R_i, denoted by circles, located around the scattering object. The forward problem consists of calculating the scattered fields $\mathbf{E}_{scat}$ from a known parameter distribution $\varepsilon(\mathbf{r})$, while the inverse problem consists of obtaining the unknown distribution $\varepsilon(\mathbf{r})$ from the measured fields $\mathbf{E}_{scat}$.*

(ii)   *Ill-conditioning*: Solving the inverse problem is generally an unstable procedure, i.e., without some type of regularization procedure, small errors due to the discretization of the problem or noise in the measurement data can lead to large discrepancies in the obtained solutions.

(iii)   *Non-linearity*: As noted, the inverse problem is nonlinear even if the forward problem is linear (such as for Maxwell's equations in linear media), which prevents the use of inversion techniques tailored for linear problems.

(iv)   *Resolution limits*: In general, if there is no a priori information available about a given inverse problem, then there is a limited maximum spatial resolution that any solution method can attain, which is related to the diffraction limit. Factors affecting the ultimate resolution are the frequency of operation, frequency bandwidth, and the aspect angle that the transmitters and receivers comprise with respect to the scattering object. Resolution limits are also closely related to the three other challenges pointed out above.

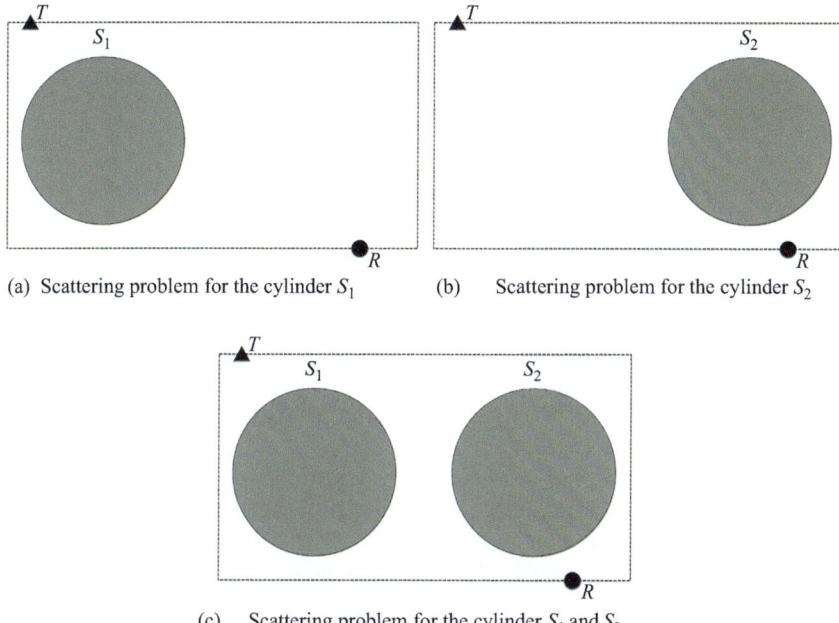

(a) Scattering problem for the cylinder S_1 (b)   Scattering problem for the cylinder S_2

(c)   Scattering problem for the cylinder S_1 and S_2

Figure 3.2   Schematic representation of the nonlinearity of the inverse scattering problem. While the relationship between incident and scattered fields is linear, the relationship between the material distribution and scattered field clearly is not linear, as the scattered field in the presence of the two cylinders S_1 and S_2 is not simply the sum of the scattered fields from either cylinder by itself; due to mutual scattering effects arising from the interaction between the two cylinders.

3.2 Traditional approaches

Before discussing the application of deep learning techniques to the problem of inverse scattering, it is useful to briefly review in a cursory fashion some traditional approaches used to solve this problem. Broadly speaking, many of the traditional approaches can be classified into two main types: approximate linearized solutions and iterative methods.

3.2.1 Traditional approximate solutions

One of the methodologies when trying to solve an inverse scattering problem is to make an approximation to the forward problem, which in turn also simplifies the inverse problem. One such approximation is to consider the incident field as propagating according to geometrical optics, i.e., that it travels in straight rays across the scatterer.

This is a good approximation when the incident wave frequency is high (or, in a particle picture, when the incident photons are very energetic and hence only very weakly affected by the scatterer). In this geometrical optics regime, a two-dimensional scatterer $s(x,y)$ can be reconstructed from its Radon transform [1]

$$p(r,\theta) = \iint s(x,y)\delta(r - x\cos\theta - y\sin\theta)dxdy, \tag{3.3}$$

where $s(\mathbf{r}) = k(\mathbf{r})/\omega$, $\delta(\,\cdot\,)$ is the Dirac delta function, r is the perpendicular distance from a line to the origin (usually assumed near the center of the object), and θ is the angle formed by the distance vector [1]. The reconstruction of $s(x,y)$ from its projections $p(r,\theta)$ is called back-projection and is of particular interest in medical science for its application to X-ray tomography, but it has less applicability to more general, nonlinear inverse scattering problems.

Other approximations can be achieved by substituting the total field $\mathbf{E}$ in the integrand of (3.2) for an approximate field with specific characteristics that simplify the problem. One such approximation, called the Born approximation [1], consists of approximating the (unknown) field in the integrand (3.2) by the (known) incident field such that

$$\mathbf{E}_{scat}(\mathbf{r}) = \int_V \overline{\mathbf{G}}(\mathbf{r},\mathbf{r}') \cdot O(\mathbf{r}')\mathbf{E}_{inc}(\mathbf{r}')d\mathbf{r}'. \tag{3.4}$$

This results in the scattered field being a linear functional of the material parameters, thus simplifying the forward problem and consequently the inverse problem. It can be shown that under the Born approximation the scattered field for a two-dimensional problem can be written in the far-field as

$$\mathbf{E}_{scat}(\rho) \approx i\frac{e^{ik_0(\rho_T+\rho_R)}}{8\pi k_0\sqrt{\rho_T\rho_R}}\tilde{O}(k_0(\hat{\rho}_R + \hat{\rho}_T)) \tag{3.5}$$

where $\rho_R\hat{\rho}_R$ is the position vector associated with the receiver location, $\rho_T\hat{\rho}_T$ is the position vector associated with the transmitter location, and $\tilde{O}$ is the Fourier transform of the function $O(\mathbf{r}) = k^2(\mathbf{r}) - k_0^2$, which is the contrast that we seek. Since in general we typically have some control over both the transmitter and receiver in an inverse problem, we are able to obtain the reconstruction of the object by calculating the inverse Fourier transform of $\tilde{O}$. The limitation of this method is that the obtained function $\tilde{O}$ is only known in Fourier space over a circle of radius at most $2k_0$, since $\hat{\rho}_R$ and $\hat{\rho}_T$ are unit vectors. Therefore the reconstructed $O(\mathbf{r})$ obtained from this method is band-limited to $|\mathbf{k}| < k_0$. Here we considered only a far-field solution. A more general derivation can be performed that takes into account near-field effects on the source and/or transmitter, but it does not change the fundamental conclusions; see [1] for a more in-depth discussion of these aspects.

Another useful approximation, called the Rytov approximation [1], consists of making the assumption that the field in the integrand can be expressed via a phase correction with respect to the incident field, such that

$$\mathbf{E}_{scat}(\mathbf{r}) = \int_V \overline{\mathbf{G}}(\mathbf{r}, \mathbf{r}') \cdot O(\mathbf{r}') \mathbf{E}_{inc}(\mathbf{r}') e^{\psi(\mathbf{r}')} d\mathbf{r}'. \tag{3.6}$$

With this assumption, the forward problem can be solved via a series expansion of the phase function $\psi(\mathbf{r})$. When applied to the inverse problem, a similar procedure can be applied to arrive at a similar result to (3.5), with the main difference being that this approximation is valid for higher frequency cases than the Born approximation.

Both the Born and Rytov approximations are accurate for weak-contrast scatterers, that is, for cases where the variation in the permittivity values across the scattering object is sufficiently small. Moreover, the Born approximation is more accurate for lower frequencies (or, equivalently, for electrically small scatterers), while the Rytov approximation is accurate at higher frequencies. Both methods are somewhat restricted in solving more general types of inverse scattering problems; nevertheless, both the Born and Rytov solutions are useful in many cases of practical interest. In addition, the solutions provided by such approximations can serve as good initial guesses for more general, iterative inverse scattering methods, described next, and for the deep learning-based approaches described later in the chapter.

3.2.2 Traditional iterative methods

Iterative methods are a powerful traditional approach to solving inverse problems. In general, an iterative method consists of an algorithm where we start with an initial guess (e.g., generated by applying the Born approximation to the problem), and then iteratively refine the guess through an optimization procedure that minimizes the error between the computed and measured field results. To exemplify the procedure, we will briefly describe the so-called Distorted Born Iterative Method [1]. For simplicity, we will describe the algorithm in its basic form, although several other variants of the basic algorithm exist. The method starts by invoking the Born approximation,

$$\mathbf{E}_{scat}(\mathbf{r}) = \int_V \overline{\mathbf{G}}(\mathbf{r}, \mathbf{r}') \cdot O(\mathbf{r}') \mathbf{E}_{inc}(\mathbf{r}') d\mathbf{r}'. \tag{3.7}$$

Furthermore, the assumption is made that the incident field originates from a point dipole source $\mathbf{p}$ at $\mathbf{r}''$ so that

$$\mathbf{E}_{inc}(\mathbf{r}') = \overline{\mathbf{G}}(\mathbf{r}', \mathbf{r}'') \cdot \mathbf{p}. \tag{3.8}$$

Therefore

$$\mathbf{E}_{scat}(\mathbf{r}) = \int_V \mathbf{M}(\mathbf{r}, \mathbf{r}', \mathbf{r}'') O(\mathbf{r}') d\mathbf{r}' \tag{3.9}$$

where $\mathbf{M}(\mathbf{r}, \mathbf{r}', \mathbf{r}'') = \overline{\mathbf{G}}(\mathbf{r}, \mathbf{r}') \cdot \overline{\mathbf{G}}(\mathbf{r}', \mathbf{r}'') \cdot \mathbf{p}$. Then, a functional can be defined as

$$I = \delta \int_V |O(\mathbf{r}')|^2 d\mathbf{r}' + \sum_{\mathbf{r},\mathbf{r}''} \left| \mathbf{E}_{scat}(\mathbf{r}) - \int_V \mathbf{M}(\mathbf{r}, \mathbf{r}', \mathbf{r}'')O(\mathbf{r}')d\mathbf{r}' \right|^2 \tag{3.10}$$

where the first term is the norm of $k^2(\mathbf{r}) - k_0^2$, the contrast we seek, the second term is the L^2 norm error between the measured scattered field and the approximate solution of the linearized Born approximation, and δ is a tuning parameter. The sum runs over all observation points (receiver positions) $\mathbf{r}$ and transmitter positions $\mathbf{r}''$. The solution for the contrast is found by minimizing this functional. Note that the first term of the functional is included as a regularization term to yield a unique solution and mitigate the ill-conditioning of the underlying inverse scattering problem. To solve this minimization problem numerically, the contrast function can next be expanded as

$$O(\mathbf{r}) = \sum_n a_n b_n(\mathbf{r}), \tag{3.11}$$

where $b_n(\mathbf{r})$ is a suitable, known basis functions for the contrast and a_n are unknown amplitudes. Then, by substituting (3.11) in (3.10),

$$I = \delta \sum_{n,m} a_n a_m^* B_{mn} + \sum_k \left| E_k - \sum_n a_n L_{kn} \right|^2, \tag{3.12}$$

where

$$B_{mn} = \int_V b_n(\mathbf{r}')b_m^*(\mathbf{r}')d\mathbf{r}', \tag{3.13a}$$

$$\mathbf{E}_k = \mathbf{E}_{scat}(\mathbf{r}_k), \tag{3.13b}$$

$$\mathbf{L}_{kn} = \int \mathbf{M}(\mathbf{r}_k, \mathbf{r}', \mathbf{r}_n)b_n(\mathbf{r}'), \tag{3.13c}$$

Minimization of the functional I gives

$$0 = \delta \sum_n a_n B_{mn} - \sum_k \mathbf{E}_k \cdot \mathbf{L}_{km}^* + \sum_k \sum_n a_n \mathbf{L}_{kn} \cdot \mathbf{L}_{km}^*, \tag{3.14}$$

which can be rewritten in matrix form as

$$0 = \delta[B] \cdot \mathbf{a} - [C] + [P] \cdot \mathbf{a}, \tag{3.15}$$

with solution

$$\mathbf{a} = [P + \delta B]^{-1} \cdot [C]. \tag{3.16}$$

Substituting these amplitudes back in (3.11) provides an approximate solution for the contrast.

The iterative nature of the method comes from taking the newly calculated $k(\mathbf{r})$ as the updated "background medium" $k_1(\mathbf{r})$. Next, a forward problem for this new background medium can be obtained, and the procedure described above can be repeated to calculate a new value for the contrast given by $O(\mathbf{r}) = k^2(\mathbf{r}) - k_1^2(\mathbf{r})$. This process is repeated until a configuration for $k(\mathbf{r})$ is found that produces scattered field data sufficiently close to the measurement data (following some residual error requirement), at which stage the inverse problem has been solved.

In addition to the traditional methods to solve nonlinear inverse problems discussed above, a myriad other types of methods also exist, such as Bayesian approaches [4,5], compressive sensing strategies [6], level set methods [7,8], multiresolution methods [9], and domain derivatives [10] to mention just a few. We will not delve into these other types of methods as they are beyond the scope of this chapter. Rather, we shall focus next on free-space inverse scattering methods based on artificial neural networks in general, and deep learning in particular.

3.3 Artificial neural networks applied to inverse scattering

The methods described in the preceding section provide solutions to the inverse scattering problem, but they are not without disadvantages. Specifically, approximate solutions suffer from having limited ranges of applicability, such as only being valid for weak scatterers or certain frequency ranges, and are not suited for solving more complex inverse scattering scenarios with strong nonlinearities. Iterative methods, on the other hand, suffer from the large cost of having to repeatedly compute the solution of the forward problem in each iteration before the solution to the inverse problem can be obtained, which can be a burden for complicated scattering scenarios and severely limits their applicability to problems that need to be solved "on-line," that is, applications where it is important to have rapid solutions.

Artificial neural networks are a natural fit to combat those disadvantages. Originally inspired by attempts to model how a biological brain works, artificial neural networks have evolved to be powerful general tools for modeling non-linear processes and find applications in several areas such as robotics, control systems, chemistry, pattern recognition, and finances. We will not discuss in detail here the historical development and the myriad strategies used to construct artificial neural networks, since these topic are better treated in other chapters of this volume. For our purposes here, it suffices to say that an artificial neural network is a collection of "neurons," which are connected computational units that mimic their biological namesake by being able to receive and transmit signals to other neurons connected to it, and this signal can be modified by the strength of the connection between neurons. Generically, the operation of an artificial neural network consists of introducing some sort of input signal that travels through the layers of connected neurons until it results in some sort of output at the end of the network. The strength of the connection between individual neurons is determined during the "training" stage of the network with a set of known input–output pairs through which the network is able to reproduce the known output results from the known input data. The specific function of a artificial neural network depends on the structure, or architecture, of the neuron connections, as well as how the network is trained. Determining the best architecture and training strategies for solving a given problem are two of the main focuses of artificial neural network research. In what follows, we refer to artificial neural networks simply as neural networks for short.

Neural networks are particularly suitable to solve inverse scattering problems because they can extract a suitable model for the nonlinear relationship between the measured data and the underlying material and geometric properties of the scattering object from a set of training data. These learned models can then solve new instances of the inverse problem with high confidence and accuracy, and at much less cost than iterative methods since the computational burden is shifted to the training process instead. It is typical for neural networks to have a layered structure. This facilitates the training and the optimization for the connection strengths. Deep learning refers to the process of constructing neural networks with many hidden layers (in excess of the input and output layers) and applying them in practice.

Though the history of neural networks in general, and deep learning in particular, is somewhat long and complex [11–13], there has been a recent major uptake of interest in the application of neural networks and deep learning to electromagnetic problems [14–20] and inverse problems in particular [21–24]. Some of these applications are discussed in other chapter of this book. This explosion of interest has been fueled in part by the increased capacity of modern computers to store and process large amounts of data. This enlarged capacity also allows for more complex neural networks (with more complex architectures such as many hidden layers and larger number of interactions), which are capable of solving increasingly more complex problems, including nonlinear inverse scattering, at a feasible cost.

3.4   Shallow network architectures

The first attempts for incorporating neural networks to solve the inverse scattering problem employed "shallow networks," that is, networks with only one or very few hidden layers. Since small networks are less capable of reconstructing the contrast distribution in detail, these applications were concerned with parametric reconstruction instead, that is, recovering a small set of fixed parameters from the problem, such as assuming the scattering objects are conducting cylinders and recovering their position and radius. The limitations of small networks also restrict their application to black-box approaches, that is, attempts that try to directly reconstruct the set of parameters being studied from the measured data without taking into account any underlying structure of the problem being solved. For the purpose of the discussion in this chapter, these initial attempts can be succinctly described as implementing a mapping of the form

$$\mathbf{q} = \mathscr{N}(\mathbf{E}_R), \tag{3.17}$$

where $\mathbf{q}$ represents the (small) set of parameters used to describe the scattering object(s), $\mathcal{N}(\,\cdot\,)$ is the nonlinear mapping corresponding to the shallow network model implementation, and $\mathbf{E}_R$ is the scattered field sampled at one or more receivers. For example, a two-layer feed-forward network is used in [25] to detect the center location, radius, and (homogeneous) dielectric permittivity of a cylinder embedded in a domain of fixed size. This problem is a simplified version of the general inverse scattering problems, since only cylindrical objects are assumed. As another example, a feed-forward network based on radial basis functions is used in [26] to solve a similar problem in the context of medical imaging, aimed at detecting proliferated bone marrow inside the bone of the lower part of a leg. While constructing the model, the problem is simplified by assuming both the leg and the proliferated marrow to be perfectly cylindrical in shape; furthermore the permittivities of all tissues are assumed known a priori. This allows for the parameter set $\mathbf{q}$ to simply correspond to the unknown location and radius of the proliferated marrow. A similar problem is also solved via a radial basis function architecture in [27], but now the scattering object is a conducting cylinder embedded in free space; the network again extracts the location and radius of the cylinder. Furthermore, a two-layer feed-forward architecture is used in [28] to extract the conductivity of a cylinder embedded in lossy media from 16 samples of the scattered field. The cylinder is considered to be embedded in a second, homogeneous cylinder of known material properties, so that the geometrical parameters are known and the only unknown is the conductivity of the inner cylinder.

The similarities between these examples consist of the limitation that the shallow network imposes on the model, restricting the sought-after output to be a small set of parameters, which also requires several assumptions to be made about the problem and therefore limits the versatility and generalization capabilities of the models. Deep learning architectures, on the other hand, are less prone to such limitations, and models can be created to directly retrieve the contrast distribution in its entirety.

3.5 Black-box approaches

Similar to shallow network architectures, black-box approaches based on deep learning can be described in a generic form by the mapping

$$\varepsilon(\mathbf{r}) = \mathcal{N}(\mathbf{E}_R) \tag{3.18}$$

where the main difference is that instead of a small output set of parameters $\mathbf{q}$ describing the scattering objects, the model now attempts to reconstruct the (discretized) distribution $\varepsilon(\mathbf{r})$ of the material parameters directly in the so-called pixel space or "pixel base." In this case, the spatial distribution of the permittivity and/or conductivity is discretized on a computational grid of pixels (or voxels in the three-dimensional case), with each pixel being assigned a separate value. In this way, the material distribution can be accurately reconstructed provided the computational pixel base is discretized in a fine enough mesh to capture the spatial variations of the actual material distribution at the appropriate spatial scale.

A very popular convolutional network architecture is the U-net [29], particularly in the field of image segmentation. Because the inverse scattering problem can be interpreted as a visual recognition task for the "image" made by the distribution of the material parameters, the U-net can be particularly effective at handling inverse scattering problems. The basic structure of the U-net contains a contracting path that consists of successive convolutions and down-sampling operations and an expansive path that performs successive convolutions and up-sampling, as shown in Figure 3.3. Skip connections between the contracting and expansive paths allow the network to combine feature information learned during the down-sampling with high-resolution spatial information. This enables the U-net to simultaneously learn both small-scale and large-scale features.

A direct inversion scheme using a convolutional U-net architecture that models (3.18) and extracts the values of the contrast defined as $\chi(\mathbf{r}) = \varepsilon_r(\mathbf{r}) - 1$ as a discretized distribution on a two-dimensional domain is proposed in [30]. The authors compare this direct inversion scheme with more robust alternatives, and show that it is only able to reconstruct simple profiles that do not exhibit sharp boundaries. In addition to this direct approach, the authors also explore the application of the U-net for solving the inverse problem in conjunction with what they call the Backpropagation Scheme (BPS) and the Dominant Current Scheme (DCS). For the BPS, an approximated induced current is calculated from the measured scattered field instead of the total field that appears in (3.2), i.e.

$$\mathbf{I}_{BPS}(\mathbf{r}') = O(\mathbf{r}')\mathbf{E}_{scat}(\mathbf{r}'), \tag{3.19}$$

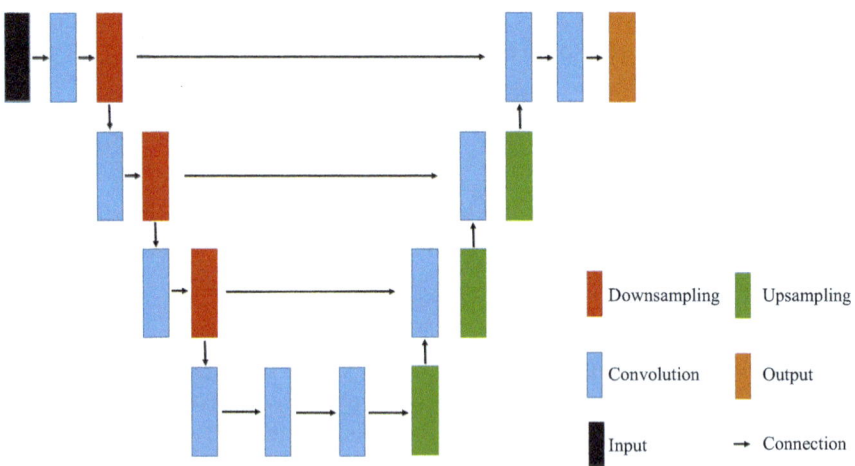

Figure 3.3 *Schematic representation of a generic U-net architecture. Each box represents a distinct operation as labeled in the legend, and the custom of depicting them in this U-shaped layout gives the architecture its name.*

which can then be used to calculate the total field and then solve a least-squares problem to obtain the approximate contrast

$$\chi(n) = \frac{\sum_p \mathbf{I}_p(n) \cdot \mathbf{E}_p^*(n)}{\sum_p \|\mathbf{E}_p(n)\|^2}, \tag{3.20}$$

where the summation p is taken over possible multiple incidences and the index n refers to the domain pixelization. This approximate contrast is then used as the input for the U-net to recover the actual reconstruction of the contrast. For the DCS, a similar approach of using an approximate contrast as input is used, but this contrast is obtained from considering an induced current formed from combining dominant singular value modes and low-frequency Fourier modes together to generate a "dominant induced current" which is then used to obtain the contrast. These two methods are able to extract the contrast distribution, even those with sharp boundaries, with better accuracy than traditional iterative methods.

An application of the U-net in the context of scattering from nanostructures is explored in [31], where the authors approach the problem in a similar manner as the direct inversion scheme of [30] but use the scattered magnetic field instead of the electric field. The construction of the network architecture and its application are otherwise similar, and for the considered case the U-net-based architecture is able to achieve high accuracy reconstructions at three times the speed of the traditional direct solution method employed as a comparison.

A study on the effect of using only real-valued data as inputs to the U-net architecture or including complex-valued data is performed in [32]. The authors test two architectures, the first separating the real and imaginary parts of the inputs and the second using the full complex-valued input, and apply them to the same direct inversion, backpropagation, and dominant current schemes presented in [30]. They show that the complex-valued architectures have better generalization capabilities than their real-valued counterparts, while having faster or similar convergence times during training. Despite this result, several of the other works discussed thorough this chapter use the method of separating the real and imaginary parts of inputs since this choice leads to a simpler implementation.

A similar approach of using a U-net architecture to create an image of the material distribution from the scattered field data collected at a limited number of receivers appears in [33], but the authors add additional U-net blocks to refine the image and obtain a higher-quality image reconstruction. The first refinement comes from using an U-net structure to obtain the approximated contrast in (3.20); the same derivation is performed, but instead of manually calculating the approximated contrast to use as input, the first network block learns the relationship between measured fields and this first approximation from the training data. The second refinement comes from using several U-net blocks cascaded together to achieve image-to-image transformations that gradually enhance the quality of the reconstruction. A similar two-step approach has also been proposed in [34], where the authors additionally use complex-valued networks directly.

A U-net architecture is also used to extract the material distribution of the scattering object in [35]. Instead of using the measured scattered fields as inputs to the U-net, the authors first calculate a backprojected approximation and use that as the input in the same way as (3.19). One salient feature of this work is that the authors also apply the network to experimental data while being trained only on a simulated training set, and show that the network is still accurate; this is an important feature as it is much easier and cost-effective to build a training set from simulation data (or some combination of simulation and experimental data) than from solely experimental results. It should be noted that the approach of using a backprojected approximation as the input to a U-net-based architecture is also used in [36], in the context of X-ray computed tomography. The authors test their approach on both synthetic and real data and show that their results have comparable accuracy to iterative reconstruction methods while having much faster computation times.

The application of U-net inverse scattering is extended to the three-dimensional case in [37]. Additionally, instead of using the scattered field as the input to their 3D U-net, the authors first calculate the Born approximation for their scattering problem to be used as the input, instead of the measured scattered fields. As a refinement to the input, the authors also apply a Monte Carlo approach to refine the initial images obtained by the Born approximation and sharpen the boundaries, so that the network input is of higher quality. The authors test their proposed method on inhomogeneous test scatterers and compare the reconstruction results with a state-of-the-art iterative method, showing that the deep learning approach is more accurate and efficient than the traditional alternative.

A similar approach of applying a convolutional network to extract the material distribution from a first guess is proposed in [38]. To obtain the first guess, the authors use what they call a "nonlinear mapping module" (NMM), which is an extreme learning machine [39] with one input layer, two hidden layers, and a preliminary imaging layer. This network has low training costs and maps the measured scattered fields into a preliminary distribution of the dielectric contrast. This preliminary distribution is then used as the input to the second part of the network, which the authors call an "image-enhancing module" (IEM) and is a convolutional network composed of an encoder, decoder, and pixel classifier; the encoder contracts the input data while the decoder synthesizes the previously encoded features, which then go through the pixel classifier for full-resolution segmentation. The primary advantage of this approach is that the first guesses provided by the extreme learning machine are more robust than those provided by approximate solutions, therefore the overall accuracy and efficiency of this approach is higher than those that depend on first guesses obtained by backpropagation or Born approximation. The authors further extend their proposed methodology to three-dimensional problems in [40]. For that, the authors first recognize that their original formulation is not suited for three-dimensional problems due to the higher computational costs in training the so-called nonlinear mapping module when shifting from a two-dimensional to a three-dimensional problem. To counteract the increased dimensionality of the problem, the authors adopt a "semi-join" approach, i.e., the extreme learning machine that encompasses the first module of their network has the nodes in their hidden layer not fully connected to the output layer, which lessens the inner matrix dimensions of the network and thus lowers the memory storage requirements and convergence rate of the training process.

3.5.1 Approaches for phaseless data

An important subset of inverse scattering problems are the so-called phaseless data problems, where the inverse problem to be solved is similar as before but we only have access to phaseless information of the scattered fields, i.e., from the generic field expression

$$\mathbf{E}_{scat} = \mathbf{E}_{amp}(\mathbf{r}, t)e^{j\phi(\mathbf{r}, t)} \tag{3.21}$$

where $\mathbf{E}_{amp}(\mathbf{r}, t)$ is a real-valued amplitude, the measurement has access only to $\mathbf{E}_{amp}(\mathbf{r}, t)$ measured at the receiver locations. This kind of limitation is relevant for some practical applications where measuring the phase information is limited due to hardware constraints in the measuring instruments.

A U-net-based architecture to solve two-dimensional phaseless inverse problems is proposed in [41]. Inspired by [30], the authors propose and compare three approaches: a direct inversion scheme, a dominant induced current scheme, and a contrast source inversion scheme. In the direct inversion scheme, the measured amplitude data is mapped directly to the contrast distribution through the U-net as a true black-box operation. In the dominant induced current method, an induced current is first calculated through an optimization procedure using only the few dominant modes of the singular value decomposition of the Green's function operator. This induced current can be used to solve an optimization problem similar to (3.20) and obtain an approximation for the contrast that is used as the input of the U-net. Finally, in the contrast source inversion approach, the dominant current is calculated via a Fourier-based expansion. Much like in [30], the authors find that the direct inversion scheme is only able to reconstruct relatively simpler images (scatterers), while the two methods that first generate a rough guess from the data through the dominant induced current method and then refine the resulting image with the U-net are more robust and yield good generalization properties.

An alternative solution using the U-net architecture is proposed in [42], where the authors separate the inverse scattering problem into a two-step procedure were first the phase information is extracted from the phaseless data and then the reconstruction is carried out. Both steps are carried out by U-net modules; the first step takes as input the square of the amplitude of the measured field and generates as output the amplitude and phase of the scattered field, which is then used as input for the second module to generate a reconstruction of the contrast distribution. Differently from most other applications of the U-net to the inverse scattering problem, the network takes as input the field values everywhere instead of a first guess of the contrast distribution.

A convolutional neural network architecture is also applied for phase recovery and phaseless data reconstruction in [43]. In this work, the authors consider specifically biomedical applications. The authors use a backpropagated intensity image separated into real and imaginary parts as the input to the convolutional neural network, which is then trained to provide the amplitude and phase reconstructions as its output. The authors apply their network to breast tissue, pap smear, and blood smear examples, and show that the network is able to recover the phase information and to provide image reconstruction with good accuracy and short computation time after training.

3.5.2 Application in electrical impedance and capacitance tomography

In some applications, electromagnetic inverse problems result from capacitance measurements rather than wave scattering measurements. In the former case, instead of transmitter and receiver antennas, a set of capacitive electrodes is placed at the boundary of a region of interest. The permittivity distribution in this region can then be approximately reconstructed based on the mutual capacitance data measured among all the electrode pair combinations. This particular sensing modality is typically denoted as electrical capacitance tomography. Electrical impedance tomography is another similar type of electromagnetic inverse problem, where the conductivity distribution in a region of interest is obtained from current-to-voltage (impedance) measurements between electrodes placed at the boundary. Many typical electrical capacitance tomography settings can be characterized as free-space. Electrical impedance tomography, on the other hand, is invariably not a free-space technique as it requires the presence of conduction currents from the transmitter to the receiver. Nevertheless, since these two tomography techniques are closely related mathematically, we discuss them together in this section.

The fundamental equations of the electrical impedance tomography problem are given by

$$\nabla \cdot \sigma(\mathbf{r})\nabla u(\mathbf{r}) = 0 \qquad \text{in } \Omega \tag{3.22a}$$

$$\sigma(\mathbf{r}) \, \hat{n} \cdot \nabla u(\mathbf{r}) = \varphi(\mathbf{r}) \quad \text{on } \partial\Omega \tag{3.22b}$$

where u is the electric potential inside the domain Ω, σ is the electric conductivity, which are the unknown material parameters, $\hat{n}$ is the unit normal vector to domain boundary, and φ is the boundary voltage. Both electrical impedance tomography problems and electrical capacitance tomography solutions are traditionally developed in two-dimensional settings, with $\mathbf{r} = (x, y)$; however, it should be noted that interest in three-dimensional tomography problems have increased in recent years due to marked developments in sensor hardware [44,45].

A deep learning U-net architecture is applied as a post-processing step in [46] to recover the sharpness in object boundaries obtained from the so-called D-bar, a traditional non-iterative method to obtain images in electrical impedance tomography. By identifying the spatial variables (x, y) with a corresponding point in the complex plane $z = x + iy$, the D-bar method consists of transforming the conductivity equation into a Schrödinger-like equation

$$\left(-\nabla^2 + q(z)\right) \tilde{u}(z) = 0 \tag{3.23}$$

where ∇^2 is the Laplace operator, $q(z) = \sigma^{-\frac{1}{2}}(z)\nabla^2\sigma^{\frac{1}{2}}(z)$ and $\tilde{u} = \sigma^{\frac{1}{2}}u$. Then, a nonphysical complex scattering variable $k = k_1 + ik_2$ can be introduced, and Complex Geometric Optics (CGO) solutions of the form $\tilde{u}(k, z) \approx \psi(k, z)e^{ikz}$ can be sought by solving the following D-bar equation

$$\bar{\partial}_k\psi(k,z) = \frac{1}{4\pi\bar{k}}\mathbf{t}(k)e^{-i(kz+\bar{k}\bar{z})}\bar{\psi}(k,z), \tag{3.24}$$

where the overbar denotes complex conjugation and

$$\mathbf{t}(k) = \int e^{i\bar{k}\bar{z}}q(z)\psi(k,z)dz \tag{3.25}$$

is the (non-linear) scattering transform of $\psi(k,z)$. By taking the asymptotic approximation $\psi(k,z) = 1$, a "Born approximation" for the scattering transform is obtained as

$$\mathbf{t}^{Born}(k) = \int e^{i\bar{k}\bar{z}}q(z)dz = \hat{q}(-2k_1, 2k_2), \tag{3.26}$$

where the hat denotes the usual Fourier transform. It can also be written as

$$\mathbf{t}^{Born}(k) = \int e^{i\bar{k}\bar{z}}(\Lambda_\sigma - \Lambda_1)e^{ikz}dz, \tag{3.27}$$

where Λ_γ is the voltage-to-current density map for the problem given a conductivity distribution γ; that is, Λ_1 is the reference data and Λ_σ is the measured data for the unknown conductivity σ. The usual D-bar method consists of solving the D-bar equation (3.24) with the Born approximation given by the measured data in (3.27), and then recovering the conductivity via the inverse relation

$$\sigma(z) \approx [\psi(z,0)]^2. \tag{3.28}$$

The application of the U-net replaces the step of solving (3.24); the network is trained with the input data in (3.27) and the ground truth as output. The authors show that the U-net is able to reconstruct test images after training with marked improvement on image quality. This is because the original D-bar method suffers from blurring effects from the Born approximation that are not present in the U-net results. The network is also effective when tested on experimental data even when trained on simulated data only.

A dominant-current scheme is proposed in [47] to solve the electrical impedance tomography problem, similar to that introduced in [30]. The authors examine both an iterative-based method and a neural network method. For the latter, a U-net architecture is used to extract the image reconstruction from the dominant currents calculated from the impedance tomography data. In their study, the authors test the network with both simulated and experimental data.

As noted above, a closely related inverse problem is that of electrical capacitance tomography, where instead of using the current-to-voltage measurements at the boundary of the domain of interest, the material parameters are obtained from capacitance measurements between a number of electrodes arranged around the region of interest. The fundamental equation of the electrical capacitance tomography problem is given by

$$\nabla \cdot \varepsilon(\mathbf{r})\nabla u(\mathbf{r}) = -\rho(\mathbf{r}) \tag{3.29}$$

where u is the electric potential, ε is the (unknown) permittivity distribution, and ρ is the charge distribution in the domain. The mutual capacitance between a pair of electrodes indexed as i and j is given by

$$C_{ij} = \frac{1}{V_{ij}} \oint_{\Gamma_j} \varepsilon(\mathbf{r})\nabla u(\mathbf{r}) \cdot \hat{n} dS \tag{3.30}$$

where V_{ij} is the potential difference between the electrodes and S_j is a closed surface (or path in two-dimensions) encircling the sensing electrode.

The forward and inverse problems in electrical capacitance tomography are solved in [48] using a feed-forward and in [49] using a Hopfield neural network architecture. In particular, the feed-forward network is trained to solve the forward problem, and the Hopfield network reconstructs the permittivity distribution from the output of the feed-forward network. While the feed-forward network is an example of a black-box implementation to solve the forward problem, the use of the Hopfield network to solve the inverse problem has similarities to the application of an iterative method, which provides a bridge into the next section where we discuss neural network architectures that are more closely related to iterative solutions.

3.6   Learning-augmented iterative methods

In this section, we will showcase some methods that employ deep learning algorithms to enhance or otherwise augment the capabilities of traditional iterative methods, either by substituting parts of a traditional iterative algorithm, e.g. the forward solver, with a surrogate learning-based model, or by applying learning-based algorithms to the entire inversion process.

The relationship between deep neural networks and traditional iterative methods to solve inverse scattering problems is elucidated in [50]. Specifically, the iterative equation for updating the contrast $\chi(\mathbf{r}) = \varepsilon_r(\mathbf{r}) - 1$ can be given in general form as

$$\chi(k+1) = \arg\min_{\chi} \left(\sum_n \|\delta\mathbf{E}_{scat}^{(n)} - \mathbf{J}_{(k)}^{(n)}\delta\chi\|_2^2 + \mathcal{R}(\chi) \right), \tag{3.31}$$

where δ denotes a difference between the quantity and its calculated value at the current iteration step, $\mathbf{J}$ denotes the Jacobian matrix of $\mathbf{E}_{scat}$ with respect to $\chi(k)$, the superscript n denotes multiple incidences, and $\mathscr{R}(\chi)$ denotes a regularization term. By considering a sparse transformation operator $\mathbf{D}$ and the regularization $\mathscr{R}(\chi) = \|D\chi\|_1$, the authors show that the iterative equation can be rewritten as

$$\mathbf{D}\chi(k+1) = \mathscr{S}\left(\mathbf{P}_k \cdot \chi(k) + \mathbf{b}_k\right), \tag{3.32}$$

where $\mathscr{S}(\,\cdot\,)$ is an element-wise soft-threshold function and

$$\mathbf{P}_k = \mathbf{D} - \mathbf{D}\left[\sum_n (\mathbf{J}_{(k)}^{(n)})^H \mathbf{J}_{(k)}^{(n)}\right]^{\dagger} \sum_n \mathbf{J}_{(k)}^{(n)} \overline{\mathbf{G}}_d \mathbf{E}_{(k)}^{(n)}, \tag{3.33a}$$

$$\mathbf{b}_k = \mathbf{D}\left[\sum_n (\mathbf{J}_{(k)}^{(n)})^H \mathbf{J}_{(k)}^{(n)}\right]^{\dagger} \sum_n \mathbf{J}_{(k)}^{(n)} \mathbf{E}_{scat}^{(n)} \tag{3.33b}$$

are terms related to the Jacobian $\mathbf{J}$, the scattered and total fields $\mathbf{E}_{scat}$ and $\mathbf{E}_{(k)}$, and the discretized Green's function $\overline{\mathbf{G}}_d$. The important connection to be made is based on the recognition of how (3.32) resembles the structure of a fully-connected deep neural network where $\mathbf{P}_k$ and $\mathbf{b}_k$ are the weight matrix and bias, respectively, k denotes the layer index, and $\mathbf{D}^H \mathscr{S}(\,\cdot\,)$ is the activation function. With this established connection between a deep neural network architecture and iterative solutions for inverse scattering problems, the authors propose a complex-valued network named DeepNIS to solve the inverse problem. Their proposed architecture consists of using cascaded convolutional neural network modules to map the input given by taking the back-propagation approximation to the real contrast distribution. Further studies on the performance and stability properties of the proposed architecture are carried out in [51].

The connection between iterative methods and deep neural networks is also directly exploited in [52], where the authors propose a convolutional network to solve the inverse problem and make a correspondence between specific steps in the optimization process used to solve the problem iteratively and associated layers in the neural network. Specifically, the authors define a Lagrangian optimization function for the inverse problem and apply the alternating direction method of multipliers (ADMM) as proposed in a previous related work [53]. In this context, the solution of the inverse problem is split into four optimization problems to solve the induced current $\mathbf{J}_{ind}$, the total field $\mathbf{E}_{tot}$ in the domain of interest, the sought-after contrast distribution χ, and an auxiliary variable $\mathbf{y}$ necessary for the Lagrangian method. These four optimization problems are then solved by convolutional layers in sequence, with knowledge from the optimization process used to guide the parameter choice of the networks. The effectiveness of the network is illustrated via simulated and experimental results.

The equation used for updating the contrast χ (3.31) can be linearized through the Born approximation as a means to simplify the problem. In this vein, a network-based regularizer was proposed in [54]. Specifically, the proposed regularizer is given by

$$\mathcal{R}(\chi_k) = \chi_k - \mathcal{N}(\chi_k), \tag{3.34}$$

where $\mathcal{N}(\cdot)$ is a denoising subnetwork implemented via a straightforward convolutional network with successive convolution, batch normalization, and rectified linear unit layers. The authors are able to establish a correspondence between the update rule for the regularizer and a deep convolutional network where each iteration can be interpreted as a single layer of the network. To reduce the model complexity, the operator $\mathcal{N}(\cdot)$ is fixed for every layer, which simplifies the training procedure. The authors compare their algorithm with the subspace optimization method and show that the proposed network achieves comparable results in terms of accuracy in less computational time. The authors then refine the algorithm presented in [55] by changing the network $\mathcal{N}(\cdot)$ in the regularizer term to be implemented via a generative adversarial network. The structure of $\mathcal{N}$ consists of two sub-networks: a generator network $\mathcal{G}$ and a discriminator network $\mathcal{D}$. The generator network $\mathcal{G}$ is similar to the one used in [54], but its output is fed as an input to the discriminator network $\mathcal{D}$ that compares it to the provided ground truth. The authors test their algorithm and evaluate its performance against both traditional iterative methods and their previous algorithm, showing that the integration of generator and discriminator networks allows the model to learn to reconstruct challenging profiles with a modest number of training samples.

A large part of the computational burden when using iterative methods for the solution of inverse scattering problems is due to the need for repeatedly solving the forward problem during the optimization process; for this reason, methodologies that quickly and efficiently solve the forward problem can be incorporated into an iterative inverse problem solution to achieve better results than traditional methods. To that end, a cascaded end-to-end convolutional network is proposed in [56] to solve the direct problem by learning the mapping between the incident fields and/or material contrast and the induced currents on the object, therefore allowing for the direct calculation of the scattered fields in an efficient manner. Specifically, the forward problem can be solved by discretizing (3.2) in a grid of M cells (or pixels) as

$$\mathbf{E}_{scat} = [G_s] \cdot \mathbf{J}_{ind}, \tag{3.35}$$

where $\mathbf{E}_{scat}$ are the scattered fields measured at N_r different receiver positions, $[G_s]$ is a matrix representation for the discretization of the Green's function operator on the given grid, and the induced current $\mathbf{J}_{ind}$ is given by

$$\mathbf{J}_{ind} = ([I] - [\chi] \cdot [G_d])^{-1} \cdot [\chi] \cdot \mathbf{E}_{inc}, \tag{3.36}$$

where $[\chi]$ is the matrix containing the contrast at each cell, $[I]$ denotes the identity matrix, and $[G_d]$ is another discretization of the Green's function operator. The difference between $[G_s]$ and $[G_d]$ is that the former represents the discretized operator related to the scattered field at the receiver locations and thus is a $N_r \times M$ matrix, where N_r is the number of receivers, while $[G_d]$ is the general, cell-based discretization and has size $M \times M$. To avoid repeatedly solving (3.36), the authors first estimate a "dominant current" $\mathbf{J}^+$ given by

$$\mathbf{J}_+ = \sum_j \frac{\psi_j^H \cdot \mathbf{E}_{scat}}{\sigma_j} \phi_j, \tag{3.37}$$

where σ_j, ψ_j, and ϕ_j are the jth singular values, left-, and right-singular vectors of $[G_s]$, respectively, the H superscript denotes conjugate transpose, and j runs from 1 to L, where L is a small number of dominant singular values. Then, a dominant electric field can be defined as

$$\mathbf{E}_+ = \mathbf{E}_{inc} + [G_d] \cdot \mathbf{J}_+, \tag{3.38}$$

and the goal is to solve the network equation

$$\mathbf{J}_{ind} = \mathcal{N}(\mathbf{J}_+, \mathbf{E}_+) \tag{3.39}$$

instead of solving (3.36). To this end, a cascaded convolutional architecture based on U-net blocks is used with skip connections between blocks, and all the subnetworks are trained simultaneously such that all the weights are updated dynamically with information from a combined loss function that takes into account all the stages of reconstruction. This network is shown to provide good generalization properties with simulated and experimental tests.

A generative adversarial network is proposed in [57] to achieve a similar result as proposed in [56]. In this particular work, the authors aim at replacing the solution of (3.36) by the application of a network model such that

$$\mathbf{J}_{ind} = \mathcal{N}(\chi, \mathbf{E}_{inc}) \tag{3.40}$$

is solved instead. The structure of $\mathcal{N}$ consists of two subnetworks: a generator network $\mathcal{G}$ and a discriminator network $\mathcal{D}$. The generator network $\mathcal{G}$ is a a convolutional U-net network that generates predictions for $\mathbf{J}_{in}$ based on the pix2pix architecture [58]. Importantly, all the quantities in consideration (i.e. $\mathbf{J}_{in}$, χ and $\mathbf{E}_{inc}$) have the same support in the discretized computational domain. Consequently, the application of a U-net like structure here is exactly the same procedure as described in Section 3.5. However, the generative adversarial model in this case also includes the discriminator network $\mathcal{D}$, which takes as input the predicted output from $\mathcal{G}$ as well as its original input, and evaluates the result with respect to the ground truth. The authors propose three different variants for the algorithm, with consist of using χ, $\chi \cdot \mathbf{E}_{in}$, or $\chi \oplus \chi \cdot \mathbf{E}_{in}$ as inputs to the generative network. The first variant consists of a direct inversion from the contrast function to the induced currents, the second variant consists of a Born-type approximation, and the third variant consists of providing the network with both the contrast information and the Born approximation information. The authors show that all these three variants are able to produce accurate results, but the third variant is the most robust and has the best generalization capacity, which is expected since the network receives the most amount of information as input.

A similar approach, based on the use of a neural network surrogate for the forward solver, is taken in [59]. In this case, however, the forward solver is embedded into another network that solves the inverse problem as well. The forward solver surrogate was originally proposed by the authors in [60] and can be described as establishing the following functional dependency

$$\mathbf{E}_{tot} = \mathcal{N}_1(\chi, \mathbf{E}_{inc}). \tag{3.41}$$

The surrogate is composed of a set of cascaded convolutional blocks that update the total field $\mathbf{E}_{tot}$ iteratively. The first block takes as input the incident field $\mathbf{E}_{inc}$ and the contrast guess χ, and outputs a first guess $\mathbf{E}_{tot}^1$ for the total field; subsequent blocks take the guess from the previous block as an additional input. To solve the inverse problem, the authors apply a supervised descent method solution where, starting from a guess for the material distribution χ_k, a new guess χ_{k+1} is calculated via a set of descent directions learned from the training data by minimizing the residual between the associated total field $\mathbf{E}_{tot}^k$ and the ground truth total field as

$$\chi_{k+1} = \mathcal{N}_2(\chi_k, \mathbf{E}_{obs} - \mathbf{E}_{tot}^k), \tag{3.42}$$

where χ_k and $\mathbf{E}_{obs} - \mathbf{E}_{tot}^k$ denote the material distribution and the difference between the observed field and the calculated total field, respectively, both evaluated at the present iteration k. The architecture $\mathcal{N}_2$ consists of a set of fully-connected layers that add nonlinearity to the data and then a set of convolutional layers to extract and refine the material distribution features. The final combined network consists of a cascade of forward solvers used to obtain the total field $\mathbf{E}_{tot}^k$ associated with the present material distribution χ_k, and the update networks that generate the new distribution χ_{k+1} from the old one and its associated field data, as shown in Figure 3.4. It is important to note that the forward solver network in this application is trained separately, which means that its parameters remain fixed while the inverse network is being trained or tested.

An iterative inverse solver based on projected linear Landweber (PNLW) assisted by three deep neural networks is proposed in [61]. The PNLW algorithm consists of solving the optimization problem

$$\arg\min_{\chi} \sum_f \|\mathbf{E}_{meas} - \mathbf{E}_{scat}(\chi)\|_2^2 \quad \text{such that } \|\chi\|_1 \leq L, \tag{3.43}$$

where $\mathbf{E}_{meas}$ is the measured scattered field, $\mathbf{E}_{scat}(\chi)$ is the field calculated via (3.35), L is a given hyperparameter that enforces sparsity in the material distribution being sought and the sum is performed across multiple operating frequencies f under consideration. Along with the hyperparameter L, the traditional PNLW algorithm also has a step size γ, not necessarily uniform, that controls the update of χ that must be selected. To enhance the traditional PNLW algorithm and avoid the need for user-defined hyperparameters, the author introduces two neural networks that predict optimal values for

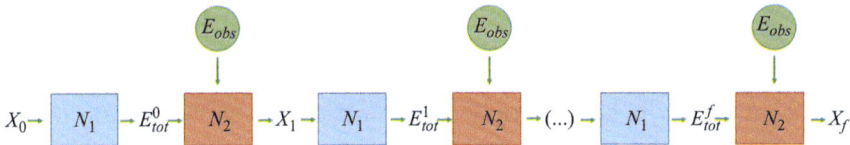

Figure 3.4   *Schematic representation of the cascaded set of networks that comprises the total inverse problem network proposed in [59]. The observed field $\mathbf{E}_{obs}$ is used as input for all instances of the $\mathcal{N}_2$ sub-network, and all instances of the $\mathcal{N}_1$ sub-network have their parameters frozen while the network is being trained. The first guess χ_0 is obtained by applying the inverse of the Frechét derivative of the forward problem to the observed data and is done offline prior to training or testing, and χ_f denotes the final output of the network.*

L and γ, as well as a third neural network that refines the output given by PNLW such that

$$L = \mathcal{N}_1(\mathbf{E}_{meas}), \tag{3.44a}$$

$$\gamma_k = \mathcal{N}_2(\chi_{k-1}), \tag{3.44b}$$

$$\chi_{out} = \mathcal{N}_3(\chi_{PNLW}). \tag{3.44c}$$

In the above, $\mathcal{N}_1$ is a fully-connected network that takes as input the measured field and outputs the prediction for the threshold L. $\mathcal{N}_2$ is a multilayer convolutional network that takes as input the current material distribution guess χ_{k-1} and outputs the step γ_k for the next PNLW iteration. Finally, $\mathcal{N}_3$ is a U-net-based architecture that takes the final material distribution given by PNWL, χ_{PNLW}, and refines it into the final output χ_{out}. The networks $\mathcal{N}_1$ and $\mathcal{N}_2$ are trained with data from applications of the traditional PNLW algorithm to a set of training cases, and the refining network $\mathcal{N}_3$ is trained with the output from these test cases and the ground truth associated with them. When compared against the traditional PNLW, the machine-learning-assisted PNLW algorithm shows comparable or better accuracy while having comparable or better running times.

3.7   Non-iterative learning methods

In this section, we will showcase methods that develop a learning-based solution for the inverse scattering problem that is neither iterative in nature nor a straight black-box solution. These methods take advantage of the underlying structure of the problem in some way to provide a direct solution.

The relationship between the training of a deep neural network and the iterative process for solving an inverse problem was also discussed in [62] in the context of certain acoustic (as opposed to elastic) seismic inverse problems. The acoustic setting differs from the electromagnetic in that acoustic waves can be treated as scalar waves instead of vector waves. However, most of the challenges in solving the inverse problem are similar in both scenarios and do not depend on the specific type of wave being studied. Particularly, the author points out the similarity between the process of updating the solution of the inverse problem iteratively and the process of training a deep neural network, which is also in a sense an iterative process of update based on optimizing some specific error measure. In the case of deep networks, the error measure is the cost function with respect to the variables in the network model, while in an iterative algorithm it is usually the 2-norm of the difference between the measured data and the data that is iteratively passed through the forward model. By describing the update equations for the forward model with a recurrent neural network and defining the material properties as the trainable weights, the task of finding out the unknown material parameters becomes equivalent to that of training the network weights. The same approach was independently proposed in [63], also in the context of seismic waveform inversion.

Inspired by these developments, a recurrent network architecture that models the electromagnetic forward scattering problem is proposed in [64]. The recurrent cell has the electromagnetic fields $\mathbf{E}$ and $\mathbf{H}$ as internal states and incorporates update equations based on the finite-difference time-domain method [65], which are deterministic and known a priori. The authors consider a 2D TM problem, for which the standard finite-difference time-domain update equations in an heterogeneous medium are given by

$$\mathbf{H}_x^{n+\frac{1}{2}} = C_p \cdot \mathbf{H}_x^{n-\frac{1}{2}} - C_q \cdot \nabla_y \mathbf{E}_z^n, \tag{3.45a}$$

$$\mathbf{H}_y^{n+\frac{1}{2}} = C_p \cdot \mathbf{H}_y^{n-\frac{1}{2}} - C_q \cdot \nabla_x \mathbf{E}_z^n, \tag{3.45b}$$

$$\mathbf{E}_z^{n+1} = C_a \cdot \mathbf{E}_z^n + C_b \cdot (\nabla_x \mathbf{H}_y^{n+\frac{1}{2}} - \nabla_y \mathbf{H}_x^{n+\frac{1}{2}} - \mathbf{J}_{source}^{n+\frac{1}{2}}), \tag{3.45c}$$

where

$$C_p = \frac{\frac{\mu}{\Delta_t} - \frac{\sigma_m}{2}}{\frac{\mu}{\Delta_t} + \frac{\sigma_m}{2}}, \tag{3.46a}$$

$$C_q = \frac{1}{\frac{\mu}{\Delta_t} + \frac{\sigma_m}{2}}, \tag{3.46b}$$

$$C_a = \frac{\frac{\varepsilon}{\Delta_t} - \frac{\sigma}{2}}{\frac{\varepsilon}{\Delta_t} + \frac{\sigma}{2}}, \tag{3.46c}$$

$$C_b = \frac{1}{\frac{\varepsilon}{\Delta_t} + \frac{\sigma}{2}} \tag{3.46d}$$

are coefficients related to the (inhomogeneous) material distribution.* In addition, ∇_x, ∇_y are discretized differential operators associated with the spatial increments Δ_x, Δ_y on a rectangular computational grid, Δ_t is the temporal discretization, and the superscripts denote the time step index. The authors compare this with the structure of a recurrent neural network, which can be written in a succinct generic form as

$$h_t = \mathcal{H}(x_t, h_{t-1}), \tag{3.47a}$$

$$y_t = \mathcal{F}(h_t), \tag{3.47b}$$

where h_t is the internal state of the network at time t, x_t is the input at time t, y_t is the output, and $\mathcal{H}(\cdot)$ and $\mathcal{F}(\cdot)$ are the network structures. The input x_t can be associated with the sources J_{source}, the internal state h_t with the fields H_x, H_y and E_z at a given time step, and the output y_t with the measured field at a given point in the computational domain. With this in mind, the structure of $\mathcal{F}(\cdot)$ is simply a sampling operator at the receiver locations, and $\mathcal{H}(\cdot)$ is given by the update equations (3.45). By introducing the material parameters in (3.46) as trainable weights, the inversion problem is recast as the training problem, and the network completes the inversion procedure once training is complete and the (reconstructed) material parameters that produce fields with best match to the original data are found.

It should be pointed out that a similar relationship between wave dynamics and recurrent neural networks is recognized in [66] but explored in a different application context: to solve the problem of classifying vowels from audio samples. Instead of traditional classification architectures, the authors implement the classification as an artificial medium that routes the different vowel waveforms into different receivers. The role of the neural network is to extract the material distribution for the routing medium that allows for the correct classification of the training dataset. Of note, the authors characterize the medium by its wave speed rather than by permittivity value as done in most other works surveyed in this chapter.

Physics-informed neural network approaches have been successfully applied to inverse scattering problems in the context of nano-optics and photonic metamaterial applications as well. For example, in reference [67], the authors use a simple feed-forward architecture but incorporate constraints from the partial differential equation under study and its boundary conditions into the loss function of the network. Considering a forward problem given by a generic partial differential equation such as

$$f\left(u(\mathbf{x}); \frac{\partial}{\partial \mathbf{x}} u(\mathbf{x}); \lambda\right) = 0, \tag{3.48}$$

where λ is the parameter sought as the solution of the inverse problem, and considering a feed-forward architecture given by

$$\hat{u} = \mathcal{N}(\mathbf{x}), \tag{3.49}$$

---

*Note that, to simplify the notation, the spatial dependency of the material parameters μ, ε, σ, and σ_m is suppressed. The symbol σ_m denotes magnetic conductivity.

then a loss function used to train the network can be written as

$$\mathcal{L}(\theta, \lambda) = w_f \mathcal{L}_f + w_i \mathcal{L}_i + w_b \mathcal{L}_b, \tag{3.50}$$

where θ are the network parameters, w_f, w_i, w_b are weights that control the relative importance of the various loss terms,

$$\mathcal{L}_f \sim \sum \| f(\mathbf{x}; \frac{\partial}{\partial \mathbf{x}} \hat{u}; \lambda) \|_2^2 \tag{3.51}$$

is the physical constraint that ensures the surrogate solution $\hat{u}$ obeys the original partial differential equation,

$$\mathcal{L}_i \sim \sum \| \hat{u}(\mathbf{x}) - u(\mathbf{x}) \|_2^2 \tag{3.52}$$

is the matching constraint that ensures the difference between the surrogate solution and the training dataset is small, and

$$\mathcal{L}_b \sim \sum \| \mathcal{B}(\hat{u}, \mathbf{x})) \|_2^2 \tag{3.53}$$

is the boundary constraint that ensures the surrogate solution obeys the appropriate boundary conditions given by $\mathcal{B}(\cdot)$ for the given problem. The authors showcase the method in the homogenization problem of a finite-size metamaterial, which corresponds to finding the unknown effective permittivity ε_{eff} that produces the same macroscopic field as the metamaterial under investigation. To generate the training data, the authors use a finite-element simulation, and to validate their results they run a new simulation with the ε_{eff} parameter extracted by the network. The authors show that good agreement is found between the effective parameter description and the original metamaterial structure. The authors also illustrate their method for inverse Mie scattering and invisible cloak designs, showing that the network is capable of retrieving the appropriate material distribution in the cases considered.

3.8 Closing remarks

This chapter surveyed applications of deep learning techniques to free-space inverse scattering. We have divided some of the most popular deep learning approaches to inverse scattering problems into three broad categories: black-box type of solutions, learning-based approaches to augment otherwise (traditional) iterative inverse scattering algorithms, and non-black-box, non-iterative learning approaches. This classification is not very sharp and has been used here simply to facilitate the presentation. Given the chapter length limitations and since the application of deep learning techniques for free-space inverse scattering is a quickly evolving area of research, this survey has been necessarily non-exhaustive. With the steady progress in computational hardware capabilities for processing of large amounts of data, it is expected that much progress will be made on this topic in the coming years to exploit the increasing availability of large training data sets. In addition, the use of very large numbers of hidden layers and unconventional neural network architecture is poised to open new research vistas and capabilities.

References

[1] Chew WC. *Waves and Fields in Inhomogeneous Media*. New York, NY: Wiley-IEEE Press; 1995.

[2] Tarantola A. *Inverse Problem Theory and Methods for Model Parameter Estimation*. New York, NY: Society for Industrial and Applied Mathematics; 2005.

[3] Chen X. *Computational Methods for Electromagnetic Inverse Scattering*. New York, NY: Wiley-IEEE Press; 2017.

[4] Schmidt DM, George JS, Wood CC. Bayesian inference applied to the electromagnetic inverse problem. *Human Brain Mapping*. 1999;7(3): 195–212.

[5] Fouda AE, Teixeira FL. Bayesian compressive sensing for ultrawideband inverse scattering in random media. *Inverse Problems*. 2014;30(11):114017.

[6] Oliveri G, Poli L, Anselmi N, *et al*. Compressive sensing-based born iterative method for tomographic imaging. *IEEE Transactions on Microwave Theory and Techniques*. 2019;67(5):1753–1765.

[7] Dorn O, Lesselier D. Level set methods for inverse scattering. *Inverse Problems*. 2006;22(4):R67–R131.

[8] Guo R, Jia Z, Song X, *et al*. Application of supervised descent method to parametric level-set approach. In: *2019 IEEE International Conference on Computational Electromagnetics (ICCEM)*; 2019. pp. 1–2.

[9] Zhong Y, Salucci M, Xu K, *et al*. A multiresolution contraction integral equation method for solving highly nonlinear inverse scattering problems. *IEEE Transactions on Microwave Theory and Techniques*. 2020;68(4):1234–1247.

[10] Hagemann F, Arens T, Betcke T, *et al*. Solving inverse electromagnetic scattering problems via domain derivatives. *Inverse Problems*. 2019;35(8):084005.

[11] Schmidhuber J. Deep learning in neural networks: an overview. *Neural Networks*. 2015;61:85–117.

[12] LeCun Y, Bengio Y, Hinton G. Deep learning. *Nature*. 2015;521(7553): 436–444.

[13] Goodfellow I, Bengio Y, Courville A. *Deep Learning*. London: MIT Press; 2016.

[14] Jiang C, Zhang H, Ren Y, *et al*. Machine learning paradigms for next-generation wireless networks. *IEEE Wireless Communications*. 2016;24(2):98–105.

[15] Zhu XX, Tuia D, Mou L, *et al*. Deep learning in remote sensing: a comprehensive review and list of resources. *IEEE Geoscience and Remote Sensing Magazine*. 2017;5(4):8–36.

[16] Massa A, Oliveri G, Salucci M, *et al.* Learning-by-examples techniques as applied to electromagnetics. *Journal of Electromagnetic Waves and Applications*. 2018;32(4):516–541.

[17] Massa A, Marcantonio D, Chen X, *et al.* DNNs as applied to electromagnetics, antennas, and propagation: a review. *IEEE Antennas and Wireless Propagation Letters*. 2019;18(11):2225–2229.

[18] Travassos XL, Avila SL, Ida N. Artificial neural networks and machine learning techniques applied to ground penetrating radar: a review. *Applied Computing and Informatics*. 2020;17(2);296–308.

[19] Campbell SD, Jenkins RP, O'Connor PJ, *et al.* The explosion of artificial intelligence in antennas and propagation: How deep learning is advancing our state of the art. *IEEE Antennas and Propagation Magazine*. 2020;63(3):16–27.

[20] Wiecha PR, Arbouet A, Girard C, *et al.* Deep learning in nano-photonics: inverse design and beyond. *Photonics Research*. 2021;9(5):B182–B200.

[21] McCann MT, Jin KH, Unser M. Convolutional neural networks for inverse problems in imaging: a review. *IEEE Signal Processing Magazine*. 2017;34(6):85–95.

[22] Lucas A, Iliadis M, Molina R, *et al.* Using deep neural networks for inverse problems in imaging: beyond analytical methods. *IEEE Signal Processing Magazine*. 2018;35(1):20–36.

[23] Chen X, Wei Z, Li M, *et al.* A review of deep learning approaches for inverse scattering problems (invited review). *Progress in Electromagnetics Research*. 2020;167:67–81.

[24] Ongie G, Jalal A, Metzler CA, *et al.* Deep learning techniques for inverse problems in imaging. *IEEE Journal on Selected Areas in Information Theory*. 2020;1(1):39–56.

[25] Caorsi S, Gamba P. Electromagnetic detection of dielectric cylinders by a neural network approach. *IEEE Transactions on Geoscience and Remote Sensing*. 1999;37(2):820–827.

[26] Rekanos IT. Neural-network-based inverse-scattering technique for online microwave medical imaging. *IEEE Transactions on Magnetics*. 2002;38(2):1061–1064.

[27] Rekanos IT. On-line inverse scattering of conducting cylinders using radial basis-function neural networks. *Microwave and Optical Technology Letters*. 2001;28(6):378–380.

[28] Bermani E, Caorsi S, Raffetto M. A threshold electromagnetic classification approach for cylinders embedded in a lossy medium by using a neural network technique. *Microwave and Optical Technology Letters*. 2000;24(1):13–16.

[29] Ronneberger O, Fischer P, Brox T. U-net: convolutional networks for biomedical image segmentation. In: *International Conference on Medical Image Computing and Computer-Assisted Intervention*. New York, NY: Springer; 2015. pp. 234–241.

[30] Wei Z, Chen X. Deep-learning schemes for full-wave nonlinear inverse scattering problems. *IEEE Transactions on Geoscience and Remote Sensing*. 2018;57(4):1849–1860.

[31] Li Y, Wang Y, Qi S, *et al.* Predicting scattering from complex nano-structures via deep learning. *IEEE Access.* 2020;8:139983–139993.

[32] Pan XM, Song BY, Wu D, *et al.* On phase information for deep neural networks to solve full-wave nonlinear inverse scattering problems. *IEEE Antennas and Wireless Propagation Letters.* 2021;20(10):1903–1907.

[33] Xu K, Zhang C, Ye X, *et al.* Fast full-wave electromagnetic inverse scattering based on scalable cascaded convolutional neural networks. *IEEE Transactions on Geoscience and Remote Sensing.* 2022;60:1–11.

[34] Yao HM, Wei E, Jiang L. Two-step enhanced deep learning approach for electromagnetic inverse scattering problems. *IEEE Antennas and Wireless Propagation Letters.* 2019;18(11):2254–2258.

[35] Sun Y, Xia Z, Kamilov US. Efficient and accurate inversion of multiple scattering with deep learning. *Optics Express.* 2018;26(11):14678–14688.

[36] Jin KH, McCann MT, Froustey E, *et al.* Deep convolutional neural network for inverse problems in imaging. *IEEE Transactions on Image Processing.* 2017;26(9):4509–4522.

[37] Xiao J, Li J, Chen Y, *et al.* Fast electromagnetic inversion of inhomogeneous scatterers embedded in layered media by born approximation and 3-D U-net. *IEEE Geoscience and Remote Sensing Letters.* 2019;17(10):1677–1681.

[38] Xiao LY, Li J, Han F, *et al.* Dual-module NMM-IEM machine learning for fast electromagnetic inversion of inhomogeneous scatterers with high contrasts and large electrical dimensions. *IEEE Transactions on Antennas and Propagation.* 2020;68(8):6245–6255.

[39] Huang GB, Zhu QY, Siew CK. Extreme learning machine: theory and applications. *Neurocomputing.* 2006;70(1–3):489–501.

[40] Xiao LY, Li J, Han F, *et al.* Super-resolution 3-D microwave imaging of objects with high contrasts by a semijoin extreme learning machine. *IEEE Transactions on Microwave Theory and Techniques.* 2021;69(11):4840–4855.

[41] Xu K, Wu L, Ye X, *et al.* Deep learning-based inversion methods for solving inverse scattering problems with phaseless data. *IEEE Transactions on Antennas and Propagation.* 2020;68(11):7457–7470.

[42] Luo F, Wang J, Zeng J, *et al.* Cascaded complex U-net model to solve inverse scattering problems with phaseless-data in the complex domain. *IEEE Transactions on Antennas and Propagation.* 2021;70:6160–6170.

[43] Rivenson Y, Zhang Y, Günaydın H, *et al.* Phase recovery and holographic image reconstruction using deep learning in neural networks. *Light: Science & Applications.* 2018;7(2):17141–17141.

[44] Zhang K, Li M, Yang F, *et al.* Three-dimensional electrical impedance tomography with multiplicative regularization. *IEEE Transactions on Biomedical Engineering.* 2019;66(9):2470–2480.

[45] Chowdhury SM, Marashdeh QM, Teixeira FL. Electronic scanning strategies in adaptive electrical capacitance volume tomography: tradeoffs and prospects. *IEEE Sensors Journal.* 2020;20(16):9253–9264.

[46] Hamilton SJ, Hauptmann A. Deep D-bar: real-time electrical impedance tomography imaging with deep neural networks. *IEEE Transactions on Medical Imaging.* 2018;37(10):2367–2377.

[47] Wei Z, Liu D, Chen X. Dominant-current deep learning scheme for electrical impedance tomography. *IEEE Transactions on Biomedical Engineering.* 2019;66(9):2546–2555.

[48] Marashdeh Q, Warsito W, Fan L, *et al.* A nonlinear image reconstruction technique for ECT using a combined neural network approach. *Measurement Science and Technology.* 2006;17(8):2097.

[49] Hopfield JJ. Neural networks and physical systems with emergent collective computational abilities. *Proceedings of the National Academy of Sciences.* 1982;79(8):2554–2558.

[50] Li L, Wang LG, Teixeira FL, *et al.* DeepNIS: deep neural network for nonlinear electromagnetic inverse scattering. *IEEE Transactions on Antennas and Propagation.* 2018;67(3):1819–1825.

[51] Li L, Wang LG, Teixeira FL. Performance analysis and dynamic evolution of deep convolutional neural network for electromagnetic inverse scattering. *IEEE Antennas and Wireless Propagation Letters.* 2019;18(11):2259–2263.

[52] Liu J, Zhou H, Ouyang T, *et al.* Physical model-inspired deep unrolling network for solving nonlinear inverse scattering problems. *IEEE Transactions on Antennas and Propagation.* 2021;70:1236–1249.

[53] Liu J, Zhou H, Chen L, *et al.* Alternating direction method of multiplier for solving electromagnetic inverse scattering problems. *International Journal of Microwave and Wireless Technologies.* 2020;12(8):790–796.

[54] Zhou H, Ouyang T, Li Y, *et al.* Linear-model-inspired neural network for electromagnetic inverse scattering. *IEEE Antennas and Wireless Propagation Letters.* 2020;19(9):1536–1540.

[55] Zhou H, Zheng H, Liu Q, *et al.* Linear electromagnetic inverse scattering via generative adversarial networks. *International Journal of Microwave and Wireless Technologies.* 2021;4:1–9.

[56] Wei Z, Chen X. Physics-inspired convolutional neural network for solving full-wave inverse scattering problems. *IEEE Transactions on Antennas and Propagation.* 2019;67(9):6138–6148.

[57] Ma Z, Xu K, Song R, *et al.* Learning-based fast electromagnetic scattering solver through generative adversarial network. *IEEE Transactions on Antennas and Propagation.* 2020;69(4):2194–2208.

[58] Isola P, Zhu JY, Zhou T, *et al.* Image-to-image translation with conditional adversarial networks. In: *Proceedings of the IEEE Conference on Computer Vision and Pattern Recognition;* 2017. pp. 1125–1134.

[59] Guo R, Lin Z, Shan T, *et al.* Physics embedded deep neural network for solving full-wave inverse scattering problems. *IEEE Transactions on Antennas and Propagation.* 2021;2021:1–24.

[60] Guo R, Shan T, Song X, *et al.* Physics embedded deep neural network for solving volume integral equation: 2D case. *IEEE Transactions on Antennas and Propagation.* 2021;70:6135–6147.

[61] Desmal A. High-quality self-contained electromagnetic imaging scheme based on projected nonlinear landweber and machine learning. *IEEE Transactions on Antennas and Propagation*. 2021;70:1380–1388.

[62] Richardson A. Seismic full-waveform inversion using deep learning tools and techniques. arXiv preprint arXiv:180107232. 2018.

[63] Sun J, Niu Z, Innanen KA, *et al*. A theory-guided deep-learning formulation and optimization of seismic waveform inversion. *Geophysics*. 2020;85(2): R87–R99.

[64] Hu Y, Jin Y, Wu X, *et al*. A theory-guided deep neural network for time domain electromagnetic simulation and inversion using a differentiable programming platform. *IEEE Transactions on Antennas and Propagation*. 2022;70(1): 767–772.

[65] Taflove A, Hagness SC, Piket-May M. Computational electromagnetics: the finite-difference time-domain method. *The Electrical Engineering Handbook*. 2005:629–670.

[66] Hughes TW, Williamson IA, Minkov M, *et al*. Wave physics as an analog recurrent neural network. *Science Advances*. 2019;5(12):6946.

[67] Chen Y, Lu L, Karniadakis GE, *et al*. Physics-informed neural networks for inverse problems in nano-optics and metamaterials. *Optics Express*. 2020;28(8):11618–11633.

Chapter 4

Deep learning techniques for non-destructive testing and evaluation

*Roberto Miorelli[1], Anastassios Skarlatos[1],
Caroline Vienne[1], Christophe Reboud[1] and Pierre Calmon[1]*

4.1 Introduction

The term of Non-Destructive Testing and Evaluation (NDT&E) gathers methods and techniques aiming at assessing the material properties of media during the industrial manufacturing process (i.e., quality control, zero-defects production, etc.) of specimen and during the exploitation cycle of the manufactured specimen (i.e., integrity check, the ageing status, etc.). NDT&E is applied to test the integrity of the deployed structures in industrial domains ranging from energy (e.g., nuclear, oil & gas, powerline electric, etc.), transportation (e.g., automotive, railways, aeronautic), civil structures (e.g., bridges, buildings, etc.), manufacturing (e.g., metallurgic, food, chemical pharmaceutical, etc.) to cite the most prominent ones. The NDT&E methods are often classified by their type of energy and the associated propagation mechanisms in the investigated specimen under testing (SUT): electromagnetic- (i.e., magnetic flux density testing, eddy current testing, microwave testing, terahertz testing), infrared (i.e., infrared thermography testing), X-rays (i.e., radiography, tomography testing) and ultrasonic (i.e., acoustic-, elasto-dynamic, guided-wave propagation regime testing). This chapter focuses on problems dealing with electromagnetic-based methods and techniques.

In the last decade, NDT&E research and development communities have been trying to develop automatic inspection systems, aiming at assisting or replacing the human involvement in data analysis and thus at enhancing productivity and reducing the risk of human errors. Indeed, in NDT&E the measurements are normally composed of a large amount of data that can behave as (multimodal-) time-series and/or (multispectral-) images. Solutions to automatize the diagnostic process or at least to provide an assistance are currently under active research, as a consequence of the digitization of manufacturing processes, called "Industry 4.0". To this end, the NDT&E community is studying Artificial Intelligence (AI)-based approaches and machine learning (ML) based algorithms. Among the most promising ML algorithms,

[1]Université Paris-Saclay, CEA, List, F-91120, Palaiseau, France

the growing family of Artificial Neural Network (ANN) and in particular the Deep Neural Network (DNN) based algorithms are catching the attention of scholars and engineers (see Figure 4.1).

Some constraints make the development and the application of ML algorithms challenging in the NDT&E context. Indeed, large collections of datasets containing close-to-reality experimental data are often not available. As a matter of fact, due to industry confidentiality constraints, collaborative and open development frameworks are quite rare and bounded on very specific cases. Another limitation is the lack of normalization of the use of such algorithms. Documents of recommended practices have been released only very recently and have not been yet applied in the various sectors of industry. The proper way of comparing performance between such solutions and actual inspection procedures still remains an open question in many sectors.

This book chapter describes the use of ML methods and techniques with a focus on DL-based methods. It provides an analysis of recent contributions within the research community. Current and future trends of the application of DL algorithms are also mentioned. Moreover, even though our analysis is based on the electromagnetic methods (see Figure 4.2), we think that this contribution may partially apply to the study of other methods (i.e., ultrasound testing, structural health monitoring, acoustic emission, visual inspection, etc.). Our review considers the most significant research axes in representative industrial sectors: energy, transportation, the civil engineering, and manufacturing.

This chapter is organized as follows. In Section 4.2, we provide the principle of electromagnetic propagation and modeling with particular emphasis on applications in the quasi-static regime for layered homogeneous conductive media. In addition,

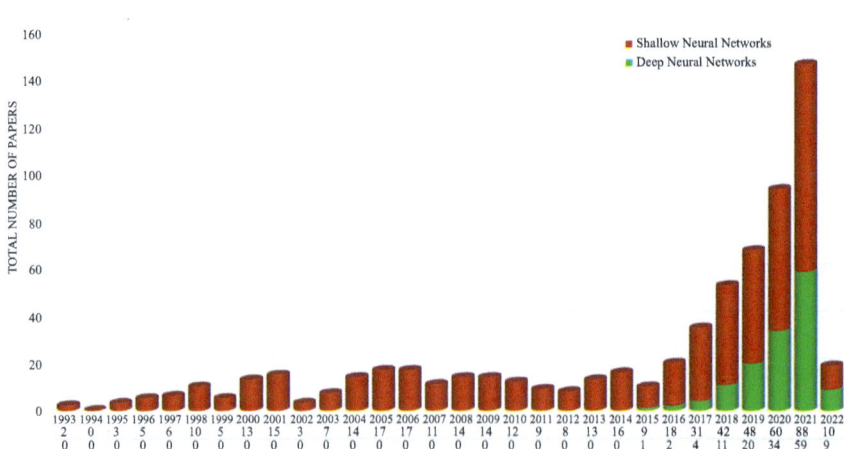

Figure 4.1   Estimated number of papers applying deep and shallow neural networks methods to electromagnetic NDT&E problems (Source: Web of Science, data updated to February 2022)

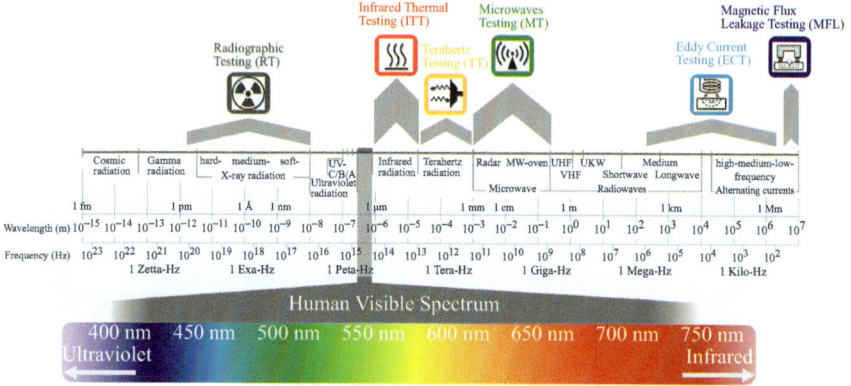

Figure 4.2   NDT&E methods as localized on the electromagnetic spectrum

some formal definitions about forward and inverse problems that are often recalled in the chapter are detailed. In Section 4.3, we review the main categories and challenges in applying deep learning methods to electromagnetic non-destructive testing signals analyzing the different kinds of signals commonly probed by the most common electromagnetic methods and techniques. In Section 4.4, we analyze some important contributions made in the field. The analysis performed has as a main purpose to shed light on the application of deep learning in the main industrial sectors concerned by electromagnetic inspection methods. In Section 4.5, we provide a brief overview of two main complementary methods for assessing the integrity of the structures. In Section 4.6, an analysis of future trends and open issues on the application of deep learning algorithms is provided. The last section is devoted to the chapter conclusion and remarks.

4.2   Principles of electromagnetic NDT&E modeling

A typical scenario for the electromagnetic inspection of a conducting and/or magnetic piece is schematically depicted in Figure 4.3. The tested piece is interacting with an incident electromagnetic field produced by a set of inducting coils, and the resulting field is sensed via a number of probes scanning the piece at the region of interest. The detection probe can be either an induction coil (with or without ferrite core), which can be designed to adapt to the specific geometrical features of the piece [1], or a magnetic field sensor, like a Hall-effect probe, a Giant Magneto-Resistance (GMR) sensor or a flux-gate, to mention the most popular ones. The measured signal carries information about the geometry and the material of the piece and it is the so-called "measurement" that will be used in the inversion phase to retrieve information about the piece.

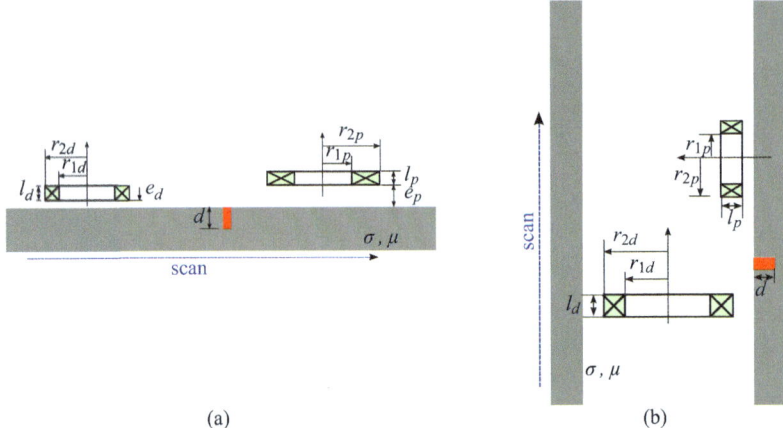

(a)   (b)

Figure 4.3   *Electromagnetic inspection of (a) planar piece and (b) tube, in driver-pickup mode. One can distinguish the piece, the driver (d) and pick-up (p) coils and the defect (in red). The eddy-current head comprising the coils is scanning the piece along the dashed line.*

To remain simple with the problem formulation, we shall restrict ourselves in the context of this article to inspection in the harmonic regime, which is the most widespread, i.e. a harmonic time dependence of the $e^{j\omega t}$, with ω being the angular frequency and $j = \sqrt{-1}$, will be assumed from this point forward for all state variables. Transient signal measurements are more suitable for particular applications [2–9].

It is convenient for both the mathematical analysis and signal interpretation purposes to decompose the measured signal into a sum of contributions, each one expressing a particular effect. Hence, the complete signal is composed of the probe response in air, the variation owing to the presence of the piece nearby the probe and finally the small signal variation sensed when scanning a flawed area of the piece. As these contributions have very different amplitudes and spatial properties, it is often much more efficient to compute them separately using perturbation approaches.

In the case of Figure 4.3 (single receiving coil), the measurement signal is proportional to the mutual impedance ΔZ_{TR}, where T stands for "transmitter" and R for receiver. The total impedance can be split into three parts, namely the mutual impedance in air $\Delta Z_{TR}^{(a)}$, the impedance change due to the piece $\Delta Z_{TR}^{(p)}$, referred to usually as the "geometry signal," and finally the impedance change owing to the presence of material defects in the illuminated zone $\Delta Z_{TR}^{(p)}$. One can thus write

$$\Delta Z_{TR}(\mathbf{r}_s, \omega) = j\omega M_{TR}^{(a)} + \Delta Z_{TR}^{(p)}(\mathbf{r}_s, \omega) + \Delta Z_{TR}^{(d)}(\mathbf{r}_s, \omega) \tag{4.1}$$

where $M_{TR}^{(a)}$ is the mutual inductance in air. Notice that both the geometry and the defect signals depend on the probe position $\mathbf{r}_s$. In the case of a magnetic field sensor, the previous splitting of the total signal in air, piece, and defect contributions also holds, where this time the complex impedance should be replaced by the magnetic field component parallel to the sensitivity direction of the magnetic sensor. In the following, we shall focus on the former case since coils are the preferred probes in the majority of practical applications. The analysis is similar for magnetic field measurements.

From the three parts in (4.1), $M_{TR}^{(a)}$ depends only upon the probe geometry and does not carry any information on the piece. It is therefore not useful for the purposes of signal processing. Besides, being constant for all scan positions, it can be easily removed by calibration. Our primary concern will thus be the calculation of the remaining two terms, namely the geometry and the defect perturbation signals.

4.2.1 Field solution for the flawless piece and calculation of the signal geometry $\Delta Z_{TR}^{(p)}$

Referring to the above introduced air-piece-defect decomposition approach, the next step in the analysis will be to calculate the response of the flawless piece, i.e. the scattering field and the probe signal owing to the piece interaction with the coil field.

The treatment of the geometry signal $\Delta Z_{TR}^{(p)}$ can provide information about the piece material and global configuration parameters such as the piece thickness and the probe lift-of. Since it depends only upon the geometry of the flawless piece, it can be often calculated by semi-analytical approaches. There is a large number of articles concerned with this calculation, all stemming in a greater of lesser extent from the seminal work of Dodd and Deeds in the 1960s [10,11] and the Auld's article on the specialization of the Lorentz reciprocity theorem [12,13]. This approach has been greatly enhanced and extended by Theodouldis and Bowler with the introduction of the Truncated Region Eigenfunctions Expansion (TREE) [14–17]. The TREE method has thus permitted the calculation of the geometry signal for canonical pieces with discontinuities like edges [18] boreholes [19–22], tubes with eccentric walls [23], etc. For more complicated pieces, one has to resort to either fully numerically techniques like the finite element method (FEM) or hybrid analytical–numerical schemes [24].

In the case of a symmetric piece, the signal geometry is also constant, and can be thus separated from the flaw signal by a simple baseline removal.

4.2.2 Defect response: calculation of the flaw signal $\Delta Z_{TR}^{(d)}$

We assume that the flawless piece is homogeneous and isotropic in the region of interest with a "base" electrical conductivity σ_b and magnetic permeability μ_b. The presence of material defects is translated to a local variation of the piece electric and magnetic properties, $\delta\sigma(\mathbf{r})$ and $\delta\mu(\mathbf{r})$, respectively. The material coefficients in the piece with the flaw can be thus written in the following way:

$$\sigma(\mathbf{r}) = \sigma_b + \delta\sigma(\mathbf{r}) \tag{4.2}$$

$$\mu(\mathbf{r}) = \mu_b + \delta\mu(\mathbf{r}). \tag{4.3}$$

The interaction of the driving (primary) electric and magnetic field $\mathbf{E}_p, \mathbf{H}_p$ with the material inhomogeneities owing to the flaw can be seen as the effect of an equivalent electric and an equivalent magnetic source, which superposed to $\mathbf{E}_p$, $\mathbf{H}_p$ will yield the total field according the expressions [25–29]

$$
\mathbf{E}(\mathbf{r}) = \mathbf{E}^p(\mathbf{r}) - j\omega\mu_b \int_{V_f} \overline{\mathbf{G}}^{ee}(\mathbf{r}, \mathbf{r}') \cdot \delta\sigma(\mathbf{r}') \, \mathbf{E}(\mathbf{r}) \, dV'
$$

$$
- j\omega \int_{V_f} \overline{\mathbf{G}}^{em}(\mathbf{r}, \mathbf{r}') \cdot \delta\mu(\mathbf{r}') \, \mathbf{H}(\mathbf{r}') \, dV' \tag{4.4}
$$

$$
\mathbf{H}(\mathbf{r}) = \mathbf{H}^p(\mathbf{r}) + \int_{V_f} \overline{\mathbf{G}}^{me}(\mathbf{r}, \mathbf{r}') \cdot \delta\sigma(\mathbf{r}') \, \mathbf{E}(\mathbf{r}') \, dV'
$$

$$
- j\omega\sigma_b \int_{V_f} \overline{\mathbf{G}}^{mm}(\mathbf{r}, \mathbf{r}') \cdot \delta\mu_b(\mathbf{r}') \, \mathbf{H}(\mathbf{r}') \, dV', \tag{4.5}
$$

where $\overline{\mathbf{G}}^{ee}$, $\overline{\mathbf{G}}^{me}$, $\overline{\mathbf{G}}^{em}$, $\overline{\mathbf{G}}^{mm}$ stand for the Green's dyads of the host medium. The integration is carried out over the defect(s) support V_f.

The $\overline{\mathbf{G}}^{ee}$ and $\overline{\mathbf{G}}^{me}$ dyads are defined as the electric and magnetic field response with a unit Dirac electric current source satisfying the Helmholtz equation [25,29]

$$
\nabla \times \nabla \times \overline{\mathbf{G}}^{ee}(\mathbf{r}, \mathbf{r}') + j\omega\mu\sigma \overline{\mathbf{G}}^{ee}(\mathbf{r}, \mathbf{r}') = \bar{\mathbf{I}}\delta(\mathbf{r} - \mathbf{r}') \tag{4.6}
$$

$$
\nabla \times \nabla \times \overline{\mathbf{G}}^{me}(\mathbf{r}, \mathbf{r}') + j\omega\mu\sigma \overline{\mathbf{G}}^{me}(\mathbf{r}, \mathbf{r}') = \nabla \times [\bar{\mathbf{I}}\delta(\mathbf{r} - \mathbf{r}')] \tag{4.7}
$$

where $\delta(\mathbf{r} - \mathbf{r}')$ is the delta function, and $\bar{\mathbf{I}}$ stands for the unit tensor. $\overline{\mathbf{G}}^{mm}$ and $\overline{\mathbf{G}}^{em}$ are defined in a similar way, as the corresponding magnetic and electric field response under magnetic current excitation, and they satisfy the same equations (4.6) and (4.7), respectively.

Note that the two pairs are interrelated via the duality principle, i.e., they can be interchanged in (4.6) and (4.7) using the following rule $\overline{\mathbf{G}}^{ee} \leftrightarrow \overline{\mathbf{G}}^{mm}$ and $\overline{\mathbf{G}}^{me} \leftrightarrow \overline{\mathbf{G}}^{em}$, which together with the interchanges $\mathbf{E} \leftrightarrow \mathbf{H}$, $\sigma \leftrightarrow -j\omega\mu$ produce the same set of equations. The duality transformation constitutes hence a symmetry of (4.6) and (4.7). The detailed derivation of the Green's dyads in planar and cylindrical stratified media is given in [25,29].

The mutual impedance variation owing to the flaw is calculated using the above solution in an elegant way by application of the reciprocity theorem

$$
\Delta Z_{TR}^{(d)}(\mathbf{r}_s) = -\frac{1}{I_T I_R} \int_{V_f} \left[\delta\sigma(\mathbf{r}') \, \mathbf{E}_T(\mathbf{r}'; \mathbf{r}_s) \cdot \mathbf{E}_R^p(\mathbf{r}'; \mathbf{r}_s) \right.
$$

$$
\left. - j\omega\delta\mu(\mathbf{r}') \, \mathbf{H}_T(\mathbf{r}'; \mathbf{r}_s) \cdot \mathbf{H}_R^p(\mathbf{r}'; \mathbf{r}_s) \right] dV' \tag{4.8}
$$

where $\mathbf{E}_R^p$ and $\mathbf{H}_R^p$ stand for the electric and the magnetic field in the flawless medium that would be produced is the receiver coil that would be fed with current I_R. $\mathbf{E}_T$, $\mathbf{H}_T$

is the field solution obtained by (4.6) and (4.7) with the transmitting coil being active and fed with current I_T. Notice the functional dependence of the field terms from the probe position $\mathbf{r}_s$ denoting that one has to consider a different field solution per scan point. The angular frequency dependence of all variables is implied.

In the case of direct magnetic field observations, (4.8) should be replaced by a calculation of the magnetic field at the probe position, namely (4.7) using the suitable expressions for the Green's dyads $\overline{\mathbf{G}}^{em}$ and $\overline{\mathbf{G}}^{mm}$*.

Conductive, non-magnetic medium with volumetric flaws

Equations (4.4)–(4.8) address the most general case of a defect inside a conducting and magnetic medium. However, in practical applications this general case concerns only ferritic steels since steel is the only ferromagnetic material of industrial interest. For the rest of workpieces the magnetic contribution due to the permeability difference is negligible, i.e., $\delta\mu = 0$ and (4.4)–(4.8) specialize to the following relations for the state equation

$$\mathbf{E}(\mathbf{r}) = \mathbf{E}^p(\mathbf{r}) - j\omega\mu_b \int_{V_f} \overline{\mathbf{G}}^{ee}(\mathbf{r},\mathbf{r}') \cdot \delta\sigma(\mathbf{r}') \, \mathbf{E}(\mathbf{r}') \, dV' \tag{4.9}$$

and the reciprocity theorem

$$\Delta Z_{TR}^{(d)}(\mathbf{r}_s) = -\frac{1}{I_T I_R} \int_{V_f} \delta\sigma(\mathbf{r}') \, \mathbf{E}_R^p(\mathbf{r}';\mathbf{r}_s) \cdot \mathbf{E}_T(\mathbf{r}';\mathbf{r}_s) \, dV'. \tag{4.10}$$

Magnetic medium with volumetric flaws

This case concerns magnetic pieces with a local variation of the permeability value, the same time that its conductivity remains constant (i.e. $\delta\sigma = 0, \delta\mu \neq 0$, and the problem reduces to (4.5) with solely magnetic contributions. Practically this case is met in the inspection of ferromagnetic specimens using static magnetic fields, a technique known as Magnetic Flux Leakage (MFL). Since $\omega \rightarrow 0$ in this limiting case (4.5) does not provide an adequate description any more. Indeed (4.5) is derived using the Faraday induction law. To address the static problem, one must devise an alternative integral equation derived by the magnetostatic equations.

A similar problem arises when calculating the magnetic flux concentration in inductors with ferrite cores. This time is the base conductivity that goes to zero $\sigma_b \rightarrow 0$ since ferrites are electrical insulators, and the integral term in (4.5) vanishes requiring again special formulation valid for the magnetostatic regime. A treatment of the core problem using a dedicated integral equation formalism can be found in [27]. The case of a pure magnetic flaw will not be examined any further.

---

*For a multilayer medium, like the ones considered in this class of problems, the Green's dyads expressions are different when source $\mathbf{r}'$ and $\mathbf{r}$ are located in different layers. This is the case for the observation equation, where the source (defect) lies in the medium whereas the observation is carried out in the air.

Conductive, non-magnetic medium with thin flaws

In non-magnetic media, a further simplification is also possible, when the thickness of the defect is negligible with respect to the other dimensions and with respect to the skin-depth in the material. This is the case of thin cracks, which is a very common category of material defects comprising the Stress-Corrosion Cracking (SCC) and the Fatigue Crack (FC) mechanisms. The appropriate formalism for the modeling of cracks in infinite medium has been introduced by Bowler *et al.* [30–34] and has been extended in the recent literature by Theodoulidis and Miorelli *et al.* [35–37]. Further developments have addressed the cases of finite media accounting end-effect such as plate edges by Theodoulidis and Bowler [38], boreholes by Pipis *et al.*, Skarlatos and Theodoulidis [22,39], and tube edges [40].

Using the fact that the normal current component at the surface of the crack must vanish, which in its turn is translated to vanishing normal electric field, (4.9) reduces to

$$\mathbf{n} \cdot \mathbf{E}^p(\mathbf{r}) = j\omega\mu_b \int_{S_f} \mathbf{n} \cdot \overline{\mathbf{G}}^{ee}(\mathbf{r}, \mathbf{r}') \cdot \mathbf{n}\, p(\mathbf{r}')\, dS' \tag{4.11}$$

where S_f is the crack surface and p expresses the electric dipole distribution over S_f, defined as

$$p(\mathbf{r}) = \lim_{\Delta x \to 0} \delta\sigma(\mathbf{r})\,\mathbf{n} \cdot \mathbf{E}(\mathbf{r})\,\Delta x \tag{4.12}$$

with Δx being the crack opening and $\mathbf{n}$ the unit normal to S_f.

Notice the simplification achieved when moving from (4.9), which is a vector Fredholm integral equation of the second kind, to (4.11), a first-order scalar Fredholm equation. The reciprocity relation is also simplified accordingly

$$\Delta Z_{TR}^{(d)}(\mathbf{r}_s) = -\frac{1}{I_T I_R} \int_{S_f} \mathbf{n} \cdot \mathbf{E}_R^p(\mathbf{r}'; \mathbf{r}_s)\, p_T(\mathbf{r}'; \mathbf{r}_s)\, dS'. \tag{4.13}$$

The interpretation of the T, R indices remains the same as above.

4.2.3 *Examples*

The application of the integral method approach for the calculation of the defect response will be illustrated via two examples.

The first example deals with the signature of a circumferential defect in a ferromagnetic tube obtained using a Remote-Field Eddy-Current (RFEC) probe. The problem configuration is depicted in Figure 4.4a.

The probe consists of a 15 mm long transmitting coil and an axial gradiometer with two coils connected in differential mode. Both receiving coils have 5 mm thickness and are located in the remote field region. The considered defect is a 50% thick (percentage with respect to the tube wall) and 5 mm wide inner groove. The results of the integral method presented above are compared against FEM simulations and measurements in Figure 4.4. The illustrated curves stand for the complex

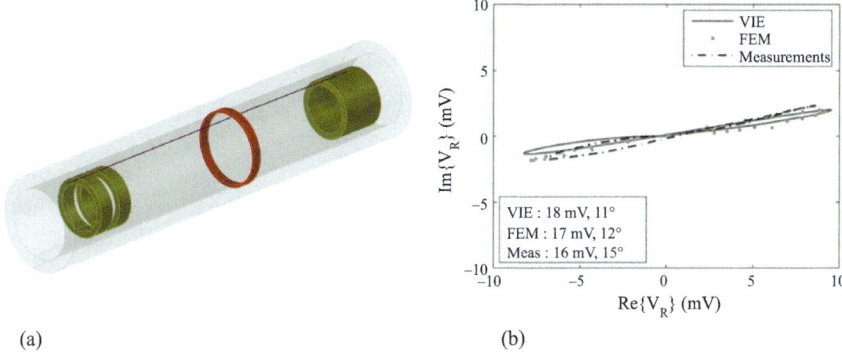

Figure 4.4   *Eddy-current inspection of a ferromagnetic tube using a REFC probe.*
              (a) Piece geometry and probe. The red ring stands for the defect. (b)
              Comparison of the simulation results obtained using the integral
              method approach and the FEM method with measurements
              (experimental data courtesy of Chen et al. [41,42]) (copyright IEEE).

plane representation of the gradiometer signal as function of the probe position. This
is a very common representation in Eddy Current Testing (ECT) applications since
the form and the angle of the curves provide direct information about the defect
features.

The second example concerns a fastener inspection affected by a narrow crack.
This case is met in the aeronautical industry, more precisely in the eddy-current
testing of fuselage fasteners. Notice that the riveted structures are regions prone
to the appearance of cracks owing to mechanical stress concentrations there. The
considered set-up is shown in Figure 4.5.

In this specific configuration, the fastener hole can be either considered as a
large defect which is addressed by means of the integral equation (4.9), a solution
proposed in [43], or alternatively as integrated part of the geometry, in which case one
needs to construct the appropriate Green's dyad that takes into account all interfaces
of the piece (horizontal interfaces and hole surface) as done in [39]. In the former
case, the Green's dyad calculation is more straightforward, however the discretization
of the defect has a negative impact to the computational burden. The latter approach
is computationally more efficient, yet one has to cope with the construction of the
appropriate Green's function, which is a hard problem.

4.2.4   Inverse problems by means of optimization and machine learning approaches

In the previous sections, we provided an overview of some computational methods that
can be used to address the direct problem, that is, to calculate the probe response for a
given configuration of inspection. This *forward model* will from now on correspond
to a function $f(\mathbf{x}|\mathbf{r}_s)$ that calculates the measured signals with respect to the probe

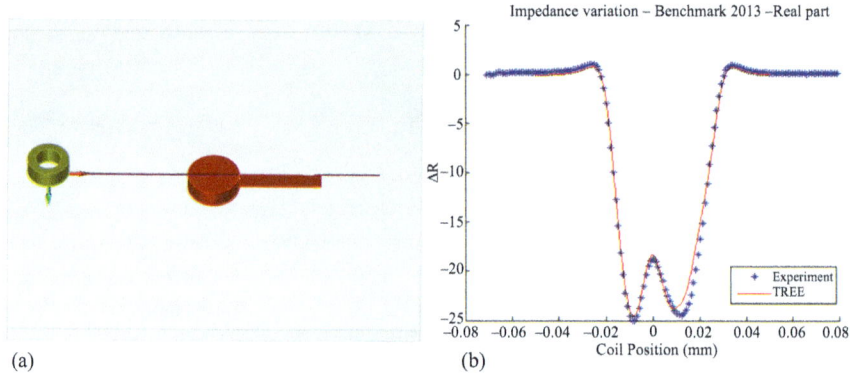

*Figure 4.5   ECT of fastener with a narrow crack. (a) Problem geometry. (b)
          Simulation vs experimental results (real part). The signal asymmetry is
          due to the presence of the crack (copyright IEEE).*

position $\mathbf{r}_s$, with $\mathbf{x}$ being some set of parameters representing the features of the geometry which we wish to estimate (conductivity, permeability, crack dimensions, crack position, etc.).

In a typical optimization approach, the forward model f is evaluated in a loop and compared with measurements $\mathbf{y}$ to recover an estimate of $\mathbf{x}$ by solving a minimization problem [44],

$$f_{obj}^{-1} := \underset{x}{\text{argmin }} mis \{f(\mathbf{x}|\mathbf{r}_s), \mathbf{y}\} + R(\mathbf{x}) \qquad (4.14)$$

where $mis: Y \times Y \to R^+$ is an appropriate measure of discrepancy in the data domain, and $R : X \to R^+$ is a regularization functional that incorporates our prior knowledge of $\mathbf{x}$. For a nonlinear inverse problem, (4.14) is usually solved by iterative methods, and in some cases it can be converted to an approximate direct inversion model. The minimizer of (4.14) is the solution provided by the objective function approach. Some typical implementations of the misfit function comprise the L^2 norm of the (normalized) difference between simulated and measured data or some kind of energy functional.

Alternatively, machine learning algorithms can be used for solving inverse problems in NDT&E. The learning approach consists first in collecting a sufficiently large amount of measured data $\mathbf{y}$, or synthetic data $f(\mathbf{x}|\mathbf{r}_s)$ and the corresponding values of parameters $\mathbf{x}$, forming a so-called training set of N pairs $\{(x_n y_n)\}$, $n = 1, ..., N$. This training set is then used to fit an ML model $\mathcal{F}_\theta (\mathbf{y})$ able to estimate $\mathbf{x}$, with $\theta \in \Theta$ the specific parameters of the ML model. In case of deep learning methods, $\mathcal{F}_\theta (\mathbf{y})$ corresponds to the architecture employed for solving the problem parametrized by

$\theta \in \Theta$. The deep learning architecture parameters are fitted to the training set during the so-called training phase (or stage) that is performed off-line by solving the optimization problems [44]

$$f^{-1}(\mathbf{y}) := \mathscr{F}_\theta(\mathbf{y}), \text{with } \theta \text{ being } \underset{\theta \in \Theta}{\operatorname{argmin}} \sum_{n=1}^{N} mis\{\mathscr{F}_\theta(\mathbf{y}_n), \mathbf{x}_n\} + R(\theta) \quad (4.15)$$

where *mis*: $X \times X \to R^+$ is a suitable measure of the mismatch in the parameter space and $R : \Theta \to R^+$ is used to regularize the solution and enhance the model generalization capabilities (i.e., avoid overfitting). Different metrics can be used to assess model mismatch depending on the learning task objective (e.g., classification or regression tasks) and the architecture employed [45]. The minimization of (4.15) is obtained through back-propagation algorithms by using a broad set of efficient minimizers (e.g., Adam, AdaGrad, RMSProp, SGD) [45]. Therefore, prediction (also called test phase) can be performed in almost real time just by evaluation of the model on an unknown set of measurements $\mathbf{y}_{test}$ such that $\hat{\mathbf{x}}_{test} = \mathscr{F}_\theta(\mathbf{y}_{test})$. It is worth mentioning that deep learning approaches can also be used to perform forward modeling tasks: in this case the model learns to generate signals $\mathbf{y}$ from a set of parameters $\mathbf{x}$.

4.3 Applications of deep learning approaches for forward and inverse problems in NDT&E

In NdT&E research and development community, numerical simulations have been historically used to design probes, inspection set-ups and assess inspection performance minimizing as much as possible time-consuming and expensive experiments. More recently, simulations have been widely exploited in order to carry out very computational demanding calculations involving statistical and sensitivity analysis studies. In this framework, ML algorithms have been employed in order to build surrogate models (also called metamodels) to speed-up otherwise computationally infeasible studies. NDT&E scholars refer to such ML paradigm as model- or physics-driven approach in contrast to the data-driven approach where ML algorithms are fit directly to measured data.

In NDT&E, model-driven ML approaches exploit the knowledge on the problem coming from simulations in order to design a suitable numerical experiment to be used for training supervised classification and regression algorithms. Once the algorithm is trained, then its performance is evaluated on a meaningful test set. Depending on the situation, the test set can be purely numerical, experimental, or a mix of the two. The performance of such a ML schema on the experimental test set can be affected by the level of agreement between experimental acquisitions and simulated data. Such an agreement depends on two main uncertainty factors, the epistemic and the aleatoric uncertainties [46,47]. The epistemic uncertainty factor can be reduced by designing a suitable ML schema or by increasing the number of simulated samples. The aleatoric uncertainty cannot be reduced since it is intrinsic to the experimental set-up. Among common sources of aleatoric uncertainty, one can mention experimental noise, probe ageing and misplacement, lack of knowledge on specimen characteristics and defect(s) morphology, etc. These uncertainties may greatly impact the ability of a trained ML model to be applied to real experimental data.

Data-driven approaches are widely employed by ML signal and image processing communities as it has access to large of real (e.g., recorded audio signals, images, etc.) open access datasets counting, very often, more than hundreds of thousand of samples. Unfortunately, in developing model-based ML strategies tackling NDT&E inspection problems for forward and inverse tasks, one needs to face two main issues. First, very few open access experimental datasets are available, thus it is difficult to establish common benchmarks to test and improve the state-of-the-art of ML-based strategies developed. Second, probed data are very often inspection and case dependent. Indeed, even in a pure data-driven approach, the aleatoric uncertainties on a given inspection problem may lead to poor generalization capabilities of the ML algorithms developed on unseen test samples (i.e., same inspection problem but an unseen experimental set-ups). Furthermore, in NDT&E acquisitions, the large majority of probed signals and/or images concerns healthy specimens, whereas to detect flaws one needs to have a lot of signals coming from flawed specimen. In classification tasks, imbalanced datasets between healthy and unhealthy specimens or between flaw types are very common and must be properly handled in the training phase. In addition, training data are often partially labeled or not labeled at all (we have the **y** data but not the corresponding values of **x** like the probe(s) position, the defect(s) geometry, the specimen characteristics, etc.). This limits the options of algorithms to so-called semi-supervised ones, which have lower performance than their supervised counterparts.

The NDT&E research community has attempted to mitigate the drawbacks associated to model-driven and data-driven ML schema by adopting different strategies. The NDT&E researchers have tried to inject physics-based knowledge in order to tailor a specific ML schema. Toward this end, specific features engineering techniques have been applied on probed signals in order to promote descriptors minimizing the aleatoric uncertainty contribution and thus enhancing the ML model generalizations performance. Furthermore, the joint use of synthetic data (which are cheaper to generate and are always labeled) and experimental data (which are similar to the test data that will be used in the end) in training sets is currently a hot research topic.

In electromagnetic NDT&E, one can distinguish two main categories of probed data. The first category gathers signals that behave like time-series signals such as scanning signal with respect to probe(s) displacement (e.g., eddy current testing acquisitions in 2D symmetrical problems) or time (e.g., pulsed eddy current testing signals for a given probe position). The second category collects all signals that can be seen as 2D cartographies (e.g., eddy current testing acquisitions in 3D problems) or 1D probe(s) displacement and a succession of time steps (e.g., a pulsed eddy current acquisition). For both categories, very often, probed signals are complex-valued and both real and imaginary parts are analyzed as the informative contents. Furthermore, multi-static probe arrays and multi-frequency acquisitions have to be considered to fulfill the inspection protocols, so the images analysed can be seen as multi-spectral ones. That is, typical electromagnetic NDT&E acquired signals behave as tensors with typical orders between 2 and 4. It is worth to be mentioned that, thanks to the high flexibility in designing DL architecture, different sources and/or different extractions (also called ways in the signal processing community) of the same data can be merged, mixed, or exploited smoothly, making use of DL methods a very convenient and flexible tool for the extraction of features and fusion in NDT&E.

4.3.1 Most relevant deep learning architecture in NDT&E

A deep learning architecture [45] consists in a chain of mathematical operations established between inputs and outputs, the so-called layers. That is, the layers perform transformation on inputs in order to extract the most meaningful features and perform the final tasks (e.g., regression, classification, etc.). Each layer is composed by arithmetic units, called neurons, that enable the mathematical transformations. In the most common deep learning architecture, the output of one layer is fed to the next layer neuron through a linear combination of weights and biases θ (i.e., see (4.15)). On these linear operations is applied an element-by-element non-linear transformation through the use the so-called activation functions (or layer) aiming at handling non-linear behaviour in mapping two successive layers. The most common activation functions are the sigmoid, Rectified Linear Unit (ReLU), Leaky-ReLU, softmax, etc. The use of a particular activation function depends on the task associated to the layer to which it is attached.

From a general point of view, the connections between two layers identify the architecture type, e.g., Fully Connected Neural Network (FCNN), Multi-Layer Perceptron (MLP), Convolutional Neural Network (CNN), Long Short Time Memory Recurrent Neural Network (LSTM-RNN) just to cite the most prominent ones. Furthermore, provided a given family of layers (e.g., CNN, FCNN, etc.), different architecture topologies (e.g., encoder–decoder, U-Net, etc.) can be obtained by connecting the different layers together in order to solve the problem at hand. Furthermore, the deep learning architecture can also be classified with respect to the machine learning task to be handled. That is, one can divide the architectures by considering supervised, semi-supervised, and unsupervised learning paradigms. The most used DL architectures that we will study in this chapter belong to the supervised learning framework aiming at solving regression and classification tasks based on labelled datasets. Nevertheless, unsupervised learning (i.e., no labels are attached to the data) based on the use of use generative models such as Variational AutoEncoder (VAE) and Generative Adversarial Network (GAN) is becoming more and more common to solve specific tasks in the NDT&E research community. The semi-supervised learning approach is also studied to enhance the DL model accuracy when a small amount of labeled data is available.

4.4 Application of deep learning to electromagnetic NDT&E

One of the first attempts in using machine learning algorithms based on the shallow neural network in the context of eddy current testing can go back to the middle of the 1990s [48–56]. Meanwhile, researchers in NDT&E studied the use of kernel machines such as support vector machines, kernel ridge regression, and Gaussian process regression, algorithms along with the use of feature extraction and feature selection techniques such as principal component analysis, partial least square, and locally linear embedding [57–61]. More recently, pushed by the large leap forward in performance of deep learning methods obtained in image and signal processing, the NDT&E community is actively developing and adapting deep learning architectures to handle classification and regression problems based on NDT&E inspected signals (see Figure 4.1).

The actual research of ML tools applied to electromagnetic NDT&E is trying to propose solid and reliable solutions to support and automatize the decision processes (i.e., defect(s) detection, localization, sizing, Remaining Useful Life (RUL), etc.) during acquisitions. In NDT&E, different levels of automation are envisaged. The integration of decisions between human supervision and machine learning algorithms is performed based on different contributions on the final outcomes. Referring to [62–65], the automation levels of NDT&E inspection systems can be divided into five levels where the increasing contribution of ML algorithms impacts the final decision ranging from a mild NDT&E operator assistance to a fully automatic system (see Figure 4.6). The higher the level the most involved and complex the ML algorithms are. The complexity of the algorithm developed should also be accounted for, in view of deploying the algorithms on embedded measurements systems that need to fulfill CPU efficiency constraint and traceability of the deployed algorithm, too. In the following, we provide a systematic analysis of deep learning methods applied to industrial sectors where electromagnetic-based NDT&E is widely employed.

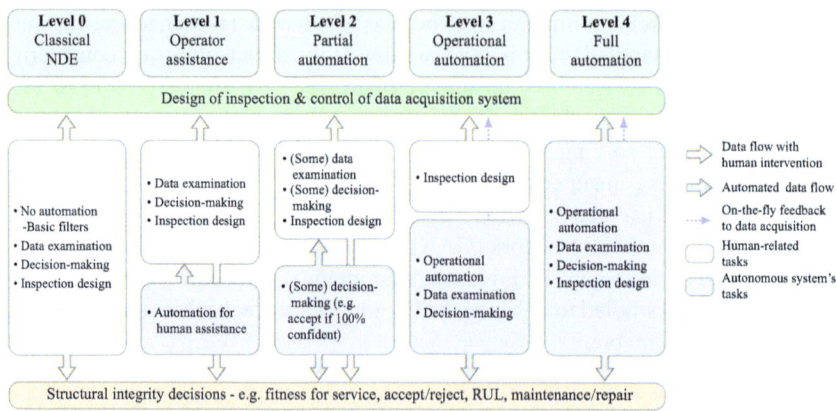

Figure 4.6   *Flow chart describing the integration between human and machine learning-based algorithms in performing decisions and tasks in NDT&E [65] (CC BY 4.0 license)*

4.4.1   *Deep learning in electromagnetic NDT&E applied to the energy sector*

The energy sector gathers a very broad set of industrial sectors ranging from nuclear energy to renewable energies (e.g., eolian, solar, etc.) and oil & gas. For all these industrial sectors, electromagnetic NDT&E is widely employed. In nuclear industry, periodic inspection of Nuclear Power Plants (NPP) Steam Generation Tubes (SGTs) is carried out with eddy current testing methods, by using of different probe(s) arrangements, and inspection protocols depending on the inspected part under test (e.g., U-bended SGT part, straight part, transition zone, near support plate, etc.). The use of AI and ML based analysis of inspection data is actually a very active research topic aiming at speeding-up the analysis of the very large quantity data acquired (e.g., a typical NPP is composed of hundreds of SGTs). The perspective, in the near future, of deployment of new array probes for these applications will increase considerably the amount of data to be analyzed, forcing the current organization, partly or exclusively based on manual analysis by experts, to adapt. This makes this topic quite strategic. In this context, the development of support tools for helping NDT&E engineers decisions are under study for reducing as much as possible the human analyses errors (e.g., the so-called human factor) when repetitive and long analysis are performed.

The use of deep convolutional neural network for defect detection has been proposed by Zhu *et al.* [66] based on multi-frequency ECT acquisitions performed in SGT. The data-driven schema developed involves the use of robust principal component analysis to properly detect the regions of interest. Thus, the convolutional neural network model proposed computes both the probability associated to the tested samples along with the epistemic uncertainty associated to the detected class (see Figure 4.7). Such an approach is supposed to be widely exploited by the NDT&E community in the near future along with the possibility to embed the explainability of a deep learning method. In [67], the authors studied the performance of a deep neural network for defect classification based on the use of two different ECT probes

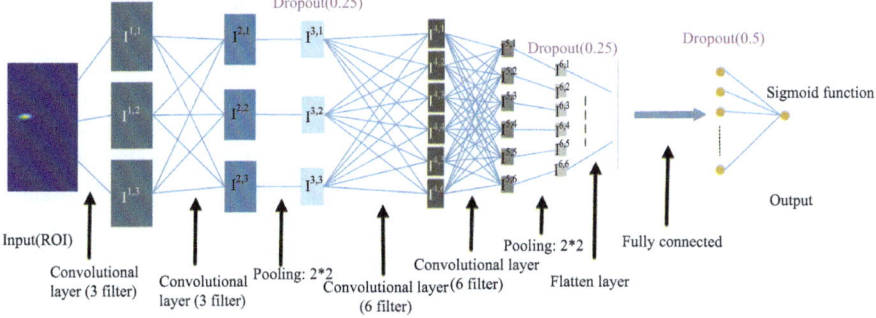

Figure 4.7   Deep convolutional neural network architecture for defect detection in SGT tube based on ECT signals enabling prediction uncertainty estimations [66] (copyright Elsevier)

(i.e., pancake coil and +Point probe) signals. The analysis performed showed that the neural network schema adopted was able to classify, with good performance, the longitudinal, circumferential, and no-defect classes. Li *et al.* [68] proposed a first attempt to crack profile reconstruction based on the use of multi-frequency ECT signals. In particular, a C-scan ECT signal was used as input to a tailored encoder–decoder deep neural network developed for this purpose loosely inspired by deep convolutional generative adversarial network architectures developed by the image and signal processing communities. The results obtained, based on numerical datasets only, were quite promising in view of an extension to more challenging problems, e.g., involving experimental signals. In [69], a set of deep residual convolutional neural networks has been tested for crack depth classification based on massive set of acquisitions performed on steel plate containing 20 machined slot defects. The study showed that the considered architectures were capable to distinguish the different defects classes with good accuracy.

In Oil and Gas (O&G) and petrochemical industries, the use of ECT method, based on both time-harmonic excitation and Pulsed Eddy Current Testing (PECT), is widely used for inspecting the presence of corrosion in pipelines. Detection, localization, and sizing (mainly in term of corrosion thickness) are the main outcomes expected by the analysis of the data acquired. The use of machine learning algorithms is expected to provide many advantages in data analysis. For instance, in PECT the interaction of a broadband signal (e.g., pulse wave form) with the SUT produces specific signatures in time and space (i.e., probe position). Unlike more common time-harmonic excitation ECT, the analysis of PECT signals needs to account for the time dimension since a certain amount of information (e.g., material characteristics, defect properties, etc.) is embedded in the SUT feedback when the PECT excitation decay. However, a time dimension composed of hundred or even thousand of samples adds to 2D mechanical scan, making the problem computationally more demanding from the learning perspective. In order to efficiently handle such large amount of data, a feature extraction of probed signals has been performed by employing principal component analysis [70]. Based on an experimental dataset, the extracted features were employed as input to a deep neural network able to account for the variations due to different acquisition temperatures. The outcomes of the architecture have been used to predict both probe position and defect geometrical characteristics (see Figure 4.8). In [71], a 1D convolutional neural network has been designed in order to perform defect classification and regressions (i.e., defect height estimation) simultaneously based on a set of A-scan PECT experimental acquisitions. The obtained results that have been compared with the state-of-the-art machine learning algorithms (e.g., Gaussian process, support vector machine, decision tree, etc.) showed a great improvements obtained by the deep learning schema proposed. In the noteworthy work [72], Dang *et al.* proposed a deep neural network + involving CNN and long-short time memory architectures for characterizing the multiphase flow (e.g., oil and water percentage) for industrial applications showing the capability to accurately measure the volume fraction of water and the total flow velocity.

Deep neural network have also been applied to magnetic flux leakage (MFL) acquisitions in large pipeline loop, where artificial natural corrosion defects were

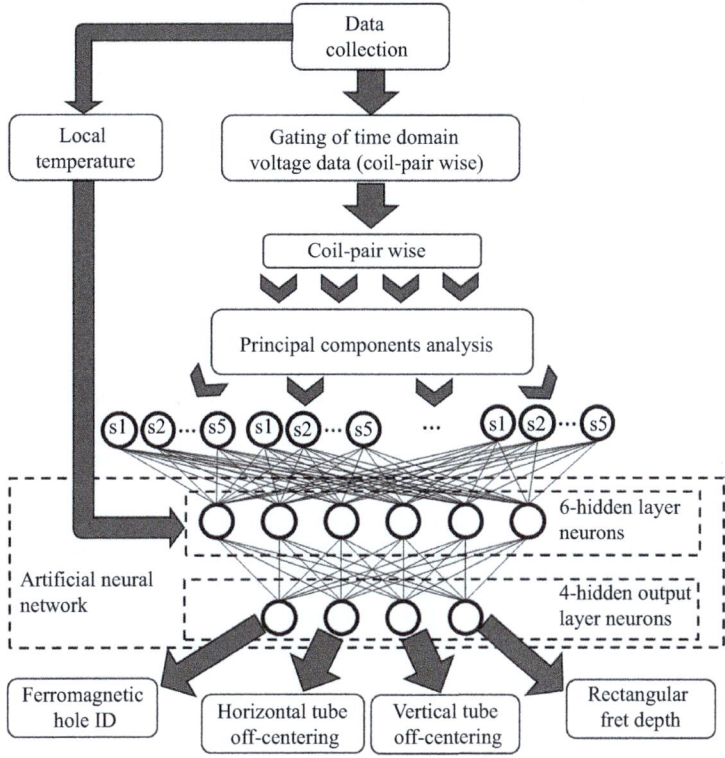

Figure 4.8   Deep neural network schema proposed in [70] to perform a regression task based on pulsed eddy current signals

present in a pipe [73,74]. The dataset obtained from in situ measurements has been augmented by simulations based on rectangular shaped artificial defects. After that, both the datasets have been used for training a specific type of CNN, called visual transformation convolutional neural network (VT-CNN), for the estimation of defect length, width and depth. The results obtained by the VT-CNN showed an higher accuracy compared to CNN without VT layer, provided a minor computational burden in training and testing phases. Sun *et al.* [75] proposed a physics-informed deep neural network architecture, called DfedResNet, to tackle the problem of estimation of defect(s) size parameters based on MFL acquisitions. The main features of such a deep CNN were the possibility of combining engineered features associated to MFL acquisitions (i.e., the physics-informed part [46,47]) along with the spatial patterns of MFL images. Furthermore, saliency maps analysis has also been investigated for enhancing the interpretability of the deep learning schema proposed. The regression results obtained by the DfedResNet showed large improvements compared to support vector machine and VT-CNN schema. In [76], Le *et al.* proposed a convolutional neural network-based surrogate model aiming at speeding up the computational time of magnetic field distribution based on the exploitation of finite-element simulations. Toward this end, a U-Net like convolutional neural network was designed considering as input two images represented by the permeability and current distributions versus the magnetic field maps (*x*- and *z*-components). The results obtained showed the capability of the deep neural network architecture to accurately predict the numerical model test data in a wide set of scenarios (e.g., different flaw type, permeability values, and current distributions).

4.4.2 Applications to the transportation and civil infrastructures sectors

Eddy current testing methods and techniques have been widely used in the past for detecting defects in aeronautic and aerospace industrial sectors, thanks to the possibility to inspect rapidly (i.e., without needing to remove coating and/or fasteners) large airplane parts, fastener sites, and bolt holes. Typically, corrosion like defects and micro cracks nearby fastener sites are the most critical defects to be detected in order to extend the operational life of airplanes without harming the residual life of the structure. The application of deep neural network in such industrial domain started in the beginning of this century. The use of two hidden layers feed forward neural network applied to defect classification based on PECT signals exploiting different feature engineering methods was studied in [50,77]. Further studies on the use of ML algorithms for defect detection in multilayered metallic structures based on time domain (pulsed) ECT have been presented in [78] for the detection of second-layer crack(s). More specifically, C-scan PECT data have been acquired on a structure composed by bolt hole with and without defect(s), bolt hole with counter sink with and without defect(s), hole with titanium and ferrous fastener with and without defect(s). The detection performance of established machine learning algorithms such as random forest, gradient boosting, and support vector machine have been assessed against deep learning methods based on long short-term memory (LSTM) recurrent neural network (RNN) and multilayer perceptron (MLP) algorithms [45]. The results obtained showed that random forest and gradient boosting have an edge in performance compared to deep neural network methods once applied on raw PECT data. In [79], time harmonic ECT has been used for defect classification based on experimental measurements in titanium plate based on tailored CNN network architecture dealing with the small amount experimental data available. The promising classification results obtained by the CNN architecture have also been compared with more established learning algorithms such as deep belief network, stacked autoencoder, and support vector machine, showing that CNN was able to achieve the best results based on the test data.

The use of ECT method coupled with deep learning has also been proposed for detecting anomalies on railways in order to exploit the strength of ECT compared to visual inspection methods (e.g., robustness with respect to environmental conditions, high speed acquisitions, etc.). In [80], the alternating current field measuring technique has been used in order to perform experimental measurements on a calibration block containing clusters of cracks and two hidden layers of multilayer perceptron neural network have been used along a Bayesian regularization method for back-propagation schema. Simulations have been used to build the training set. The results obtained showed a good generalization capability on the studied DL schema in retrieving the equivalent length of the cluster of cracks on both simulated and experimental test sets. In [81], a data-driven convolutional neural network architecture has been developed in order to exploit the ECT acquisitions post-processed through wavelet-based algorithms (i.e., continuous wavelet transform) and perform classification of surface breaking and superficial anomalies in rails (i.e., weld, squat, and joint anomalies). The CNN architecture has been trained with data processed by wavelet power spectrum transform and the obtained classification results have been compared with a large set of classification methods involving logistic regression, ensemble methods, quadratic discriminant analysis, etc. The CNN results showed an edge in performance compared to the other ML methods studied.

In [82], a Deep Belief Network (DBN) [45] composed by a set of stacked restricted Boltzman machines trained in an unsupervised fashion (i.e., see Figure 4.9) has been applied in order to extract the most meaningful set of features from ECT signals measured on titan plates where slots and holes machined defects were embedded. The features extracted by DBN in an unsupervised way have been fed to vector valued Least Square Support Vector Machine (LS-SVM) algorithm in order to perform the defect characterization (i.e., defect(s) sizing) tasks. Thanks to the use of the DBN, the proposed method does not require any feature engineering stage before providing data to LS-SVM algorithms. The results obtained by the proposed learning schema have been compared with principal component analysis and Boltzman machine feature extraction algorithms. These comparisons showed that the higher accuracy was obtained by the DBN and LS-SVM approach. In [83], the use of two chained artificial neural network architectures has been developed in order to perform classification of defect depth and width based on MFL experimental measurements on steel wire ropes. The developed schema was based on the use

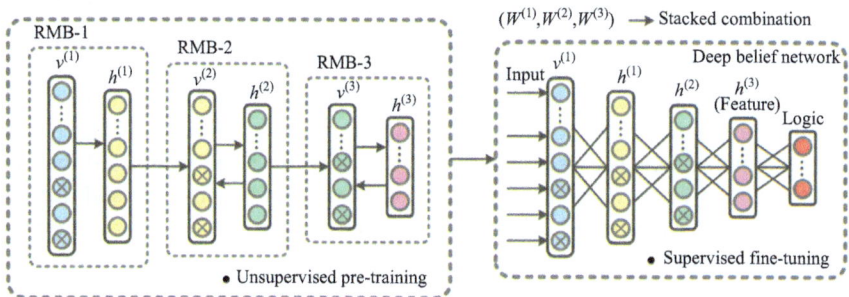

Figure 4.9   Deep belief network schema developed for unsupervised feature extraction based on ECT measurements in titanium plates [82] (CC BY 4.0 license)

of engineered features extracted from acquisitions. The obtained results showed the capability of the developed methodology to diagnose defects in wire ropes with good accuracy.

In civil infrastructures industrial sectors, Ground Penetrating Radar (GPR) is widely applied to detect, classify, and possibly localize buried objects under the soils or pavements, in reinforced concrete structures and masonry (e.g., landmines, pipes, voids, cracks, rebars). Deep convolutional neural network for detecting buried explosives based on GPR B-scans measurements has been proposed in [84]. The schema proposed is based on a pre-processing step aiming at detecting and selecting a suitable region of interest from the original B-scan before performing the training phase. The results obtained, compared with different state-of-the-art algorithms for anomalies detection, showed that the CNN approach has the capability to outperform other traditional post-processing algorithms. In [85], experimental B-scan images have been decomposed into smaller patches containing buried landmine signatures (i.e., echos hyperbolas) and ground signatures before training a convolutional neural network targeting landmines detection. High detection accuracy in detecting landmine has been observed on the test set data for the three tested architectures compared to histogram oriented gradients detection procedure. In [86–88], different 3D convolutional neural network architectures have been developed in order to jointly use the information contents of B-, C-, and D-scans GPR experimental acquisitions targeting objects detection. The results obtained showed an enhanced accuracy in classification results compared to classical approaches based on the use of B-scans only training input data. Automatic GPR signatures detection have been studied by several scholars for identify and classify echos hyperbolas in measured data. More into details, mask and faster Region-based Convolutional Neural Network (R-CNN) based architectures have been employed to detect hyperbolas performing semantic segmentation of echoes in B-scan measurements [89–93]. In [94,95], different versions of YOLO architectures have been studied to tackle the hyperbolas identification and defect localization tasks based on B-scans acquisitions. A worthy mention on the use of deep learning methods applied to GPR data is about the estimation of permittivity characteristics and electromagnetic waves velocity estimations in complex soils. Researchers are investigating the use of deep generative models in order to perform a pixel-wise reconstruction of electromagnetic characteristic of soils based on encoder–decoder, U-Net like and generative adversarial network architectures [96]. Furthermore, tailored convolutional neural network architectures proposed in [97–100] showed high capability in retrieving permittivity characteristics as well as velocity of electromagnetic in complex soils (i.e., see Figure 4.10). It is worth to be mentioned that the previously cited works rely on simulated results for 2D problems only and that the performance of the DNN model developed is directly linked to the number of training samples considered, i.e., a large or very large training set is needed. Furthermore, a noteworthy result targeting forward and inverse modeling based on the use of deep fully connected neural network has been shown in [101,102]. In these works Giannakis *et al.* employed DL schemas to infer rebars positions and size in concrete based on the use of fully numerical training sets.

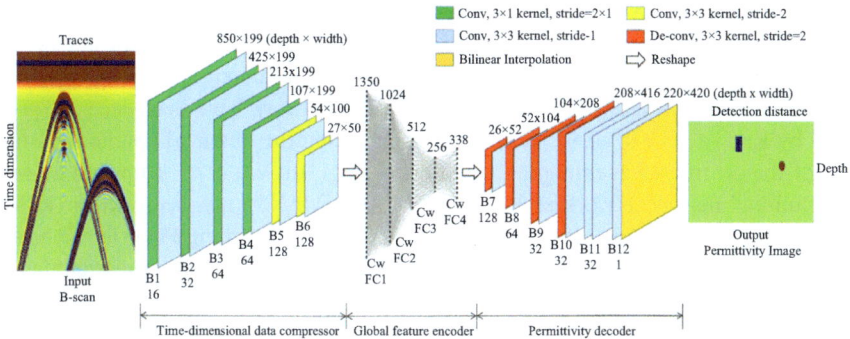

Figure 4.10   Convolutional neural network used for predicting ground permittivity characteristics based on GPR B-scan signals [98] (copyright IEEE)

4.4.3   Applications to the manufacturing and agri-food sectors

The Industry 4.0 paradigm, roughly consisting in global digitization and exploitation of data coming from manufacturing processes, for the next generation production and manufacturing plants aims at bringing the third industrial revolution in our society. This production paradigm is expected to extensively exploit the emerging technologies boosted by the most recent AI advances. In this regard, the wide use of robotics and numerically controlled procedures will be used for enhancing the production outcomes, lowering the production time and costs, and increasing the production quality (i.e., the so-called zero-defects production). Toward this end, production chains are becoming more and more connected through a broad variety of possibly heterogeneous sensors aiming at collecting the largest set of information to control the productions factors (e.g., quality of manufactured specimens, devices, etc.). The integration, connection, and exploitation of such heterogeneous set of information are among the highest challenges of AI-based algorithms for the next decade.

Microwave-based NDT&E method is applied in different manufacturing sectors for checking the quality of products during the fabrication process for dielectric of weakly conductive materials. In [103,104], a convolutional neural network approach aims at retrieving the moisture density in porous foam inspected by a microwave tomography system. Toward this end, numerical simulations have been used in order to generate the training, validation, and test sets. The real and imaginary parts of S-parameters have been fed to a CNN composed by two CNN blocks and a fully connected layer that, once reshaped, provided a vertical slice of the moisture density and/or the permittivity map(s). Microwave non-destructive testing and shallow neural networks have been jointly applied more than 20 years ago in the agri-food industrial sector [105] to establish the moisture content in wheat. More recently, Ref. [106] proposed a deep neural network architecture aiming at handling multi-frequency sweep microwave measurements in order to predict the moisture measurement of sweet corn. A noteworthy end-to-end experimental set-up, exploiting both CPU or FPGA hardware, based on the use of microwaves and deep neural network learning schema has been developed for detecting contaminant in food jars [107]. Microwave inversion of dielectric rods in complex geometry in a fully data-driven approach exploiting convolutional neural network and recurrent neural network have been recently proposed by Ran *et al.* [108,109] successfully comparing the obtained results with the state-of-the-art microwave imaging techniques. In [110], Wu *et al.* proposed a deep convolutional neural network architecture, called VMFNet (i.e., see Section 4.11), aiming at performing damage detection on curved Radar Absorbing Materials (RAM). The designed network exploits the inputs coming from visual and microwave images of curved RAMs through two distinct backbone convolutional neural network, thus the extracted features have concatenated in order to perform detection. The results obtained by the proposed network showed large improvements in detecting cracks compared to state-of-the art algorithms in computer vision (e.g., YOLOv4, Faster R-CNN, EfficientNet). Rohkohl *et al.* [111] studied a deep learning schema aiming at perform weld inspection of electric contact in battery cell manufacturing based on eddy current testing. In particular, the authors proposed to train DNN model based on

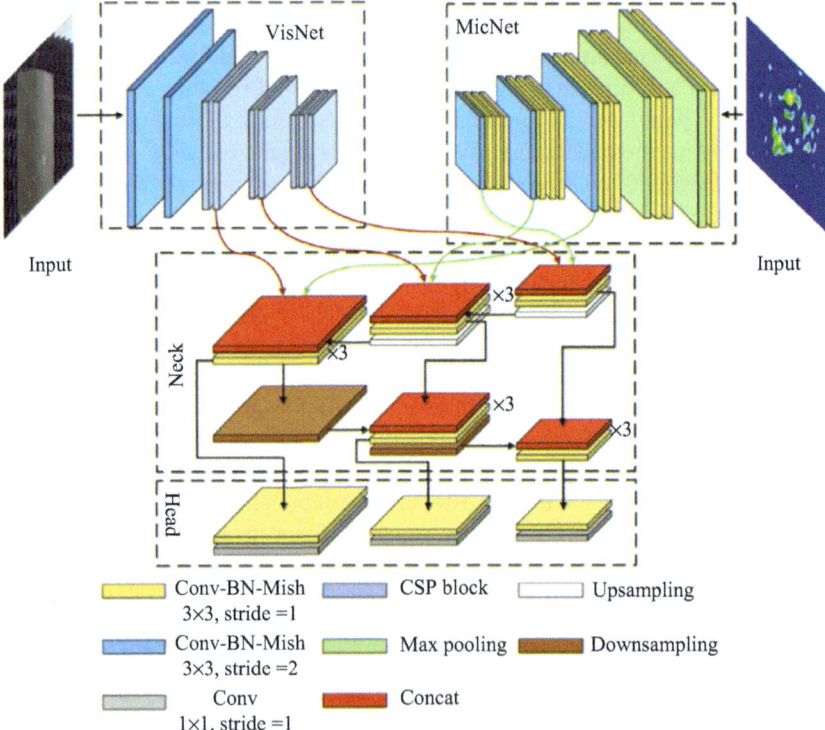

Figure 4.11   Deep convolutional neural network applied to defect detection in radar absorbing materials based on visual and microwave features exploitation [110] (CC BY 4.0 license)

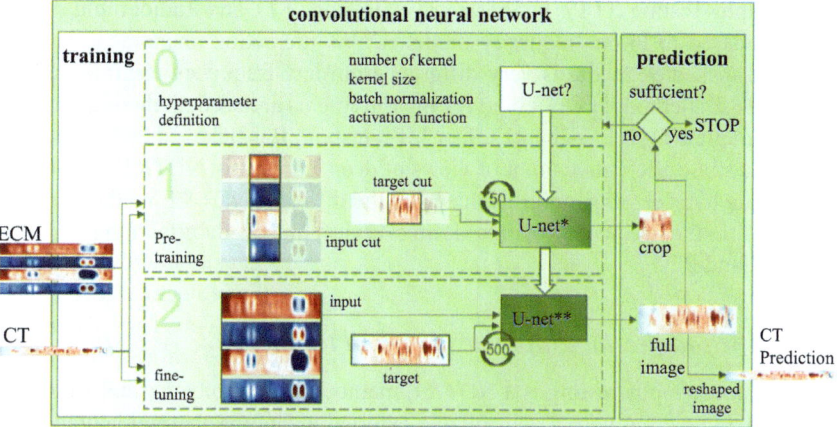

Figure 4.12   *Deep learning schema used to enhanced the interpretation of Eddy Current Measurements (ECM) based on the use of X-ray computer tomography (CT) data [111,112] (CC BY 4.0 license)*

ECT signals in order to predict results from a reference method such as radiography testing in order to enhance the interpretation of ECT acquisitions. A convolutional neural network with U-Net like architectures has been developed in order to encode ECT into RT tomography map. Generated results based on the ECT tests are converted to cone beam tomography results that can be easily interpreted by humans without performing any expensive and time consuming ionizing RT tomography acquisitions.

Automatic characterization of magnetic properties and the design of innovative materials based on the used of machine learning are catching the interest of academic researchers and development engineers in the manufacturing industry. In [113], Elman neural network was used for the identification of non-linear hysteresis model parameters. More recently, in [114], a recurrent neural network model was used to accurately predict the behavior of the hysteresis loops in ferromagnetic materials under a limited amount of measurement data available. In Maciusowicz *et al.* [115], magnetic Barkhausen noise measurements have been exploited in order to predict grain orientation in a ferrosilicon alloy for electrical steel. More into details, short-time Fourier transform has been applied on magnetic Barkhausen noise measurements, then the obtained signal maps have been fed to a specifically designed convolutional neural network. The obtained results showed the possibility to correctly classify the grains orientation angles for the experimental set-up considered.

4.5   Applications to higher frequency NDT&E methods

Accordingly to the schema displayed in Figure 4.2, the highest frequencies of the electromagnetic spectrum are taken by three widely used NDT&E methods: Infrared

Thermography Testing (ITT), Terahertz waves Testing (TT), and Radiography Testing (RT). Compared to lower-frequency methods, the interaction phenomena between source and inspected media is very different from low-frequency electromagnetic-based NDT&E method. ITT, TT, and RT measurements are used methods along with electromagnetic NDT&E such as ECT or MFL testing. It is believed that this section can provide some alternative point of views of deep learning in NDT&E and suggest possible data hybridization and fusion across the different NDT&E methods treated in this work. In the following, we provide an overview on how deep learning is applied to ITT, TT, and RT.

4.5.1 Infrared thermography testing and terahertz wave testing

Infrared thermography testing is a NDT&E method that exploits thermal signatures of the SUT interacting with a controllable external excitation thermal source (eddy current induction, lamp flash, laser, etc.). In ITT, the interactions between thermal source and SUT are ruled by the convection and conduction equations that make ITT a complementary method to ECT and MFL. Indeed, ITT is one among the best suited methods for fast inspection of composite material and surface breaking defects in conducting media, masonry, and concrete structures [116].

The thermal signature emitted by the SUT is collected by infrared cameras and lenses sensing rays within the infrared spectrum. Depending on the considered problem, ITT collected signals behaves as order-1 tensor if pixel-wise time-dependent measurements are considered. Order-2 and order-3 tensors are considered when acquisitions behave as images or video sequences, respectively. In this framework, deep learning techniques issued from image and signal processing communities have been adapted to ITT data. In particular, for order-1 tensor data are exploited when a limited amount of data is available. In [117–119], temperature time series signals are analyzed employing recurrent neural networks, where the temperature signatures were processed by employing a long-short memory recurrent neural network to classify different defects that typically affect Carbon Fiber Reinforced Plastic (CFRP) material in assembled structures. In [120,121], a data-drive approach has been used to train a 1D CNN architecture in order to perform pixel-wise pristine versus damage classification of CFRP material based on temperature signature with respect to time. The prediction results were subsequently concatenated in order to obtain an binarized image of the whole specimen under testing. The classification results have been compared with classic ITT signal processing state-of-the-art techniques showing promising improvements.

In Xie *et al.* [122], an AutoEncoder (AE) based architecture has been studied in order to extract the meaningful set of hidden features associated to an experiment temperature time-series back-wall cracks; the proposed methodology showed to be able to enhance the quality of ITT images. Deep neural network architectures based on CNN and/or LSTM-RNN have been developed based on order-2 and order-3 tensor ITT inputs signals in order to gather and mix both spatial and temporal features [123–125] (i.e., see Figure 4.13). Subsequently, the joint exploitation of both spatial and temporal features information allowed to increase the classification and regression performance

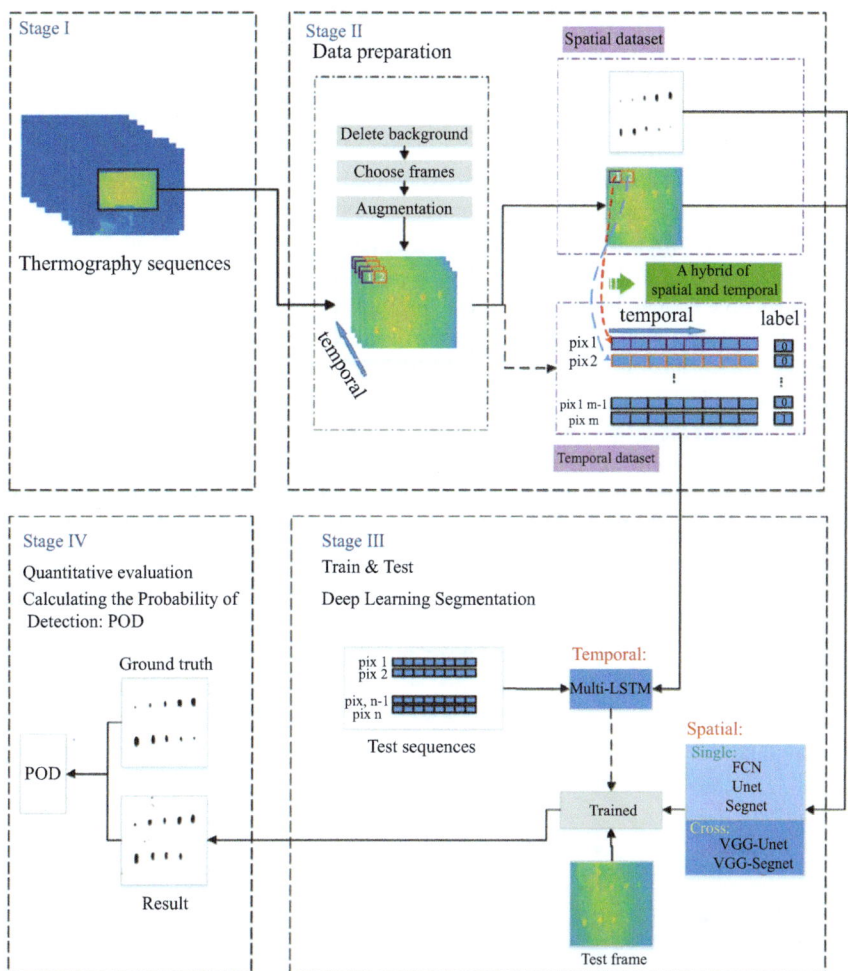

Figure 4.13 Example of spatial and/or time feature based deep neural networks applied to ITT. A defect detection schema based on deep neural network architecture based on fully-connected network, VGG architecture and LSTM-RNN is displayed [123] (copyright Elsevier).

compared to state-of-the-art approaches. Deep neural network architecture has been also used in ITT to provide fast and reliable image segmentation [126,127] to improve the image analysis stage. In [126], authors proposed a DNN architecture based on convolutional layers and inception modules chained for tackling the problem of automatic segmentation of cracks profile in concrete structures. The inputs to the network are composed by hyper-spectral image gathering both visible and infrared spectra information the classification outputs obtained can be also interpreted in terms of probabilities being concrete or crack pixels.

The use of deep learning methods applied to ITT is also raising the interest of the experimental physics research community historically involved in numerical simulations of highly complex problems in a vast set of research domains. Very recently, the use of DL algorithms has been extensively used in order to model complex interactions between plasma and confinement barriers in fusion reactors. That is, ITT is used for in-service monitoring of plasma facing components to detect unexpected hot points to be handled instantaneously in order to avoid fusion reactor damages. Simulated heat flux images have been used to predict plasma parameters based on a set of six different deep neural networks involving feed-forward neural network to deep Inception ResNet [128]. In [129], a generative adversarial network framework has been developed for enhancing the defect detection capabilities based on ITT measurements performed on composite fibre reinforced plastic plate. In [130], deep residual network has been used for deblurring purposes based on ITT image acquisitions.

Terahertz wave testing is used in NDT&E to perform contactless measurements of millimeter an submillimeter electromagnetic waves interacting with the SUT. The use of TT is gaining particular attention in NDT&E research community for inspecting dielectric materials such as glass fiber reinforced plastic, glass fiber composite, ceramics, plastic materials, and in food industry. In [131], Wang *et al.* proposed an experimental validation of two different deep learning architecture for pixel-wise defect depth classification. That is, a bidirectional LSTM-RNN and a 1D-CNN were fed with time-domain signals or spectral signals and the results obtained by the two different architectures as well as the different input signals have been compared showing

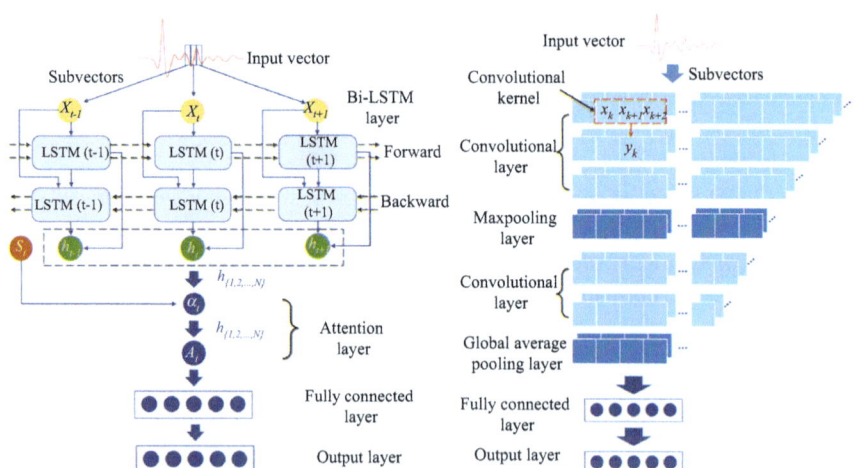

Figure 4.14   Deep neural network architecture as applied to terahertz testing acquisitions. On the left a bidirectional LSTM-RNN and 1D-CNN architecture are sketched, respectively [131] (copyright Elsevier).

an edge in performance by the 1D-CNN regarding the input signals considered. In Zhang *et al.* [132], time domain TT measurements have been used in order to perform air gap thickness measurements in insulation equipment. Different ML classification methods have been trained and tested and compared on the experimental signals. That is, the performance results have been studied based on a deep learning schema composed by a chain of combination of CNN, residual shrinkage network and fully connected blocks as backbone for terahertz waveform feature extraction purpose with a classification block provided by a Bayes classifier, a softmax layer or a support vector machine. The best results were obtained by using the support vector machine layer. Very recently, the use of deep learning for super-resolution purposes based on terahertz imaging images is attracting the attention of many researchers in the NDT&E field. In particular, in [131,133,134], tailored deep convolutional neural network architectures have been proposed to enhance the resolution of terahertz images based on measurements performed on different kind of structures. In [135,136], the super-resolution task has been tackled by considering generative adversarial network adapted to a dataset of experimental terahertz images.

4.5.2 Radiographic testing

X-rays are a form of electromagnetic radiation of extremely short wavelength, ranging from 10^{-12} to 10^{-8} meter, that have the ability to penetrate the matter. The inspection of the internal structure of an object through X-ray testing consists in passing an X-ray beam through this object and recording its attenuation on a receptor. With digital radiography, a 2D grey-level projection image is acquired from the transmitted X-ray beam. Due to the similarities between the X-ray images and the visual ones, all modern computer vision techniques have been naturally applied to X-ray testing [137]. In particular, DL approaches have been employed to target real-time detection and automatic classification of encountered flaws, contaminants or threats for different industrial applications such as quality control of welds, inspection of automotive and aeronautics parts, and food products or baggage screening. In this section, we present the most relevant applications of DL algorithms to RT data in different industrial sectors (defect detection in welds and casting parts, contaminants in food industry, threats in baggage screening).

Defects detection in weld inspection, casting and assembled parts
X-ray quality inspection of welds (e.g., pipes in NPP), casting light-alloy parts (e.g., wheel rims, steering knuckles, and steering gear boxes) and assembled (e.g., composite structures) parts is commonly used in the nuclear, naval, chemical, automotive, and aeronautical industries for ensuring the safety and the quality of the parts. Traditionally, radiographic images are manually inspected by human experts in order to detect and characterize potential defects. However, this task requires experienced inspectors and is time-consuming. In order to avoid the effect of human factors, to cope with the throughput of the production and analysis pipeline and to improve detection accuracy, fully automated inspection systems are deployed.

In the last decades, several works have focused on the automatic detection and identification of the most common welding defects and deep learning approaches have recently been applied to this task. In [138], the implementation of the automatic defect detection relies on three steps: the segmentation of the weld area, the application of a classification model on patches of the image, and the detection of defects in the entire weld area using a sliding window algorithm. The classification model is constructed by stacking several sparse auto-encoders performing unsupervised learning and one softmax classifier using supervised learning. The proposed algorithm is applied on the public database GDXray [139] that includes a weld dataset of 88 images taken by the BAM Federal Institute for Materials Research and Testing. Several experiments are implemented including extracting the features using SAE and examining the classification accuracy under different parameters of the model. The overall method is illustrated in Figure 4.15. With this approach, defects can be accurately detected but not classified.

Wang *et al.* [140] propose a method to identify three types of welding defects (blowhole, underfill, or incomplete penetration) and their locations in X-ray images by using a pre-trained RetinaNet-based CNN and develop a dataset constituted of 6,714 labeled images. Mean average precision (mAP) ratings are 0.76, 0.79, and 0.92 for the defect types. Yang *et al.* [141] proposed an improved CNN model based on LeNet-5, whose architecture consists of 7 layers (excluding the input layer), in which the layers 1, 3, and 5 are convolution layers, and the 2 and 4 are down-sampling layers. The CNN X-ray input is fed with 60×60 patches taken from the radiographic images and is shown to outperform LeNet-5, ANN, and SVM methods in terms of recognition accuracy.

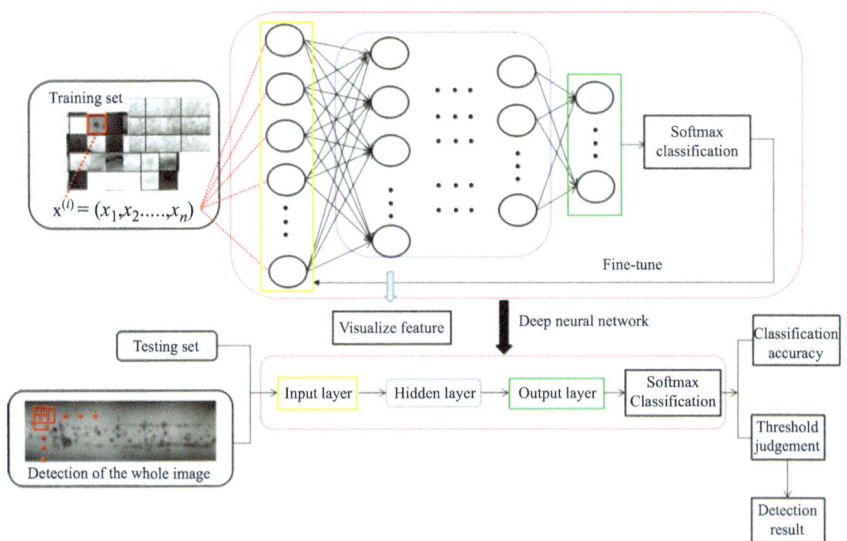

Figure 4.15   Illustration of the method proposed in [138] for defect detection in weld images (CC-BY 3.0 license)

Ferguson *et al.* [142] investigate the potential of different CNN architectures to localize casting defects in X-ray image. By decoupling the feature extraction layer from the object detection architecture, they studied three object detection architectures, namely Faster R-CNN, R-FCN (Region-based Fully Convolutional Networks [143]), and SSD (Single Shot Multibox Detector) with different feature extractors (VGG-16 and ResNet-101). Using an adapted version of the Faster R-CNN architecture, they achieve a mAP value of 0.921 on the GDXray dataset, constituted of 2727 X-ray images. For a similar use case, Du *et al.* [144] proposed Feature Pyramid Network (FPN) as the defect detection framework, which proved to be better suited for detecting small defects than Faster R-CNN, with a 40.9% improvement of the mAP. In the final regression and classification stage, RoIAlign (see Figure 4.16) indicated apparent accuracy improvement in bounding boxes location compared with RoI pooling, which could increase accuracy by 23.6% under Faster R-CNN.

In general the proposed detection techniques cannot classify a lot of defect types with high accuracy and do not consider the scale variation among different defect categories. Moreover, a lack of datasets, especially due to the lack of defective radiographic images, is noted and justifies to investigate specific data augmentation techniques, transfer learning approaches and generative adversarial networks.

Contaminants and threats detection in food industry and baggages

Ensuring contaminant-free products is a major concern in food industry, especially with the development of high-speed and fully automated production lines. X-ray inspection offers today the most effective way to detect and eliminate products containing foreign elements such as glass fragments, stones, metal pieces, or organic external elements such as insects or wood chips. For the task of contaminants detection, unsupervised learning approaches could be preferred because they can learn only with contaminant-free images, much easier to record in industrial environment. However, because defective product images in the context of contaminant detection are only slightly different from legitimate ones, they cannot be well separated through

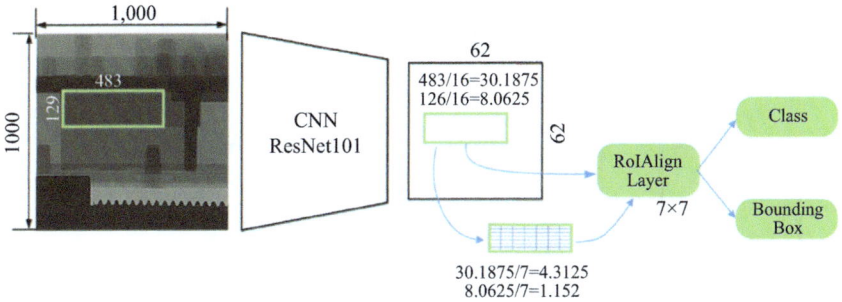

Figure 4.16   Faster R-CNN with RoIAlign applied to defect detection in X-ray images of casting parts [144] (copyright Elsevier)

one-class classification (OCC). For this reason, object class segmentation (OCS) is more commonly used for industrial flaw detection.

Bergmann *et al.* [145] propose a multi-object, multi-defect dataset of RGB camera images for anomaly detection and an evaluation of multiple OCS methods for unsupervised anomaly detection. Based on their results, Kim *et al.* [146] found OCS method not suitable for the contaminants detection in heterogeneous food items and proposed a supervised learning approach with a reduced dataset of industrial abnormal data. This database was augmented in a cut-paste manner using 500 images of different food product to create various backgrounds and 50 images of three types of contaminant without background (see Figure 4.17). The test data were constructed from defective product X-ray images collected in the field. By predicting the test data with the object detection network YOLOv4 (see Figure 4.18), trained on the augmented data, normal and defective products were classified with at least 94% accuracy for all foods.

Another use of industrial X-ray imaging concerns baggage screening, largely deployed for maintaining security at airports and other public spaces. In this field, screening is still very often realized by a human operator but due to the complexity of the image, containing lots of items overlapping, and the limited decision time, the performance of the control is not optimal. For this reason, several works have been dedicated to automatic threat detection. A thorough survey of this literature is reported in [147] based on 213 relevant references, among which 36 were identified as using deep-learning algorithms and categorized as supervised (classification, detection and segmentation) and unsupervised (anomaly detection) approaches.

The performances of different supervised approaches, applied on a same input X-ray image, are illustrated in Figure 4.19. For more details on the algorithms pipelines implemented in these different works, we refer the reader to [147].

Overall, despite promising results, the automated X-ray baggage screening remains an open question with a main limitation due to the lack of large unbiased datasets. There is also a lower detection accuracy in highly complex scenes and with

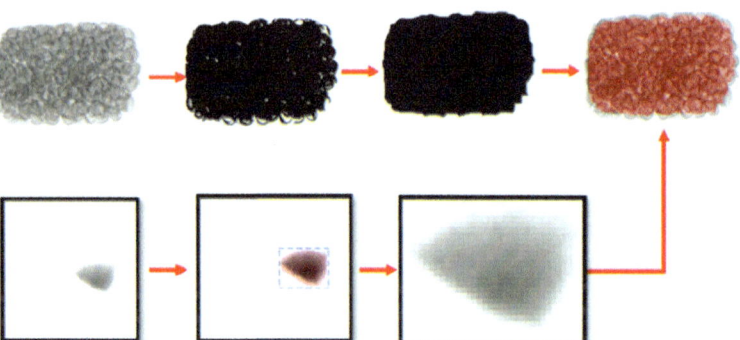

Figure 4.17   Augmentation of the dataset by merging contaminants alone images in different backgrounds [146] (CC-BY 4.0 license)

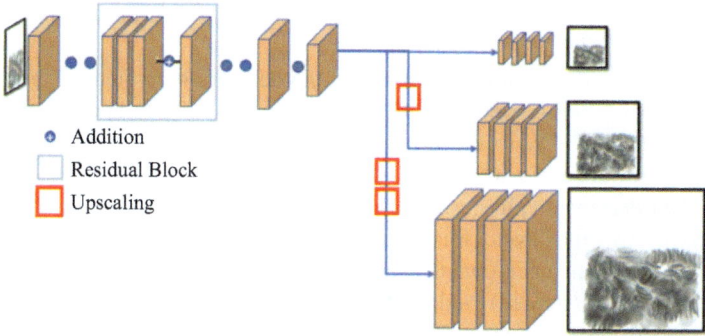

Figure 4.18   Basic architecture of YOLOv3 applied to food package analysis in [146] (CC-BY 4.0 license)

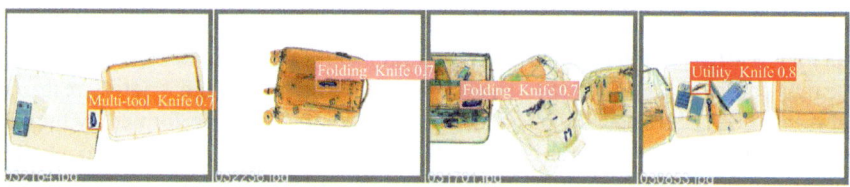

Figure 4.19   Results of DL tasks applied to X-ray baggage screening, the detection is performed with YOLOv5 on the GDXray dataset

thin objects such as sharps or knives with a critical role played here by multi-view X-ray systems, because there is a higher probability to have uninformative views of the threat with such objects. However, modern X-ray detection systems provide two orthogonal views or even four different angles of the same object, potentially offering clearer perspectives of an object occluded in the first view. Here again, the datasets of multi-view X-ray imagery are scarce. Unsupervised anomaly detection approaches exhibit lower performances and could be improved.

4.6   Future trends and open issues for deep learning algorithms as applied to electromagnetic NDT&E

The broad survey performed in this chapter shed some light on the future trends in the application of DL algorithms in NDT&E. Active reach efforts are currently focused on hybridization of numerical solvers and deep learning algorithms, on the possibility to perform data fusion of NDT&E acquisitions for enhancing predictions accuracy, in the assessment and propagation of uncertainties in predictions, the interpretability and the explainability of deep learning decisions in the NDT&E.

Embedding the physics knowledge in deep learning architecture for electromagnetic NDT&E

The use of full data-driven approaches makes the deep learning methods behave like black-box models. In the context of NDT&E, the possibility to add either physical meaning or explainability to the DL model developed is an active field of study and research. Toward this end, the use of numerical solvers is envisaged in order to hybridize data-driven and physics-driven approaches or by conditioning the learning process by injecting the physics knowledge directly into the DL architecture [93,148–152]. In both cases, the main goal is to end up with gray-box DL models that allow to better understand the (trained) models characteristics and discard unrealistic solutions from the physical point of view. Furthermore, the use of numerical solvers can be exploited to enhance the generalization capabilities of DL models when measurement data are subject to uncertainties.

Embedding explainability and interpretability in deep learning decision for electromagnetic NDT&E

The possibility to explain the ML models and the DL models specifically, also called explainable artificial intelligence (XAI), is a very active topic within the deep learning research community [153–155]. The XAI will be one of the most active research fields in NDT&E. Indeed, the possibility to link the prediction performed by the DL algorithms to the input features can lead, from the empirical point of view, to a better understanding and interpretation of the DL model mechanisms. In this framework, model-agnostic XAI algorithms are among the most suitable candidates [156,157].

Quantify and propagate the uncertainties in deep learning-based model for electromagnetic NDT&E

The application of DL to NDT&E acquisitions needs to account requirements in terms of prediction accuracy and robustness with respect to the test data provided in the online phase. That is, the estimation of probabilities in predictions (e.g., classification classes) as well as the uncertainties associated to the predictions need to be considered [46,47]. Bayesian inference applied to deep learning exploiting deep Bayesian Neural Network (BNN) architectures can be used to estimate the DL model uncertainties also called epistemic uncertainties [66,158]. The identification of the epistemic uncertainties enables as a consequence to estimate the aleatoric uncertainties linked to the intrinsic variability expressed by the measured data (e.g., measurement noise, uncertainties on probe position, probe ageing, specimen characteristics).

Data fusion based on deep learning algorithms applied to electromagnetic NDT&E

Deep learning methods are suitable to handle multiple structured input features during the learning process (e.g., multi-spectral images). In the context of electromagnetic NDT&E, such a feature allows, for instance, to handle directly complex valued signals in ECT. In the same manner researchers in NDT&E are trying to improve the prediction performances through the fusion of features coming from multiple channels data (e.g., multi-static probe signals, multiple frequencies). In view of deployment of DL algorithms in interconnected manufacturing systems, such a homogeneous data fusion process is supposed to be extended to non-electromagnetic NDT&E methods targeting an heterogeneous data fusion process. In this sense, a noteworthy result has been recently proposed in the biomedical context [159].

Open issues about the application of deep learning algorithms to electromagnetic NDT&E

It is well known that, in order to achieve good performance in terms of accuracy and robustness, DL algorithms need a suitable amount of data that is often larger than the one needed for other ML-based algorithms e.g., kernel machines. In NDT&E research community, due to secrecy and security issues, the availability of open dataset containing close-to-reality labeled data for developing and benchmark DL algorithms on a statistically meaningful set of samples is very scarce. In fact, most of the available acquisitions have been performed on specimen that do not account realistic inspection set-up or defect typologies, thus the direct use of these datasets for training DL algorithms can be applied by the whole community on narrow and case-dependent scenarios. Nevertheless, the joint use of numerical simulations and a small amount of labeled experimental data is believed to mitigate such systematic lack of realistic data through the use of semi-supervised and generative deep learning models.

Certification of NDT&E algorithms

Another great challenge to tackle in order to largely deploy such solutions in industry is the certification of DL-based algorithms and their inclusions in the norms ruling its various sectors. Many aspects have indeed to be taken into considerations. First, the metrics used to compare performance between such diagnostic algorithms and existing procedures should be established, depending on the level of autonomy given to the algorithm: assistance of the operator to highlight suspect regions of the analyzed data, proposition of diagnostic based on classification or fully automatic diagnostic. Then, when considering algorithms that update their learning with respect to incoming data, the question of how to ensure that the performance level is at least the same when adding some data should be answered to. Finally, another point is to manage the robustness of such algorithms (and possibly their recertification) to changes in the inspection conditions, like changes in environmental factors or replacement of parts of the acquisition chain due to some failures.

4.7 Conclusion and remarks

In this chapter, we analyzed some applications of deep learning methods to electro-magnetic NDT&E and tried to show how deep neural networks can be adapted to different scenarios involving electromagnetic probing waves ranging from the quasi-static regime to microwave. In particular, CNN have been deeply exploited when the treated signals behave "as images" such as in the case of ECT and MFL inspections where real and imaginary parts of the impedance variation as well as the magnetic flux density are probed. Furthermore, time domain signals as in PECT or GPR mea-surements have been addressed, too, by employing LSTM-RNN and/or through CNN explicitly adapted for the purpose (e.g., pixel-wise inversion). Our analysis underlined that specifically tailored deep neural architectures have obtained a better prediction performances than pre-trained networks based on state-of-the-art architectures. In fact, the systematic lack of large shared datasets containing labeled measurements of realistic acquisitions makes it difficult to properly benchmark and improve such backbone architectures. Moreover, the difficulties in collecting labeled measurements for defect parameters (e.g., the defect geometry) downsize the practical applications of deep learning models mostly to classification problems.

The survey performed in this chapter has also highlighted that the application of deep learning in NDT&E is also going toward the acceleration of numerical forward solvers for NDT&E modeling and simulations in a fully model-driven approach. It is believed that the ability of DL methods to handle problems having large cardinality (e.g., NDT&E parameters such as large number of defect classes, and defect geometry description) will boost the research and its application to time consuming statistical studies (see, e.g., [160,161]). Moreover, our analysis showed that the use of numerical solvers proves useful in designing the most suitable DL schemas as well as in improving the prediction accuracy when a low amount of measurements is available. Finally, a large amount of works in the literature showed that exploitation of deep learning algorithms directly on embedded systems (e.g., FPGA hardware) is already possible without an appreciable degradation in prediction performance.

4.8   Acknowledgments

The authors are grateful to Dominique Lesselier for all the wise advices, helpful suggestions, and insightful comments concerning this manuscript. We deeply believe that this chapter improved a lot thanks to him.

References

[1]   Marchand B, Decitre JM, Sergeeva-Chollet N, *et al*. Development of flexible array eddy current probes for complex geometries and inspection of magnetic parts using magnetic sensors. *AIP Conference Proceedings*. 2013;1511(1):488–493.

[2]   Xie S, Chen Z, Takagi T, *et al*. Efficient numerical solver for simulation of pulsed eddy-current testing signals. *IEEE Transactions on Magnetics*. 2011;47(11):4582–4591.

[3]   Xie S, Chen Z, Takagi T, *et al*. Development of a very fast simulator for pulsed eddy current testing signals of local wall thinning. *NDT & E International*. 2012;51:45–50.

[4]   Xie S, Chen Z, Chen HE, *et al*. Sizing of wall thinning defects using pulsed eddy current testing signals based on a hybrid inverse analysis method. *IEEE Transactions on Magnetics*. 2013;49(5):1653–1656.

[5]   Bowler JR, Johnson M. Pulsed eddy-current response to a conducting half-space. *IEEE Transactions on Magnetics*. 1997;33(3):2258–2264.

[6]   Fu F, Bowler JR. Transient eddy-current driver pickup probe response due to a conductive plate. *IEEE Transactions on Magnetics.* 2006;42(8): 2029–2037.

[7]   Theodoulidis T. Developments in calculating the transient eddy-current response from a conductive plate. *IEEE Transactions on Magnetics.* 2008;44(7):1894–1896.

[8]   Theodoulidis T, Skarlatos A. Efficient calculation of transient eddy-current response from multilayer cylindrical conductive media. *Philosophical Transactions of the Royal Society A.* 2020;378:20190588.

[9]   Skarlatos A, Theodoulidis T, Poulakis N. A fast and robust semi-analytical approach for the calculation of coil transient eddy-current response above planar specimens. *IEEE Transactions on Magnetics.* 2022;58(9):1–9.

[10] Dodd CV, Deeds WE. Analytical solutions to eddy current probe coil problems. *Journal of Applied Physics.* 1968;39(6):2829–2838.

[11] Dodd CV, Cheng CC, Deeds WE. Induction coils coaxial with an arbitrary number of cylindrical conductors. *Journal of Applied Physics.* 1974;45(2):638–647.

[12] Auld BA. Theoretical characterization and comparison of resonant-probe microwave eddy-current testing with conventional low-frequency eddy-current methods. In: G. Birnbaum and G. Free, American Society for Testing and Materials, editors, *Eddy-Current Characterization of Material and Structures.* vol. 12; 1981. p. 332–347.

[13] Auld BA, Muennemann F, Winslow DK. Eddy current probe response to open and closed surface flaws. *Journal of Nondestructive Evaluation.* 1981;2(1): 1–21.

[14] Theodoulidis TP, Kriezis EE. *Eddy Current Canonical Problems (with Applications to Nondestructive Evaluation).* Forsyth, GA: Tech Science Press; 2006.

[15] Theodoulidis TP, Bowler JR. The truncated region eigenfunction expansion method for the solution of boundary value problems in eddy current nondestructive evaluation. *AIP Conference Proceedings.* 2005;760(1):403–408.

[16] Burke SK, Bowler JR, Theodoulidis TP. An experimental and theoretical study of eddy-current end effects in finite rods and finite length holes. *AIP Conference Proceedings.* 2006;820(1):361–368.

[17] Bowler JR, Theodoulidis TP. Eddy currents induced in a conducting rod of finite length by a coaxial encircling coil. *Journal of Physics D: Applied Physics.* 2005;38(16):2861–2868.

[18] Theodoulidis TP, Bowler JR. Eddy current coil interaction with a right-angled conductive wedge. *Proceedings of the Royal Society A.* 2005;461(2062): 3123–3139.

[19] Theodoulidis TP, Bowler JR. Impedance of an induction coil at the opening of a borehole in a conductor. *Journal of Applied Physics.* 2008;103(2): 024905-1–024905-9.

[20] Skarlatos A, Theodoulidis T. Solution to the eddy-current induction problem in a conducting half-space with a vertical cylindrical borehole. *Proceedings of the Royal Society.* 2012;468(2142):1758–1777.

[21]　Skarlatos A, Theodoulidis T. Analytical treatment of eddy-current induction in a conducting half-space with a cylindrical hole parallel to the surface. *IEEE Transactions on Magnetics*. 2011;47(11):4592–4599.

[22]　Skarlatos A, Theodoulidis T. Calculation of the eddy-current flow around a cylindrical through-hole in a finite-thickness plate. *IEEE Transactions on Magnetics*. 2015;51(9):6201507.

[23]　Skarlatos A, Theodoulidis T. Impedance calculation of a bobbin coil in a conductive tube with eccentric walls. *IEEE Transactions on Magnetics*. 2010;46(11):3885–3892.

[24]　Skarlatos A. A mixed spatial-spectral eddy-current formulation for pieces with one symmetry axis. *IEEE Transactions on Magnetics*. 2020;56(9):1–11.

[25]　Chew WC. *Waves and Fields in Inhomogeneous Media*. New York, NY: Wiley-IEEE Press; 1995.

[26]　de Hoop AT. *Handbook of Radiation and Scattering of Waves*. Delft: Academic Press; 1995.

[27]　Sabbagh HA. A model of eddy-current probes with ferrite cores. *IEEE Transactions on Magnetics*. 1987;23: 1888–1904.

[28]　Abubakar A, van den Berg PM. Iterative forward and inverse algorithms based on domain integral equations for three-dimensional electric and magnetic objects. *Journal of Computational Physics*. 2004;195:236–262.

[29]　Skarlatos A, Pichenot G, Lesselier D, *et al*. Electromagnetic modeling of a damaged ferromagnetic metal tube by a volume integral equation formulation. *IEEE Transactions on Magnetics*. 2008;44:623–632.

[30]　Bowler JR. Eddy-current interaction with an ideal crack. I. The forward problem. *Journal of Computational Physics*. 1994;75(12):8128–8137.

[31]　Bowler JR. Eddy-current interaction with an ideal crack. II. The inverse problem. *Journal of Computational Physics*. 1994;75(12):8138–8144.

[32]　Bowler JR, Harfield N. Evaluation of probe impedance due to thin-skin eddy-current interaction with surface cracks. *IEEE Transactions on Magnetics*. 1998;34(2):515–523.

[33]　Yoshida Y, Bowler JR. Thin-skin eddy-current interaction with semielliptical and epicyclic cracks. *IEEE Transactions on Magnetics*. 2000;36(1):281–291.

[34]　Bowler JR, Harfield N. Vector potential integral formulation for eddy-current probe response to cracks. *IEEE Transactions on Magnetics*. 2000;36(2): 461–469.

[35]　Theodoulidis T. Developments in efficiently modelling eddy current testing of narrow cracks. *NDT & E International*. 2010;43(7):591–598.

[36]　Theodoulidis T, Poulakis N, Dragogias A. Rapid computation of eddy current signals from narrow cracks. *NDT & E International*. 2010;43(1):13–19.

[37]　Miorelli R, Reboud C, Theodoulidis T, *et al*. Efficient modeling of ECT signals for realistic cracks in layered half-space. *IEEE Transactions on Magnetics*. 2013;49(6):2886–2892.

[38]　Bowler JR, Theodoulidis TP. Boundary element calculation of eddy currents in cylindrical structures containing cracks. *IEEE Transactions on Magnetics*. 2009;45:1012–1015.

[39] Pipis K, Skarlatos A, Theodoros T, *et al.* ECT-signal calculation of cracks near fastener holes using an integral equation formalism with dedicated Green's kernel. *IEEE Transactions on Magnetics.* 2016;52(4):6200608.

[40] Pipis K, Skarlatos A, Theodoulidis T, *et al.* Impedance of an induction coil accounting for the end-effect in eddy-current inspection of steam generator tubes. In: H. Kikuchi, N. Yusa, T. Uchimoto, editors. *Electromagnetic Non-destructive Evaluation* (XIX), vol. 41 of Studies in Applied Electromagnetics and Mechanics. Amsterdam: IOS Press; 2016. p. 237–244.

[41] Rebican M, Chen Z, Yusa N, *et al.* Investigation of numerical precision of 3-D RFECT signal simulations. *IEEE Transactions on Magnetics.* 2005;41:1968–1971.

[42] Chen Z, Rebican M, Miya K, *et al.* Three-dimensional simulation of remote field ECT using the Ar method and a new formula for signal calculation. *Res Nondestr Eval.* 2005;16:35–53.

[43] Miorelli R, Reboud C, Theodoulidis T, *et al.* Coupled approach VIM–BEM for efficient modeling of ECT signal due to narrow cracks and volumetric flaws in planar layered media. *NDT & E International.* 2014;62:178–183.

[44] Chen X, Wei Z, Li M, *et al.* A review of deep learning approaches for inverse scattering problems (invited review). *Progress in Electromagnetics Research.* 2020;167:67–81.

[45] Goodfellow I, Bengio Y, Courville A. *Deep Learning.* London: MIT Press; 2016.

[46] Hüllermeier E, Waegeman W. Aleatoric and epistemic uncertainty in machine learning: an introduction to concepts and methods. *Machine Learning.* 2021;110(3):457–506.

[47] Gawlikowski J, Tassi CRN, Ali M, *et al.* A survey of uncertainty in deep neural networks. arXiv preprint arXiv:210703342. 2021.

[48] Coccorese E, Martone R, Morabito FC. A neural network approach for the solution of electric and magnetic inverse problems. *IEEE Transactions on Magnetics.* 1994;30(5):2829–2839.

[49] Elshafiey I, Udpa L, Udpa S. Application of neural networks to inverse problems in electromagnetics. *IEEE Transactions on Magnetics.* 1994;30(5):3629–3632.

[50] Lingvall F, Stepinski T. Automatic detecting and classifying defects during eddy current inspection of riveted lap-joints. *NDT & E International.* 2000;33(1):47–55.

[51] Chady T, Enokizono M, Sikora R. Neural network models of eddy current multi-frequency system for nondestructive testing. *IEEE Transactions on Magnetics.* 2000;36(4):1724–1727.

[52] Chady T, Enokizono M, Sikora R, *et al.* Natural crack recognition using inverse neural model and multi-frequency eddy current method. *IEEE Transactions on Magnetics.* 2001;37(4):2797–2799.

[53] Yusa N, Cheng W, Chen Z, *et al.* Generalized neural network approach to eddy current inversion for real cracks. *NDT & E International.* 2002;35(8):609–614.

[54] Wrzuszczak M, Wrzuszczak J. Eddy current flaw detection with neural network applications. *Measurement*. 2005;38(2):132–136.

[55] Chady T, Lopato P. Flaws identification using an approximation function and artificial neural networks. *IEEE Transactions on Magnetics*. 2007;43(4): 1769–1772.

[56] Rosado LS, Janeiro FM, Ramos PM, *et al.* Defect characterization with eddy current testing using nonlinear-regression feature extraction and artificial neural networks. *IEEE Transactions on Instrumentation and Measurement*. 2013;62(5):1207–1214.

[57] Bilicz S, Lambert M, Gyimóthy S. Kriging-based generation of optimal databases as forward and inverse surrogate models. *Inverse Problems*. 2010;26(7):074012.

[58] Douvenot R, Lambert M, Lesselier D. Adaptive metamodels for crack characterization in eddy-current testing. *IEEE Transactions on Magnetics*. 2011;47(4):746–755.

[59] Bernieri A, Betta G, Ferrigno L, *et al.* Multifrequency excitation and support vector machine regressor for ECT defect characterization. *IEEE Transactions on Instrumentation and Measurement*. 2013;63(5):1272–1280.

[60] Salucci M, Anselmi N, Oliveri G, *et al.* Real-time NDT-NDE through an innovative adaptive partial least squares SVR inversion approach. *IEEE Transactions on Geoscience and Remote Sensing*. 2016;54(11):6818–6832.

[61] Ahmed S, Reboud C, Lhuillier PE, *et al.* An adaptive sampling strategy for quasi real time crack characterization on eddy current testing signals. *NDT & E International*. 2019;103:154–165.

[62] Aldrin JC, Lindgren EA, Forsyth DS. Intelligence augmentation in non-destructive evaluation. In: *AIP Conference Proceedings*. vol. 2102. AIP Publishing LLC; 2019. p. 020028.

[63] EASA. *Artificial Intelligence Roadmap – A Human-Centric Approach to AI in Aviation*. EASA; 2020.

[64] EASA. *First Usable Guidance for Level 1 Machine Learning Applications*. EASA; 2021.

[65] Cantero-Chinchilla S, Wilcox PD, Croxford AJ. Deep learning in automated ultrasonic NDE – developments, axioms and opportunities. *NDT and E International*. 2022;131:102703.

[66] Zhu P, Cheng Y, Banerjee P, *et al.* A novel machine learning model for eddy current testing with uncertainty. *NDT & E International*. 2019;101:104–112.

[67] Park J, Han SJ, Munir N, *et al.* MRPC eddy current flaw classification in tubes using deep neural networks. *Nuclear Engineering and Technology*. 2019;51(7):1784–1790.

[68] Li S, Anees A, Zhong Y, *et al.* Crack profile reconstruction from eddy current signals with an encoder–decoder convolutional neural network. In: *2019 IEEE Asia-Pacific Microwave Conference (APMC)*; 2019. p. 96–98.

[69] Meng T, Tao Y, Chen Z, *et al.* Depth evaluation for metal surface defects by eddy current testing using deep residual convolutional neural networks. *IEEE Transactions on Instrumentation and Measurement*. 2021;70:1–13.

[70]   Buck JA, Underhill PR, Morelli JE, *et al*. Simultaneous multiparameter measurement in pulsed eddy current steam generator data using artificial neural networks. *IEEE Transactions on Instrumentation and Measurement*. 2016;65(3):672–679.

[71]   Fu X, Zhang C, Peng X, *et al*. Towards end-to-end pulsed eddy current classification and regression with CNN. In: *2019 IEEE International Instrumentation and Measurement Technology Conference (I2MTC)*; 2019. p. 1–5. ISSN: 2642-2077.

[72]   Dang W, Gao Z, Hou L, *et al*. A novel deep learning framework for industrial multiphase flow characterization. *IEEE Transactions on Industrial Informatics*. 2019;15(11):5954–5962.

[73]   Feng J, Li F, Lu S, *et al*. Injurious or noninjurious defect identification from MFL images in pipeline inspection using convolutional neural network. *IEEE Transactions on Instrumentation and Measurement*. 2017;66(7):1883–1892.

[74]   Lu S, Feng J, Zhang H, *et al*. An estimation method of defect size from MFL image using visual transformation convolutional neural network. *IEEE Transactions on Industrial Informatics*. 2018;15(1):213–224.

[75]   Sun H, Peng L, Huang S, *et al*. Development of a physics-informed doubly fed cross-residual deep neural network for high-precision magnetic flux leakage defect size estimation. *IEEE Transactions on Industrial Informatics*. 2022;18(3):1629–1640.

[76]   Le M, Pham CT, Lee J. Deep neural network for simulation of magnetic flux leakage testing. *Measurement*. 2021 Jan;170:108726.

[77]   Liu Z, Forsyth D, Lepine B, *et al*. Investigations on classifying pulsed eddy current signals with a neural network. *Insight-Non-Destructive Testing and Condition Monitoring*. 2003;45(9):608–614.

[78]   Liu Y, Liu S, Liu H, *et al*. Pulsed eddy current data analysis for the characterization of the second-layer discontinuities. *Journal of Nondestructive Evaluation*. 2019;38(1):1–8.

[79]   Deng W, Bao J, Ye B. Defect image recognition and classification for eddy current testing of titanium plate based on convolutional neural network. *Complexity*. 2020;2020:e8868190.

[80]   Rowshandel H, Nicholson G, Shen J, *et al*. Characterisation of clustered cracks using an ACFM sensor and application of an artificial neural network. *NDT & E International*. 2018;98:80–88.

[81]   Alvarenga TA, Carvalho AL, Honorio LM, *et al*. Detection and classification system for rail surface defects based on eddy current. *Sensors*. 2021;21(23):7937. Number: 23. Publisher: Multidisciplinary Digital Publishing Institute.

[82]   Bao J, Ye B, Wang X, *et al*. A deep belief network and least squares support vector machine method for quantitative evaluation of defects in titanium sheet using eddy current scan image. *Frontiers in Materials*. 2020;7:322.

[83]   Kim JW, Park S. Magnetic flux leakage sensing and artificial neural network pattern recognition-based automated damage detection and quantification for wire rope non-destructive evaluation. *Sensors*. 2018;18(1):109.

[84] Besaw LE, Stimac PJ. Deep convolutional neural networks for classify-
      ing GPR B-scans. In: *Detection and Sensing of Mines, Explosive Objects,
      and Obscured Targets XX*. vol. 9454. International Society for Optics and
      Photonics; 2015. p. 945413.
[85] Lameri S, Lombardi F, Bestagini P, *et al*. Landmine detection from GPR
      data using convolutional neural networks. In: *2017 25th European Signal
      Processing Conference (EUSIPCO)*. New York, NY: IEEE; 2017. p. 508–512.
[86] Kim N, Kim S, An YK, *et al*. Triplanar imaging of 3-D GPR data
      for deep-learning-based underground object detection. *IEEE Journal of
      Selected Topics in Applied Earth Observations and Remote Sensing*.
      2019;12(11):4446–4456.
[87] Kang MS, Kim N, Lee JJ, *et al*. Deep learning-based automated underground
      cavity detection using three-dimensional ground penetrating radar. *Structural
      Health Monitoring*. 2020;19(1):173–185.
[88] Khudoyarov S, Kim N, Lee JJ. Three-dimensional convolutional
      neural network-based underground object classification using three-
      dimensional ground penetrating radar data. *Structural Health Monitoring*.
      2020;19(6):1884–1893.
[89] Lei W, Hou F, Xi J, et al. Automatic hyperbola detection and fitting in GPR
      B-scan image. *Automation in Construction*. 2019;106:102839.
[90] Gao J, Yuan D, Tong Z, *et al*. Autonomous pavement distress detection using
      ground penetrating radar and region-based deep learning. *Measurement*.
      2020;164:108077.
[91] Feng J, Yang L, Wang H, *et al*. GPR-based subsurface object detection and
      reconstruction using random motion and depthnet. In: *2020 IEEE Inter-
      national Conference on Robotics and Automation (ICRA)*. New York, NY:
      IEEE; 2020. p. 7035–7041.
[92] Liu H, Lin C, Cui J, *et al*. Detection and localization of rebar in concrete by
      deep learning using ground penetrating radar. *Automation in Construction*.
      2020;118:103279.
[93] Hou F, Lei W, Li S, *et al*. Improved Mask R-CNN with distance guided inter-
      section over union for GPR signature detection and segmentation. *Automation
      in Construction*. 2021;121:103414.
[94] Li Y, Zhao Z, Luo Y, *et al*. Real-time pattern-recognition of GPR images with
      YOLO v3 implemented by Tensorflow. *Sensors*. 2020;20(22):6476.
[95] Li S, Gu X, Xu X, *et al*. Detection of concealed cracks from ground pen-
      etrating radar images based on deep learning algorithm. *Construction and
      Building Materials*. 2021;273:121949.
[96] Alvarez JK, Kodagoda S. Application of deep learning image-to-image trans-
      formation networks to GPR radargrams for sub-surface imaging in infrastruc-
      ture monitoring. In: *2018 13th IEEE Conference on Industrial Electronics and
      Applications (ICIEA)*. New York, NY: IEEE; 2018. p. 611–616.
[97] Liu B, Ren Y, Liu H, *et al*. GPRInvNet: deep learning-based ground-
      penetrating radar data inversion for tunnel linings. *IEEE Transactions on
      Geoscience and Remote Sensing*. 2021.

[98]   Ji Y, Zhang F, Wang J, *et al*. Deep neural network-based permittivity inversions for ground penetrating radar data. *IEEE Sensors Journal*. 2021;21(6):8172–8183.

[99]   Leong ZX, Zhu T. Direct velocity inversion of ground penetrating radar data using GPRNet. *Journal of Geophysical Research: Solid Earth*. 2021;e2020JB021047.

[100]   Feng J, Yang L, Wang H, *et al*. Subsurface pipes detection using DNN-based back projection on GPR data. In: *2021 IEEE Winter Conference on Applications of Computer Vision (WACV)*. New York, NY: IEEE; 2021. p. 266–275.

[101]   Giannakis I, Giannopoulos A, Warren C. A machine learning-based fast-forward solver for ground penetrating radar with application to full-waveform inversion. *IEEE Transactions on Geoscience and Remote Sensing*. 2019;57(7):4417–4426.

[102]   Giannakis I, Giannopoulos A, Warren C. A machine learning scheme for estimating the diameter of reinforcing bars using ground penetrating radar. *IEEE Geoscience and Remote Sensing Letters*. 2020;18(3):461–465.

[103]   Lähivaara T, Yadav R, Link G, *et al*. Estimation of moisture content distribution in porous foam using microwave tomography with neural networks. *IEEE Transactions on Computational Imaging*. 2020;6:1351–1361.

[104]   Yadav R, Omrani A, Link G, *et al*. Microwave tomography using neural networks for its application in an industrial microwave drying system. *Sensors*. 2021;21(20):6919.

[105]   Bartley PG, Nelson SO, McClendon RW, *et al*. Determining moisture content of wheat with an artificial neural network from microwave transmission measurements. *IEEE Transactions on Instrumentation and Measurement*. 1998;47(1):123–126.

[106]   Zhang J, Du D, Bao Y, *et al*. Development of multifrequency-swept microwave sensing system for moisture measurement of sweet corn with deep neural network. *IEEE Transactions on Instrumentation and Measurement*. 2020;69(9):6446–6454.

[107]   Ricci M, Štitić B, Urbinati L, *et al*. Machine-learning-based microwave sensing: a case study for the food industry. *IEEE Journal on Emerging and Selected Topics in Circuits and Systems*. 2021;11(3):503–514.

[108]   Ran P, Qin Y, Lesselier D, *et al*. Subwavelength microstructure probing by binary-specialized methods: contrast source and convolutional neural networks. *IEEE Transactions on Antennas and Propagation*. 2020;69(2):1030–1039.

[109]   Ran P, Chen S, Serhir M, *et al*. Imaging of sub-wavelength micro-structures by time reversal and neural networks, from synthetic to laboratory-controlled data. *IEEE Transactions on Antennas and Propagation*. 2021;69(12):8753–8762.

[110]   Wu X, Wei X, Zhang L, *et al*. VMFNet: visual-microwave dual-modality real-time target detection model for detecting damage to curved radar absorbing materials. *Optics Express*. 2021;29(15):23182–23201.

[111]   Rohkohl E, Kraken M, Schönemann M, *et al.* How to characterize a NDT method for weld inspection in battery cell manufacturing using deep learning. *The International Journal of Advanced Manufacturing Technology.* 2022; 1–15.

[112]   Rohkohl E, Kraken M, Schönemann M, *et al.* How to develop a NDT method for weld inspection in battery cell manufacturing using deep learning. 18 June 2021, DOI:10.21203/rs.3.rs-572568/v1.

[113]   Xu R, Zhou M. Elman neural network-based identification of Krasnosel'skii–Pokrovskii model for magnetic shape memory alloys actuator. *IEEE Transactions on Magnetics.* 2017;53(11):1–4.

[114]   Grech C, Buzio M, Pentella M, *et al.* Dynamic ferromagnetic hysteresis modelling using a preisach-recurrent neural network model. *Materials.* 2020;13(11):2561.

[115]   Maciusowicz M, Psuj G, Kochmański P. Identification of grain oriented SiFe steels based on imaging the instantaneous dynamics of magnetic Barkhausen noise using short-time fourier transform and deep convolutional neural network. *Materials.* 2022;15(1):118.

[116]   Vavilov V, Burleigh D. *Infrared Thermography and Thermal Nondestructive testing.* New York, NY: Springer; 2020.

[117]   Hu C, Duan Y, Liu S, *et al.* LSTM-RNN-based defect classification in honeycomb structures using infrared thermography. *Infrared Physics & Technology.* 2019;102:103032.

[118]   Fang Q, Maldague X. A method of defect depth estimation for simu-lated infrared thermography data with deep learning. *Applied Sciences.* 2020;10(19):6819. Number: 19. Publisher: Multidisciplinary Digital Publishing Institute.

[119]   Wang Q, Liu Q, Xia R, *et al.* Defect depth determination in laser infrared thermography based on LSTM-RNN. *IEEE Access.* 2020;8:153385–153393.

[120]   Cao Y, Dong Y, Cao Y, *et al.* Two-stream convolutional neural network for non-destructive subsurface defect detection via similarity comparison of lock-in thermography signals. *NDT & E International.* 2020;112:102246.

[121]   Marani R, Palumbo D, Galietti U, *et al.* Deep learning for defect charac-terization in composite laminates inspected by step-heating thermography. *Optics and Lasers in Engineering.* 2021;145:106679.

[122]   Xie J, Xu C, Chen G, *et al.* Improving visibility of rear surface cracks during inductive thermography of metal plates using autoencoder. *Infrared Physics & Technology.* 2018;91:233–242.

[123]   Luo Q, Gao B, Woo WL, *et al.* Temporal and spatial deep learning network for infrared thermal defect detection. *NDT & E International.* 2019;108:102164.

[124]   Yang J, Wang W, Lin G, *et al.* Infrared thermal imaging-based crack detection using deep learning. *IEEE Access.* 2019;7:182060–182077.

[125]   Müller D, Netzelmann U, Valeske B. Defect shape detection and defect reconstruction in active thermography by means of two-dimensional con-volutional neural network as well as spatiotemporal convolutional LSTM network. *Quantitative InfraRed Thermography Journal.* 2020;1–19.

[126]   Jang K, Kim N, An YK. Deep learning-based autonomous concrete crack evaluation through hybrid image scanning. *Structural Health Monitoring.* 2019;18(5–6):1722–1737.

[127]   Wu H, Zhang H, Hu G, *et al.* Deep learning-based reconstruction of the structure of heterogeneous composites from their temperature fields. *AIP Advances.* 2020;10(4):045037.

[128]   Blatzheim M, Böckenhoff D, & W7-X Team, Max Planck Institute for Plasma Physics, Max Planck Society. Neural network regression approaches to reconstruct properties of magnetic configuration from Wendelstein 7-X modeled heat load patterns. *Nuclear Fusion.* 2019;59(12):126029.

[129]   Liu K, Li Y, Yang J, *et al.* Generative principal component thermography for enhanced defect detection and analysis. *IEEE Transactions on Instrumentation and Measurement.* 2020;69(10):8261–8269.

[130]   Jiang H, Chen F, Liu X, *et al.* Thermal wave image deblurring based on depth residual network. *Infrared Physics & Technology.* 2021;117:103847.

[131]   Wang Q, Liu Q, Xia R, *et al.* Automatic defect prediction in glass fiber reinforced polymer based on THz-TDS signal analysis with neural networks. *Infrared Physics & Technology.* 2021;115:103673.

[132]   Zhang Z, Peng G, Tan Y, *et al.* THz wave detection of gap defects based on convolutional neural network improved by residual shrinkage network. *CSEE Journal of Power and Energy Systems.* 2020.

[133]   Long Z, Wang T, You CW, *et al.* Terahertz image super-resolution based on a deep convolutional neural network. *Applied Optics.* 2019;58(10): 2731–2735.

[134]   Lei T, Tobin B, Liu Z, *et al.* A terahertz time-domain super-resolution imaging method using a local-pixel graph neural network for biological products. *Analytica Chimica Acta.* 2021;1181:338898.

[135]   Lei T, Li Q, Sun DW. A Dual AE-GAN guided THz spectral Dehulling model for mapping energy and moisture distribution on sunflower seed kernels. *Food Chemistry.* 2021;131971.

[136]   Zhang Z, Zhang L, Chen X, *et al.* Modified generative adversarial network for super-resolution of terahertz image. In: *2020 International Conference on Sensing, Measurement & Data Analytics in the Era of Artificial Intelligence (ICSMD).* New York, NY: IEEE; 2020. p. 602–605.

[137]   Mery D. X-ray testing by computer vision. In: *2013 IEEE Conference on Computer Vision and Pattern Recognition Workshops*; 2013. p. 360–367.

[138]   Hou W, Wei Y, Guo J, *et al.* Automatic detection of welding defects using deep neural network. *Journal of Physics: Conference Series.* 2018;933:012006.

[139]   Mery D, Riffo V, Zscherpel U, *et al.* GDXray: the database of X-ray images for nondestructive testing. *Journal of Nondestructive Evaluation.* 2015;34.

[140]   Wang Y, Shi F, Tong X. In: *A Welding Defect Identification Approach in X-ray Images Based on Deep Convolutional Neural Networks*; 2019. p. 53–64.

[141]   Yang N, Niu H, Chen L, *et al.* X-ray weld image classification using improved convolutional neural network. vol. 1995; 2018. p. 020–035.

[142]   Ferguson M, Ak R, Lee YTT, *et al*. Automatic localization of casting defects with convolutional neural networks. In: *2017 IEEE International Conference on Big Data (Big Data)*; 2017. p. 1726–1735.

[143]   Dai J, Li Y, He K, *et al*. R-FCN: object detection via region-based fully convolutional networks. In: *Proceedings of the 30th International Conference on Neural Information Processing Systems. NIPS'16*. Red Hook, NY: Curran Associates Inc.; 2016. p. 379–387.

[144]   Du W, Shen H, Fu J, *et al*. Approaches for improvement of the X-ray image defect detection of automobile casting aluminum parts based on deep learning. *NDT & E International*. 2019;107:102144.

[145]   Bergmann P, Fauser M, Sattlegger D, *et al*. MVTec AD – a comprehensive real-world dataset for unsupervised anomaly detection. In: *2019 IEEE/CVF Conference on Computer Vision and Pattern Recognition (CVPR)*; 2019. p. 9584–9592.

[146]   Kangjik K, Hyunbin K, Junchul C, *et al*. Real-time anomaly detection in packaged food X-ray images using supervised learning. *Computers, Materials & Continua*. 2021;67(2):2547–2568.

[147]   Akcay S, Breckon T. Towards automatic threat detection: a survey of advances of deep learning within X-ray security imaging. *Pattern Recognition*. 2022;122:108245.

[148]   Raissi M, Perdikaris P, Karniadakis GE. Physics-informed neural networks: a deep learning framework for solving forward and inverse problems involving nonlinear partial differential equations. *Journal of Computational Physics*. 2019;378:686–707.

[149]   Shukla K, Jagtap AD, Blackshire JL, *et al*. A physics-informed neural network for quantifying the microstructural properties of polycrystalline nickel using ultrasound data: a promising approach for solving inverse problems. *IEEE Signal Processing Magazine*. 2021;39(1):68–77.

[150]   Guo R, Shan T, Song X, *et al*. Physics embedded deep neural network for solving volume integral equation: 2D case. *IEEE Transactions on Antennas and Propagation*. 2021.

[151]   Guo L, Li M, Xu S, *et al*. Electromagnetic modeling using an FDTD-equivalent recurrent convolution neural network: accurate computing on a deep learning framework. *IEEE Antennas and Propagation Magazine*. 2021;2–11.

[152]   Guo R, Lin Z, Shan T, *et al*. Physics embedded deep neural network for solving full-wave inverse scattering problems. *IEEE Transactions on Antennas and Propagation*. 2021;1–1.

[153]   Adadi A, Berrada M. Peeking inside the black-box: a survey on explainable artificial intelligence (XAI). *IEEE Access*. 2018;6:52138–52160.

[154]   Gunning D, Aha D. DARPA's explainable artificial intelligence (XAI) program. *AI Magazine*. 2019;40(2):44–58.

[155]   Arrieta AB, Díaz-Rodríguez N, Del Ser J, *et al*. Explainable Artificial Intelligence (XAI): concepts, taxonomies, opportunities and challenges toward responsible AI. *Information Fusion*. 2020;58:82–115.

[156] Ribeiro MT, Singh S, Guestrin C. "Why should i trust you?" Explaining the predictions of any classifier. In: *Proceedings of the 22nd ACM SIGKDD International Conference on Knowledge Discovery and data Mining*; 2016. p. 1135–1144.

[157] Lundberg SM, Lee SI. A unified approach to interpreting model predictions. In: Guyon I, Luxburg UV, Bengio S, *et al.* (eds.), *Advances in Neural Information Processing Systems*. Curran Associates, Inc.; 2017, pp. 4765–4774.

[158] Wei Z, Chen X. Uncertainty quantification in inverse scattering problems with Bayesian convolutional neural networks. *IEEE Transactions on Antennas and Propagation*. 2020;69(6):3409–3418.

[159] Qin Y, Ran P, Rodet T, *et al.* Breast imaging by convolutional neural networks from joint microwave and ultrasonic data. *IEEE Transactions on Antennas and Propagation*. 2022;1–1.

[160] Aldrin JC, Oneida EK, Shell EB, *et al.* Model-based probe state estimation and crack inverse methods addressing eddy current probe variability. In: *AIP Conference Proceedings*, vol. 1806. AIP Publishing LLC; 2017. p. 110013.

[161] Cai C, Miorelli R, Lambert M, *et al.* Metamodel-based Markov-Chain-Monte-Carlo parameter inversion applied in eddy current flaw characterization. *NDT & E International*. 2018;99:13–22.

Chapter 5
Deep learning techniques for subsurface imaging

Rui Guo[1], Maokun Li[1] and Aria Abubakar[2]

Geophysical subsurface imaging is an effective tool for understanding the Earth's internal structures. It is widely used in ecological and hydrological applications, oil and gas industry, etc. The exploration is usually done with remote sensing tools. Sensors record the secondary field induced by distant targets. One needs to see through countless rocks with various properties to distinguish the buried targets. Many obstacles exist in data collection. For example, the sources and receivers may be insufficient to illuminate the region of interest. The energy of the scattered field may attenuate to be undetectable, and the measuring environment can be very noisy.

Due to the imperfect measurement, subsurface data inversion has strong non-uniqueness, i.e., multiple models may fit the same field data. To obtain a geologically reasonable model, one needs to constrain the inversion with prior knowledge. Additionally, multi-physics measurements such as gravity, seismic, and electromagnetic (EM) data can be jointly used for comprehensively understanding the domain of investigation. The large-volume data, large-scale survey domain, as well as the complex numerical modeling process, make geophysical inversion computationally expensive. It is also common to repeat multi-physics inversion many times to obtain a reliable geological model, which aggravates the computational burden.

The past few years have witnessed a popularity of deep learning (DL) techniques thanks to the development of modern computing hardware, big data storage and efficient computing methods. Geophysical inversion can benefit from this progress. The advanced computing power permits fast and high-quality imaging with large-volume datasets. Besides, experience of interpretation can be trained into a DL model, which enables the fusion of prior knowledge and inversion seamlessly.

This chapter aims to overview the frontiers of DL as applied to subsurface imaging. The detailed implementations will not be introduced; instead, we hope to provide readers with a broad view of DL as applied to geophysical inversion. We mainly focus on EM methods, the depth of investigation ranging from hundreds to thousands of meters. After briefly reviewing the history of learning-based inversion, we show state-of-the-art techniques in applying DL in EM inversion, including purely data-driven approaches, physics-embedded data-driven approaches and learning-assisted physics-driven approaches. We also present several DL-based methods for seismic

[1]Department of Electronic Engineering, Tsinghua University, Beijing, China
[2]Schlumberger, Houston, TX, USA

data inversion that may benefit the EM community. We further discuss different approaches of constructing training datasets. At last, we conclude the state of DL in EM subsurface imaging and show outlooks.

5.1   Introduction

Geophysical EM inversion recovers resistivity from the measured electric (E-) and magnetic (H-) fields. The transmitters and receivers are generally placed on the Earth's surface or in boreholes. The skin depth of EM waves is [1]

$$\delta \approx 503\sqrt{\rho/f}, \tag{5.1}$$

where ρ is the electrical resistivity and f is the frequency. To illuminate the region inside Earth, low-frequency EM waves are adopted. In this case, the wave number,

$$k = 2\pi f \sqrt{\mu \varepsilon}, \tag{5.2}$$

with μ and ε being the permeability and the permittivity, and are nearly zero. Therefore, EM fields propagate in Earth mainly by diffusion. The lower frequency and conductivity (reciprocal of resistivity), the deeper we can see. On the other hand, the resolution decreases with the frequency since EM waves are insensitive to targets much smaller than the wavelength. For high-quality imaging, prior information are needed to reduce the non-uniqueness.

Classical physics-driven methods simulate the EM response to find the model that fit the observed data. Bayesian and deterministic inversion are two mainstream frameworks. In Bayesian inversion, the posterior distribution of the model parameter $\mathbf{m}$ given the observed data $\mathbf{d}_{obs}$, i.e.,

$$\pi(\mathbf{m}|\mathbf{d}_{obs}) \propto \pi(\mathbf{d}_{obs}|\mathbf{m})\pi(\mathbf{m}) \tag{5.3}$$

needs to be computed, where $\pi(\mathbf{d}_{obs}|\mathbf{m})$ is the likelihood function and $\pi(\mathbf{m})$ is the prior distribution of model parameters. By properly constructing $\pi(\mathbf{m})$ and sampling probability density function, Bayesian inversion can flexibly incorporate prior information into the model reconstruction. However, this process involves a large number of calls of forward modeling, which is prohibitively expensive for 2D or 3D inversion.

Deterministic approaches use gradient-based methods to seek a solution that minimizes the cost function, for instance,

$$L(\mathbf{m}) = \|F(\mathbf{m}) - \mathbf{d}_{obs}\|^2 + \lambda\phi(\mathbf{m}), \tag{5.4}$$

where $F(\cdot)$ is the forward modeling function and $\phi(\cdot)$ is the regularization function weighted by λ. The prior knowledge of model parameters can be incorporated into the inversion by regularizations. The deterministic inversion has been widely applied in industry. However, the reconstructed models are in general locally optimal. Furthermore, it is difficult to integrate the experience from interpretation into inversion by regularizations. In practice, the resistivity models are usually progressively refined by repeating inversion and interpretation several times.

Machine learning (ML) provides an alternatives method for geophysical inversion. The process is compelled by data, assuming that the inverse operator, Θ_I, can be obtained through offline training. The models will be predicted with the trained operator, i.e.,

$$\mathbf{m} = \Theta_I(\mathbf{d}_{obs}).$$ (5.5)

ML is not new in geoscience. In 1990, Raiche A overviewed the use of neural networks (NNs) for geophysical inversion [2], indicating that the NN-based inversion should learn from experience, be robust to noise, and be able to infer structures not contained in the training set. In 1991, McCormack MD pointed out the limitations of NNs [3], i.e., slow learning rates, nonoptimal training results, imprecise numerical answers, and black-box property. Nevertheless, early work on the application of NNs in EM inversion proved its feasibility [4,5].

Van der Baan M and Jutten C introduced most recent progress of the NN theory and optimization to the geophysics community in 2000 [6]. They provided the strategies of choosing NN structures, input parameters, and training algorithms. The NNs at this stage can handle more parameters than in the 1990s, but the input and output still need to be preprocessed for dimensionality reduction. The authors concluded that NNs are too expensive to be of real value in geophysics, which is consistent with the popular perception of ML in this decade [7]. Despite the criticisms, research on this topic has not stopped. NN-based inversion of EM data for 1D, 2D or 3D models was performed in [8–11]. Aside from NNs, the support vector machines also showed effectiveness in localization of subsurface objects [12,13].

Current research frontiers extend to various DL-based inversion. Graphics processing units (GPUs), stochastic gradient descent methods, advanced deep NN structures, as well as user-friendly DL frameworks, allow us to train NNs with billions of parameters at a reasonable cost. Advanced NNs such as the convolutional neural networks (CNNs), recurrent neural networks (RNNs) and generative adversarial networks (GANs) have succeed in computer vision (CV) and natural language processing (NLP). Some considered as limitation decades ago have been overcome. However, the theory of DL-based inversion still needs to be studied. Different from CV or NLP where NNs are used as the approximation of a process difficult to model, data inversion is the reverse process of wave propagation that has a clear physical model. Geophysicists are familiar with the governing equations of waves and fields and have physics intuition. Such domain knowledge should be combined with DL to provide faithful answers. Toward this goal, researchers have studied many ways of DL-based data inversion, including end-to-end NNs, physics-embedded NNs and DL-assisted cost function inversion. Various NN architectures as well as training methods are explored. In the following, we provide an overview.

5.2 Purely data-driven approach

ML can be grouped into three types:

Supervised learning: Machine is expected to learn the input-output mapping from a labeled training dataset. For each labeled sample, the answer of the input is also given. It includes two types of problems: classification and regression.

Unsupervised learning: No labeled data are given for training. The machine finds optimal representations or unknown patterns in the dataset using self-organized manners. It can be used for clustering and dimensionality reduction.

Reinforcement learning: The machine interacts with the environment by taking actions that will, in turn, feed punishments or rewards back to the machine. The machine maximizes the future rewards it receives during the lifetime of the task. A well-known example is the AlphaGo [14].

In data inversion, the input of **supervised learning** can be the field data, such as the measured E-field, H-field, or apparent resistivity, while the output is generally the subsurface resistivity model. The **unsupervised learning** usually involves forward modeling in the loss function of training. The study on **reinforcement learning** for inversion is still under progress.

This section focuses on the methods without physics constraint where NNs are expected to learn the data-to-model mapping through big data training. We call them purely data-driven approaches. The architectures are mainly based on modified CNNs, RNNs, or GANs.

5.2.1   Convolutional neural network

Compared with fully connected NNs, the CNN greatly reduces the number of trainable parameters by shared weights and sparse connections [15,16] (Figure 5.1). It has been widely used in image classification, segmentation, deblurring, etc. [17–20]. Its ability of learning spatially local correlation in natural images shows potentials for data inversion.

CNNs can be applied with the input being EM data and the output being the resistivity. In [21], the author applies 1D CNN to invert frequency domain EM data. The input is six response values received in two antenna modes with three offsets, while the output is the discrete conductivity represented by twelve values. Both synthetic and field data are tested on the trained CNN. In [22], the authors perform transient EM (TEM) data inversion with 1D CNN, where the input is the time domain response and the output is a 300-layer resistivity model. Synthetic and field data test show that the CNN is more robust to noise, tends to generate higher resolution models and runs much faster than the Gauss-Newton method.

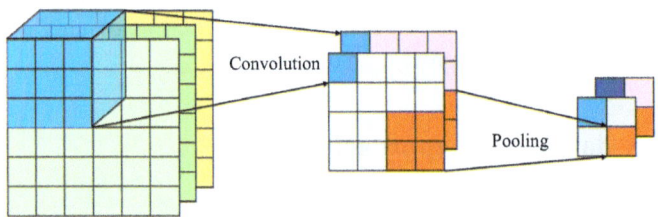

Figure 5.1   Schematic of the CNN

Similar conclusions are also drawn in [23–26] through inverting 1D audio-magnetotelluric (AMT), airborne EM (AEM), TEM, marine controlled source EM (CSEM), and borehole EM data. In addition, the authors find that increasing the data volume improves the accuracy of prediction, such as improving the number of frequencies [25] and transmitting–receiving configurations [26].

Uncertainty analyses of the inversion can be performed [22,25,27]. In [22,25], the results are predicted with a series of CNNs selected at the consecutive epochs around the epoch with the optimal trained parameters. For example, if the CNN is well trained at the Kth epoch, the CNNs trained at previous ten epochs can be selected to generate several realizations of the output, see Figure 5.2. The approach significantly reduces the risk of overfitting. In [27], Bayesian CNN is applied to predict the epistemic and aleatoric uncertainty of 1D inversion. Both resistivity model and aleatoric uncertainty are the outputs of the network. The CNN predicts many times with the random dropout of network parameters. Statistical analysis can be made on these predictions. However, the authors claim that this method cannot guarantee robustness when the test data are beyond the scope of the training data.

Two- and three-dimensional (2D and 3D) inversions are also performed. CSEM imaging for CO_2 monitoring is achieved by fully CNN in [28]. It is shown that the CNN is robust to normally distributed noise, and capable of predicting accurate results even when the receiver grid in the test (32×32) is different from that in the training (64×64), thanks to the CNN's strong learning ability in spatial correlation. In [29], resistive salt bodies are predicted from marine EM data. The CNN can reconstruct salts buried in inhomogeneous medium although the background of training samples is simple. In [30], the 2D and 3D resistivity distribution for reservoir monitoring is recovered by 2D and 3D CNNs.

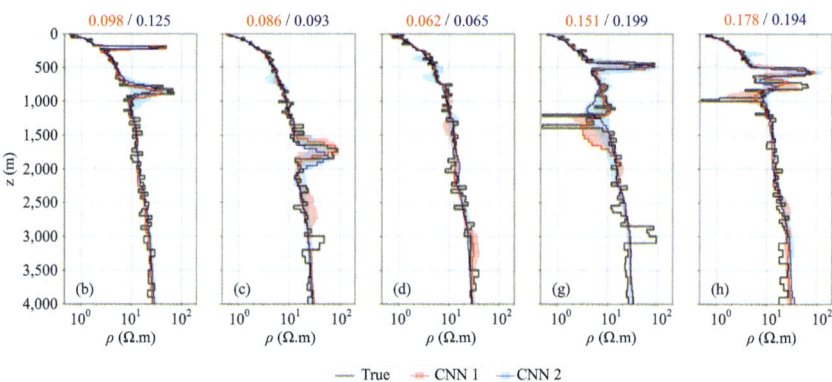

Figure 5.2   *Uncertain analysis of two CNNs in [25]. The solid lines are the predictions of well trained CNNs and the shadings show the range of uncertainty.*

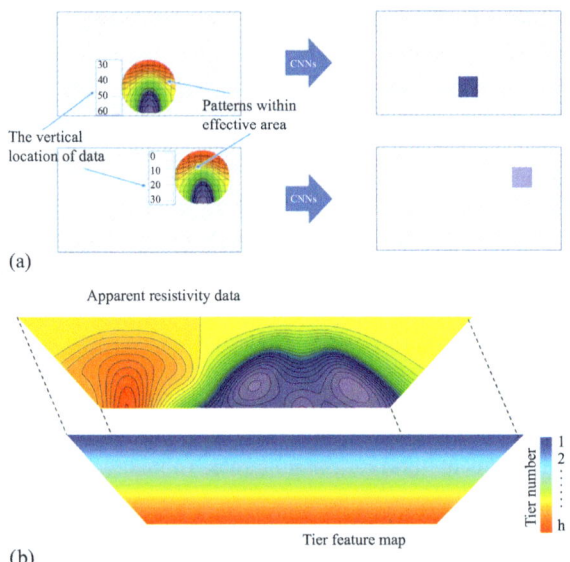

(a)

(b)

Figure 5.3   Incorporating depth information into the CNN [31]. (a) The potential ambiguity of CNNs during training. (b) The input data after concatenating the tier feature map.

Modifying the NN architecture based on geophysical domain knowledge can improve the prediction accuracy. In [31], the mapping from the apparent resistivity (input) to the resistivity model (output) is built. The authors find that only using the local spatial features of apparent resistivity as the input will cause ambiguity in the depth domain, see Figure 5.3(a). Therefore, a tier feature map containing depth information is provided to the CNN, see Figure 5.3(b). Furthermore, a depth weighting function and a smooth constraint are introduced to the training loss for obtaining reasonable resistivity models.

5.2.2   Recurrent neural network

RNNs are suitable for sequential data prediction such as speech, text, and video [32,33]. As shown in Figure 5.4, the output at the current step is determined by not only the current input but also the state of the previous step. Popular architectures include bidirectional RNN [34], long short-term memory (LSTM) [35], and gated recurrent units (GRUs) [36]. The RNN is attractive for inversion because it can describe the dependence of resistivity in spatial domain. For example, the reconstruction of deep earth resistivity using late-time EM response will be affected by the shallow resistivity inferred from early-time response. Such characteristics can be captured by the RNN.

In [37], 1D inversion of airborne TEM data is achieved by a hybrid CNN-RNN. As shown in Figure 5.5, the time domain response and the flight altitude are input into four CNNs to extract features, and then the LSTMs use these features to predict

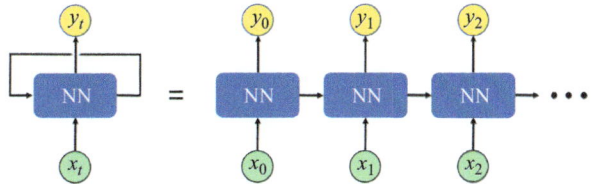

Figure 5.4   Schematic of the RNN

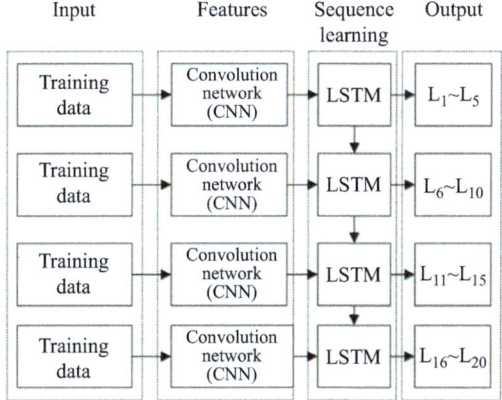

Figure 5.5   The RNN structure proposed in [37]. L_m represents the resistivity of the mth layer.

resistivity of different layers sequentially. In the field data test, the imaging results are slightly different from the Occam inversion. However, inverting a test line with 1,500 stations takes 1 s, implying the feasibility of real-time data processing for large TEM surveys.

RNNs can be used to increase the horizontal continuity in 2D inversion. The EM data along the test line may be separated into segments, and sequentially used as the input to predict part of the underground resistivity [38]. With a similar idea, bidirectional LSTM is applied to enhance the spatial continuity in consecutive and long survey line ground penetrating radar (GPR) imaging [39].

5.2.3   Generative adversarial network

The GAN contains a generator and a discriminator: the generator is trained to produce an output that fools the discriminator, while the discriminator is trained to distinguish the candidate produced by the generator, see Figure 5.6. The discriminator makes sure that the generated output has the same statistics as the training set.

In [40], the loss function of the CNN for electrical resistivity inversion is composed of the total variation loss, mean squared error loss and adversarial loss, see Figure 5.7(a). The first term constrains the sharpness of the resistivity model.

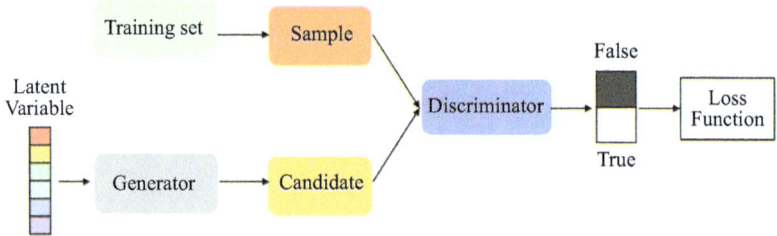

Figure 5.6   Schematic of the GAN

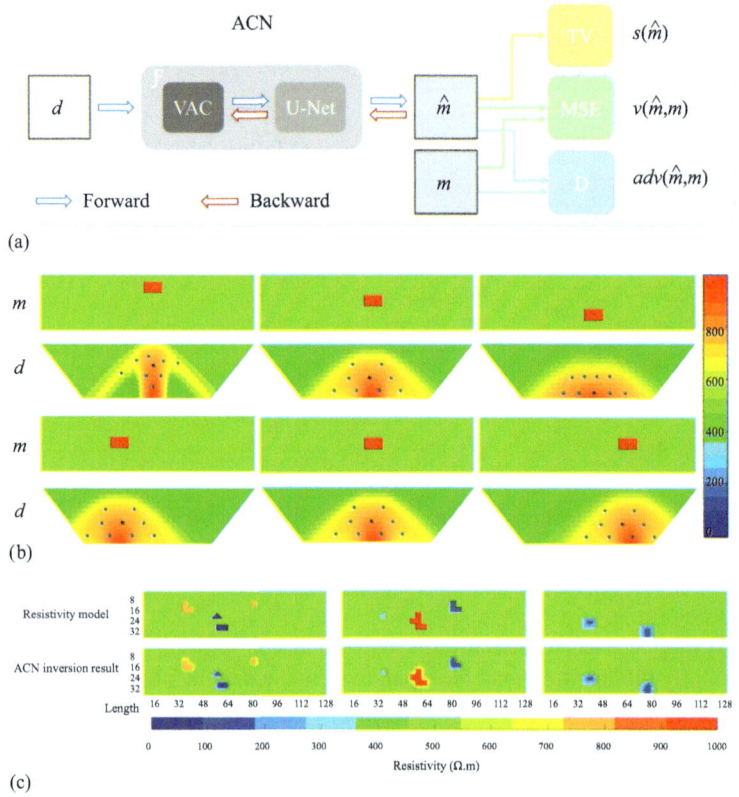

Figure 5.7   The adaptive CNN for electrical resistivity inversion [40]. (a) Training of the CNN. (b) The adaptively learned convolution positions represented by dots. (c) Ground truths and inversion results.

The second term minimizes the distance between predictions and labels. The adversarial loss is added by a discriminator to constrain the statistics of the output, which ensures that the inverted model looks realistic. To alleviate the ambiguity conventional CNNs may face when targets are located at different depths, the authors apply active convolution units (ACUs) [41] to adaptively learn the features in depth. In an ACU, the convolution does not have a fixed shape of the receptive field, instead, the positions for convolutions are learnable, see Figure 5.7(b). The combined use of above schemes reduces the ill-posedness of inversion and produces models with high resolution, see Figure 5.7(c).

5.3 Physics embedded data-driven approach

Although a NN can approximate any continuous function [42], many researchers have realized its difficulty in high-dimensional data inversion. Due to the large uncertainties of 2D or 3D models, the training dataset for optimizing the NN parameters is usually incomplete. The trained NN may be unstable when facing new data, and it is difficult to know the boundary of accurate prediction. To address these issues, strategies that incorporate physics law into ML are proposed.

5.3.1 Supervised descent method

The supervised descent method (SDM) was first proposed by Xiong X and De la Torre F for solving nonlinear least squares problems in computer vision [43]. Instead of learning an end-to-end mapping, it iteratively learns a set of linear descent directions for optimizing the least squares problem. In [44], SDM was first applied to data inversion.

The data misfit function of inversion can be

$$L(\mathbf{m}) = \|\mathbf{d}_{\text{obs}} - F(\mathbf{m})\|^2, \tag{5.6}$$

where $\mathbf{m}$ is the unknown to be recovered, $\mathbf{d}_{\text{obs}}$ is the observed field data and $F(\cdot)$ is the forward modeling function. In the Gauss–Newton method, the model update at the kth step is computed by

$$\Delta\mathbf{m}_k = (\mathbf{J}^T\mathbf{J})^{-1}\mathbf{J}^T(\mathbf{d}_{\text{obs}} - F(\mathbf{m}_{k-1})), \tag{5.7}$$

where $\mathbf{J}$ is the Fréchet derivative of F at $\mathbf{m}_{k-1}$, and $\cdot^T$ represents the transpose of a matrix. Notice that $\mathbf{J}$ only contains the local property of F and the computations of $\mathbf{J}$ and $F(\mathbf{m})$ are usually expensive.

SDM aims to learn the descent direction $\mathbf{K}$ rather than computing $\mathbf{J}$ and $(\mathbf{J}^T\mathbf{J})^{-1}\mathbf{J}^T$ online. With an iterative manner, the loss function for training is

$$L(\mathbf{K}_k) = \sum_{i=1}^{N} \|\Delta\mathbf{m}_{i,k} - \mathbf{K}_k(F(\mathbf{m}_{i,T}) - F(\mathbf{m}_{i,k}))\|^2, \tag{5.8}$$

where $\Delta\mathbf{m}_{i,k} = \mathbf{m}_{i,T} - \mathbf{m}_{i,k}$, with $\mathbf{m}_{i,k}$ being the ith model in the kth step, and $\mathbf{m}_{i,T}$ represents the ith training model. In online data inversion, the model can be updated by

$$\mathbf{m}_{k+1} = \mathbf{m}_k + \mathbf{K}_k(\mathbf{d}_{\text{obs}} - F(\mathbf{m}_k)). \tag{5.9}$$

Note that the model is updated by the production of the learned descent directions and the computed data residual. Since the latter term contains forward modeling, the prediction can obey physical law. More details about training and prediction can be found in [45].

In [44], both model- and pixel-based 1D TEM inversions are achieved by SDM. In the model-based inversion, the unknowns include the thickness and resistivity, while in the pixel-based inversion, the unknown is the discrete resistivity defined on fixed layers. It is shown that SDM can interpolate new models that are not contained in the training set. This work validate the feasibility of SDM in reconstructing inhomogeneous medium.

The generalizability is further validated by 2D MT inversion [46]. With the online regularization and restart scheme, SDM is able to predict inhomogeneous medium while the training models are quite simple, see Figure 5.8. In addition, the trained descent direction can be modified according to different prior knowledge, which follows the concept of transfer learning.

In [47], the logging-while-drilling (LWD) inverse problem is solved by SDM. The authors use the real-time feedback from the downhole to update the descent direction, which achieves higher accuracy than traditional gradient inversion. Authors in [48] develop SDM for inverting logging data in anisotropic formations. In addition, SDM is applied in electrical-source airborne transient electromagnetic (GREATEM) data inversion [49] and unexploded ordnance detection [50].

5.3.2   *Physics embedded deep neural network*

In this part, we first introduce a physics constrained deep neural network (DNN) developed by Jin Y *et al.* for solving LWD problems [51]. The architecture of the NN is inspired by the auto-encoder that widely applied in unsupervised learning. A vanilla auto-encoder comprises an encoder and a decoder, the parameters in which are all trainable. In [51], the encoder part maps data to model, which is considered as the inverse function of forward modeling F^{-1}. The decoder part, however, is explicitly represented by numerical simulation $F : \mathbf{m} \mapsto \mathbf{d}$.

The loss function for training consists of model misfit and data misfit, i.e.,

$$L = \alpha L_{\text{model}} + \beta L_{\text{data}}, \tag{5.10}$$

with

$$L_{\text{model}} = \|\mathbf{m} - \mathbf{m}_T\|^2, \tag{5.11}$$

$$L_{\text{data}} = \|F(\mathbf{m}) - \mathbf{d}_T\|^2, \tag{5.12}$$

where $\mathbf{m}_T$ and $\mathbf{d}_T$ is the training model and the corresponding data, respectively. The training of the NN requires backpropagating the gradients of the data misfit. The first-order gradient of data misfit can be computed by

$$\nabla_{\text{dl}}\Theta = \frac{\partial L}{\partial \Theta} = \frac{\partial L}{\partial \mathbf{d}} \frac{\partial \mathbf{d}}{\partial \mathbf{m}} \frac{\partial \mathbf{m}}{\partial \Theta} \tag{5.13}$$

with Θ representing the network parameters. The authors estimate the Jacobian $\partial \mathbf{d}/\partial \mathbf{m}$ via the finite difference method.

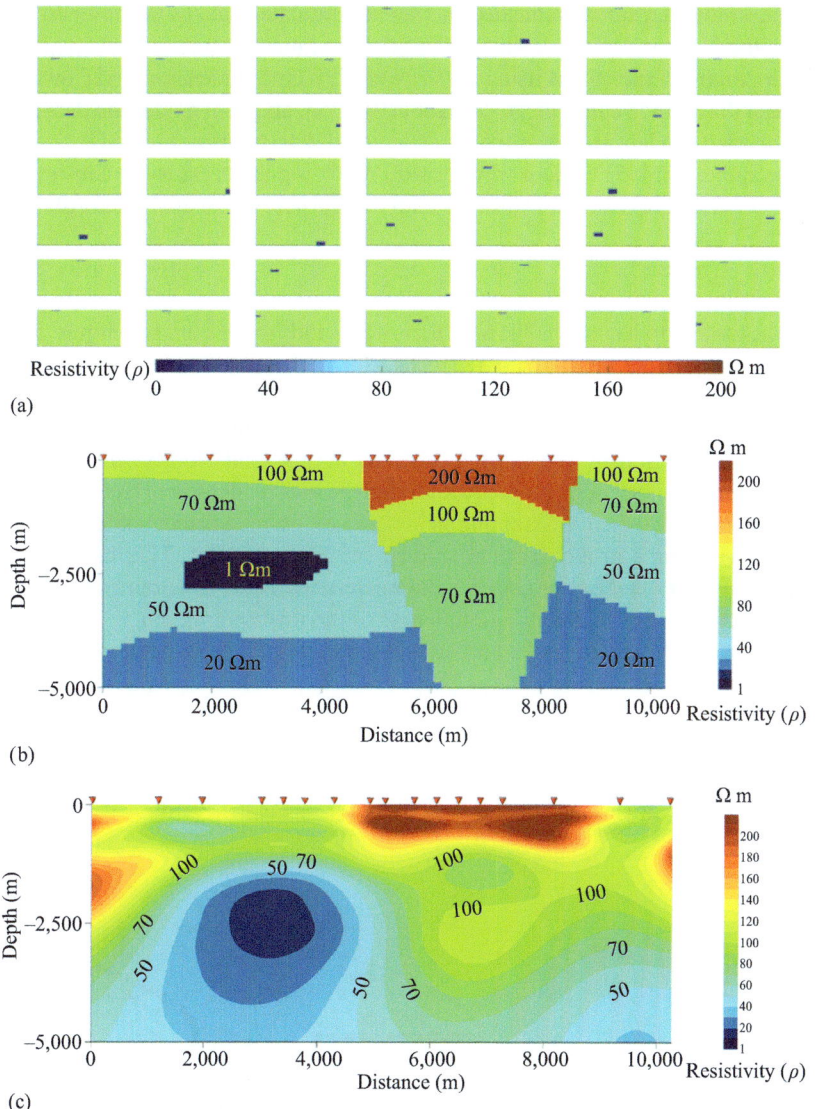

*Figure 5.8   MT inversion with SDM in [46]. (a) The resistivity models for training.
(b) The test model. (c) The reconstructed model.*

The constraint in the data domain makes the training less prone to overfitting, hence the NN can achieve higher accuracy than training with only model misfit. However, the training process is more expensive since the Jacobian needs to be backpropagated.

Since the data residual indicates whether the simulated data fit the observed data, utilizing the feedback of data residual can generate more accurate models. The same research group rewrites SDM using an RNN in [52]. One-dimensional MT inversion shows that the method can achieve fast reconstruction, and both model and data misfit of the inverted models are lower than conventional deterministic methods.

The authors of [53] explain why forward modeling is necessary to constrain the training process. For example, consider a forward process with analytical solutions:

$$m \triangleq F(p) = p^2, \tag{5.14}$$

where the inversion has two solutions: $p = +\sqrt{m}$ and $p = -\sqrt{m}$, see Figure 5.9(a). The authors construct the training set so that for each sample $(m, \sqrt{m})$, there exists another one $(m, -\sqrt{m})$. For each pair of the samples in this form, the point that simultaneously minimizes the distance between both solutions is zero. When the training is supervised by the label $\pm\sqrt{m}$, the output tends to be zero, shown in Figure 5.9(b). The non-uniqueness in the training dataset renders incorrect predictions. On the other hand, when the training is supervised by p^2, correct solutions can be predicted with some proper regularizations, see Figure 5.9(c). This mathematical example illustrates that optimizing NN parameters with data constraint can alleviate the ambiguity brought by the non-uniqueness of the inverse problem.

Computing the Jacobian of a numerical solver and backpropagating it in training is expensive. Therefore, the authors propose to approximate the forward function using a DNN Θ_F. The training is to find the optimal forward solver Θ_F^* and inversion solver Θ_I^* such that the data misfit is minimal:

$$(\Theta_F^*, \Theta_I^*) = \arg \min_{\Theta_F, \Theta_I} \left\{ \|(\Theta_F \circ \Theta_I)(\mathbf{d}) - \mathbf{d}\|^2 + \|\Theta_F(\mathbf{m}) - \mathbf{d}\|^2 \right\}, \tag{5.15}$$

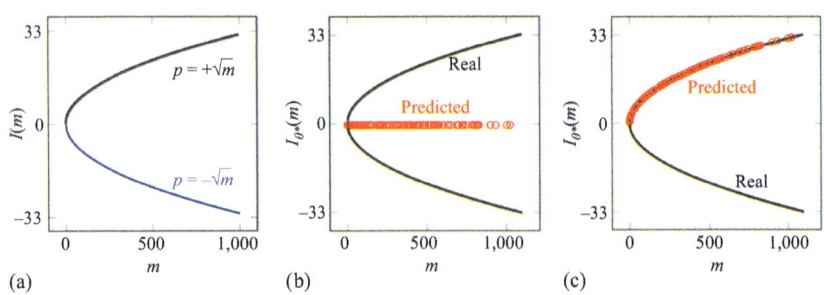

Figure 5.9 *Incorporating forward modeling in DNN can reduce the non-uniqueness of the inverse problem [53]. (a) Inverse operator with two branches. (b) Predictions of the DNN using the loss function based on the model misfit. (c) Predictions of the DNN using the loss function based on the data misfit.*

where $\Theta_F \circ \Theta_I$ constitutes an auto-encoder. It is possible to decompose the above training into two steps: (1) training the forward solver Θ_F^* and (2) training the inverse operator Θ_I^*, which can be written as

$$\Theta_F^* = \arg\min_{\Theta_F} \|\Theta_F(\mathbf{m}) - \mathbf{d}\|^2, \tag{5.16}$$

$$\Theta_I^* = \arg\min_{\Theta_I} \|(\Theta_F \circ \Theta_I)(\mathbf{d}) - \mathbf{d}\|^2. \tag{5.17}$$

With such scheme, the authors improve the accuracy of the DNN-based LWD inversion presented in [26].

Based on the above insights, a 2.5D DL scheme for LWD data inversion is achieved in [54]. The inversion workflow contains three DL modules: one fault detection module and two inversion modules trained with or without the existence of fault. The fault detection module is first applied to determine the presence or absence of fault planes. Then one of the inversion modules is selected to predict the parameters describing the geometry and resistivity of the subsurface model. This work provides a new view in data-driven inversion: the heavy task of reconstructing complex models can be decomposed into several parts that are affordable to train in practice.

Deep reinforcement learning is first applied to EM inversion in [55], where a trained agent predicts the actions of updating resistivity according to the fitness of data. In the following, we first introduce deep Q-learning.

The "Q" in Q-learning stands for quality. Q-value refers to the expected total reward in the future. The Q-value at state s, taking an action a, is the immediate reward $R(s, a)$ plus the highest Q-value possible from the next state s':

$$Q(s,a) = R(s,a) + \gamma \max_a Q'(s',a) \tag{5.18}$$

where γ is the discount rate. The optimal Q-values are found by trial-and-error according to the following update scheme [56]:

$$\text{New } Q(s,a) \leftarrow Q(s,a) + \alpha \left[R(s,a) + \gamma \max_a Q'(s',a) - Q(s,a) \right], \tag{5.19}$$

where α is the learning rate. Once a Q-value table is established, the optimal strategy can be obtained by selecting a series of actions that have the highest Q-values.

In conventional approach, establishing a Q-value table is computationally intensive. To reduce the computing burden, DNNs are applied to replace the process. The neural network is called deep Q network (DQN), which is validated in [57]. The input of a DQN is the state, while the output is the Q-value. The authors of [55] analogize MT inversion to the deep Q-learning: (1) the set of resistivity and the index of layers can be considered as states. (2) Adding or subtracting the resistivity can be considered as actions. (3) The environment is the fixed setting of layers. In the training, the agent traverses from the first to the last layer, disturbing the resistivity of each layer according to the policy that the maximal Q-value is selected at a probability of 90%. The perturbation of each layer is called an update. In every update, the agent will receive a positive reward when the data misfit decreases, otherwise be punished with a negative reward. It will circularly loop over all layers until the data fitness reaches a predefined level.

The authors perform 300 times inversion, during which the Q-values keep updating. In each time of inversion, the initial model is the homogeneous background. As the Q-value increases, the training loss decreases, and the agent updates models more efficiently. After the 300 times of inversion are completed, statistical analysis is performed over all inverted models, and the resistivity with the highest probability in each layer is chosen as the final model. This method provides a new perspective in evaluating uncertainty.

5.4   Learning-assisted physics-driven approach

Although physics can be incorporated in the training process, the learned inverse operator is unavoidably biased towards the training samples. This is a good thing when prior knowledge is reliable and can be represented by sufficient training data. However, such ideal scenario seldom happens in industrial applications. Researchers have found that data-driven approaches display larger data misfits than conventional physics-driven methods [46,51,58]. Based on these concerns, another viewpoint is that DL can help the reconstruction escape local minima, but the accuracy of inversion should be controlled by physics-driven approaches.

In deterministic inversion, the cost function can be written as

$$L(\mathbf{m}) = \|\mathbf{W}_d \left(\mathbf{d}_{\text{obs}} - F(\mathbf{m})\right)\|^2 + \lambda \|\mathbf{W}_m \left(\mathbf{m} - \mathbf{m}_{\text{ref}}\right)\|^2, \tag{5.20}$$

where the first term measures the data misfit, the second term is the regularization weighted by coefficient λ, $\mathbf{W}_d$, and $\mathbf{W}_m$ are weighting matrices, and $\mathbf{m}_{\text{ref}}$ is the prior model. Conventionally, the prior model is determined based on the estimation of the survey field. Gradient descent methods can be used to minimize the cost function, but the solution is usually not globally optimal.

On the other hand, the DL-based inversion predicts the model $\mathbf{m}_p$ from the observed data $\mathbf{d}_{\text{obs}}$ using a trained neural network Θ, i.e., $\mathbf{m}_p = \Theta(\mathbf{d}_{\text{obs}})$. The prediction does not rely on initial models and can generate the optimal solution when the network is correctly trained. However, the statistics of the training data and real data are hardly the same, which limits the generalizability of the DL approach.

Colombo *et al.* propose physics-driven deep learning inversion (PhyDLI), which benefits from the pseudo-stochastic sampling of the model and data spaces through training, the nonlinear mapping represented by DNNs, and the powerful local optimization capability of gradient based L2 method [58]. Ideally, the DNN may help the deterministic inversion avoid local minima, and the deterministic inversion in turn expands the distribution of the model space in training.

The PhyDLI links the conventional L2 and DL inversion in the model space. Given the observed data, the trained NN first predicts reference models for L2 inversion. Note that the statistics of the training models is not required to be the same as the real case. With the reference model, L2 inversion will be performed for the field dataset. The resultant models and the corresponding simulated data is then added to the training dataset for further retraining the NN. After several iterations, the two

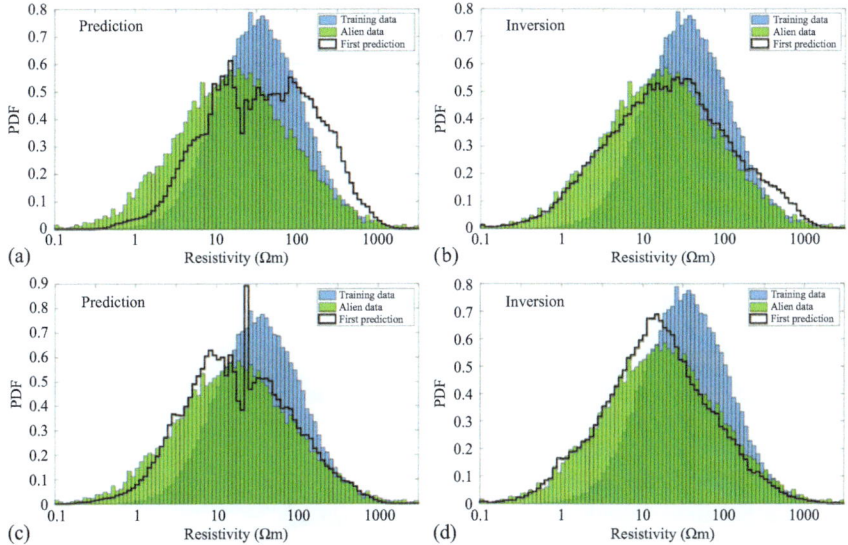

Figure 5.10   Different stages of the PhyDLI with the training data [58]. (a) Distribution of the initial prediction. (b) Distribution of the first inversion. (c) Distribution of the second prediction after retraining. (d) Final distribution.

procedures tend to output a common model that satisfies both prior knowledge and data fitness.

The method is demonstrated by 1D helicopter TEM inversion. Two sets of resistivity models are generated with different Gaussian distributions. The set called "Training" is used for training-validation-test, while the set called "Alien" represents the field model with unknown distribution. The distributions of resistivity are shown as blue and green histograms in Figure 5.10. Within two loops, the distribution of the inverted resistivity gradually moves from the Training set to the Alien set, showing the effectiveness of the method. It has also been adopted for 2D MT inversion [59].

Another work surrogates the computation of forward problem and Jacobian matrix with DNN solvers [60]. The inversion is performed in the deterministic framework, but the Jacobian matrix is predicted using a DNN call "jNet." The input of the DNN consists of the resistivity vector $\mathbf{m}$, the offset of transmitter and receiver coils, and the index t in $\mathbf{m}$ on which the partial derivative will be calculated. The output is the partial derivative $\partial \mathbf{d}/\partial m_t$.

During inversion, the EM response can be either computed by a surrogate network "fNet," or by a fullwave numerical solver "fFull." The authors investigate four schemes with different hybridization: fFull–jFull (jFull is numerically computing the Jacobian), fFull–jNet/jFull, fFull–jNet, and fNet–jNet, where fFull–jNet/jFull means changing to fFull–jFull after fFull–jNet reaches a certain misfit. The performance is shown in Table 5.1. The fNet–jNet computes the fastest but with a higher misfit. The fFull–jNet finds a good balance between speed and accuracy. After the jNet is

Table 5.1   Comparison of the four schemes in [60].

Inversion Method	# of iterations	Total inversion time (minutes)	Total misfit ϕ (m)
fFull-jFull	6	357	0.2284
fFull-jNet/jFull	7	84	0.2289
fFull-jNet	6	27	0.2304
fNet-jNet	6	9	0.2322

The dataset is a 3D volume with 2,822 soundings

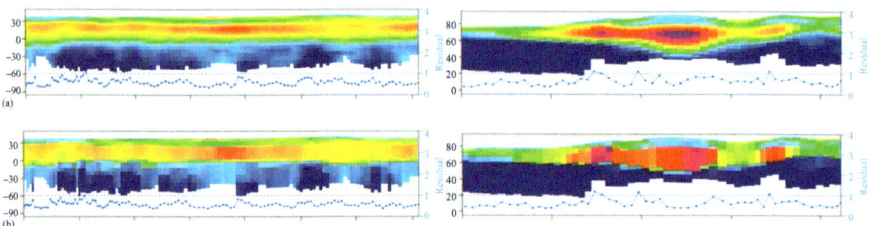

Figure 5.11   Field data test with the predicted Jacobian [60]. (a) Inversion with the smooth constraint. (b) Inversion with sharp boundary constraint.

trained, it can be flexibly used to inverted models with smooth or sharp boundary constraint. The field tests in Figure 5.11 validate that the predicted Jacobian can be readily employed in practice.

5.5   Deep learning in seismic data inversion

Aside from EM methods, seismic exploration is another effective method to identify underground structures. Seismic waves can be excited by artificial sources, such as exploding dynamite or air guns, and they propagate at different speeds in different lithologies. For example, the velocity in the soil, clay and limestone is 200–800 m/s, 1,800–2,400 m/s, and 3,200–5,500 m/s, respectively. The seismic signals detected by surface stations can be inverted to the distribution of subsurface velocity.

There are similarities between EM and seismic methods. Both inverse problems are highly nonlinear and ill-posed, and the computations are time and memory consuming. Many DL techniques have been applied to address these challenges in seismic inversion. In this section, we will introduce some works that may be instructive for EM inversion.

5.5.1   Inversion with unsupervised RNN

Sun J *et al.* set up the forward modeling of the acoustic wave propagation in an RNN architecture [61] and conclude that training such a network and updating its weights is equivalent to gradient-based seismic full waveform inversion (FWI). Specifically, the time domain wave equation in 2D acoustic media with constant density is

$$\nabla^2 u(\mathbf{r}, t) = \frac{1}{v^2(\mathbf{r})} \frac{\partial^2 u(\mathbf{r}, t)}{\partial t^2} + s(\mathbf{r}, t)\delta(\mathbf{r} - \mathbf{r_s}) \qquad (5.21)$$

where ∇^2 is the spatial Laplacian operator, u is the pressure or displacement, v is the velocity, s is the source function, $\mathbf{r}$ is the position, $\mathbf{r_s}$ is the source location and t is the time coordinate. Using the second-order finite difference in time, the wavefield at time $t + \Delta t$ can be represented by

$$u(\mathbf{r}, t + \Delta t) = v^2(\mathbf{r})\Delta t^2 \left[\nabla^2 u(\mathbf{r}, t) - s(\mathbf{r}, t)\delta(\mathbf{r} - \mathbf{r_s})\right] + 2u(\mathbf{r}, t) - u(\mathbf{r}, t - \Delta t).$$

$$(5.22)$$

Therefore, the wave propagation can be recurrently modeled given the source s and the wavefield at two previous time steps. The authors unfold this process on an RNN, the recurrent unit of which is shown in Figure 5.12.

Both forward modeling and inversion can be implemented with the RNN. In the forward problem, the wavefield is obtained by forward propagating the source s (in black) and velocity v (in purple) without training. In the inverse problem, the

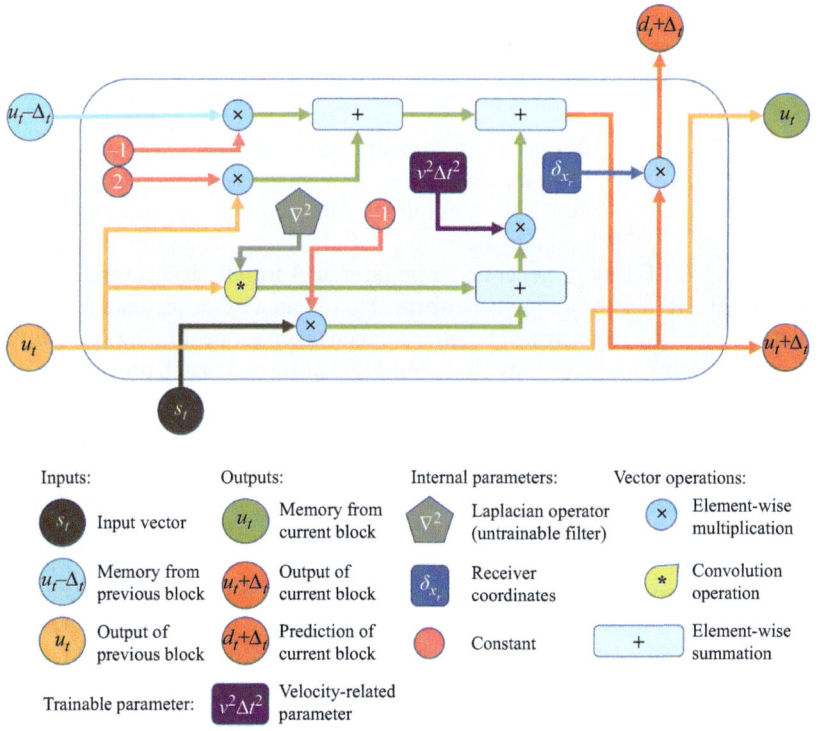

Figure 5.12   The recurrent unit of the waveform RNN in [61]

velocity is set as trainable parameters, and the loss function for training is the data misfit. With the unsupervised approach, the velocity is recovered to honor seismic data. Results show that training with adaptive moment (Adam) optimizer achieves faster convergence and lower data misfit than the nonlinear conjugate gradient (CG) and limited-memory Broyden–Fletcher–Goldfarb–Shanno (L-BFGS) optimizations in conventional FWI.

EM fields can also be simulated by finite-difference time-domain (FDTD) method. Efforts have been done in [62,63], where EM scattering and inverse scattering problems are solved by RNN. We have not found related research in subsurface EM exploration. Rigorously speaking, this method is not a data-driven approach, but rather a fast computation technique within the DL framework. The inversion can be as reliable as conventional methods. The EM community will also benefit from the computing resources powered by GPUs, advanced optimization algorithms and user-friendly software packages.

5.5.2 Low-frequency data prediction

Inversion with only high-frequency (HF) data is easy to be trapped into local minima (cycle-skipping). Low-frequency (LF) data play an important role in stabilizing the process. However, low-frequency components less than 5 Hz are of poor quality in most collected seismic dataset. To address this issue, efforts in extrapolation of LF data were made in [64,65]. Predicting LF data from HF data was also performed in [66–68]. In the following, we introduce the progressive transfer learning strategy [67] for LF data prediction.

The FWI with the LF data prediction strategy consists of two parts: DL and physics-based inversion. First, an arbitrary velocity model is selected to generate LF and HF data, which are taken as the training set. After training, the measured HF data are input into the DNN to predict inaccurate initial LF data. Then FWI is performed on the predicted LF data to generate an initial model for HF data inversion. Based on this model, HF data inversion is performed to obtain a more accurate model. The simulated data of the reconstructed model are then taken as the training set to further refine the DNN. In each progressive iteration, HF data inversion will provide the DNN with richer information about the subsurface. Furthermore, the DL module predicts LF data more accurately as the training data update, so that the FWI can generate more and more reliable models. Through this strategy, the DL module and the FWI module interact and complement with each other to help FWI escape from local minima. In experiments, it shows that the scheme can achieve the same level of accuracy as wide band FWI , much better than inverting only with HF data.

The same strategy is exploited in the time domain FWI [68]. The authors use a 2D CNN to learn the patterns in shot gathers. The predicted LF data can progressively converge to the ground truth. With this scheme, the authors reconstruct the Marmousi model [69] and extend it to invert weak elastic data with acoustic FWI.

Compared with the sequential strategy that first predicting LF data and then performing FWI, we tentatively think that the iterative method is more adaptable in practice. Signals processed by DL cannot well honor wave propagation, hence may lead to implausible inversion results. The proposed strategy integrates signal processing and data inversion together and improves the performance of both as a whole system, which is illuminating for EM data inversion.

5.5.3 Physically realistic dataset construction

The training dataset plays an important role in data-driven inversion. Collecting real geological models and the corresponding seismic data can be quite expensive. On the other hand, synthetic models and data are usually too simple to represent real cases. To improve the generalizability of data-driven inversion, Feng *et al.* propose a style transform method for creating physically realistic subsurface velocity models from natural images [70].

Generating realistic geologic images is similar to generating realistic images that contain certain art styles in computer vision, if we consider the velocity model as an image and geologic features as the styles of the image. Using image style transfer one can find a composite image y whose style is similar to a style image y_s and its content is similar to a content image y_c.

The styles of an image can be measured by Gram matrix $G_j(x)_{mn}$ [71]

$$G_j(x)_{mn} = \sum_p \phi_j(x)_{mp}\phi_j(x)_{np}, \tag{5.23}$$

where $\phi_j(x)$ is the activations at the jth layer of the network ϕ for input x, m and n are feature maps in layer j. The styles are similar if the style loss is low:

$$L_{style} = \sum_{j \in S} \frac{1}{U_j} \left\| G_j(y) - G_j(y_s) \right\|^2, \tag{5.24}$$

where S is a set of NN layers used in style reconstruction, and U_j is the number of units in layer j.

The contents of two images are similar if their high-level features extracted by a DNN are close. The content loss is defined as

$$L_{content} = \sum_{j \in C} \frac{1}{U_j} \left\| \phi_j(y) - \phi_j(y_c) \right\|^2 \tag{5.25}$$

where C is a set of NN layers used in content reconstruction.

The illustration of the style-transform network is shown in Figure 5.13. The image transform network (bottom) is an auto-encoder trained to transform natural images to velocity models. The feature models are extracted from a pre-trained VGG-16 network (top) [20]. The style loss is defined as the Gram matrix difference between the features of natural images and geology models in the top four layers, while the content loss is the mean squared error between the features of natural images and geology models in the second latent layer. The total loss is defined as

$$L_{trans} = \alpha_{style} L_{style} + \alpha_{content} L_{content} \tag{5.26}$$

where α_{style} and $\alpha_{content}$ are the coefficients for style and content reconstruction. The effect of the style coefficient is shown in Figure 5.14.

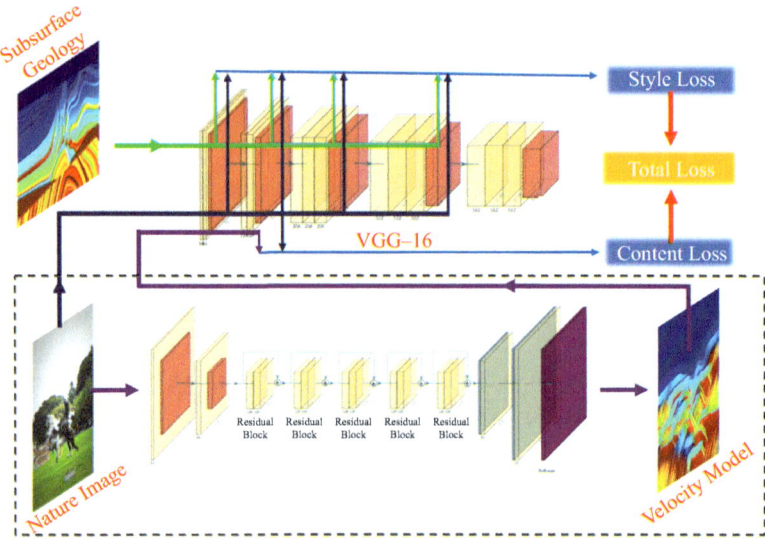

Figure 5.13   *A schematic illustration of the style-transform network [70]*

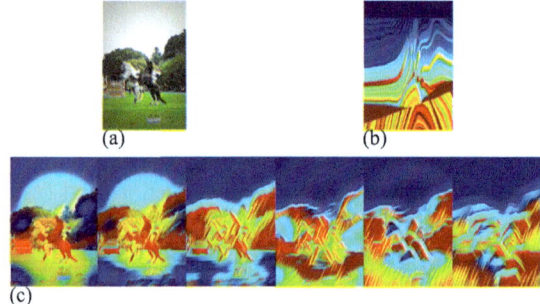

Figure 5.14   *The effect of the style coefficient [70]. (a) Content image. (b) Style image. (c) The output images with the style weight increasing from left to right.*

With the proposed method, the authors use 67,000 natural images as the content images and the Marmousi model as the style image to construct physically realistic subsurface velocity models. Training on this dataset, data-driven FWI shows good generalizability. The lack of reliable training set is also a problem in EM inversion. In the future, we may apply the method to generate more reasonable resistivity models for training.

5.5.4   *Learning the optimization*

Learning optimization algorithms is usually referred to "metalearning," or "learning to learn," in the ML community. In [72], the authors make LSTM network take the gradient of the misfit function as the input, and output the model update, which is called ML-descent. The purpose is to allow the machine to learn how to better update models in FWI, thereby to accelerate the convergence.

With LSTM network, the model update Δx_{t+1} is predicted from the gradient g_t, a state variable c_{t+1} and a hidden state variable h_{t+1}:

$$(\Delta x_t, c_{t+1}, h_{t+1}) = \Theta_{LSTM}(g_t, c_t, h_t). \tag{5.27}$$

To reduce the complexity of the LSTM, the model is updated coordinate-wise. Different coordinates (pixels) share the same RNN parameters for updating, but the unique behavior is controlled by separate hidden stable variables.

In the training, it is computationally expensive if one uses the gradient of the loss function in FWI as the input. Considering that any nonlinear inverse problems can be locally linearized, the authors use a quadratic loss function f as

$$f = \|Wx - b\|^2, \tag{5.28}$$

where the matrix W and vector b are randomly selected from a Gaussian distribution. The training loss function is defined as

$$L = \sum_t f(x_t). \tag{5.29}$$

For instance, each function f is optimized for 100 iterations, and every 20 iterations, the summation of the misfit will be evaluated to update the LSTM parameters. The authors run 100 epochs with each epoch containing 2,000 quadratic functions to train the network. It is verified that the trained optimizer outperforms other state-of-the-art algorithms.

To reduce the number of unknowns in FWI, the authors further use variational auto-encoder (VAE) [73] to compress model parameters. Consequently, a velocity model of size 40×40 is represented by 9 variables in the latent layer of VAE. Compared with the classic auto-encoder, VAE has a more continuous latent space and is more stable when computing its Fréchet derivatives.

Supposing that the unknown (latent variables) to be recovered is z_m, the gradient of the loss function w.r.t. z_m is computed by

$$\frac{\partial L}{\partial z_m} = \frac{\partial L}{\partial m} \frac{\partial m}{\partial z_m} \tag{5.30}$$

where $\partial m / \partial z_m$ can be obtained using the automatic differentiation. The use of VAE largely reduces the computational burden and makes the ML-descent applicable in FWI. The trained optimizer yields the best result compared with RMSprop [74], Adam, stochastic gradient descent (SGD), and Nesterov accelerated-gradient (NAG) methods.

This method and SDM have some similarities in both learning model update directions. They provide us another perspective of applying DL in inversion: learning the optimization process rather than learning an end-to-end mapping. By monitoring the data misfit during inversion, they can be more reliable than purely data-driven approaches, but at the expense of the computational cost.

5.5.5 Deep learning constrained traveltime tomography

Traveltime tomography is another method that determines the subsurface velocity. It only inverts the arrival times of seismic waves, and the most commonly used is the first arrival time. It provides a smooth velocity model that can be further used as the initial model of FWI. Traveltime tomography is also nonlinear and ill-posed, hence prior information should be incorporated to stabilize the inversion.

In [75], the authors propose to constrain the traveltime tomography with the prior model generated by GAN. The inversion is still in the deterministic framework, but the inverted model is made to have structural similarities with the prior model.

The cost function for inversion is written as

$$L(\mathbf{m}) = \|\mathbf{d}_{\mathrm{obs}} - F(\mathbf{m})\|^2 + \alpha \|\nabla^2 \mathbf{m}\|^2 + \|u(\mathbf{m}, \mathbf{m}_{\mathrm{GAN}})\|^2, \tag{5.31}$$

where u is the cross-gradient function [76] that adds structural similarity constraint and $\mathbf{m}_{\mathrm{GAN}}$ is the output of the GAN. The cross-gradient regularization is a weak constraint imposed on the gradient of the image rather than its values, which has been widely adopted for multi-physics joint inversion:

$$u(\mathbf{m}_T, \mathbf{m}_G) = \nabla \mathbf{m}_T(x, y) \times \nabla \mathbf{m}_G(x, y), \tag{5.32}$$

In the k-th iteration,

$$\mathbf{m}_{\mathrm{GAN}}^k = \Theta_{\mathrm{GAN}}(\mathbf{m}^{k-1}). \tag{5.33}$$

The inverted model will obtain clear boundaries with the constraint.

Although both the input and output of the GAN are velocity models, this work is instructive for DL-based joint inversion. For instance, when the training set contains various geophysical attributes, one can utilize a DNN to learn implicit relationships among them. In the following, we will introduce several works on DL-based joint inversion.

5.6 Deep learning in multi-physics joint inversion

A subsurface model has a variety of geophysical properties. In order to characterize underground structures, various exploration methods need to be used to estimate these properties. However, these methods have different resolutions and sensitivities, and the inverted models from different physical data in the same survey field may be different. This brings challenges in geophysical data interpretation. Since the EM and seismic data contain complementary information, jointly inverting the EM and seismic data can highlight abnormalities through mutually constraining the model space.

In [77], the authors perform the joint inversion for audio-magnetotelluric (AMT) and seismic first arrival traveltime data with deep learning. The resistivity–velocity correlations and the structural similarity are learned by a DNN. Assuming that the forward problem of AMT and seismic traveltime is $\mathbf{d} = F(\boldsymbol{\rho})$ and $\mathbf{t} = G(\mathbf{s})$, where $\boldsymbol{\rho}$ and $\mathbf{s}$ is the resistivity and slowness (reciprocal of velocity), respectively, the objective function of the joint inversion is

$$\begin{aligned} L_{Joint}(\boldsymbol{\rho}, \mathbf{s}) = {}& \alpha_D^{\rho} \|F(\boldsymbol{\rho}) - \mathbf{d}_{\mathrm{obs}}\|^2 + \lambda_R^{\rho} \|\boldsymbol{\rho} - \Theta_{s2\rho}(\mathbf{s})\|^2 + \lambda_{\nabla}^{\rho} L_{\nabla}(\boldsymbol{\rho}) \\ & + \alpha_D^s \|G(\mathbf{s}) - \mathbf{t}_{\mathrm{obs}}\|^2 + \lambda_R^s \|\mathbf{s} - \Theta_{\rho 2s}(\boldsymbol{\rho})\|^2 + \lambda_{\nabla}^s L_{\nabla}(\mathbf{s}), \end{aligned} \tag{5.34}$$

where α_D^ρ and α_D^s are the normalized coefficients, $\mathbf{d_{obs}}$ and $\mathbf{t_{obs}}$ are the observed AMT and traveltime data, L_∇ is the regularization function of the smoothness, $\Theta_{s2\rho}$ or $\Theta_{\rho2s}$ is the DNN that projects slowness to resistivity or resistivity to slowness, respectively, λ_R^ρ, λ_∇^ρ, λ_R^s, and λ_∇^s are regularization coefficients.

Using the iterative method to alternatingly update ρ and $\mathbf{s}$, we have

$$L_{AMT}(\boldsymbol{\rho}_k) = \alpha_D^\rho \|F(\boldsymbol{\rho}_k) - \mathbf{d_{obs}}\|^2 + \lambda_R^\rho \|\boldsymbol{\rho}_k - \Theta_{s2\rho}(\mathbf{s}_{k-1})\|^2 + \lambda_\nabla^\rho L_\nabla(\boldsymbol{\rho}_k), \quad (5.35)$$

$$L_{TT}(\mathbf{s}_k) = \alpha_D^s \|G(\mathbf{s}_k) - \mathbf{t_{obs}}\|^2 + \lambda_R^s \|\mathbf{s}_k - \Theta_{\rho2s}(\boldsymbol{\rho}_{k-1})\|^2 + \lambda_\nabla^s L_\nabla(\mathbf{s}_k), \quad (5.36)$$

The workflow of the joint inversion algorithm is shown in Figure 5.15.

Two DNNs are trained to learn the mappings from velocity to resistivity and vice versa. The training models are generated according to the prior knowledge of resistivity–velocity relationship. For example, the log data from two wells can be

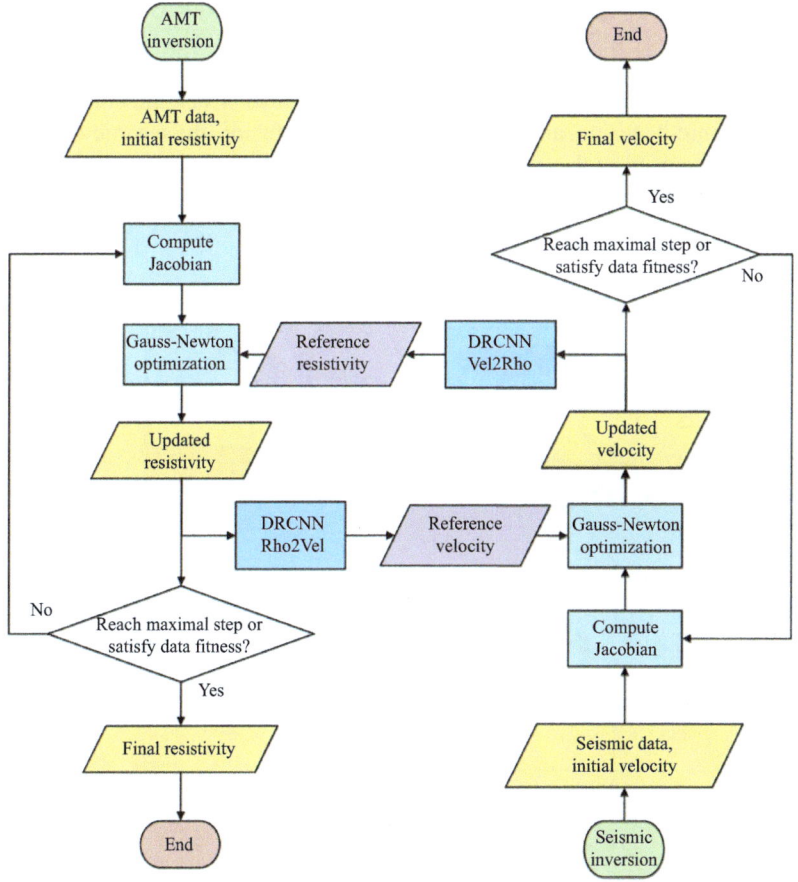

*Figure 5.15   Workflow of the joint inversion with DL constraint [77]. The DRCNN
is the abbreviation of deep residual convolutional neural networks.*

used to derive resistivity and velocity, and their relationships can be grouped based on their depth, see Figure 5.16. The DNNs simultaneously learn the mappings between resistivity and velocity in different depths by designing the training set according to Figure 5.16(b). On the contrary, conventional regression methods are difficult to describe this relationship with mathematical equations. In Figure 5.17, we show the true resistivity and velocity models and the recovered models with different inversion schemes. The DL constrained joint inversion produces the most reasonable models, and [77] shows that the prior joint distribution of resistivity and velocity can be well preserved in the inverted models.

Based on the similar idea, DL-based attribute fusion schemes are applied to joint inversion [78,79]. Instead of training two separate networks for attribute mapping, a single DNN is trained to jointly interpret resistivity and velocity models. The inputs are separately inverted resistivity and velocity models, while the labels are true resistivity and velocity models. Given inverted models, the DNN can output reference models taking into account the correlations between resistivity and velocity.

A more aggressive data-driven approach for seismic and CSEM joint inversion is performed in [80]. The network is trained to predict the outlines of salts from seismic gather shots and CSEM data. The authors investigate three NN architectures, namely early fusion, middle fusion, and late fusion. In the early fusion, the seismic and CSEM data are concatenated as the input. In the middle fusion, the features extracted from seismic and CSEM data are concatenated in different levels. In the late fusion, the

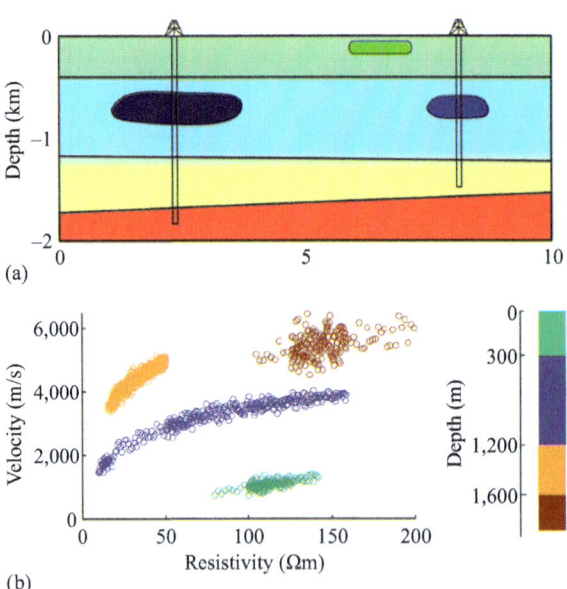

Figure 5.16   An example shown in [77]. (a) Profile of the geology model and the two wells. (b) Resistivity–velocity distribution derived from well logs.

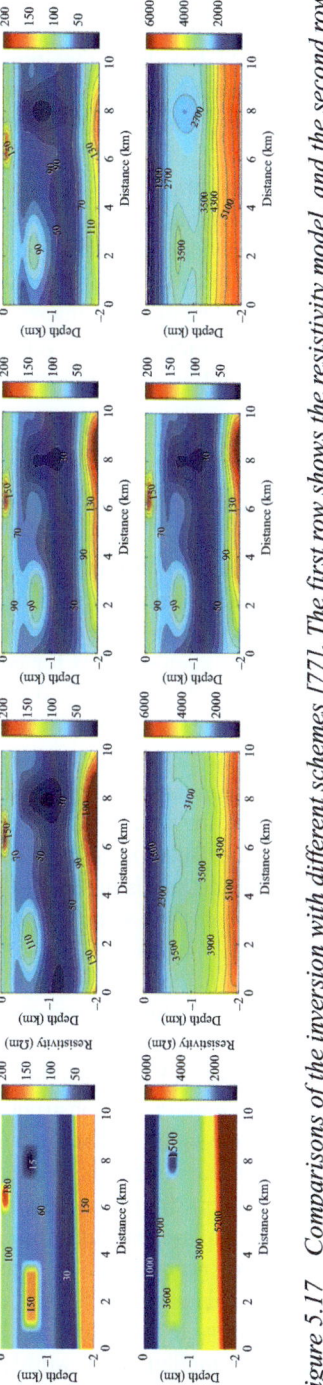

Figure 5.17   Comparisons of the inversion with different schemes [77]. The first row shows the resistivity model, and the second row shows the velocity model.

extracted features of seismic and CSEM data are concatenated near the end of the NN. The authors find that the middle fusion architecture provides the best results.

The predictions from unseen CSEM data shows that jointly using the seismic and EM data can improve the imaging accuracy. Top boundaries can be delineated with the information contained in seismic data, and lower boundaries are well bounded by EM data.

Compared with the DL constrained physics-driven approach, this method does not require adjusting the weights of seismic and EM data. The imaging process is governed by training data, thus less dependent on subjective judgements. In addition, the online inversion is much faster. However, it is still an open question whether the balance of multi-physics data should be made by interpreters or neural networks. Concerns will be raised when the sensitivities or noise levels of two methods are quite different. DL-based joint inversion is still at an early stage and more investigations are needed in future work.

5.7   Construction of the training dataset

The training dataset plays a key role in data-driven methods. In computer science, there are a number of public datasets that elaborately collected or designed for ML tasks. Researchers can develop new algorithms, validate the performance of the algorithm based on these datasets. Unfortunately, such datasets for geophysical inversion are still quite rare mainly due to commercial copyrights, diverse scenarios and large uncertainties of underground structures. Therefore, constructing a proper dataset is also a concern when developing DL-based inversion algorithms.

The most straightforward way is randomly sampling the model space. This is applicable in 1D inversion. For instance, the thickness of layers and the corresponding resistivity can be sampled within a predefined range [23]. To generate geophysically reasonable models, a pseudo-random method is applied [37]: the resistivity of the next layer has a probability (40%) of being the same as the current layer; otherwise it takes a random value within a predefined range. Furthermore, the labels can be the inverted models from conventional methods [60], which ensures that the training process is only supervised by resolvable resistivity structures and therefore eliminates the non-uniqueness of the inversion.

Random sampling is inapplicable for high dimensional inversion due to the curse of dimensionality. Prior knowledge is required to construct a reasonable dataset. In reservoir monitoring [30], the reservoir saturation at a given time can be regularly updated with data from production and new wells, hence various fluid flow realizations through time can be performed by a dynamic fluid flow simulator. In [28,38,80,81], the scenarios are inverting abnormalities in homogeneous or layered medium. Especially, Bang *et al.* [38] generate training models according to the sensitivity of data, and evaluate the distribution of training and field data using t-distributed stochastic neighbor embedding (t-SNE) algorithm [82]. Despite these datasets have validated the algorithms, they are still simple to be widely applied in real-world applications. Techniques in natural image generation may provide new ideas of constructing datasets [70].

The requirement for training dataset in the physics embedded data-driven and learning assisted physics-driven approaches are not as strict as in purely data-driven approaches. For instance, after learning the descent directions of optimization through simple models [46,83] or quadratic functions [72], complex models can be accurately recovered. In [58,67,68,77], the probability distributions of training models are not necessarily the same as test cases, since the DL process only provides assistance to the deterministic inversion.

When constructing the training dataset, factors such as the prior knowledge, the extent to which physical laws are incorporated, the completeness of training models and the data sensitivity should all be evaluated. Besides, the time of generating a dataset is a critical issue. We also look forward to more open-source datasets in this community.

5.8 Conclusions and outlooks

Geophysical EM inversion has followed the developments of DL closely. Advanced learning concepts and algorithms as well as NN structures have been exploited for subsurface imaging. The prior knowledge extracted from big data can largely improve the accuracy and speed of model reconstruction. Experience of joint interpretation can be fused into joint inversion through DL. Literatures also show that hybridizations of wave physics and DL can reduce the burdens in training and enhance the reliability of inversion.

Despite the promising results, limitations still exist in real-world applications. First, due to the diversity of underground structures and the diversity in the survey design, training a global surrogate for inversion is difficult. The current applications mainly focus on training local surrogates. There is still a long way to achieve the ideal status, i.e., training once and predicting all. The price of setting up a DL procedure should be carefully evaluated in practice. Second, the reliability of NNs will always be a concern before its interpretability is solved. Simply using NNs for regression may fail in predicting out-of-distribution data. This limits the industrial applications especially when the cost of decision is expensive, e.g., whether to drill a well. Third, datasets for training are not always available. A reliable geophysical model is not easy to obtain, and the imaging often suffers from non-unique explanations of different interpreters. Synthetic models and data, however, do not provide enough complexity comparable to real-world scenarios. Fourth, DL techniques provide new perspectives of multi-physics data integration for joint inversion, but it should be investigated more in the future.

The last few decades have witnessed several winters of artificial intelligence. However, it has never changed the way of our life and industry as much as it does today. Applications of DL in subsurface imaging require not only developments in hardwares and softwares but also the domain knowledge in geophysics and geology. Current studies have validated the feasibility. In the future, we may explore the Earth more deeply and accurately through the integration of data science, computer science and geoscience.

References

[1]   Nabighian MN. *Electromagnetic Methods in Applied Geophysics: Volume 1, Theory*. Houston, TX: Society of Exploration Geophysicists; 1988.

[2]   Raiche A. A pattern recognition approach to geophysical inversion using neural nets. *Geophysical Journal International*. 1991;105(3):629–648.

[3]   McCormack MD. Neural computing in geophysics. *The Leading Edge*. 1991;10(1):11–15.

[4]   Poulton MM, Sternberg BK, Glass CE. Location of subsurface targets in geophysical data using neural networks. *Geophysics*. 1992;57(12): 1534–1544.

[5]   Zhang Y, Paulson K. Magnetotelluric inversion using regularized Hopfield neural networks. *Geophysical Prospecting*. 1997;45(5):725–743.

[6]   Van der Baan M, Jutten C. Neural networks in geophysical applications. *Geophysics*. 2000;65(4):1032–1047.

[7]   Dramsch JS. 70 years of machine learning in geoscience in review. *Advances in Geophysics*. 2020;61:1.

[8]   El-Qady G, Ushijima K. Inversion of DC resistivity data using neural networks. *Geophysical Prospecting*. 2001;49(4):417–430.

[9]   XU HL, WU XP. 2-D Resistivity inversion using the neural network method. *Chinese Journal of Geophysics*. 2006;49(2):507–514.

[10] Ho TL. 3-D inversion of borehole-to-surface electrical data using a back-propagation neural network. *Journal of Applied Geophysics*. 2009;68(4):489–499.

[11] Neyamadpour A, Abdullah WW, Taib S. Inversion of quasi-3D DC resistivity imaging data using artificial neural networks. *Journal of Earth System Science*. 2010;119(1):27–40.

[12] Bermani E, Boni A, Kerhet A, *et al.* Kernels evaluation of SVM-based estimators for inverse scattering problems. *Progress in Electromagnetics Research*. 2005;53:167–188.

[13] Lizzi L, Viani F, Rocca P, *et al.* Three-dimensional real-time localization of subsurface objects: — from theory to experimental validation. In: *2009 IEEE International Geoscience and Remote Sensing Symposium*, vol. 2; 2009. pp. II–121–II–124.

[14] Silver D, Huang A, Maddison CJ, *et al.* Mastering the game of Go with deep neural networks and tree search. *Nature*. 2016;529(7587):484–489.

[15] LeCun Y, Bengio Y. Convolutional networks for images, speech, and time series. *The Handbook of Brain Theory and Neural Networks*. 1995;3361(10):1995.

[16] Deng J, Dong W, Socher R, *et al.* Imagenet: a large-scale hierarchical image database. In: *2009 IEEE Conference on Computer Vision and Pattern Recognition*. New York, NY: IEEE; 2009. pp. 248–255.

[17] Krizhevsky A, Sutskever I, Hinton GE. Imagenet classification with deep convolutional neural networks. *Advances in Neural Information Processing Systems*. 2012;25:1097–1105.

[18]   Badrinarayanan V, Kendall A, Cipolla R. Segnet: a deep convolutional encoder–decoder architecture for image segmentation. *IEEE Transactions on Pattern Analysis and Machine Intelligence.* 2017;39(12):2481–2495.

[19]   Xu L, Ren JS, Liu C, *et al.* Deep convolutional neural network for image deconvolution. *Advances in Neural Information Processing Systems.* 2014;27:1790–1798.

[20]   Simonyan K, Zisserman A. Very deep convolutional networks for large-scale image recognition. arXiv preprint arXiv:14091556. 2014.

[21]   Moghadas D. One-dimensional deep learning inversion of electromagnetic induction data using convolutional neural network. *Geophysical Journal International.* 2020;222(1):247–259.

[22]   Wu S, Huang Q, Zhao L. Convolutional neural network inversion of airborne transient electromagnetic data. *Geophysical Prospecting.* 2021;69(8-9): 1761–1772.

[23]   Liu Z, Chen H, Ren Z, *et al.* Deep learning audio magnetotellurics inversion using residual-based deep convolution neural network. *Journal of Applied Geophysics.* 2021;188:104309.

[24]   Noh K, Yoon D, Byun J. Imaging subsurface resistivity structure from airborne electromagnetic induction data using deep neural network. *Exploration Geophysics.* 2020;51(2):214–220.

[25]   Puzyrev V, Swidinsky A. Inversion of 1D frequency-and time-domain electromagnetic data with convolutional neural networks. *Computers & Geosciences.* 2021;149:104681.

[26]   Shahriari M, Pardo D, Picón A, *et al.* A deep learning approach to the inversion of borehole resistivity measurements. *Computational Geosciences.* 2020;24(3):971–994.

[27]   Oh S, Byun J. Bayesian uncertainty estimation for deep learning inversion of electromagnetic data. *IEEE Geoscience and Remote Sensing Letters.* 2021;19:8010705.

[28]   Puzyrev V. Deep learning electromagnetic inversion with convolutional neural networks. *Geophysical Journal International.* 2019;218(2):817–832.

[29]   Oh S, Noh K, Yoon D, *et al.* Salt delineation from electromagnetic data using convolutional neural networks. *IEEE Geoscience and Remote Sensing Letters.* 2019;16(4):519–523.

[30]   Colombo D, Li W, Sandoval-Curiel E, *et al.* Deep-learning electromagnetic monitoring coupled to fluid flow simulators. *Geophysics.* 2020;85(4): WA1–WA12.

[31]   Liu B, Guo Q, Li S, *et al.* Deep learning inversion of electrical resistivity data. *IEEE Transactions on Geoscience and Remote Sensing.* 2020;58(8): 5715–5728.

[32]   Lipton ZC, Berkowitz J, Elkan C. A critical review of recurrent neural networks for sequence learning. arXiv preprint arXiv:150600019. 2015.

[33]   Hochreiter S, Schmidhuber J. LSTM can solve hard long time lag problems. In *Advances in Neural Information Processing Systems.* Cambridge MA: MIT Press; 1997. pp. 473–479.

[34] Schuster M, Paliwal KK. Bidirectional recurrent neural networks. *IEEE Transactions on Signal Processing*. 1997;45(11):2673–2681.

[35] Hochreiter S, Schmidhuber J. Long short-term memory. *Neural Computation*. 1997;9(8):1735–1780.

[36] Chung J, Gulcehre C, Cho K, *et al.* Empirical evaluation of gated recurrent neural networks on sequence modeling. arXiv preprint arXiv:14123555. 2014.

[37] Li J, Liu Y, Yin C, *et al.* Fast imaging of time-domain airborne EM data using deep learning technology. *Geophysics*. 2020;85(5):E163–E170.

[38] Bang M, Oh S, Noh K, *et al.* Imaging subsurface orebodies with airborne electromagnetic data using a recurrent neural network. *Geophysics*. 2021;86(6):E407–E419.

[39] Wang J, Liu H, Jiang P, *et al.* GPRI2Net: a deep-neural-network-based ground penetrating radar data inversion and object identification framework for consecutive and long survey lines. *IEEE Transactions on Geoscience and Remote Sensing*. 2021;60:5106320.

[40] Liu B, Guo Q, Wang K, *et al.* Adaptive convolution neural networks for electrical resistivity inversion. *IEEE Sensors Journal*. 2020;21(2): 2055–2066.

[41] Jeon Y, Kim J. Active convolution: Learning the shape of convolution for image classification. In: *Proceedings of the IEEE Conference on Computer Vision and Pattern Recognition*; 2017. pp. 4201–4209.

[42] Hornik K, Stinchcombe M, White H. Multilayer feedforward networks are universal approximators. *Neural Networks*. 1989;2(5):359–366.

[43] Xiong X, De la Torre F. Supervised descent method and its applications to face alignment. In: *Proceedings of the IEEE Conference on Computer Vision and Pattern Recognition*; 2013. pp. 532–539.

[44] Guo R, Li M, Fang G, *et al.* Application of supervised descent method to transient electromagnetic data inversion. *Geophysics*. 2019;84(4):E225–E237.

[45] Guo R, Li M, Yang F, *et al.* First arrival traveltime tomography using supervised descent learning technique. *Inverse Problems*. 2019;35(10):105008.

[46] Guo R, Li M, Yang F, *et al.* Application of supervised descent method for 2D magnetotelluric data inversion. *Geophysics*. 2020;85(4):WA53–WA65.

[47] Hu Y, Guo R, Jin Y, *et al.* A supervised descent learning technique for solving directional electromagnetic logging-while-drilling inverse problems. *IEEE Transactions on Geoscience and Remote Sensing*. 2020;58(11):8013–8025.

[48] Hao P, Sun X, Nie Z, *et al.* A robust inversion of induction logging responses in anisotropic formation based on supervised descent method. *IEEE Geoscience and Remote Sensing Letters*. 2022;19:8011505.

[49] Lu S, Liang B, Wang J, *et al.* 1-D inversion of GREATEM data by supervised descent learning. *IEEE Geoscience and Remote Sensing Letters*. 2022;19:8007305.

[50] Xie W, Zhang X, Mu Y, *et al.* A subsurface targets' classification method utilizing gradient learning technique. *IEEE Geoscience and Remote Sensing Letters*. 2022;19:3000305.

[51] Jin Y, Shen Q, Wu X, *et al.* A physics-driven deep-learning network for solving nonlinear inverse problems. *Petrophysics—The SPWLA Journal of Formation Evaluation and Reservoir Description.* 2020;61(01):86–98.

[52] Jin Y, Hu Y, Wu X, *et al.* RNN-based gradient prediction for solving magnetotelluric inverse problem. In: *SEG International Exposition and Annual Meeting.* OnePetro; 2020.

[53] Shahriari M, Pardo D, Rivera JA, *et al.* Error control and loss functions for the deep learning inversion of borehole resistivity measurements. *International Journal for Numerical Methods in Engineering.* 2021;122(6): 1629–1657.

[54] Noh K, Pardo D, Torres-Verdín C. 2.5 D deep learning inversion of LWD and deep-sensing EM measurements across formations with dipping faults. *IEEE Geoscience and Remote Sensing Letters.* 2022;19:8023805.

[55] Wang H, Liu Y, Yin C, *et al.* Stochastic inversion of magnetotelluric data using deep reinforcement learning. *Geophysics.* 2021;87(1):1–52.

[56] Sutton RS, Barto AG. *Reinforcement Learning: An Introduction.* London: MIT Press; 2018.

[57] Mnih V, Kavukcuoglu K, Silver D, *et al.* Playing atari with deep reinforcement learning. arXiv preprint arXiv:13125602. 2013.

[58] Colombo D, Turkoglu E, Li W, *et al.* Physics-driven deep-learning inversion with application to transient electromagnetics. *Geophysics.* 2021;86(3): E209–E224.

[59] Alyousuf T, Li Y. Inversion using adaptive physics-based neural network: application to magnetotelluric inversion. In: *First International Meeting for Applied Geoscience & Energy.* Houston, TX: Society of Exploration Geophysicists; 2021. pp. 1455–1459.

[60] Asif MR, Bording TS, Maurya PK, *et al.* A Neural network-based hybrid framework for least-squares inversion of transient electromagnetic data. *IEEE Transactions on Geoscience and Remote Sensing.* 2022;60:4503610.

[61] Sun J, Niu Z, Innanen KA, *et al.* A theory-guided deep-learning formulation and optimization of seismic waveform inversion. *Geophysics.* 2020;85(2): R87–R99.

[62] Guo L, Li M, Xu S, *et al.* Study on a recurrent convolutional neural network based FDTD method. In: *2019 International Applied Computational Electromagnetics Society Symposium—China (ACES)*, vol. 1. New York, NY: IEEE; 2019. pp. 1–2.

[63] Hu Y, Jin Y, Wu X, *et al.* A theory-guided deep neural network for time domain electromagnetic simulation and inversion using a differentiable programming platform. *IEEE Transactions on Antennas and Propagation.* 2022;70(1):767–772.

[64] Li YE, Demanet L. Full-waveform inversion with extrapolated low-frequency data. *Geophysics.* 2016;81(6):R339–R348.

[65] Hu W, Chen J, Liu J, *et al.* Retrieving low wavenumber information in FWI: an overview of the cycle-skipping phenomenon and solutions. *IEEE Signal Processing Magazine.* 2018;35(2):132–141.

[66]   Jin Y, Hu W, Wu X, *et al.* Learn low wavenumber information in FWI via deep inception based convolutional networks. In: *2018 SEG International Exposition and Annual Meeting*. OnePetro; 2018.

[67]   Hu W, Jin Y, Wu X, *et al.* Progressive transfer learning for low-frequency data prediction in full waveform inversion. *Geophysics*. 2021;86(4): 1–82.

[68]   Zhao T, Abubakar A, Cheng X, *et al.* Augment time-domain FWI with iterative deep learning. In: *SEG Technical Program Expanded Abstracts 2020*. Houston, TX: Society of Exploration Geophysicists; 2020. pp. 850–854.

[69]   Brougois A, Bourget M, Lailly P, *et al.* Marmousi, model and data. In: *EAEG Workshop-Practical Aspects of Seismic Data Inversion*. European Association of Geoscientists & Engineers; 1990. p. cp–108.

[70]   Feng S, Lin Y, Wohlberg B. Multiscale data-driven seismic full-waveform inversion with field data study. *IEEE Transactions on Geoscience and Remote Sensing*. 2021;60:1–14.

[71]   Gatys LA, Ecker AS, Bethge M. Image style transfer using convolutional neural networks. In: *Proceedings of the IEEE Conference on Computer Vision and Pattern Recognition*; 2016. pp. 2414–2423.

[72]   Sun B, Alkhalifah T. ML-descent: an optimization algorithm for full-waveform inversion using machine learning. *Geophysics*. 2020;85(6):R477–R492.

[73]   Kingma DP, Welling M. Auto-encoding variational Bayes. arXiv preprint arXiv:13126114. 2013.

[74]   Ruder S. An overview of gradient descent optimization algorithms. arXiv preprint arXiv:160904747. 2016.

[75]   Li Z, Jia X, Zhang J. Deep learning guiding first-arrival traveltime tomography. In: *SEG International Exposition and Annual Meeting*. OnePetro; 2019.

[76]   Gallardo LA, Meju MA. Joint two-dimensional DC resistivity and seismic travel time inversion with cross-gradients constraints. *Journal of Geophysical Research: Solid Earth*. 2004;109(B3):B03311.

[77]   Guo R, Yao HM, Li M, *et al.* Joint inversion of audio-magnetotelluric and seismic travel time data with deep learning constraint. *IEEE Transactions on Geoscience and Remote Sensing*. 2021;59(9):7982–7995.

[78]   Hu Y, Jin Y, Wu X, *et al.* Deep learning enhanced joint geophysical inversion for crosswell monitoring. In: *2021 United States National Committee of URSI National Radio Science Meeting (USNC-URSI NRSM)*. New York, NY: IEEE; 2021. pp. 101–102.

[79]   Zhou H, Guo R, Tao D, *et al.* Joint inversion of audio-magnetotelluric and seismic travel time data using attribute fusion based on deep learning. In: *2021 International Applied Computational Electromagnetics Society (ACES-China) Symposium*. New York, NY: IEEE; 2021. pp. 1–2.

[80]   Sun Y, Denel B, Daril N, *et al.* Deep learning joint inversion of seismic and electromagnetic data for salt reconstruction. In: *SEG Technical Program Expanded Abstracts 2020*. Houston, TX: Society of Exploration Geophysicists; 2020. pp. 550–554.

[81] Liu W, Xi Z, Wang H, *et al.* Two-dimensional deep learning inversion of magnetotelluric sounding data. *Journal of Geophysics and Engineering.* 2021;18(5):627–641.

[82] Van der Maaten L, Hinton G. Visualizing data using t-SNE. *Journal of Machine Learning Research.* 2008;9(11):2579–2605.

[83] Guo R, Lin Z, Shan T, *et al.* Physics embedded deep neural network for solving full-wave inverse scattering problems. *IEEE Transactions on Antennas and Propagation.* 2021;122:106686.

Chapter 6

Deep learning techniques for biomedical imaging

Yuan Fang[1], Kazem Bakian-Dogaheh[1] and Mahta Moghaddam[1]

Electromagnetic phenomena are at the heart of the majority of medical imaging techniques, such as computed tomography (CT) scan, magnetic resonance imaging (MRI), microwave imaging (MWI), and electrical impedance tomography (EIT). These medical imaging techniques provide a non-invasive approach that is capable of generating images of complex biological structure for clinical diagnosis and follow-up treatment procedures.

In the last decade, computational image reconstruction algorithms have benefited from the rapid evolution of deep learning methods, particularly for high- and super-resolution medical imaging studies. The vastly increasing medical imaging databases, in turn, have accelerated the development of more advanced deep learning neural networks. Synergizing data-driven deep learning neural networks with legacy physics-based imaging algorithms in recent years shows promising enhancement in the performance of medical imaging methods.

This chapter starts by reviewing the current state-of-art deep learning methods used in medical imaging approaches. We then introduce the typical physics-based medical imaging techniques with a brief comparison of these techniques, followed by the progress of machine learning and deep learning methods in the next section (Section 6.1). Then in Section 6.2, we will illustrate the physics of electromagnetic medical imaging techniques and their related physical imaging methods (Section 6.2). In Section 6.3, we discuss the commonly used deep learning neural networks with their applications in medical imaging. The recent studies on synergizing learning-assisted and physics-based imaging methods are discussed in Section 6.4, followed by a summary in Section 6.5.

6.1 Introduction

Electromagnetic (EM) waves across the electromagnetic spectrum and with different energy levels extensively in medical imaging applications [1]. For example, computed

[1]Microwave Systems, Sensors and Imaging Lab (MiXIL), Part of Ming Hsieh Department of Electrical Engineering, University of Southern California, USA

tomography (CT) scans measure the X-ray attenuation through the imaging area. At a high energy level, X-rays are excited within the ionizing radiation range of 300 PHz–30 EHz. In the frequency range between 0.3 GHz and 3 GHz, microwave imaging (MWI) can be applied to reconstruct dielectric images from scattered EM signals due to dielectric contrast between targets and background. At lower frequency bands, such as within 3–300 MHz range, the magnetic resonance imaging (MRI) technique utilizes a strong magnetic field to measure detailed longitudinal and spin-lattice relaxation time data at each image pixel. Lastly, most electrical impedance tomography systems provide an impedance distribution of the target area from measured electrical voltage and current at frequencies lower than 3 MHz.

CT and MRI are relatively mature among the medical imaging techniques above, and commercial imaging systems are widely used in most hospitals and clinics. The physics of these imaging techniques determines the advantages of CT and MRI usage in different medical imaging applications. CT scans show better image contrasts of the bones and sclerotic lesions, and they are commonly used in initial staging and re-evaluation after treatment. When detecting and treating soft-tissue diseases is of interest, the image resolution of CT scans are not as good as those of MRI, which makes MRI often used in stroke detection and cancer diagnosis.

However, there are shortcomings associated with CT and MRI. In order to obtain high-resolution CT scan images, the equipment needs sufficient contrast agents, and patients undergo through a certain amount of radiation exposure during the repeated CT scans [2]. Although MRI does not impose radiation risk, high magnetic field may create excessive SAR levels, also the magnetic field could react in patients with artificial metallic implants [3]. Furthermore, and from a logistical stand point, MRI equipment is usually more expensive than CT scan systems, which limits the resources per capita [4]. In the last two decades, MWI and EIT have gained increasing attention as alternative imaging modalities that can potentially overcome some of the CT and MRI aforementioned challenges [2–7]. MWI and EIT systems provide portability with less cost than most CT and MRI systems. Because of the low radiation risk, the medical MWI and EIT systems are much safer than CT systems. Therefore, multiple preclinical MWI systems have been successfully demonstrated in medical imaging applications, including the brain, breast, and forearm imaging, or those of for the diagnosis and treatment of stroke, cancer, and fractures [3,5–7]. Furthermore, commercial EIT systems have also been utilized in applications such as detecting pulmonary emboli, monitoring heart function and blood flow, lung monitoring, and breast cancer detection [8–12].

The physics of CT or MR image reconstruction are fairly straight forward signal processing based techniques and are well-established methods that date back to the 1990s. [2,13]. On the contrary, the MWI and EIT image formation is based on an inverse-forward solver scheme with a cost function derived from Maxwell's equations. Therefore, these physics-based imaging algorithms are often more complicated than those of CT and MRI, particularly for 3D image reconstruction. Conventional inverse solvers based on the Newton, Gauss-Newton, and conjugate gradient methods have been routinely applied for solving physics-based inverse problems. More advanced inverse solvers, such as the contrast source inversion (CSI), Born iterative method (BIM), and distorted Born iterative method (DBIM), have been developed and extensively analyzed for the MWI and EIT applications [8,13–20]. Nevertheless, the achieved imaging resolution of the physics-based algorithms is limited by the measurement configuration and system performance (details are in Section 6.2).

As a branch of artificial intelligence, machine learning (ML) algorithms have attracted increasing attention in medical imaging due to their pattern recognition and extraction capability for computer-aided detection (CADe) and diagnosis (CADx). In the last decade, machine learning researchers have focused on developing new algorithms such as K means, Bayesian learning, AdaBoost algorithms, and support vector machine (SVM), to assist diagnosis through quantitatively distinguishing different tissues and segmenting the malignant area from healthy tissues in MRI and CT images [21–24]. Besides the ML applications for CT scans and MRI, the data-driven ML models are also introduced as the prior knowledge of physics-based inverse solvers to improve the imaging quality of the MWI and EIT [25–28].

To increase the robustness of learning-assisted medical imaging, deep learning (DL) has received the bulk of research interest since the late 2010s. Unlike machine learning methods that utilize manually designed features from input data, deep learning artificial neural networks (ANN) uncover relationships between data characteristics in a more intuitive way than human neural networks. As the ML area emerged in the computer vision field, DL methods, including the fully connected neural network (FCN), convolutional neural network (CNN), recurrent neural network (RNN), and generative adversarial network (GAN), have achieved great success in both segmentation of MRI and CT images, and have shown great promise in assistance with physics-based MWI and EIT inversion. More recently, the hybridization of both physics-based and learning MWI and EIT models has shown promising potential as an advanced EM imaging method in medical imaging [26–29].

6.2   Physics of medical imaging

In a CT scan, as X-rays propagate through the patient, their intensity decreases due to the absorption of multiple tissues. This attenuation can be represented by Lambert–Beer equation: $-ln(I/I0) = \sum_i \rho_i x_i$, where ρ_i, x_i are the attenuation coefficient and propagation length in ith material, respectively. The image reconstruction algorithms include matrix inversion, iterative Radon transformation, and backprojection [30].

MRI systems utilize a fixed longitudinal magnetic field and an oscillating transverse magnetic field to excite the nuclear magnetic resonances and measure the magnitude and phase of magnetic signal experienced by the nuclei during its relaxation time, which can be described by the Bloch equations. Fourier transformations are then applied to transfer the acquired signal from the frequency domain to k space for displaying the corresponding intensity levels as gray-shade images. The imaging is achieved due to different hydrogen nuclei relaxation times during the magnetization in human tissues. Two types of relaxation time determine the most common MRI sequences (1) the time when the longitudinal magnetization of the excited proton has recovered to equilibrium (T1), and (2) the time when transverse magnetization has decayed to equilibrium (T2). By changing the repetition time between successive pulse sequences (TR) and the time to echo between the excited magnetic pulse and the receipt of the echo signal (TE), MRI systems can generate T1-weighted (short TR and TE) images and T2-weighted (long TR and TE) images [31].

Although MRI and CT scans transmit and receive electromagnetic waves to form their corresponding images, their imaging principles are not based on EM theory or Maxwell's equations. This section mainly focuses on explaining the medical imaging techniques particularly for MWI and EIT, which are developed from the basics of electromagnetic theory. MWI and EIT are relatively new compared with CT and MRI techniques. Unlike commercial CT, MRI, and EIT systems, most MWI system prototypes are still in laboratory test settings. Details of CT and MRI imaging principles can be found in many textbooks such as [30,31].

6.2.1   Maxwell's equations

Maxwell's equations describe the relations between electromagnetic sources and the fields generated by them in the presence of media. EM sources may be impressed and conduction electric current density $\mathbf{J}_i$ and $\mathbf{J}_c$ (A/m^2) and impressed magnetic current density $\mathbf{M}_i$ (V/m^2). The time-harmonic electric field $\mathbf{E}$ (V/m) and magnetic field $\mathbf{H}$ (A/m) in Maxwell's equations can be written as [32] (time-harmonic term $e^{j\omega t}$):

$$\nabla \times \mathbf{E} = -\mathbf{M}_i - j\omega\mathbf{B} \tag{6.1}$$

$$\nabla \times \mathbf{H} = \mathbf{J}_i + \mathbf{J}_c + j\omega\mathbf{D} \tag{6.2}$$

$$\nabla \cdot \mathbf{D} = q_{ev} \tag{6.3}$$

$$\nabla \cdot \mathbf{B} = q_{mv} \tag{6.4}$$

where electric and magnetic flux density $\mathbf{D}$ (C/m^2) and $\mathbf{B}$ (W/m^2) in isotropic media are defined as: $\mathbf{D} = \varepsilon\mathbf{E}$, $\mathbf{B} = \mu\mathbf{H}$. ε and μ are the media permittivity and permeability. The quantities q_{ev} (C/m^3) and q_{mv} (W/m^2) are electric and magnetic charge density, respectively. Although magnetic charges q_{mv} and impressed magnetic current densities $\mathbf{M}_i$ have not been physically observed in the real world, they have been introduced to balance Maxwell's equations [32].

By introducing Ohm's law, $\mathbf{J}_c = \sigma\mathbf{E}$. Equation (6.2) can be rewritten as:

$$\nabla \times \mathbf{H} = \mathbf{J}_i + j\omega\tilde{\varepsilon}\mathbf{E} \tag{6.5}$$

where $\tilde{\varepsilon} = \varepsilon + \sigma/(j\omega)$.

6.2.2   Formulations of EIT

The EIT technique constructs the impedance profile of biological tissues, which consists of the resistance and reactance. Fluids manifest the resistance characteristic, while the cell membranes act as defective capacitors [14]. Since the EIT is operated in a low-frequency regime, the effect of the impressed electric current source $\mathbf{J}_i$ is negligible. In the absence of magnetic source $\mathbf{M}_i$, Maxwell's equations (6.1) and (6.2) within the EIT regime can be written as [14]:

$$\nabla \times \mathbf{E} = 0 \tag{6.6}$$

$$\nabla \times \mathbf{H} = \mathbf{J}_i + \sigma \mathbf{E} \tag{6.7}$$

For EIT, $\mathbf{J}_i$ can be set as zero since there is no current source inside the imaging domain [13–15].

Defining the electric potential u through $\mathbf{E} = -\nabla u$, we notice the left-hand term of equation (6.7) is divergence free, and equation (6.7) can be written as Poisson's equation [33]:

$$\nabla \cdot [\sigma(\mathbf{r})\nabla u(\mathbf{r})] = 0, \mathbf{r} \in \Gamma \tag{6.8}$$

where Γ is the imaging domain.

The electric potential u satisfies the Neumann boundary condition (6.9) and Dirichlet boundary condition (6.10):

$$-\sigma(\mathbf{r})\frac{\partial u(\mathbf{r})}{\partial n} = j_n(\mathbf{r}), \mathbf{r} \in \partial\Gamma \tag{6.9}$$

$$u(\mathbf{r}) = 0, \mathbf{r} \in S_{ref} \tag{6.10}$$

S_{ref} is the surface of the reference (ground) electrode and $\partial\Gamma$ is the boundary surface of the imaging domain, where $j_n(\mathbf{r})$ is zero on $\partial\Gamma$ except at the excited nth electrode location [33].

A typical EIT system has N_t electrodes, in which one electrode is set as the reference (ground) electrode. When a total electric current I flows in the source electrode located at $\mathbf{r}_{tx}$, the current density $j_n(\mathbf{r}, \mathbf{r}_{tx})$ at $\mathbf{r}_{tx}$ satisfies the equations (6.11) and (6.12) due to the conservation of charge [16,33].

$$\int_{S_t} j_n(\mathbf{r}, \mathbf{r}_{tx}) ds = -I \tag{6.11}$$

$$\int_{S_{ref}} j_n(\mathbf{r}, \mathbf{r}_{tx}) ds = I \tag{6.12}$$

Note that $j_n(\mathbf{r}, \mathbf{r}_{tx}) = 0$ when $\mathbf{r}$ is on the rest of $\partial\Gamma$ and S_t is the source electrode surface. A finite-element method (FEM) forward solver can be used to construct the stiffness matrix $(S(\mathbf{u}, v))$, where v is the unknown array of conductivity (σ) distribution. The image reconstruction can be achieved by using (6.8)–(6.12) equations to form a least-square cost function for the minimization in the form of $\min(||S(\mathbf{u}, v)v - \beta||)$, where β is the forcing vector generated from the multiplication of testing (weighting function) measured voltage or current signal array of the electrodes [8,13–16].

The minimization in these works is conducted through direct inverse solvers, such as singular value decomposition (SVD) [14] or iterative inverse solvers, such as the Gauss-Newton method [8,14,16]. Bayford has summarized common direct and iterative inverse solvers in [14]. In [8], the L_2-norm cost function with multiplicative regularization was proposed for 3D EIT reconstruction. Results evaluation shows an acceptable performance for edge-preserving and noise. Gupta recently used the combined $L_1 - L_2$-norm cost function to solve the current density impedance imaging with a sparse reconstruction inverse problem [34]. The proposed method obtains high-resolution images with sharp edges, even in the presence of noise.

As introduced in the previous section, MWI generates dielectric images through the inverse scattering theory. Ybarra *et al.*, proposes a method using the similar "scattering" method to allow the Born-approximation based method to be used in EIT [33]. The method starts by designating a the reference (background) conductivity $\sigma_b(\mathbf{r})$ and the electric potential $u_b(\mathbf{r}, \mathbf{r}_{tx})$ before the EIT operation on patient. By multiplying $u_b(\mathbf{r}, \mathbf{r}_a)$ ($\mathbf{r}_a$ at another different location away from $\mathbf{r}_{tx}$) with equation (6.8), we will have:

$$\int_{\Gamma} u_b(\mathbf{r}, \mathbf{r}_a) \nabla \cdot [\sigma(\mathbf{r}) \nabla u(\mathbf{r}, \mathbf{r}_{tx})] \, dv = 0 \tag{6.13}$$

Notice that $\sigma_b(\mathbf{r})$ and $u_b(\mathbf{r}, \mathbf{r}_{tx})$ still satisfy Poisson equation and two boundary conditions:

$$\nabla \cdot [\sigma_b(\mathbf{r}) \nabla u_b(\mathbf{r}, \mathbf{r}_{tx})] = 0 \tag{6.14}$$

$$-\sigma_b(\mathbf{r}) \frac{\partial u_b(\mathbf{r}, \mathbf{r}_{tx})}{\partial n} = j_n(\mathbf{r}, \mathbf{r}_{tx}), \mathbf{r} \in \partial \Gamma \tag{6.15}$$

$$u_b(\mathbf{r}) = 0, \mathbf{r} \in S_{ref} \tag{6.16}$$

Integrating equation (6.13) by parts yields:

$$\int_{\partial \Gamma} u_b(\mathbf{r}, \mathbf{r}_a) \sigma(\mathbf{r}) \frac{\partial u(\mathbf{r}, \mathbf{r}_{tx})}{\partial n} ds - \int_{\Gamma} \sigma(\mathbf{r}) \nabla u_b(\mathbf{r}, \mathbf{r}_a) \nabla u(\mathbf{r}, \mathbf{r}_{tx}) dv = 0 \tag{6.17}$$

Similarly, when multiplying equation (6.14) with $u(\mathbf{r}, \mathbf{r}_a)$, we have:

$$\int_{\partial \Gamma} u(\mathbf{r}, \mathbf{r}_a) \sigma_b(\mathbf{r}) \frac{\partial u_b(\mathbf{r}, \mathbf{r}_{tx})}{\partial n} ds - \int_{\Gamma} \sigma_b(\mathbf{r}) \nabla u(\mathbf{r}, \mathbf{r}_a) \nabla u_b(\mathbf{r}, \mathbf{r}_{tx}) dv = 0 \tag{6.18}$$

From Neumman boundary conditions in (6.9) and (6.15), the terms $\sigma_b(\mathbf{r}) \partial u_b(\mathbf{r}, \mathbf{r}_{tx})/\partial n$ and $\sigma(\mathbf{r}) \partial u(\mathbf{r}, \mathbf{r}_{tx})/\partial n$ can be replaced by $-j_n(\mathbf{r}, \mathbf{r}_{tx})$. By subtracting (6.18) from (6.17) the "scattering" potential can be written as follows:

$$u(\mathbf{r}_{tx}, \mathbf{r}_a) - u_b(\mathbf{r}_{tx}, \mathbf{r}_a) = \frac{1}{I} \int_{\Gamma} \sigma(\mathbf{r}) \nabla u_b(\mathbf{r}, \mathbf{r}_a) \nabla u(\mathbf{r}, \mathbf{r}_{tx}) - \sigma_b(\mathbf{r}) \nabla u(\mathbf{r}, \mathbf{r}_a) \nabla u_b(\mathbf{r}, \mathbf{r}_{tx}) dv$$

$$\tag{6.19}$$

Since most EIT systems operate at low frequency and the imaging domain of EIT is less than 1 m [8,13–16], the term $k_b L(\sigma - \sigma_b)$ is much less than 1. Hence, the first-order Born approximation validity satisfies and $u_b \approx u$. Equation (6.19) can be written as:

$$\Delta u(\mathbf{r}_{tx}, \mathbf{r}_a) \approx \frac{1}{I} \int_\Gamma \Delta \sigma(\mathbf{r}) \nabla u_b(\mathbf{r}, \mathbf{r}_a) \nabla u(\mathbf{r}, \mathbf{r}_{tx}) dv \tag{6.20}$$

$$\Delta u(\mathbf{r}_{tx}, \mathbf{r}_a) = u(\mathbf{r}_{tx}, \mathbf{r}_a) - u_b(\mathbf{r}_{tx}, \mathbf{r}_a) \tag{6.21}$$

$$\Delta \sigma(\mathbf{r}) = \sigma(\mathbf{r}) - \sigma_b(\mathbf{r}) \tag{6.22}$$

The integral term can be calculated through either the numerical method such as FEM [33], or the Faddeev Green's function [35,36]. Equation (6.21), can be used to construct the cost function for the Born-approximation-based methods. We will introduce these Born-approximation-based methods in the later subsection to demonstrate the unity of these inverse methods in both EIT and MWI applications.

6.2.3 Formulations of MWI

In the absence of magnetic sources (i.e., $\mathbf{M}_i = 0$, $q_{mv} = 0$), we introduce the magnetic vector potential $\mathbf{A}$ and define $\mathbf{B} = \nabla \times \mathbf{A}$. The Maxwell's equation (6.1) can be rewritten as:

$$\nabla \times (\mathbf{E} + j\omega \mathbf{A}) = 0 \tag{6.23}$$

The $\mathbf{E} + j\omega \mathbf{A}$ is an irrotational vector, hence can be written as:

$$\mathbf{E} = -\nabla u - j\omega \mathbf{A} \tag{6.24}$$

where u is the electric potential. By choosing the Lorenz gauge $\nabla \cdot \mathbf{A} + j\omega\mu\varepsilon u = 0$, (6.24) is [32]:

$$\mathbf{E} = -\nabla u - j\omega \mathbf{A} = -j\omega \left[\mathbf{A} + \frac{1}{k^2} \nabla(\nabla \cdot \mathbf{A}) \right] \tag{6.25}$$

For a point electric dipole $\mathbf{J}$, the magnetic vector potential is defined as [32,37]:

$$\mathbf{A}(\mathbf{r}) = \mu \int_V \mathbf{J}(\mathbf{r}')g(\mathbf{r}, \mathbf{r}')d\mathbf{r}' \tag{6.26}$$

where $g(\mathbf{r}, \mathbf{r}')$ is the scalar Green's function between source location $\mathbf{r}'$ and receiver location $\mathbf{r}$. In homogeneous medium, $g(\mathbf{r}, \mathbf{r}') = e^{-jk_b|\mathbf{r}-\mathbf{r}'|}/(4\pi|\mathbf{r} - \mathbf{r}'|)$.

Now we define the electric and magnetic fields without imaging objects as the background fields $\mathbf{E}^b$, $\mathbf{H}^b$, and the electric and magnetic fields with imaging objects as the total fields $\mathbf{E}$, $\mathbf{H}$. The fields satisfy the Maxwell's equations:

$$\nabla \times \mathbf{E}^b = -\mathbf{M} - j\omega\mu_b\mathbf{H}^b \tag{6.27}$$

$$\nabla \times \mathbf{H}^b = \mathbf{J} + j\omega\tilde{\varepsilon}_b\mathbf{E}^b \tag{6.28}$$

$$\nabla \times \mathbf{E} = -\mathbf{M} - j\omega\mu\mathbf{H} \tag{6.29}$$

$$\nabla \times \mathbf{H} = \mathbf{J} + j\omega\tilde{\varepsilon}\mathbf{E} \tag{6.30}$$

where material properties of the background are denoted as $\mu_b, \tilde{\varepsilon}_b$ and properties of background and objects are $\mu, \tilde{\varepsilon}$. Subtracting (6.29) and (6.30) from (6.27) and (6.28) results in scattered field:

$$\nabla \times \mathbf{E}^s = -j\omega(\mu\mathbf{H} - \mu_b\mathbf{H}^b) = -j\omega(\mu - \mu_b)\mathbf{H} - j\omega\mu_b\mathbf{H}^s \tag{6.31}$$

$$\nabla \times \mathbf{H}^s = j\omega(\tilde{\varepsilon}\mathbf{E} - \tilde{\varepsilon}_b\mathbf{E}^b) = j\omega(\tilde{\varepsilon} - \tilde{\varepsilon}_b)\mathbf{E} + j\omega\tilde{\varepsilon}_b\mathbf{E}^s \tag{6.32}$$

where $\mathbf{E}^s = \mathbf{E} - \mathbf{E}^b$ and $\mathbf{H}^s = \mathbf{H} - \mathbf{H}^b$.

By defining volume equivalent electric current $\mathbf{J}_{eq} = j\omega(\tilde{\varepsilon} - \tilde{\varepsilon}_b)\mathbf{E}$ and magnetic current source $\mathbf{M}_{eq} = j\omega(\mu - \mu_b)\mathbf{H}$, (6.31) and (6.32) can be written as

$$\nabla \times \mathbf{E}^s = -\mathbf{M}_{eq} - j\omega\mu_b\mathbf{H}^s \tag{6.33}$$

$$\nabla \times \mathbf{H}^s = \mathbf{J}_{eq} + j\omega\tilde{\varepsilon}_b\mathbf{E}^s \tag{6.34}$$

In most MWI applications, both background and object materials are non-magnetic ($\mu = \mu_b = \mu_0 \approx 4\pi \times 10^{-7}$ H/m). Therefore, scattered electric field $\mathbf{E}^s$ and magnetic field $\mathbf{H}^s$ can be viewed as excited from equivalent volume electric current $\mathbf{J}_{eq}$.

Now replacing $\mathbf{E}$ in (6.25) with scattered electric field $\mathbf{E}^s$, and using volume equivalent source $\mathbf{J}_{eq}$ to represent the magnetic vector potential $\mathbf{A}$ in (6.26). The scattered electric field at any observation location $\mathbf{r}$ can be written as [32]:

$$\mathbf{E}^s(\mathbf{r}) = k_b^2 \int_V \chi(\mathbf{r}')\left(\bar{\bar{\mathbf{I}}} + \frac{1}{k_b^2}\nabla\nabla\cdot\right)g(\mathbf{r},\mathbf{r}')\mathbf{E}(\mathbf{r}')d\mathbf{r}', \mathbf{r}' \in V \tag{6.35}$$

where $\mathbf{E}(\mathbf{r}')$ is the total field at the object location $\mathbf{r}'$. The electric contrast $\chi(\mathbf{r}')$ between background and object material at location $\mathbf{r}'$ is:

$$\chi(\mathbf{r}') = \frac{\tilde{\varepsilon}(\mathbf{r}')}{\tilde{\varepsilon}_b} - 1 \tag{6.36}$$

$\tilde{\varepsilon}(\mathbf{r}')$ and $\tilde{\varepsilon}_b$ are the complex permittivity of the object at location $\mathbf{r}'$ and the background.

The operators with the scalar Green's function inside the integral can be replaced by the dyadic Green's function $\bar{\bar{\mathbf{G}}}$, and (6.35) can be written as [32]:

$$\mathbf{E}^s(\mathbf{r}) = k_b^2 \int_V \chi(\mathbf{r}')\bar{\bar{\mathbf{G}}}(\mathbf{r},\mathbf{r}')\mathbf{E}(\mathbf{r}')d\mathbf{r}', \mathbf{r}' \in V \tag{6.37}$$

This is the scattered electric field volume integral equation (EFVIE) that is commonly used in microwave imaging. The key element of 6.37 is the background electric dyadic Green's function $\bar{\bar{\mathbf{G}}}$. When the background is a homogeneous or planarly/cylindrically/spherically layered medium, the detailed expression of $\bar{\bar{\mathbf{G}}}$ is discussed in [37]. However, Green's function can only be numerically calculated for most inhomogeneous media.

In most MWI system prototypes, vector network analyzers (VNA) are integrated with the antenna and array systems to measure the signal as a scalar quantity, known as scattering parameters (S-parameter) [3,5,17–19], which are voltage ratios. It can be shown that the scattered S-parameter measurement $S_{n,m}^s$ between transmitter (Tx) port m and receiver (Rx) port n can be rewritten as the following:

$$S_{n,m}^s = k_b^2 \int_V \chi(\mathbf{r}')\mathbf{G}_{n,m}(\mathbf{r}') \cdot \mathbf{E}(\mathbf{r}')dv', \mathbf{r}' \in V \tag{6.38}$$

where the scattered S-parameter can be derived by $S_{n,m}^s = S_{n,m}^t - S_{n,m}^b$. Superscripts t and b separately indicate the total and the background S-parameters, respectively. The waveport vector Green's function (WVGF) $\mathbf{G}_{n,m}(\mathbf{r}')$ is [17–19]:

$$\mathbf{G}_{n,m}(\mathbf{r}') = \frac{\sqrt{Z_m}}{\sqrt{Z_n}} \frac{j\mathbf{E}_n^b(\mathbf{r}')}{\omega\mu_0 \int \int_{A_m} \mathbf{e}_m^t(\mathbf{r}) \times \mathbf{h}_m^t(\mathbf{r})I_0 \cdot d\mathbf{r}}, \mathbf{r} \in A_m \tag{6.39}$$

where $\mathbf{E}_n^b(\mathbf{r}')$ is the background field at location $\mathbf{r}'$ in object domain V, which is generated by the Rx port n. The quantities $\mathbf{e}_m^t$ and $\mathbf{h}_m^t$ are the electric and magnetic field mode templates (e.g., TEM mode with coaxial feeds) excited at Tx port m. I_0 is the source signal response in the frequency domain, μ_0 is the vacuum permeability, Z_m and Z_n are the Tx port m and Rx port n impedance. In the above equations, $\mathbf{E}_m^b$ and $\mathbf{E}$ are calculated by numerical methods such as the finite difference time domain (FDTD) method and FEM, and $\mathbf{e}_m^t$ and $\mathbf{h}_m^t$ are calculated by the 2D versions of these numerical methods.

Equation (6.38) is a nonlinear equation since the total field inside the imaging domain V is a function of unknown χ. Similar to the EIT method, the Born approximation can be applied in (6.38) with $\mathbf{E} \approx \mathbf{E}^b$:

$$S_{n,m}^s \approx k_b^2 \int_V \chi(\mathbf{r}')\mathbf{G}_{n,m}(\mathbf{r}') \cdot \mathbf{E}^b(\mathbf{r}')dv', \mathbf{r}' \in V \tag{6.40}$$

6.2.4 Inverse methods for EIT and MWI

Three-dimensional (3D) EIT or microwave imaging is a challenging task due to the ill-posedness, nonlinearity, and underdeterminedness nature of the 3D inverse scattering problems. As we mentioned before, various direct and iterative inverse scattering methods such as the SVD, classical Newton, Gauss–Newton, and conjugate gradient (CG) methods have been applied for 3D impedance image reconstruction in the past few decades [8,13–16]. These methods are applicable in microwave imaging [38–44]. For 3D EIT and MWI imaging with a large number of unknowns, other nonlinear iterative solvers such as the Born iterative method (BIM), distorted Born iterative methods (DBIM), or contrast source inversion (CSI) have been developed with various optimization solvers [2–7,17–20].

6.2.4.1 Cost function

Constructing the inversion scheme starts with discretizing the system equation. The system equation in EIT is represented by (6.20), where the potential difference between mth Tx electrode and nth Rx electrode ($m, n \in [1, N_t - 1]$) is contributed by a total number of Q voxels:

$$\Delta u(\mathbf{r}_m, \mathbf{r}_n) \approx \sum_{q=1}^{q=Q} \Delta \sigma_q \int_{\Gamma_q} \frac{\nabla u_b(\mathbf{r}_q, \mathbf{r}_n) \nabla u(\mathbf{r}_q, \mathbf{r}_m)}{I} dv_q \qquad (6.41)$$

where q indicates the voxel index number and $q \in [1, Q]$, and the location of the qth voxel is denoted by $\mathbf{r}_q$. Defining the total number of measurement is P, then $P = (N_t - 1) \times (N_t - 1)$.

Similarly, after discretizing the MWI system equation (6.38):

$$S_{n,m}^s \approx \sum_{q=1}^{q=Q} \chi_q \int_{V_q} k_b^2 \mathbf{G}_{n,m}(\mathbf{r}_q) \cdot \mathbf{E}(\mathbf{r}_q) dv_q \qquad (6.42)$$

For a MWI system with M Txs and N Rxs, the total number of measurement P is $M \times N$.

The above two equations can be represented as a unified equation with matrix multiplication:

$$\mathbf{b} = \mathbf{A}\mathbf{x} \qquad (6.43)$$

where the length of the measurement vector $\mathbf{b}$ is P, length of the unknown vector $\mathbf{x}$ is Q, and the size of matrix $\mathbf{A}$ is $P \times Q$. The elements of $\mathbf{b}$, $\mathbf{x}$, and $\mathbf{A}$ in EIT applications are as follows:

$$b_p = \Delta u(\mathbf{r}_m, \mathbf{r}_n), x_q = \sigma_q,$$
$$A_{p,q} = \int_{\Gamma_q} \frac{\nabla u_b(\mathbf{r}_q, \mathbf{r}_n) \nabla u(\mathbf{r}_q, \mathbf{r}_m)}{I} dv_q \qquad (6.44)$$

where p indicates pth measurement, and can be found as $p = (m - 1) \times (N_t - 1) + n$.

In the MWI application:

$$b_p = S_{n,m}^s, x_q = \chi_q,$$
$$A_{p,q} = \int_{V_q} k_b^2 \mathbf{G}_{n,m}(\mathbf{r}_q) \cdot \mathbf{E}(\mathbf{r}_q) dv_q \qquad (6.45)$$

and similarly, $p = m \times N + n$.

In (6.44) and (6.45), all elements in vector $\mathbf{b}$ are measured by EIT or MWI systems, while all elements in matrix $\mathbf{A}$ are calculated through the forward solver. To eliminate the possible discrepancy between the measurement and simulation caused by the systematic errors, both sides of matrix function (6.2.4.1) can be normalized by dividing both sides of the measured background voltage or S-parameters in case of EIT or MWI, respectively [3].

For the Born iterative method (BIM), the dielectric constant and conductivity images are reconstructed through minimizing the L_2-norm cost function:

$$F(\mathbf{x}, \mathbf{b}) = \frac{1}{2} \| \mathbf{b} - \mathbf{A}\mathbf{x} \|_2^2 + \frac{1}{2} \gamma^2 \| \mathbf{x} \|_2^2 \qquad (6.46)$$

where the second term of (6.46) is the Tikhonov regularization term to smooth the reconstructed target. The regularization term (γ) can be found through covariance matrix multiplication, and L-curve optimization [18,19,45,46].

The gradient of cost function (6.46) with respect to **x** can be written as follows:

$$\frac{\partial F(\mathbf{x}, \mathbf{b})}{\partial \mathbf{x}} = [\mathbf{A}^T \mathbf{A} + \gamma^2 \mathbf{I}]\mathbf{x} - \mathbf{A}^T \mathbf{b} \tag{6.47}$$

where superscript T represents the transpose operation.

For the distorted Born iterative method (DBIM), the cost function is updated at each iterative step [47,48]. For instance, the DBIM cost function F_D at ith iterative step is $(i \geq 2)$:

$$F_D(\delta \mathbf{x}_i, \delta \mathbf{b}_i) = \frac{1}{2} \frac{\| \delta \mathbf{b}_i - \mathbf{A}(\mathbf{x}_{i-1})\delta \mathbf{x}_i \|_2^2}{\| \mathbf{b} \|_2^2} + \frac{1}{2}\gamma^2 \frac{\| \delta \mathbf{x}_i \|_2^2}{\| \mathbf{x}_{i-1} \|_2^2} \tag{6.48}$$

The gradient of the cost function (6.48) is:

$$\frac{\partial F_D(\delta \mathbf{x}_i, \delta \mathbf{b}_i)}{\partial \delta \mathbf{x}_i} = \left[\frac{\mathbf{A}^T(\mathbf{x}_{i-1})\mathbf{A}(\mathbf{x}_{i-1})}{\| \mathbf{b} \|_2^2} + \frac{\gamma^2}{\| \mathbf{x}_{i-1} \|_2^2}\mathbf{I} \right]\delta \mathbf{x}_i - \frac{\mathbf{A}^T(\mathbf{x}_{i-1})\delta \mathbf{b}_i}{\| \mathbf{b} \|_2^2} \tag{6.49}$$

where $\delta \mathbf{b}_i$ at ith iterative step is the difference between the measured scattered data **b** and the predicted data $\mathbf{A}(\mathbf{x}_{i-1})\mathbf{x}_{i-1}$ calculated from the $(i-1)$th step. The difference of reconstructed unknown parameters between ith and $(i-1)$th steps is defined as $\delta \mathbf{x}_i = \mathbf{x}_i - \mathbf{x}_{i-1}$.

6.2.4.2   Born approximation-based algorithms

The inverse solvers are designed to find the minimum value of the cost function (6.46) through solving the matrix equation $\partial F(\mathbf{x}, \mathbf{b})/\partial \mathbf{x} == 0$. As we mentioned before, BIM and DBIM solvers are introduced when solving the nonlinear inverse problem with a large number of unknowns. The BIM and DBIM flowchart for both EIT and MWI applications is shown in Figures 6.1 and 6.2.

The constant θ in Figures 6.1 and 6.2 is the convergence threshold for the inverse solvers. Reconstructed electric and magnetic fields of the object, object's electric potential, Green's function, inverse system matrix, and model parameter at ith step are separately denoted as $\mathbf{E}^i$, $\mathbf{H}^i$, u^i, $\mathbf{G}_i$, $\mathbf{A}_i$, and $\mathbf{x}_i$. After conducting the initial Born approximation with equations (6.46) and (6.47), the BIM starts with minimizing (6.46) and solving $\partial F(\mathbf{x}, \mathbf{b})/\partial \mathbf{x} == 0$ and the DBIM starts with minimizing (6.48) and solving $\partial F_D(\delta \mathbf{x}_i, \delta \mathbf{b})/\partial(\delta \mathbf{x}_i) == 0$. In the center of the BIM and DBIM flowcharts (shaded area), local optimization methods, such as the Newton method, the SVD method, and the conjugate gradient method, are usually used for minimization [3,18].

In the flowcharts, both BIM and DBIM solvers need to update the reconstructed electric field or transmitted electric potential inside the imaging domain at each iterative step. However, the entire BIM workflow only requires one-time calculation of the background Green's function **G** or background receiving electric potential $u_b(:, \mathbf{r}_n)$ at the initial step, while each DBIM step requires extra updates of **G** or $u_b(:, \mathbf{r}_n)$. This additional calculation at each DBIM step increases the reconstruction accuracy but costs more computational resources than the BIM method.

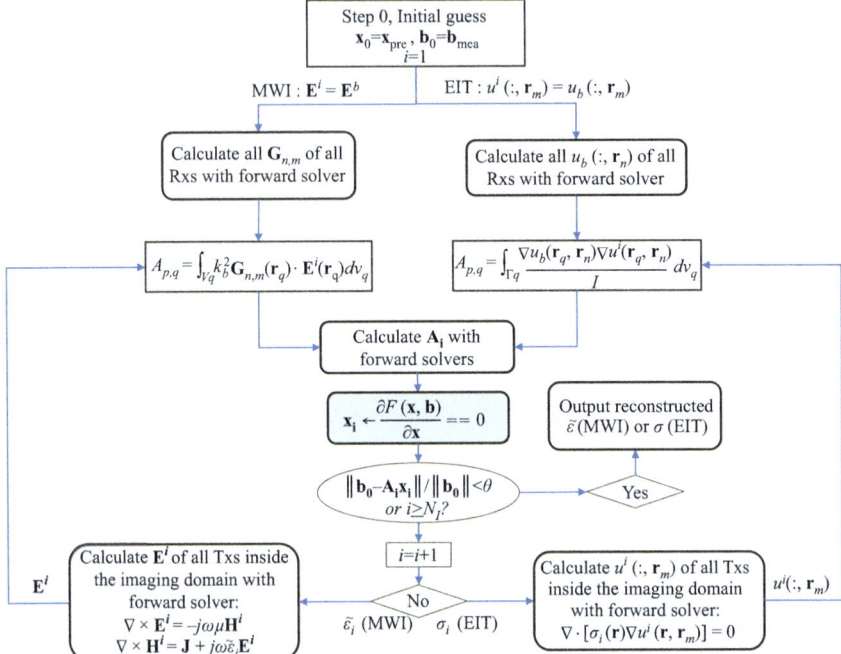

Figure 6.1   The flowchart of the BIM solver for the MWI (left-half part) and EIT (right-half part) applications

Ybarra *et al.* proposed a DBIM inverse solver for the EIT application [33]. The forward solver to update the electric potential and background electric potential is the second-order FEM method, whose accuracy level is 2-times higher than the first-order FEM. The robustness of this DBIM solver has been validated with the 2D experimental EIT system and successfully detects targets with both positive and negative conductivity contrast.

Haynes *et al.* proposed a MWI system prototype for breast imaging [5]. The system prototype is designed to work at a single frequency of 2.75 GHz. The imaging is conducted using the BIM method and can sufficiently detect the target after four iterative steps [5].

A follow-on system was designed by the same group to achieve the real-time imaging for the thermal therapy monitoring, and the system exhibits successful tracking of the dielectric variance caused by the temperature change within 2.1 s [3]. The system prototype in [3] is designed to work at 915 MHz, and the imaging is achieved through the Born approximation inversion (1st step of BIM). As Figure 6.3(c) shows, the system prototype can reconstruct the dielectric change distribution during the cooling experiment in [3].

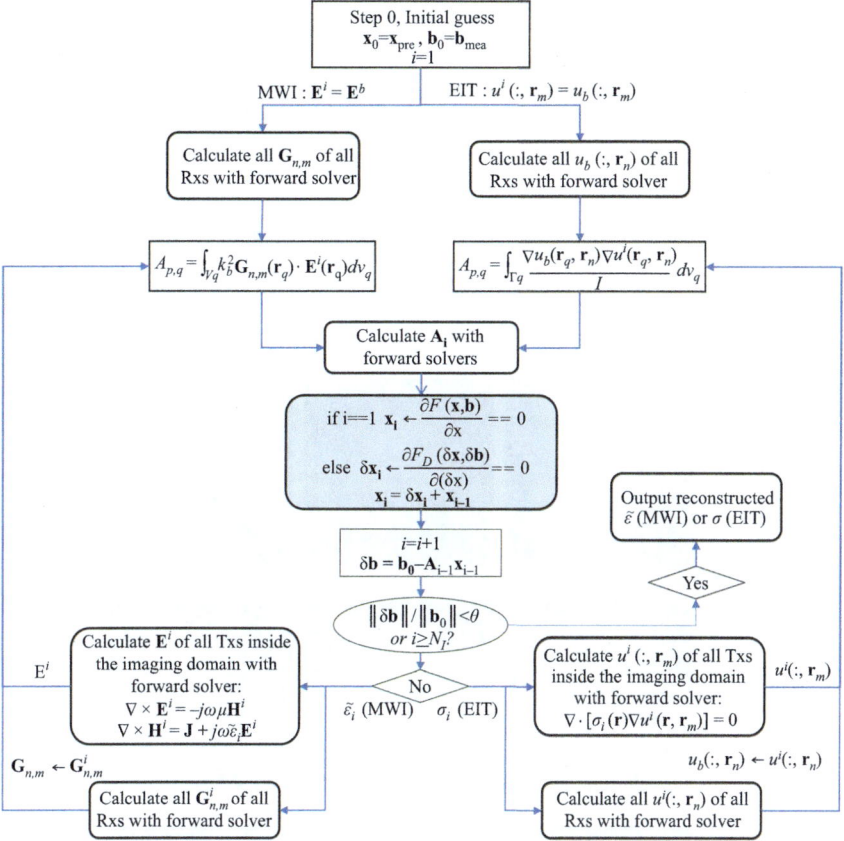

Figure 6.2   The flowchart of the DBIM solver for the MWI (left-half part) and EIT (right-half part) applications

The Green's function in [3,5] is numerically calculated in HFSS, which is based on a FEM solver. Chen *et al.* proposed the expression of a new waveport vector Green's function (WVGF), which is validated in the experimental measurement [17]. This work is further applied in the real-time DBIM inverse algorithm with GPU accelerated FDTD method [18]. Figure 6.4 demonstrates the DBIM capability to reconstruct the dielectric distribution during the simulated heating procedure with a synthetic brain-tumor phantom model [18].

To further improve the imaging resolution, Shah *et al.* implemented a combined $L_1 - L_2$ norm cost function with a level-set method in the BIM framework [19]. This new method can recover shapes, locations, and sizes of objects accurately using a small number of forward calculation steps with a significantly reduced computational cost, even in scenarios where the number of measurements is relatively small. In [19],

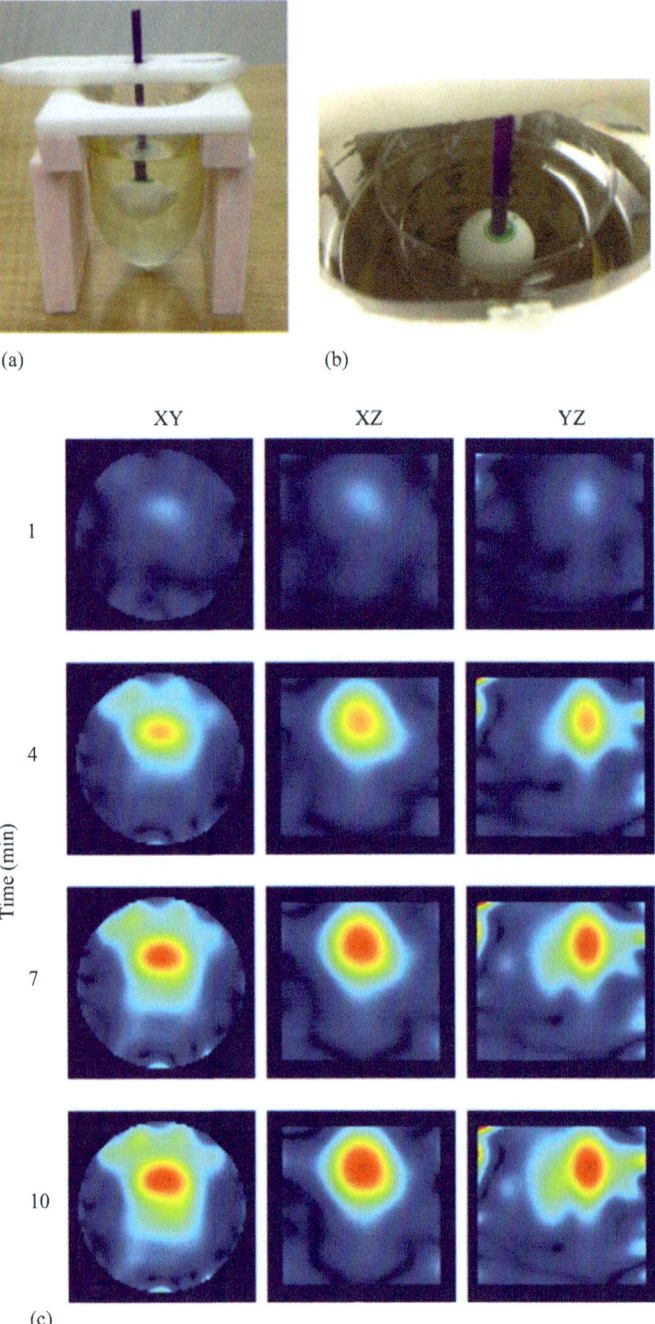

Figure 6.3 *(a) The breast phantom to test the system prototype, where the cooling water-ball target is shown in (b), and (c) is time series of dielectric variance for the phantom in (a). Phantom and heated target are taken as the background object at time zero. Cuts are through the peak contrast (figure is referred from [3]).*

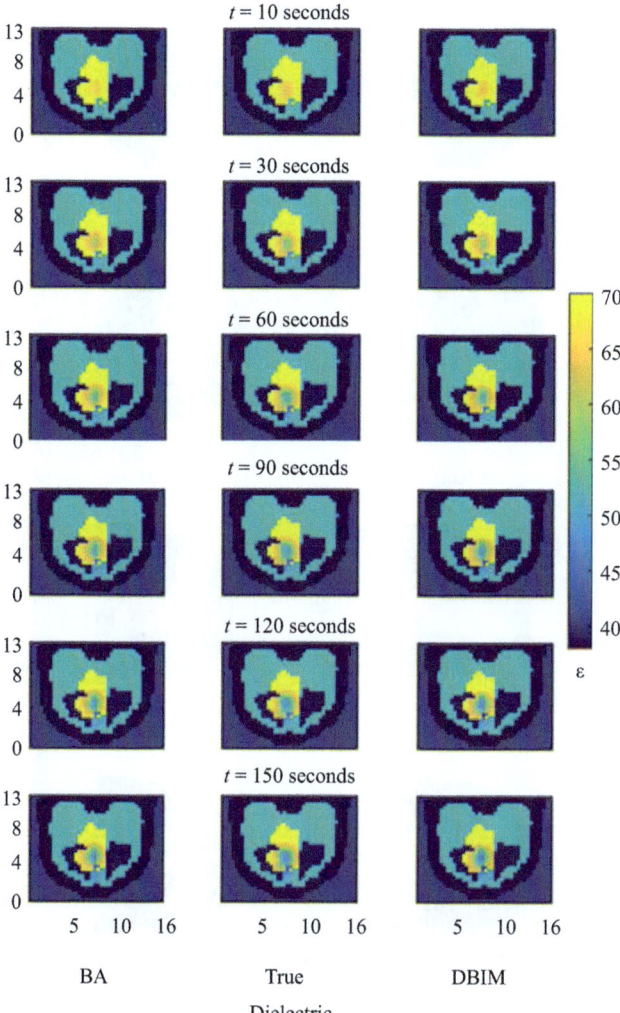

Figure 6.4 *Coronal slice of relative dielectric constant maps directly adjacent to interstitial probe over time during thermal therapy (figure is referred from [18]). The middle column is the true dielectric maps. The left and right columns are the reconstructed dielectric using BA and DBIM, respectively. Spatial units are centimeters.*

two numerical breast phantoms were tested in a simulated MWI system prototype to evaluate the performance of the inverse solver. Figure 6.5 shows the reconstructed class 2 phantom with 4 mm image resolution, and the reconstructed image error within the breast domain is less than 0.82% [19].

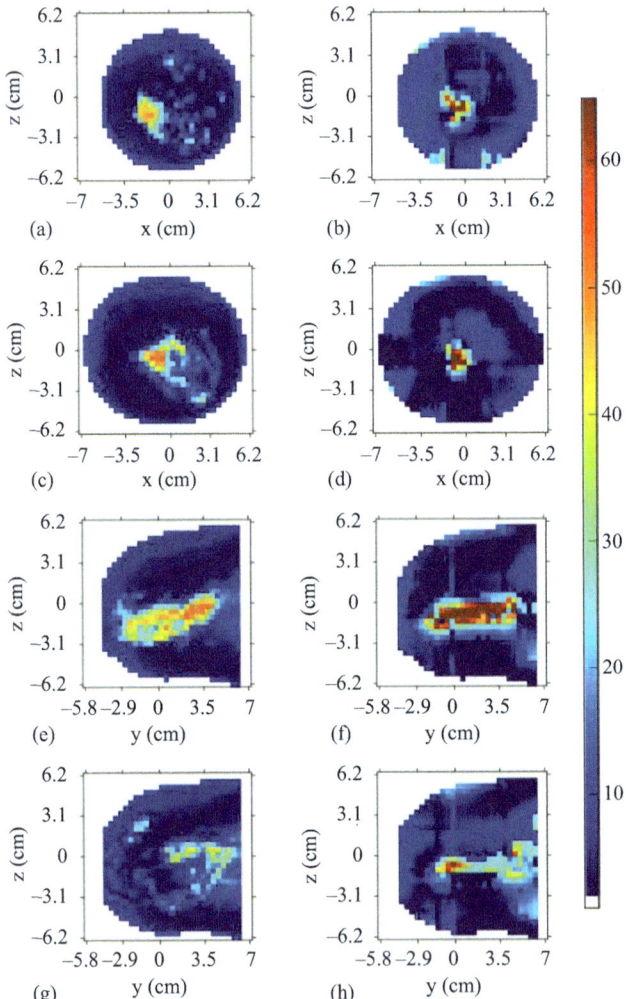

Figure 6.5   Class 2 (scattered fibroglandular) phantom that derived from the MRI [49]. Left column: actual geometry, right column: joint contrast and shape recovery (figure is referred from [19]). The top two rows show a 2D view in coronal cross section and bottom two rows show sagittal 2D view.

The authors recently have designed a new multi-frequency MWI system prototype for medical imaging and thermal therapy monitoring [50]. Figure 6.6 demonstrates the configuration of our new system. Compared with the system prototypes in [3], the new system is over four times larger. A newly designed antenna array can operate at four

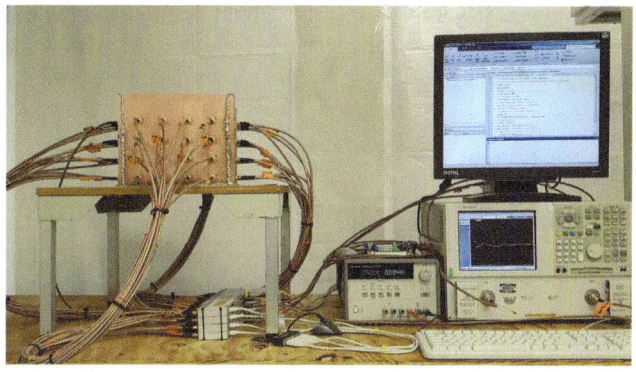

Figure 6.6   The new multi-frequency MWI preclinical system in [50]

resonant frequencies than span in the range of 0.5–3 GHz and the novel protein-based emulsion shows low loss, dielectric controllable, long-term and wide-temperature stability [50]. New multi-frequency inverse algorithms and animal phantom experiments are currently being designed for future medical imaging applications.

6.3   Deep-learning in medical imaging

We have explained the principles of common medical imaging techniques and the electromagnetic inverse algorithms used for EIT and MWI systems in the last section. The image reconstruction involves combining the physics of imaging, measured data, and prior information [21–24]. For MRI and CT scans, the majority of research interests in recent years has been concentrated on data processing and extraction of image information. As for MWI and EIT, there have been several creative methods to improve the performance of the physics-based imaging algorithms. Nonetheless, the difficulties of the inverse scattering problem, such as ill-posedness, nonlinearity, and non-uniqueness are still largely unsolvable.

The inverse scattering at the heart of physics-based imaging methods impose inherent challenges on the entire imaging framework. To address some of these challenges, machine learning (ML) assisted methods have offered promising alternatives. As the computational capabilities have increased over the years, the development of learning-based methods has enabled breakthroughs in medical imaging applications. In this section, the machine learning concepts and their most recent developments are explained at first. Then the basic architectures of common deep learning neural networks are introduced. The rest of this section will discuss recent studies of deep learning neural networks in medical imaging applications.

From the evolution of pattern recognition and computational learning theory in artificial intelligence, machine learning is aimed at constructing algorithms that can learn from data and make predictions. According to Arthur Samuel, machine learning methods "give computers the ability to learn without being explicitly programmed" [51]. The computational models generated from ML methods help researchers uncover hidden insights through learning from previous input data and their patterns or trends [52].

6.3.1   Machine learning

The critical element in ML methods is the data. Typically, the data set to construct an ML method can be split into: training and testing data sets. The training data set is further separated into estimating and validation data sets. Computational models are selected and "trained" to obtain the optimized model parameters using the estimation data set to match the target features from the validation data set as the input. This estimation continues until the trained model with the optimized model parameters can successfully predict the results.

Machine learning algorithms are usually categorized into two groups. The first one is supervised learning, where elements in the training data set are labeled, which implies each input data is related with a labeled target feature as the output. A trained computational model or a function is generated from the training data set in this learning approach. This model/function can predict the results from the new input data. Supervised learning is usually used in regression analysis and categorizing statistical characteristics. The other approach is unsupervised learning, in which no feature or output value for a decision is in the data set. The unsupervised learning method is commonly designed for clustering and anomaly detection. During unsupervised learning, the ML method is developed to find the groups with similar features or characteristics from the input data set.

In medical imaging, many ML algorithms have been developed to extract information from the images and improve image resolution. Example of ML algorithms in medical imaging [53–55]:

1. K-Nearest Neighbor (KNN): In a training sample set, each data point in the sample set corresponds to a label. When entering new data without labels, each feature of the new data is compared with features of all categories in the sample set. Then the algorithm extracts the categorical label of the data with the most similar (the nearest) features in the sample set. The KNN algorithm is suitable for data with limited number of labels with overlaps [53].

2. Naive Bayes: The Naive Bayes classifier is designed to categorize the data based on the Bayesian decision theory. The classifier is named "naive" because all features in the dataset are mutually independent and equally important. Compared with other ML algorithms, the Naive Bayes is effective even using small-size datasets and can process multiple features in one problem [54].

3. Support Vector Machine (SVM): The SVM algorithm finds the $(n-1)$-hyperplane of the n-dimensional vector that can separate two datasets with the maximum margin. This method is commonly used in high-dimension data regression and classification and can be applied in both supervised and unsupervised learning approaches. Compared with the KNN and Naive Bayes methods, the SVM can better handle the datasets in which elements have nonlinearity [54,55].

4. Random Forest: Random Forest is the extension of bootstrap aggregation (also called "bagging"). Based on the decision tree, the Random Forest randomly chooses a feature subset first, then chooses the "best" feature from this subset for splitting. If the size of the chosen subset is the same as the number of features of the input data set, the base tree construction of the random forest will be the same as the decision tree [53,54].

5. K-means: As an unsupervised learning algorithm, K-means clustering calculates the minimum square error of all K clusters in the sample set with a size of n to find the optimum clusters, which classifies all data points to the optimum clusters. Generally, this algorithm requires a sizeable computational cost, and neural networks are introduced to accelerate the learning speed [54].

Other imaging methods, such as Boosting, support vector regression (SVR), C-means, supervised descent method (SDM), and compressive sensing, are also used in some medical imaging applications [24,56–58].

In medical imaging, ML methods are mainly applied in anomaly classification (diagnosis) and imaging pattern regression (segmentation). Table 6.1 lists some of the ML studies in the last decade applied in different medical imaging techniques and their achievements. Note that we only list the machine learning methods that are not solely utilized by the deep learning neural networks in Table 6.1. In recent years, deep learning methods in medical imaging have been a frontier research area. The deep learning neural networks will be reviewed and discussed in the following subsection.

In Table 6.1, almost all applications of ML algorithms are to segment images and disease diagnosis or detection. The majority of study interests are focused on the SVM applications and improvements [59,71,82]. Most ML algorithms are used for imaging in the brain and chest (including breast and lung) area for cancer or lesion diagnosis [63,76]. Compared with a large amount of ML applications in MRI and CT, ML applications in EIT and MWI have thrived in the recent half decade. Unlike the EIT machine learning methods that combined physics-based models with ML methods to generate conductivity images, most MWI ML methods are purely data-driven methods that detect cancers or lesions from the electromagnetic field distribution [82–84].

Deep learning developments are relatively new compared to other branches of machine learning. Because of the ever-increasing evolution of computational capabilities, the recent deep learning methods can solve more complicated problems by building a more complex multi-layer artificial neural network (ANN) [86]. ANN is designed to mimic the interaction between the human neural system and real-world objects. Unlike most ML methods, the raw data input into the deep learning neural networks (DNN) based on the multi-layer ANN are not manually labeled. Features can be automatically extracted through the multiple hidden layers inside the ANN. The following subsection will introduce the commonly used DNNs in medical imaging.

Table 6.1   Machine learning methods in medical imaging (excluded deep learning method)

Imaging	ML method	Organs	Clinical purpose
MRI	SVM & CNN	Brain	Brain Tumor Segmentation [59]
	SVM	Brain	Diagnosis (Alzheimer) [60]
	SVM	Carotid	Diagnosis (Carotid Plaques) [61]
	SVM	Breast	Diagnosis (Cancer) [62]
	K-means & SVM	Brain	Brain Tissue Segmentation [63]
	Random Forest	Bone	Diagnosis (bone chondrosarcoma) [64]
	Random Forest	Heart	Heart Tumor Segmentation [65]
	K-means & C-means	Brain	Brain Tumor Segmentation [58]
	K-means	Blood	Diagnosis of Leukemia [66]
	kNN	Breast	Mammographic Microcalcifications [67]
	Naïve Bayes	Brain	Brain Tumor Segmentation [68]
PET/CT scan	SVM	breast	Diagnosis (Cancer) [69]
	SVM	Brain	Diagnosis (Alzheimer) [70]
	SVM	Liver	Diagnosis (Liver Lesion) [71]
	SVM	Lung	Diagnosis (Covid 19) [72]
	SVM	Lung	Diagnosis (Cancer) [73]
	K-means	Lung	Diagnosis (Cancer) [74]
	Random Forest	Brain	CT synthesis [75]
	Random Forest	Chest	Tissue Segmentation [76]
	K-means	Kidney	Kidney Segmentation [77]
	Naïve Bayes	Heart	Diagnosis (Coronary artery) [78]
EIT	K-means & kNN	Phantom	Image reconstruction [79]
	SVM	Brain	Diagnosis (Stroke) [80]
	kNN	Phantom	Image reconstruction [81]
	SDM	Lung	Image reconstruction [25]
MWI	SVM	Breast	Cancer Detection [82]
	SVM	Brain	Stroke Detection [83]
	SVM & kNN	Breast	Lesion Detection [84]
	K-means & C-means	Brain	Stroke Detection [85]

6.3.2   Deep learning neural networks

Deep learning is an area of machine learning that obtains featured representations from raw data through multi-step feature transformation and further input to the prediction function to get the final result. The idea behind "Deep" learning comes from multiple linear and nonlinear perceptrons in the neural network layers that are applied to automatically extract all unlabeled and hidden features from raw input data. The critical problem that a deep learning method needs to solve is the Credit Assignment Problem, which analyzes the effect of each component in the learning method on the final output result and updates the weight and learning rate from the feedback of analysis during the learning process [87].

Each inner component (layer) cannot directly obtain the supervised information, but rather from the final supervised information of the whole learning network. Nowadays, neural networks are commonly used in deep learning since the error back-propagation algorithm can be applied in this deep learning model to solve the credit assignment problem [88]. Common DNN architectures are listed below.

6.3.2.1 Fully connected neural network (FCNN)

A fully connected neural network (FCNN) (Figure 6.7) is also known as a feedforward neural network, since there is no feedback connection between the model function input and output [89]. FCNN adopts a one-way multi-layer structure, to fit a model function and is the simplest type of neural network. Signals received by each neuron in the FCNN are from the neurons in the previous layer. After processing the received signals, neurons will produce output signals to their following layer. Generally, the initial and the last layers of an FCNN are separately defined as the input and output layers. Other intermediate layers in the FCNN are the unified as "hidden layer," which can be a one-layer or a multi-layer structure.

The main algorithms used in the FCNN are the forward-propagation (FP) and back-propagation (BP) algorithms. If we assume the function of ith layer is f_i, the input data $\mathbf{X}$ and output data $\mathbf{Y}$ can be written as a nested function $\mathbf{Y} = F(\mathbf{X}) = f_N(f_{N-1}(\cdots f_2(f_1(\mathbf{X}))))$. The BP algorithm passes the information of the loss to the "back" layers by calculating the function derivative of $\mathbf{X}$ in each layer $(\partial f_i / \partial \mathbf{X})$,

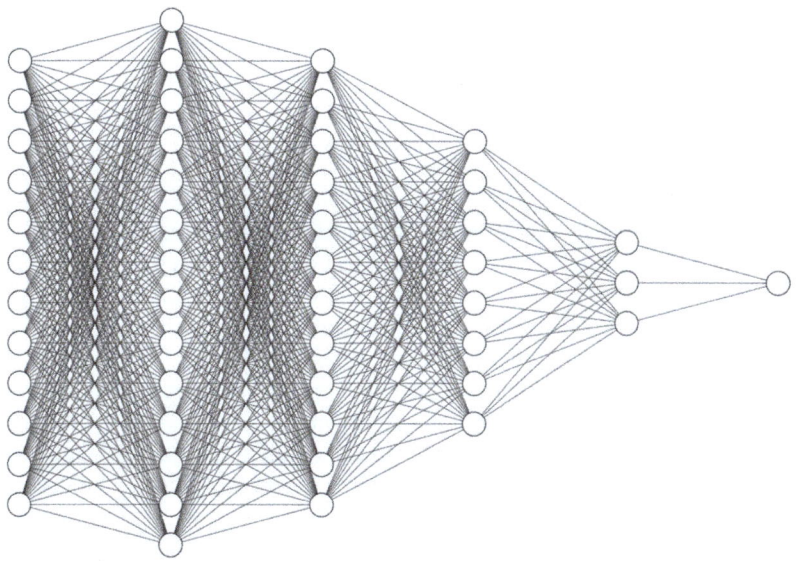

Input Layer $\in R^{12}$ Hidden Layer $\in R^{14}$ Hidden Layer $\in R^{12}$ Hidden Layer $\in R^{8}$ Hidden Layer $\in R^{3}$ Output Layer $\in R^{1}$

Figure 6.7   Structure of a typical fully connected neural network

which can be achieved through the chain rule calculation. The derivative is then used to update the weight of each layer.

The function f_i in each layer with its input $\mathbf{X}_i$ is usually written as $f_i(\mathbf{X}_i) = \sigma_i(\mathbf{W}_i\mathbf{X}_i + \mathbf{b}_i)$, where $\mathbf{W}_i$ and $\mathbf{b}_i$ are the weight matrix and bias of each layer. The activation function (σ_i) is generally taken as a nonlinear function (e.g., ReLU, Sigmoid, Logistic) to mimic the neuron's nonlinear perception behavior. A loss function (e.g., Softmax, Sigmoid cross-entropy, Euclidean) is used to evaluate the neural network's performance.

This section briefly introduces the structures and concepts of different neural networks. Details can be found in multiple textbooks [90–92].

6.3.2.2   Convolutional neural network

The large number of parameters in FCNN lead to a few shortcomings that makes training difficult and cause over-fitting. Because of these drawbacks and increasing computational cost for image processing, the convolutional layer was introduced in the neural network and evolved as the CNN [93]. A typical CNN contains the following layers: Input layer, Convolutional layer, ReLU layer, Pooling layer, Fully connected layer, and Loss layer.

The most crucial part in CNN is the convolutional layer. For two discretized signals f and g in signal processing, the convolution is defined as $(f * g)[i] = \sum_{m=0}^{m=I} f[m]g[i - m] = \sum_{m=0}^{m=I} f[m]g([i - m]_{mod(I)})$. However, the operation in the convolution layer of the CNN is "cross-correlation." When a filter F with size K is applied to a data vector $\mathbf{x}$ with length M, this operation will result in $y_m = \sum_{k=1}^{K} F_k x_{m+k-1}$. Similarly, for a 2D matrix data $\mathbf{X}$ (size: $M \times N$) with weighting matrix $\mathbf{W}$ (size: $J \times K$), the convolution is defined as $\mathbf{Y} = \mathbf{W} \otimes \mathbf{X}$, where each element of the convoluted matrix $\mathbf{Y}$ can be written as follows:

$$y_{m,n} = \sum_{j=1}^{J} \sum_{k=1}^{K} w_{j,k} x_{m+j-1, n+k-1} \tag{6.50}$$

Visually, the weighting matrix $\mathbf{W}$, or **convolutional kernel**, is sliding from left to right and from top to bottom along with the input matrix $\mathbf{X}$ with step length 1. The output element $\mathbf{Y}[m, n]$ is calculated with the window around on $\mathbf{X}[m, n]$ through the cross-correlation operation using the input data and chosen convolution kernels. In this circumstance, the output size is $(M - J + 1) \times (N - K + 1)$.

In general, parameters that define a convolutional layer are: number of convolutional kernels P, size of each kernel (J, K), zero-padding size (M_0, N_0), stride length (s_h, s_w). Once we have the output from the convolutional layer, the ReLU layer will use it as the input and generate the nonlinear relation between the convolutional and pooling layers.

Although the number of neuron connections is decreased by introducing the convolutional layer, the number of neurons for each feature is not changed significantly. The pooling layer is presented in the CNN to shrink the size of neuron groups. The pooling layer can also mitigate the sensitivity of convolutional layers to location and spatially downsample representations.

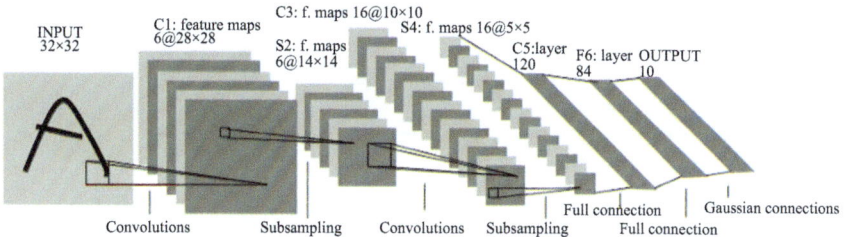

Figure 6.8   *Structure of the LeNet-5 convolutional neural network (figure is replotted from [93]). A convolutional NN, here used for digits recognition. Each plane is a feature map, i.e., a set of units whose weights are constrained to be identical [93].*

Similar to the convolutional layer, a pooling function slides along the input matrix **Y** from the convolutional layer. The pooling function is usually chosen to calculate either the maximum or the average value of the elements in the pooling window. For instance, if the pooling layer size is $U \times V$, the mean-value pooling function value at element index (α, β) of the pooling output matrix **Z** is:

$$z_{\alpha,\beta} = \frac{\sum_{j=1}^{J} \sum_{k=1}^{K} y_{\alpha+j-1,\beta+k-1}}{UV} \qquad (6.51)$$

The pooling function also needs zero padding and defining strides. Unlike the convolutional layer, the pooling layer pools each input channel separately instead of summing the inputs up over media [94]. Therefore, the number of input channels is the same as the output channels during the pooling operation.

After several convolutional and pooling layers, the neurons go through fully-connected neural layers and finally output loss values to all possible features. Figure 6.8 depicts the well-known LeNet-5 convolutional neural network for the handwritten and machine-printed character recognition [93]. This network contains two convolutional layers, two average pooling layers, two fully connected layers, and a softmax classifier in the end. Details of the example network can be found in [93,95]. This specific DNN structure has been widely used for image processing since 1998, and it is the major deep learning method used in medical imaging applications. We will discuss the medical imaging applications using CNN in the following subsection.

6.3.2.3   Recurrent neural network

Above DNNs are the feed-forward networks in which the input information begins at the input layer, passes through hidden layers, and eventually reaches the output layer. In recurrent neural networks, also known as RNNs, feedback signals are used throughout the construction process to enable the creation of internal states or memories. These memories store important information that is connected to the stimuli

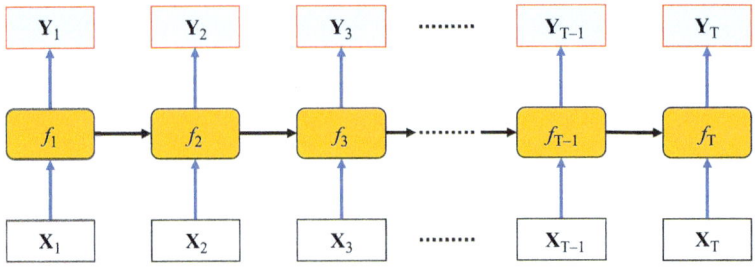

Figure 6.9   Structure of a one-layer recurrent neural network, where subscripts
$1, 2, 3, \cdots, T - 1, T$ are the time steps

received before (recurrent). Since RNN takes into account both the most recent input data and the feedback data from prior states, it is well suited for modeling sequential data (inputs are dependent on one another in a streaming pattern), which may include both temporal and spatial information [96].

The RNN is built based on the latent autoregressive model. This model uses the hidden state to store the model state of the previous step and applies the stored hidden state as the input of the model function with the input data to calculate the model state at the following step. Starting from a one-layer FCNN as an example, we know the output can be written as:

$$f = \sigma(\mathbf{W}_{xf}\mathbf{X} + \mathbf{b}_{xf}) \tag{6.52}$$

Here we rewrote the weighting matrix as $\mathbf{W}_{xf}$ and bias as $\mathbf{b}_{xf}$ to show the relationship between input data $\mathbf{X}$ and function f. The sequential data $\mathbf{X}_t$ are input at different time steps $1, 2, \cdots, t, \cdots, T$. Based on the definition of the latent autoregressive model, the output of the model f_t at time step t is generated from the input data at time step t and the last output state of the model f_{t-1}, which can be written as:

$$f_t = \sigma(\mathbf{W}_{xf}\mathbf{X}_t + \mathbf{W}_{ff}f_{t-1} + \mathbf{b}_{xf}) \tag{6.53}$$

where $\mathbf{W}_{ff}$ is the weighting matrix mapping model output from time step $(t - 1)$ to t. Bias $\mathbf{b}_{xf}$ relates to both input data and the model output at time step $(t - 1)$, and has the same size as the model output f.

The basic RNN structure is shown in Figure 6.9. In [90–92,94], more advanced RNN models are explained for different applications.

6.3.2.4   Generative Adversarial Network

The mathematical derivation and structure of GANs are complicated and beyond the scope of this book chapter. Here we only briefly introduce the basics of GANs. The detailed theory about the GAN can be found in [97] and [91].

The Generative Adversarial Network (GAN) is a relatively new deep learning network compared with CNN and RNN. It was first proposed in 2014 [97]. Studies about the GAN applications have grown substantially since 2018, particularly the medical imaging applications [98–100]. The GAN framework can be separated into a generative network and a discriminative network. The word "adversarial" is used because the generative and discriminative networks contest. The generative network generates a group of "false" datasets with similar features as the true data distribution. The discriminative network is designed to extract the true data from those "false" datasets. Each network is trained and learns from the output of the other network until the generative network can achieve acceptable accuracy. The discriminative network cannot distinguish the generative "false" dataset and the true dataset. Deconvolutional neural networks are usually used as generative networks, and CNNs are applied as discriminative networks.

6.3.3 DNN in medical imaging

In recent years, compared with the conventional ML methods such as SVM, random forest, and K-means clustering, more deep learning methods have been developed and analyzed for imaging segmentation, diagnosis, and prediction [24,96]. We demonstrate the applications of different DNNs in medical imaging in the rest of this subsection. Since an overwhelming majority of studies of deep learning applications in medical imaging are focused on MRI and CT scans, the majority of recent DNN studies that are reviewed in this section are utilized with these two medical techniques. We will briefly introduce some DNN studies in MWI and EIT, and comprehensive example results of DNN applications in MWI and EIT are discussed in Section 6.4.

6.3.3.1 FCNN in medical imaging

As the basic structure of the ANN, the FCNN has been replaced by more advanced ANNs (CNN, RNN, GAN) in modern medical imaging. However, there are still specific applications in which the FCNN exhibits a good performance. Sheikhjafari *et al.* proposed an FCNN for unsupervised image registration with MRI training data that suffer a certain level of deformation [101]. This application is helpful for realistic medical imaging since many ANNs in imaging registration use synthetic medical imaging data for training. In contrast, the clinical medical data could experience unknown deformation generated from both fixed and moving images [101]. The proposed method outperforms the GPU-accelerated mathematical imaging registration algorithm in [102].

In [103], Feng *et al.* proposed an FCNN for virtual monochromatic imaging in spectral CT. Although CNNs are more common for CT applications as it shown in the next subsection, a spectral CT image could be distorted by the non-ideal detector response. Meanwhile, it is challenging for CNN to construct a detector response function and accurate incident source spectrum [103]. This FCNN can provide accurate virtual monochromatic linear attenuation coefficients to build proper system response functions and shows outstanding performance on denoising and artifact suppression.

As for the EIT and MWI, the FCNN is more frequently applied with the physics-based inversions [26,27,104]. For example, Dachena et al. proposed a FCNN to calculate the inversion of MWI system matrix $\mathbf{A}^{-1}$. This method was used to reconstruct the 2D neck dielectric images and successfully detected the tumor from surrounding tissues [104]. Moreover, the FCNN can help reduce the matrix calculation time significantly [26] and provide preconditioned images for further reconstruction [27]. The detailed implementation of the FCNN in EIT and MWI will be explained in Section 6.4.

6.3.3.2 CNN for diagnosis, imaging segmentation, and reconstruction

Compared with the FCNN, CNN can learn hierarchical feature representations of the image and compute features from locally connected voxels, which helps the image classification and segmentation [96,105]. The architecture of CNN is also suitable for high-performance computing platforms, such as the graphics processing unit (GPU) [24]. Because of these advantages, a significantly increasing number of studies have been conducted on the topic in the last decade. In [96] and [106], dozens of CNN research papers between 2015 and 2018 have been reviewed on multiple MRI and CT imaging applications (diagnosis, detection, segmentation, registration, enhancement, and reconstruction). Yang and Yu also summarized the CNN and its extended neural networks in detection and segmentation in medical imaging [107]. In Table 6.2, we show some of the latest developments of CNN in medical imaging published just in 2021 and 2022.

In [108], a 2D-CNN based on the Keras and Tensorflow libraries with ResNet-101 was utilized to diagnose the multi-category Alzheimer's disease. The proposed CNN was trained using the Open Access Chain of Imaging Studies (OASIS) database and SegNet features with the ADNI dataset. The validation accuracy of this CNN can achieve 97% with less than 0.1 validation loss. Another new development of CNN was conducted by joint research from the United Kingdom, China, and New Zealand [109]. Using the quantum principle of entanglement on the ReLU activation function in Hilbert ReLU-based and Hilbert LReLU-based states, the authors constructed the quantum ReLU. They modified quantum ReLU activation functions for the deep CNN to diagnose Parkinson's disease and COVID-19 [109]. Although more extended computation is needed for this quantum deep learning CNN, the overall accuracy and reliability of the proposed CNN are increased by over 50% compared with the CNNs with traditional ReLU and leaky ReLU activation functions [109].

The development of CNNs benefits disease diagnosis and improves the accuracy of image segmentation. In [110], Ahmad *et al.* proposed a 3D Gaussian-weight initialization of CNN (Ga-CNN) for liver segmentation in the CT scan. Three benchmark databases (MICCAI SLiver'07, 3Dircadb01, and LiTS17) were utilized for training and testing the proposed Ga-CNN to segment the anomalies from the liver tissues with an average 96.6% segmentation accuracy [110]. The dice similarity coefficient is 1%–5% higher than the other liver segmentation DNNs [110–112]. Recently, Nvidia Corporation introduced a unified framework consisting of two architectures, UNet-Former and UNetFormer+, to segment tumors from the surrounding tissues [113]. This method is constructed with a 3D Swin Transformer-based encoder, CNN, and transformer-based decoders. These frameworks show a >90% segmentation accuracy compared with traditional CNNs in validation using the Liver databases above and the Multimodal Brain Tumor Image Segmentation Benchmark (BraTS) database from BraTs 13 to BraTs 21 [113].

The newly developed CNNs are also applicable in physics-based EIT and MWI. In [114], Biasi built an EIT-based tactile sensor. The sensors incorporate the FEM model and the CNN learning method through COMSOL and Matlab Deep Learning Toolbox to reconstruct a conductivity map from measured data. Compared with the L_2 reconstructed results (physics-based inversion only), the proposed CNN-assisted method shows a lower reconstructed RMSE value [114]. Capps et al. proposed a D-bar method to build the EIT inversion framework and use a CNN to generate reference results to fuse with the physical reconstructed results [115]. Qin *et al.* combined the 2D physics scattering EM inversion model with a multi-task CNN learning strategy for microwave breast imaging [116]. This method shows better robustness and reconstruction accuracy than the CSI-only reconstruction [116]. We will show more details of hybrid physics-based and learning-assisted MWI/EIT reconstruction through some example results provided in [26–29].

Apart from the papers introduced above, there are many papers on CNN development and applications in medical imaging [128–138]. Newly developed CNNs are applicable in a wide range of applications including image segmentation and disease diagnosis and prediction, such as COVID-19 and Alzheimer [119,120,123–126]. Refer to Table 6.2.

Even though CNNs have been primarily utilized to solve a wide range of medical imaging problems, their further applications are still limited. Due to the hierarchical feature learning, a high-dimension calculation is required in the CNN structure, which causes a high computational cost. In addition, CNN training needs a large number of training sets [144]. Recurrent Neural Network (RNN) and Generative Adversarial Network (GAN) are used in medical imaging to overcome the limitations of CNN.

6.3.3.3   Medical RNN applications

In FCNN and CNN, all input data batches are independent sequentially, which implies that the output of these neural networks depends on the input at the current state only. However, in many medical imaging applications, an imaging system with a state machine (i.e., finite state machine, FSM) to show the medical images in both current and previous states is required. Under these circumstances, the RNN is applied to explore the temporal information of the medical imaging data sequences [96,145].

In [96], RNN applications reported between 2015 and 2018 in computer-aided diagnosis (CADx), such as Detection, Prediction, Classification, and Image Reconstruction, are summarized. Azizi *et al.* proposed a hybrid CNN (ResNet) and RNN model to predict the lung cancer treatment response [146]. This method can monitor the therapy follow-up response, track the radiographic changes of tumors over time, and predict patients' survival and other clinical endpoints. In [147], the Conditional Random Field was implemented in the RNN and was in conjunction with the full CNN (V-net) for imaging segmentation. This method successfully reduced the computational complexity from $O(N^2)$ to $O(N)$ and accelerated the computational speed through hybrid CPU/GPU calculation. The algorithm was tested with both 3D MRI datasets and 2D RGB data. For breast lesion classification, an RNN structure, long

Table 6.2   CNN applications in medical imaging during 2021–2022

Purpose	Organ	Malignant	CNN net	Imaging	Accuracy
DIAG	Thyroid	Cancer	Xception	CT scan US	98% [117]
	Teeth	Lesion	YOLO-CNN	X-ray	95% [118]
	Brain	Alzheimer	Fully-CNN	MRI	>95% [119]
	Brain	Alzheimer	AlexNet GoogLENet VGG16 ResNet	MRI	94% [120]
	Brain	Cancer	GA-CNN	MRI	99% [121]
	Spine	Ossification (OPLL)	VGG16	MRI	94% [122]
	Lung	COVID-19	SARS-Net	X-ray	98% [123]
	Lung	COVID-19	VGG16 InceptionV3 Xception	X-ray	94% [124]
	Lung	COVID-19	DenseNet169 XGBoost	X-ray	98% [125]
	Lung	COVID-19	EfficientNet	X-ray	99% [126]
	Pancreas	Cancer	MBU-Net	CT scan	89% [127]
SEG	Lung	Parenchyma	UNet++	CT scan	0.98 (DSC) [128]
	Brain	Tumor	DeepMedic	MRI	0.8 (DSC) [129]
	Retinal	Vessel	T-Net	OCT	0.83 (F1) [130]
	Lung	Cancer	MA-Unet	CT scan	0.97 (MDC) [131]
	Cells	Bacteria	IRUnet	Microscopy	0.93 (DSC) [132]
	Brain Liver	Tumor Spleen	UNETR	MRI CT scan	0.84 (MDC) [133]
	Liver	Lesion	CFCN	CT scan	94% [134]
	Heart	Left ventricle	U-Net	MRI	0.93 (DSC) [135]
	Brain	Brain injury	MU-Net-R	MRI	0.9 (DSC) [136]
	Brain	Tumor	U-Net	MRI	0.88 (Dice) [137]
	Breast	Tumor	MRFE-CNN	CT scan	0.89 (Dice) [138]
RECON	Breast	Tumor	GaussNewton +U-net	US MWI	0.038 (MSE) [139]
	Breast	\	CSI +CNN	MWI	0.12 (RMSE) [140]
	Brain	Tumor	YOLOv3	MWI	0.95 (F1-score) [141]
	Brain	Stroke	FEM +Residual CNN	EIT	97% [142]
	Lung	\	FEM +VGG	CT EIT	0.08 (RMSE) [143]

DIAG: Diagnosis
SEG: Segmentation
RECON: Reconstruction
US: Ultrasound

short-term memory network (LSTM), was proposed to enable the integration of temporal components of DCE-MRI with the features extracted from the CNN (VGGNet) in cancer screening and staging [148].

Another application of RNN in medical imaging is to improve the image resolution of other ML and DL methods. Between the input and the output of the ML module (i.e., Gradient descent, Variable splitting, SVM) or DL network (i.e., FCNN, CNN), the learning rate and weight coefficients are updated iteratively. Each iterative step can be viewed as an internal state before the output. An RNN can be applied as an optimization method to enhance the reconstruction quality or accelerate the learning update speed by reducing the iterative steps [149].

Moreover, owing to the temporal processing ability of the RNN, there are some specific applications where RNN shows its unique advantages. One example is the functional MRI (fMRI), an imaging technique that can monitor the dynamic states of the brain to measure the brain activity that associates with blood flow, in which RNN is the natural DL choice for this application. Wang *et al.* proposed a 5-layer deep sparse RNN (DSRNN) incorporated with the LSTM and Gated Recurrent Units (GRU) for the brain state recognition with the fMRI [150]. This method shows at least $\geq$ 80% accuracy in recognizing multiple brain activities (working memory task, gambling, motor, language, social, relational, emotional). Dvornek *et al.* also proposed a learning-generalized RNN for the task-fMRI application [151]. Particularly, this method has higher ASD classification performance with a collaborative learning process with the FC-SVM [152]. The other suitable application for RNN is Angiography, an X-ray-based technique to monitor blood flow in blood vessels. In [153], the RNN is applied in the digital subtraction angiography (DSA) and validated that RNNs can correct commonly occurring motion artifacts or incomplete DSA acquisitions to extract accurate quantitative parameters.

Compared with the CNN applications in EIT and MWI, there are not many RNN implementations in these two areas, owing to fact that the target tissues in EIT and MWI are mostly static. But some certain situations in MWI can utilize the RNN. For instance, Geng *et al.* proposed a hybrid deep learning network with both CNN and RNN to achieve human activity recognition [154,155].

6.3.3.4 New GAN for medical imaging

Although GAN was initially proposed in 2014, it has been frequently applied in medical imaging applications since 2018 [98]. In [98] and [99], a summary of GAN utilization in medical imaging is provided. The majority of papers use GANs and related medical datasets for GAN training in MRI and CT image synthesis, reconstruction, and segmentation after 2018. Xun *et al.* reviewed more than 120 GAN-based architectures for medical imaging segmentation before September 2021 [100].

New GAN structures have been constructed to improve image resolution in recent years. A Fused Attentive Generative Adversarial Networks (FA-GAN) is proposed to achieve super-resolution in MRI [156]. Results show that the proposed FA-GAN can reconstruct MR images better than other GAN structures proposed in 2017 and 2018 [156]. Recently, a Super-resolution Optimized Using Perceptual-tuned

Generative Adversarial Network (SOUP-GAN) was constructed to improve the MRI quality through anti-aliasing and deblurring [157].

Although GAN shows powerful capability in the medical imaging application, it requires highly computational hardware to generate the training model [158], particularly for 3D medical EIT and MWI reconstructions that contain limited measurements but require dense meshes.

In this subsection, we have reviewed recent DNN developments in medical imaging, where the majority of these studies were applied in MRI and CT scans. The main reason for this situation is that most clinical and commercial medical imaging systems are MRI and CT, and there are a large amount of open-access MR- and CT-datasets to help with designing and developing the DNN. On the contrary, EIT and MWI systems are mostly preclinical, and most training and testing datasets for DNN studies in EIT and MWI are synthetically generated through simulation software [27,29]. However, the new deep learning research goal in EIT and MWI applications targets to combine the learning method within the physics-based inversion framework [26,27,29], and to incorporate with other imaging techniques to improve the learning performance [28]. The next section will explain some of the example studies to understand this combination of physics-based and learning methods.

6.4   Hybrid physics-based learning-assisted medical imaging: example studies

This section will introduce several combined physics- and learning-based imaging algorithms in the medical imaging application that are developed from the basic electromagnetic theory (Maxwell's equations). Examples of studies of the algorithm in EIT and MWI are adapted from recent publications.

6.4.1   Example 1: EIT-based SDL-assisted imaging

Zhang *et al.* suggested to use the supervised descent method (SDM) for the EIT application in thorax model imaging [25]. With the SDM, prior information about the thoracic structure can be easily integrated into the inversion through training samples. Compared with the Gauss–Newton Inversion (GNI), the SDM integrated inversion achieves better image reconstruction resolution. However, as a linear regression method, the SDM may slow down the convergence of the inversion [26]. To that end, Lin *et al.* suggested the integration of fully connected neural network (FCNN) with the SDM inversion framework [26].

The inversion scheme in [25] and [26] was based on GNI. The cost function with generalized Tikhonov regularization for GNI could be written as follows:

$$\mathscr{L}(\sigma) = \| \mathscr{F}(\sigma) - \mathbf{u}_* \|_2^2 + \alpha \| \mathbf{G}\sigma \|_2^2 \tag{6.54}$$

where α is the regularization parameter, $\mathbf{u}_*$ is the measured data, $\mathbf{G}$ is the discretized operator for the spatial derivative, and $\mathscr{F}$ is the forward solver that is constructed by the FEM method for (6.8)–(6.12).

Assuming $\mathscr{L}$ is twice differentiable, the second-order Taylor expansion of the cost function $\mathscr{L}$ can be expressed as follows:

$$\mathscr{L}(\sigma_0 + \delta\sigma) \approx \mathscr{L}(\sigma_0) + \mathbf{J}_{\mathscr{L}}(\sigma_0)^T \Delta\sigma + \frac{1}{2}\sigma_0^T \mathbf{H}_{\mathscr{L}}(\sigma_0)\Delta\sigma \tag{6.55}$$

where $\mathbf{J}_{\mathscr{L}}(\sigma_0)$ and $\mathbf{H}_{\mathscr{L}}(\sigma_0)$ are the Jacobian and Hessian matrices of $\mathscr{L}$ evaluated at σ_0. The minimum of $\mathscr{L}$ is achieved when the gradient of (6.55) with respect to $\Delta\sigma$ is set to zero, which means:

$$\Delta\sigma = -\mathbf{H}_{\mathscr{L}}(\sigma_0)^{-1}\mathbf{J}_{\mathscr{L}}(\sigma_0) \tag{6.56}$$

where $\mathbf{J}_{\mathscr{L}}(\sigma_0) = 2\mathbf{J}_{\mathscr{F}}^T(\mathscr{F}(\sigma_0) - \mathbf{u}_*) + 2\alpha\mathbf{G}^T\mathbf{G}\sigma_0$, and $\mathbf{H}_{\mathscr{L}}(\sigma_0) = 2\mathbf{J}_{\mathscr{F}}^T(\sigma_0)\mathbf{J}_{\mathscr{F}}(\sigma_0) + 2\alpha\mathbf{G}^T\mathbf{G}$. $\mathbf{J}_{\mathscr{F}}(\sigma_0)$ are respectively the Jacobian and Hessian matrices of $\mathscr{F}$ evaluated at σ_0.

In the kth GNI step, $\sigma_k = \sigma_{k-1} + \Delta\sigma_k$ though constructing $\mathbf{J}_{\mathscr{L}}(\sigma_0)$ and $\mathbf{H}_{\mathscr{L}}(\sigma_0)$, which are computationally expensive. However, under SDM learning, calculation of these matrices can be avoided since $\Delta\sigma_k$ at kth learning step can be rewritten as:

$$\Delta\sigma_k = \mathscr{M}_k(\Delta\mathbf{u}_k) \tag{6.57}$$

where $\mathbf{u}_k$ is the reconstructed electric potential co-related with the conductivity map σ_k at kth learning step, and $\Delta\mathbf{u}_k = \mathscr{F}(\sigma_{k-1}) - \mathbf{u}_*$, $\Delta\sigma_k = \sigma_{k-1} - \sigma_*$. The quantity $\mathscr{M}_k$ is the descent mapping operation learned through the SDM at the kth step. Now considering N-dimensional training datasets $\Sigma_* = [\sigma_*^1, \sigma_*^2, \cdots, \sigma_*^N]^T$ and their corresponding measured data $\mathbf{U} = [\mathbf{u}_*^1, \mathbf{u}_*^2, \cdots, \mathbf{u}_*^N]^T$ as 0th initial input of the SDM, the cost function for the SDM at kth learning step can be reorganized as follows:

$$\Gamma(\Delta\sigma_k^i) = \sum_{i=1}^{N} \| \Delta\sigma_k^i - \mathscr{M}_k(\Delta\mathbf{u}_k^i) \|_2^2 \tag{6.58}$$

where $\sigma_k = \sigma_{k-1} + \mathscr{M}_k(\Delta\mathbf{u}_k)$. Cost function in (6.58) can be written in matrix form:

$$\Gamma(\Delta\Sigma_k) = \| \Delta\Sigma_k - \mathscr{M}_k(\Delta\mathbf{U}_k) \|_F^2 \tag{6.59}$$

where $\Sigma_k = \Sigma_{k-1} + \mathscr{M}_k(\Delta\mathbf{U}_k)$.

Authors in [26] assumed that $\mathscr{M}$ is a linear function as $\mathscr{M}(\Delta\mathbf{u}) = \mathbf{M}\Delta\mathbf{u}$ and $\mathscr{M}(\Delta\mathbf{U}) = \Delta\mathbf{U} \cdot \mathbf{M}^T$, the minimization of cost function (6.59) will have a close form solution for $\mathbf{M}_k$ at kth learning step for this linear SDM (LSDM):

$$\mathbf{M}_k^T = (\Delta\mathbf{U}_k^T\Delta\mathbf{U}_k)^{-1} \cdot (\Delta\mathbf{U}_k^T\Delta\Sigma_k) \tag{6.60}$$

where the singular value decomposition (SVD) method can be applied to solve the matrix inversion above [25,26].

For the nonlinear descent direction mapping $\mathscr{M}(\Delta\mathbf{u})$, a FCNN was exploited $\mathbf{Net}_k$ to calculate the kth descent direction mapping, denoting as $\mathscr{M}_k(\Delta\mathbf{u}_k^i) = \mathbf{Net}_k(\Delta\mathbf{u}_k^i)$. The formation of $\mathbf{Net}_k(\Delta\mathbf{u}_k^i)$ can be written as:

$$\mathbf{Net}_k(\Delta\mathbf{u}_k^i) = \hat{\mathbf{y}}_k = \mathbf{W}^{[2]}\mathscr{H}\left(\mathbf{W}^{[1]}\Delta\mathbf{u}_k + \mathbf{b}^{[1]}\right) + \mathbf{b}^{[2]} \tag{6.61}$$

where $\hat{\mathbf{y}}_k$ is the output vector of $\mathbf{Net}_k$, $\mathbf{W}^{[1]}$ and $\mathbf{b}^{[1]}$ are the weight matrix and bias vector of the hidden layer, $\mathbf{W}^{[2]}$ and $\mathbf{b}^{[2]}$ are those of the output layer, and $\mathscr{H}$ is the non-linear activation function of the hidden layer (ReLU is chosen in [26]).

Although only one hidden layer is chosen to build the FCNN in the SDM framework (NN-SDM), authors in [26] shows that the performance of the chosen FCNN can better reconstruct the conductivity profile than the end-to-end neural network (E2E-NN) in the actual EIT measurement. As the results shown in Figure 6.10, the LSDM, NN-SDM, and GNI can effectively reconstruct the conductivity maps, in

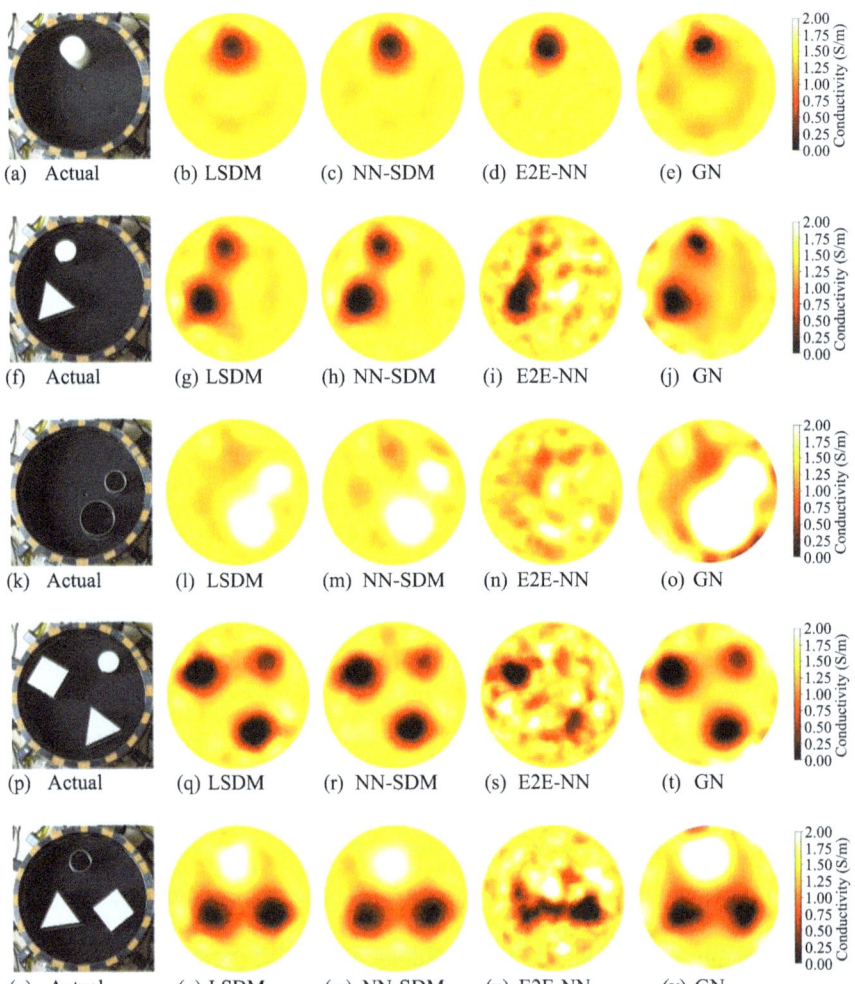

*Figure 6.10   The actual phantom model in the 2D EIT measurement (the 1st
            column figures from left) and reconstructed conductivity images by
            the LSDM, NN-SDM, E2E-NN and Gauss–Newton (GN) methods
            (images in 2nd–5th columns) (figure is referred from [26])*

which NN-SDM shows its unique capability to distinguish multiple objects with negative conductivity contrast. When predicting the conductivity map, the NN-SDM calculation is 6.5 times faster than GNI and 1.5 times faster than the LSDM.

6.4.2 Example 2: MWI(CSI)-based UNet-assisted imaging

Amer *et al.* presented a CSI cost function based on FEM construction derived from the wave equation below [159]:

$$\nabla \times \nabla \times \mathbf{E}_t^s(\mathbf{r}) - k_b^2(\mathbf{r})\mathbf{E}_t^s(\mathbf{r}) = k_b^2(\mathbf{r})\mathbf{w}_t(\mathbf{r}) \tag{6.62}$$

where subscript t is an arbitrary transmitter, $\mathbf{E}_t^s$ is the scattered field due to the presence of the object illustrated by the EM field generated from transmitter t. The contrast source variable $\mathbf{w}_t$ is defined as $\mathbf{w}_t(\mathbf{r}) = \chi(\mathbf{r})\mathbf{E}(\mathbf{r})$. Given the boundary conditions, the FEM matrix equation of (6.62) can be reorganized as follows:

$$\mathbf{H}_b[\mathbf{E}_t^s] = \mathbf{R}_b\mathbf{w}_t \tag{6.63}$$

where $\mathbf{H}_b$ is the FEM discretization matrix. The matrix $\mathbf{R}_b$ depends on the background medium properties and accounts for projecting the contrast source variables in the imaging domain onto the mesh edges in the computational domain [159].

Along the mesh edges, the scattered field is calculated as $\mathbf{E}_t^s = \mathbf{H}_b^{-1}[\mathbf{R}_b\mathbf{w}_t] = \mathscr{L}[\mathbf{w}_t]$, where $\mathbf{w}_t$ is the contrast source. With this "inverse" FEM operator $\mathscr{L}$, the cost function of the FEM-CSI is described as follows:

$$\mathscr{F}^{CSI}(\chi, \mathbf{w}_t) = \mathscr{F}^S(\mathbf{w}_t) + \mathscr{F}^D(\chi, \mathbf{w}_t) = \frac{\sum_t \| \mathbf{E}_t^{s,meas} - \mathscr{M}_{S,t}\mathscr{L}[\mathbf{w}_t] \|_S^2}{\sum_t \| \mathbf{E}_t^{s,meas} \|_S^2}$$
$$+ \frac{\sum_t \| \chi \odot \mathbf{E}_t^b - \mathbf{w}_t + \chi \odot \mathscr{M}_D\mathscr{L}[\mathbf{w}_t] \|_S^2}{\sum_t \| \chi \odot \mathbf{E}_t^b \|_S^2} \tag{6.64}$$

where operators $\mathscr{M}_{S,t}$ and $\mathscr{M}_D$ are interpolated matrix operators that transform field values calculated along the mesh edges to spatial-vector field values at receivers on surface S, and at the centers of volume meshes inside the imaging domain D. Operator $\odot$ is the Hadamard (i.e., element-wise) product [159].

At nth inversion step, the cost function is the multiplication of the MR and CSI term:

$$\mathscr{F}_n(\chi, \mathbf{w}_t) = \mathscr{F}_n^{MR}(\chi) \times \mathscr{F}^{CSI}(\chi, \mathbf{w}_t)$$
$$= \frac{1}{V} \int_D \frac{|\nabla\chi|^2 + \mathscr{F}^D(\chi_n^{CSI}, \mathbf{w}_{t,n})A^{-1}}{|\nabla\chi_{n-1}|^2 + \mathscr{F}^D(\chi_n^{CSI}, \mathbf{w}_{t,n})A^{-1}} dv \tag{6.65}$$

where V is the volume value of D, A is the mean of the facets for meshes in D. Function (6.65) can be minimized through the conjugate gradient method using Polak-Ribere search directions [159].

In 2019, Asefi and LoVetri designed an air-based MWI system prototype for breast imaging [6]. The system can detect tumors in the breast phantom using the multiplicatively regularized (MR) FEM with the contrast source inversion (FEM-CSI) algorithm. Reconstructed tumor models' size and dielectric value shows a relatively high-accuracy level. However, the anomalies associated with reconstructed images could cause misjudgment on the tumor detection. To eliminate the anomalies and increase the reconstruction accuracy, two implementations of the DNN with the CSI imaging are proposed [27,29].

Initially an implementation proposed in [29], where a CNN structure, UNet, is used to improve the imaging accuracy after CSI reconstruction. The entire combined CSI-CNN inversion framework is shown in Figure 6.11. The UNet is pre-trained with 600 breast-tumor phantom dielectric data points corresponding to their scattered field measurements. In each phantom model, tumor(s) is randomly placed at a position and grow randomly until the maximum diameter threshold is reached. The maximum diameter is also randomly chosen between 1.1 and 1.5 cm. After the CSI reconstruction is completed, the reconstructed real and imaginary dielectric images at five frequencies are input to the UNet and generate the final dielectric images.

Another DNN implementation is proposed using a magnetic-based MWI system prototype in [27]. Before the CSI reconstruction, a supervised FCNN is applied to generate the dielectric map as a preconditioner for the physics-based inverse algorithm. The FCNN is pre-trained with more than 50,000 synthetic data vectors with their measured magnetic scattered field data with the existence of both tumor and fibroglandular region $H_{fibro+tumor}^{sct}$ [160]. Each data vector contains four features of the synthetic tumor and surrounding fibroglandular region: radius, height, and the real and imaginary parts of the dielectric values. This FCNN-CSI framework is shown in Figure 6.12. With the trained FCNN, the network can generate the predicted property data vector $\mathbf{p}$ from its measured $H_{fibro+tumor}^{sct}$ as the pre-input for the CSI algorithm. Then the CSI uses this dielectric map from FCNN with the measured scattered magnetic field H_{tumor}^{sct} with the tumor only to reconstruct the dielectric images of the tumor [27].

Figure 6.11   Schematic for the proposed U-Net to reconstruct the real part of permittivity. The input to the network is the CSI reconstruction, and the network is trained to output the corresponding true 3D permittivity map (figure replotted from [29]).

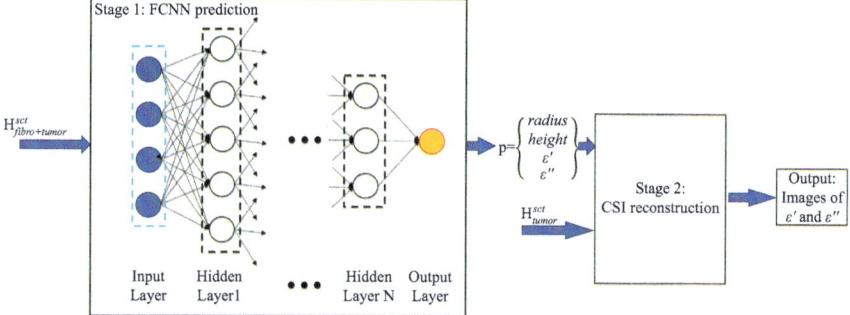

*Figure 6.12   Inversion of scattered-field data to recover prior information (vector of parameters **p**), followed by CSI reconstruction using neural network recovered prior information (figure replotted from [27])*

6.4.3   Example 3: MWI(BIM)-based CNN-assisted imaging

The incorporation of DNN to assist physics-based inverse algorithms can be categorized into three classes:

1. During the physics-based inversion, using the DNN to help calculate model parameter variance between every two iterative steps [26].
2. After the physics-based inversion, using pre-trained DNNs to improve the quality of reconstructed images [29].
3. Before the physics-based inversion, using pre-trained DNNs to generate an initial dielectric profile of both the target and surrounding environment as the pre-input of the physics-based inverse solver [27].

In this section, we will review a new perspective of the inclusion of the DNN in electromagnetic medical imaging, namely, a BIM-based and CNN-assisted imaging method that was proposed by Chen *et al*. [28]. Instead of using the dielectric image data to train a CNN as in [29], the CNN in [28] is designed with another approach by mapping the more extensive and accurate MRI database to corresponding dielectric images. Both of the geometric and biologic properties of tissues from other imaging approaches (i.e., MRI, CT scans) can be reconstructed into the dielectric image through this method in [28].

The method framework method shown in Figure 6.13 has three stages: first, the mapping function is learned from a pre-trained CNN that can transfer input MR images to dielectric images. After that, the trained CNN will be able to provide predicted dielectric scans for each new patient by applying the trained CNN to the MRI scan of that patient. In the final step of the process, the estimated dielectric maps are utilized as a refined initial guess for the iterative dielectric image reconstruction that is based on a physics model.

In comparison to the conventional initialization methods, the CNN-predicted dielectric images are much more accurate representations of the true dielectric images.

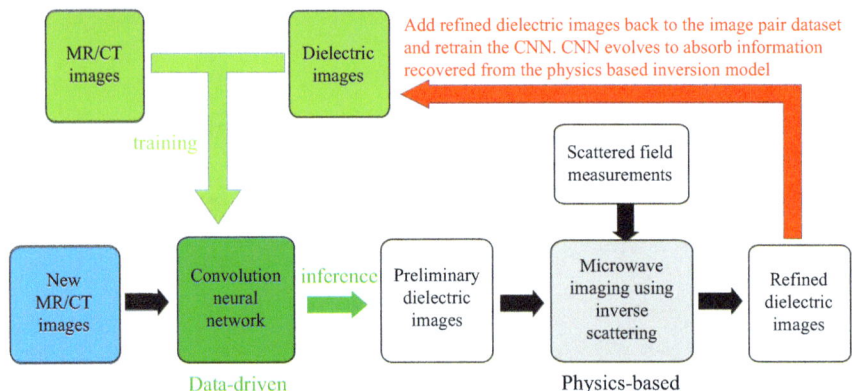

Figure 6.13   CNN assisted multi-modality dielectric imaging method description (figure is referred from [28])

It is because CNN-predicted dielectric images contain an abundance of brain structure and pixel-level tissue property information that was transferred from other imaging modalities. The CEM forward model, FDTD, provides a substantially more accurate first estimate of the total electric field by initializing with the CNN-predicted dielectric image. This method achieves a considerable reduction in the nonlinearity and ill-posedness of the MWI inverse scattering problem. Then, by performing BIM-based microwave imaging, this physics-based method recovers additional information that may have been absent from the training data and not learned by the CNN. Finally, the BIM output results can update the training of the CNN by including a fresh set of image pairings in the training database, which will include the reconstructed dielectric image together with its associated MR image. The CNN model can enhance itself to generate higher-resolution dielectric images from subsequent input MRI images. Therefore, this method can be dynamically evolved through ingesting new information retrieved from a physics-based model and incorporating that information into the CNN model.

The learning-inversion framework, significantly alleviates the nonlinearity and ill-posedness of the physics-based inversion model by the CNN-predicted dielectric images. Additionally, the physics-based inversion model will complement the CNN by recovering information that was absent in the training data and was not learned by the CNN, hence enhancing the performance of the CNN. Note that the CNN can also be used to map and ultrasound images to the dielectric images and is not specific to learning from MRI images.

6.4.3.1   Training data preparation: MR-dielectric data pairs

Before the CNN training process, the MR-dielectric data pairs need to be generated. In [28], MR segmentation is initially conducted using MR T2 images from the BRATS-15 dataset through the Statistical Parametric Mapping (SPM) tool box. Four major tissue types of the brain including soft tissues, white matter, grey matter, cerebrospinal fluid (CSF), and glioma tumor, are obtained during the process. The unified segmentation approach integrates a smooth and nonlinear registration with tissue probability maps and generates a model based on a combination of Gaussian distributions.

After the segmentation, authors assumed that both T2 pixel values and dielectric values of the input and reconstructed images are clustered and with similar statistical distribution for each tissue type [28]. Based on the assumption, they used a piecewise-linear mapping method to map the T2 pixel intensity to the dielectric distributions for different tissues. For single-frequency dielectric reconstruction where complex permittivity value is written as $\tilde{\varepsilon} = \varepsilon + \sigma/(j\omega)$, the mapping functions to map the permittivity and conductivity values to the T2 pixel intensity value at the ith image pixel are shown in equations (6.66) and (6.67) below:

$$\varepsilon(i) = 1.1\varepsilon_m + \frac{(0.9\varepsilon_m - 1.1\varepsilon_m) \times [V_{T2}(i) - (\mu_{T2} + 2\delta_{T2})]}{(\mu_{T2} - 2\delta_{T2}) - (\mu_{T2} + 2\delta_{T2})} \tag{6.66}$$

$$\sigma(i) = 1.1\sigma_m + \frac{(0.9\sigma_m - 1.1\sigma_m) \times [V_{T2}(i) - (\mu_{T2} + 2\delta_{T2})]}{(\mu_{T2} - 2\delta_{T2}) - (\mu_{T2} + 2\delta_{T2})} \tag{6.67}$$

where V_{T2} is the T2 pixel value. The mean μ_{T2} and the standard deviation δ_{T2} of Gaussian probability density functions of T2 image pixel histogram related to different brain tissue type are shown in Table 6.3.

For the multi-frequency reconstruction (0.5–2 GHz in [28]), a single-pole Debye model (6.68) is used for the complex dielectric reconstruction:

$$\tilde{\varepsilon}(\omega) = \varepsilon_\infty + \frac{\Delta\varepsilon}{1 + (j\omega\tau_d)} + \frac{\sigma_d}{j\omega\varepsilon_0} \tag{6.68}$$

where ε_∞, $\Delta\varepsilon$, and σ_d the parameters of this single-pole Debye model. The values of these parameters of each brain tissues are listed in Table 6.4.

Table 6.3　T2 pixel intensity distribution and the measured dielectric value (at 1.2 GHz) of brain tissue [28]

	μ_{T2}	δ_{T2}	ε_m	σ_m
White matter	108.82	20.68	38.07	0.69
Grey matter	145.64	39.41	51.56	1.08
CSF	244.62	72.72	68.09	2.55
Tumor	182.24	52.25	N/A	N/A

Table 6.4　Brain tissue T2 voxel intensity distribution and 0.5–2 GHz Debye model dielectric parameters (τ_d=13.27ps) [28]

	μ_{T2}	δ_{T2}	ε_∞	$\Delta\varepsilon$	σ_d
White matter	108.82	20.68	5.04	33.57	0.45
Grey matter	145.64	39.41	5.72	46.63	0.75
CSF	244.62	72.72	18.73	50	2.22

With the complex dielectric values of different types of tissues, these values generated from the patient training process are co-registered with the correlated MR T1 images and create the MR-dielectric image pairs. For single-frequency reconstruction, MR T1-ε and MR T1-σ data pairs are created, while the MR T1-ε_∞, MR T1-$\Delta\varepsilon$, and MR T1-σ_d data pairs are created for the multi-frequency reconstruction.

6.4.3.2   CNN training

The CNN structure in [28] is shown in Figure 6.14, which has three convolution layers (Conv) with parameterized ReLU (PReLU) in each layer, and a pooling layer at the end. In each convolution layer, a computation is performed to learn the weights of the kernel and determine the inner product of the local patch of the input image and the kernel. Non-linearity relation between the output of each layer and the input x_i is provided through the PReLU as $\max(0, x_i) + \alpha_i \min(0, x_i)$, where α_i is the parameter that is to be learned by the CNN. After the convolution, the intermediate image is downsampled through the pooling layer to an appropriate resolution for the BIM solver, which will physically increase the resolution through the inversion process. The loss function for the CNN is defined as:

$$L(\Theta) = \frac{1}{N} \sum_{i=1}^{N} \|F(t_i, \Theta) - d_i\|^2, \tag{6.69}$$

where d_i is the synthesized dielectric image patch, t_i is the MR T1 image patch, N is the batch size, and Θ consists of all dielectric parameters to be learned. F the mapping function to map the input MR T1 image to output its dielectric image. In [28], authors used a gradient-based method with an adaptive moment estimation (adam) using Caffe framework to achieve the minimization of the cost function. The training process uses 80% patient data of the selected BRATS datasets while the other 20% is used for testing.

6.4.3.3   MWI inversion: BIM

The synthetic MWI inversion in [28] used the same EFVIE function as that in equation (6.37). Since the dielectric images generated by the pre-trained CNN are used as the input, the background is inhomogeneous and the Green's function needs to be numerically calculated through the CEM forward solver, herein an in-house developed

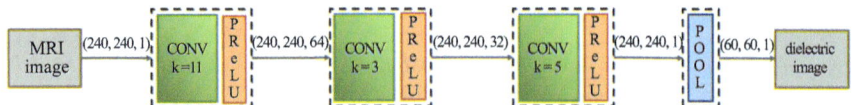

Figure 6.14   Block diagram of CNN for learning dielectric image from an MR T1 image (figure is referred from [28]). First two values within the bracket on the arrow indicate the size of the image in the x and y directions, respectively. The third value indicates number of channels/filters for the layer on the left. K denotes the filter size.

FDTD solver. Synthetic scattered electric field data are also simulated from the FDTD method. For each new patient, with the *a priori* input dielectric function and the scattered electric field, the BIM converges fast with a higher resolution than the CNN-predicted dielectric results. The BIM-refined dielectric images are then co-registered with the new patient's MR T1 image and input into the training data set. Details of the BIM is discussed in the previous Section 6.2.4.

6.4.3.4   Reconstructed results

Figure 6.15 presents the outcome of the inverse scattering image reconstruction performed on a patient (AAB). In the FDTD simulation, 32 transmitters and 32 receivers working at 1.2 GHz are placed around the brain phantom. The background medium is a coupling fluid with $\varepsilon_r = 20$ and $\sigma = 0.1$ S/m. The physics model recovers some of the contrast that is otherwise lost in the reconstructed dielectric image, which represents a substantial portion of the tumor. However, the physics model cannot recover all of the details of contrast. The fact that there were not enough learning instances of tumors included in the training data is a contributing factor in the CNN's inability to fully predict the details of the tumor area. When compared to those of the three other kinds of tissue, the number of voxels in the chosen low-grade glioma (LGG) MR datasets that correspond to tumors is considerably lower. Because of the high dielectric heterogeneity that could be in glioma tumors and the fact that only a rough

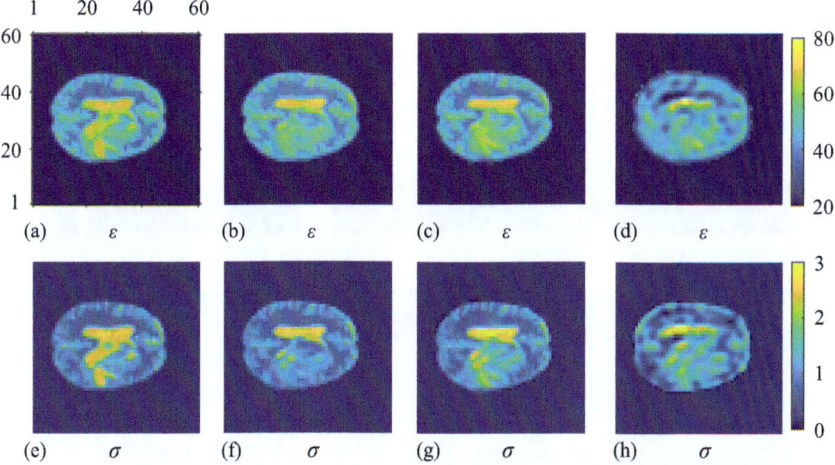

Figure 6.15   *4 mm resolution single frequency reconstruction results of patient AAB slice 90 (figure is referred from [28]). The first row is ε images and the second row is σ images. First column is the true dielectric images. Second column is CNN predicted dielectric images. Third column is the recovered images starting from CNN predicted image. Fourth column is the recovered images starting from brain phantom filled with average tissue value.*

calculation of the tumor's dielectric value can be done owing to a lack of measurement data, the CNN has very little usable information to learn about the tumor. Nevertheless, the images that were reconstructed from the pre-trained CNN are visually close to the true dielectric images. On the other hand, the images that were reconstructed in the conventional way by starting from the phantom with an average tissue value are quite unresolved and contain very little information of clinical use.

A multi-frequency reconstruction can demonstrate more information comparing to a single-frequency method. As we can see in Figure 6.16 the reconstructed image with synthetic scattered electric field data at 0.8 GHz, 1.2 GHz, 1.6 GHz, and 2 GHz from 36 transmitters and 36 receivers, the Debye parameter $\Delta\varepsilon$ in (d), (e), and (f) can show both tissue and the tumor area. The BIM reconstructed images in Figure 6.16(c),

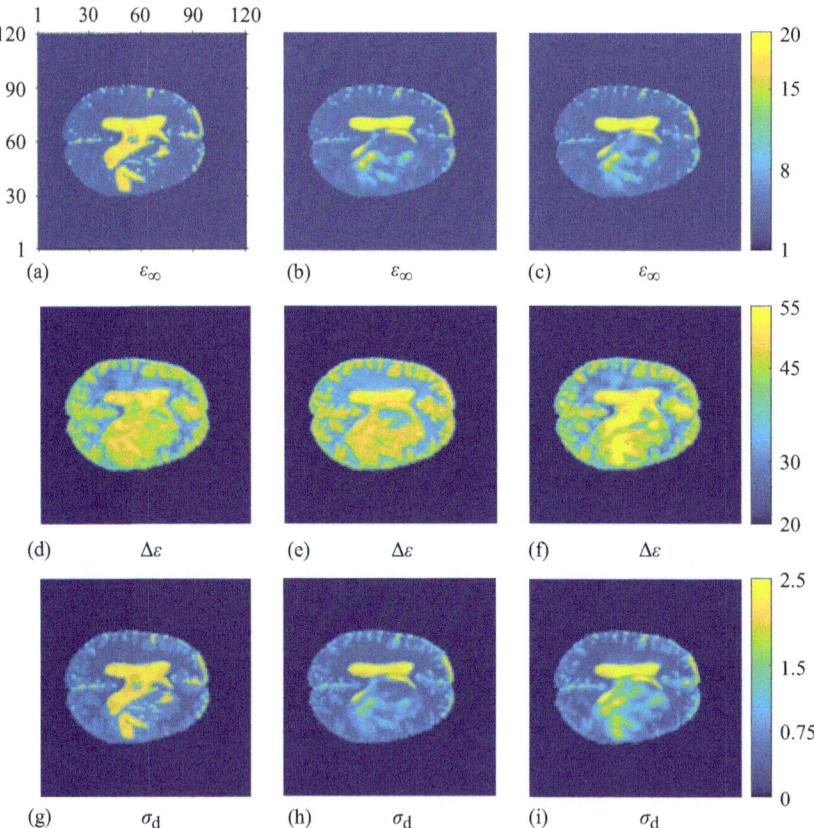

Figure 6.16   2 mm resolution 4 frequencies Debye model reconstruction results of patient AAB slice 90 (figure is referred from [28]). The first, second, and third rows are the ε_∞, $\Delta\varepsilon$, and σ_d images, respectively. The first, second, and third columns are the true, CNN predicted, and the reconstructed dielectric images.

(f), and (i) exhibit more details of the tumor and surrounding tissues using the CNN-predicted ε_∞, $\Delta\varepsilon$, and σ_d images compared with only CNN-predicted results in Figure 6.16 (b), (e), and (h). The CNN-predicted and BIM-refined results in Figure 6.16(c), (f), and (i) are closer to the ground truth images in Figure 6.16(a), (d), and (g).

6.5 Summary

In this chapter, we have discussed the physics of the common medical imaging techniques that are applied in the electromagnetic spectrum: electrical impedance tomography (EIT), magnetic resonance imaging (MRI), microwave imaging (MWI), and computed tomography (CT) scan. The EIT and MWI imaging techniques based on the electromagnetic theory and Maxwell's equations are specifically investigated with detail physics and imaging algorithms (BIM and DBIM) flowcharts. Then we explained the basic theories and structures of the machine learning and deep learning methods, and discussed their applications in medical imaging within the last decade. Particularly, we discussed the most recent deep learning network applications in medical imaging diagnosis, segmentation, and reconstruction. Furthermore, more advanced applications that combine the electromagnetic physics-based imaging methods with deep learning networks for imaging improvements are discussed in details through four recent EIT and MWI imaging studies published.

References

[1]   Ling SJ, Sanny J, Moebs W. *University Physics*. Volume 2. OpenStax, Rice University; 2021.

[2]   Shea JD, Kosmas P, Hagness SC, *et al.* Three-dimensional microwave imaging of realistic numerical breast phantoms via a multiple-frequency inverse scattering technique. *Medical Physics*. 2010;37(8):4210–4226.

[3]   Haynes M, Stang J, Moghaddam M. Real-time microwave imaging of differential temperature for thermal therapy monitoring. *IEEE Transactions on Biomedical Engineering*. 2014;61(6):1787–1797.

[4]   Suzuki Y, Kidera S. Resolution enhanced distorted Born iterative method using ROI limiting scheme for microwave breast imaging. *IEEE Journal of Electromagnetics, RF and Microwaves in Medicine and Biology*. 2021;5(4): 379–385.

[5]   Haynes M, Stang J, Moghaddam M. Microwave breast imaging system prototype with integrated numerical characterization. *International Journal of Biomedical Imaging*. 2012;2012.

[6]   Asefi M, Baran A, LoVetri J. An experimental phantom study for air-based quasi-resonant microwave breast imaging. *IEEE Transactions on Microwave Theory and Techniques*. 2019 Sep;67(9):3946–3954.

[7]   Kaye C, Jeffrey I, LoVetri J. Improvement of multi-frequency microwave breast imaging through frequency cycling and tissue-dependent mapping. *IEEE Transactions on Antennas and Propagation*. 2019;67(11):7087–7096.

[8]   Zhang K, Li M, Yang F, *et al.* Electrical impedance tomography with multiplicative regularization. In: *2017 Progress in Electromagnetics Research Symposium-Fall (PIERS-FALL)*. New York, NY: IEEE; 2017. pp. 1326–1333.

[9]   Sobota V, Suchomel J. Monitoring of pulmonary embolism using electrical impedance tomography: a case study. In: *2013 E-Health and Bioengineering Conference (EHB)*; 2013. pp. 1–4.

[10]   Zhu J, Snowden JC, Verdejo J, *et al.* EIT-kit: an electrical impedance tomography toolkit for health and motion sensing. In: *The 34th Annual ACM Symposium on User Interface Software and Technology*; 2021. pp. 400–413.

[11]   Kao TJ, Amm B, Isaacson D, *et al.* A 3D reconstruction algorithm for real-time simultaneous multi-source EIT imaging for lung function monitoring. bioRxiv. 2020. Available from: https://www.biorxiv.org/content/early/2020/05/31/2020.05.29.124222.

[12]   Putensen C, Hentze B, Muenster S, *et al.* Electrical impedance tomography for cardio-pulmonary monitoring. *Journal of Clinical Medicine*. 2019;8(8):1176.

[13]   Borcea L, Berryman JG, Papanicolaou GC. High-contrast impedance tomography. *Inverse Problems*. 1996;12(6):835.

[14]   Bayford RH. Bioimpedance tomography (electrical impedance tomography). *Annual Review of Biomedical Engineering* 2006;8:63–91.

[15]   Borcea L. Electrical impedance tomography. *Inverse Problems*. 2002 oct;18(6):R99–R136. Available from: https://doi.org/10.1088/0266-5611/18/6/201.

[16]   Cheney M, Isaacson D, Newell JC. Electrical impedance tomography. *SIAM Review*. 1999;41(1):85–101.

[17]   Chen G, Stang J, Moghaddam M. Numerical vector Green's function for S-parameter measurement with waveport excitation. *IEEE Transactions on Antennas and Propagation*. 2017 July;65(7):3645–3653.

[18]   Chen G, Stang J, Haynes M, *et al.* Real-time three-dimensional microwave monitoring of interstitial thermal therapy. *IEEE Transactions on Biomedical Engineering*. 2018;65(3):528–538.

[19]   Shah P, Chen G, Stang J, *et al.* 3-D Level set method for joint contrast and shape recovery in microwave imaging. *IEEE Transactions on Computational Imaging*. 2019 March;5(1):97–108.

[20]   Asefi M, LoVetri J. Use of field-perturbing elements to increase nonredundant data for microwave imaging systems. *IEEE Transactions on Microwave Theory and Techniques*. 2017;65(9):3172–3179.

[21]   Wang G, Ye JC, Mueller K, *et al.* Image reconstruction is a new frontier of machine learning. *IEEE Transactions on Medical Imaging*. 2018;37(6):1289–1296.

[22]   Gong K, Catana C, Qi J, *et al.* PET image reconstruction using deep image prior. *IEEE Transactions on Medical Imaging*. 2018;38(7):1655–1665.

[23] Zhang HM, Dong B. A review on deep learning in medical image reconstruction. *Journal of the Operations Research Society of China*. 2020;8(2): 311–340.

[24] Li M, Guo R, Zhang K, *et al*. Machine learning in electromagnetics with applications to biomedical imaging: a review. *IEEE Antennas and Propagation Magazine*. 2021;63(3):39–51.

[25] Zhang K, Guo R, Li M, *et al*. Supervised descent learning for thoracic electrical impedance tomography. *IEEE Transactions on Biomedical Engineering*. 2020;68(4):1360–1369.

[26] Lin Z, Guo R, Zhang K, *et al*. Neural network-based supervised descent method for 2D electrical impedance tomography. *Physiological Measurement*. 2020;41(7):074003.

[27] Edwards K, Khoshdel V, Asefi M, *et al*. A machine learning workflow for tumour detection in breasts using 3D microwave imaging. *Electronics*. 2021;10(6):674.

[28] Chen G, Shah P, Stang J, *et al*. Learning-assisted multimodality dielectric imaging. *IEEE Transactions on Antennas and Propagation*. 2020;68(3): 2356–2369.

[29] Khoshdel V, Asefi M, Ashraf A, *et al*. Full 3D microwave breast imaging using a deep-learning technique. *Journal of Imaging*. 2020;6(8):80.

[30] Leeuwen Tv, Brune C. Computed Tomography [homepage on the Internet]. GitHub; 2022 [updated 22/4/2022; cited 2022 Jun 9]. Available from: https://tristanvanleeuwen.github.io/IP_and_Im_Lectures/tomography.html.

[31] MRI Basics [homepage on the Internet]. Cleveland, OH: Case Western Reserve University; 2016 [updated 07/04/16; cited 2022 Jun 9]. Available from: https://case.edu/med/neurology/NR/MRI%20Basics.htm.

[32] Balanis CA. *Advanced Engineering Electromagnetics*. New York, NY: John Wiley & Sons; 2012.

[33] Ybarra GA, Liu QH, Ye G, *et al*. Breast imaging using electrical impedance tomography (EIT). In: J Suri, RM Rangayyan, and S Laxminarayan (eds.), *Emerging Technologies in Breast Imaging and Mammography*. American Scientific Publishers; 2006.

[34] Gupta M, Mishra RK, Roy S. Sparse reconstruction of log-conductivity in current density impedance tomography. *Journal of Mathematical Imaging and Vision*. 2020;62(2):189–205.

[35] Faddeev L. SCHRÖDINGER EQUATION. *Nine Papers on Partial Differential Equations and Functional Analysis*. 1967;65:139.

[36] Delbary F, Hansen PC, Knudsen K. Electrical impedance tomography: 3D reconstructions using scattering transforms. *Applicable Analysis*. 2012;91(4): 737–755.

[37] Chew W. *Waves and Fields in Inhomogeneous Media*. New York, NY: Springer; 1990.

[38] Gu GZ, Li DH, Qi L, *et al*. Descent directions of quasi-Newton methods for symmetric nonlinear equations. *SIAM Journal on Numerical Analysis*. 2002;40(5):1763–1774.

[39]   Abubakar A, Habashy TM, Pan G, *et al*. Application of the multiplicative reg-ularized Gauss–Newton algorithm for three-dimensional microwave imaging. *IEEE Transactions on Antennas and Propagation*. 2012;60(5):2431–2441.

[40]   Rubaek T, Meaney PM, Meincke P, *et al*. Nonlinear microwave imaging for breast-cancer screening using Gauss–Newton's method and the CGLS inversion algorithm. *IEEE Transactions on Antennas and Propagation*. 2007;55(8): 2320–2331.

[41]   Ostadrahimi M, Mojabi P, Zakaria A, *et al*. Enhancement of Gauss–Newton inversion method for biological tissue imaging. *IEEE Transactions on Microwave Theory and Techniques*. 2013;61(9):3424–3434.

[42]   De Zaeytijd J, Franchois A, Eyraud C, *et al*. Full-wave three-dimensional microwave imaging with a regularized Gauss–Newton method– theory and experiment. *IEEE Transactions on Antennas and Propagation*. 2007;55(11): 3279–3292.

[43]   Bisio I, Estatico C, Fedeli A, *et al*. Brain stroke microwave imaging by means of a Newton-conjugate-gradient method in L^p Banach spaces. *IEEE Transactions on Microwave Theory and Techniques*. 2018;66(8):3668–3682.

[44]   Bisio I, Estatico C, Fedeli A, *et al*. Variable-exponent Lebesgue-space inver-sion for brain stroke microwave imaging. *IEEE Transactions on Microwave Theory and Techniques*. 2020;68(5):1882–1895.

[45]   Oraintara S, Karl WC, Castanon DA, *et al*. A method for choosing the regularization parameter in generalized Tikhonov regularized linear inverse problems. In: *Proceedings 2000 International Conference on Image Pro-cessing* (Cat. No. 00CH37101). vol. 1. New York, NY: IEEE; 2000. pp. 93–96.

[46]   Kaltenbacher B, Kirchner A, Vexler B. Adaptive discretizations for the choice of a Tikhonov regularization parameter in nonlinear inverse problems. *Inverse Problems*. 2011;27(12):125008.

[47]   Li F, Liu QH, Song LP. Three-dimensional reconstruction of objects buried in layered media using Born and distorted Born iterative methods. *IEEE Geoscience and Remote Sensing Letters*. 2004;1(2):107–111.

[48]   Liang B, Qiu C, Han F, *et al*. A new inversion method based on distorted Born iterative method for grounded electrical source airborne transient elec-tromagnetics. *IEEE Transactions on Geoscience and Remote Sensing*. 2017; 56(2):877–887.

[49]   Burfeindt MJ, Colgan TJ, Mays RO, *et al*. MRI-derived 3-D-printed breast phantom for microwave breast imaging validation. *IEEE Antennas and Wireless Propagation Letters*. 2012;11:1610–1613.

[50]   Fang Y, Bakian-Dogaheh K, Stang J, *et al*. A versatile and shelf-stable dielectric coupling medium for microwave imaging. *IEEE Transactions on Biomedical Engineering*. Early Access, 2022.

[51]   Samuel AL. Machine learning. *The Technology Review*. 1959;62(1):42–45.

[52]   Ganapathy K, Abdul SS, Nursetyo AA, *et al*. Artificial intelligence in neurosciences: a clinician's perspective. *Neurology India*. 2018;66(4):934.

[53] Sagar M, Islam S, Ouassal H, *et al.* Application of machine learning in electromagnetics: mini-review. *Electronics.* 2021;10(22):2752.

[54] Harrington P. *Machine Learning in Action.* New York, NY: Simon and Schuster; 2012.

[55] Raschka S. *Python Machine Learning.* Birmingham: Packt Publishing Ltd; 2015.

[56] Zhang W, Zhu L, Hallinan J, *et al.* Boostmis: boosting medical image semi-supervised learning with adaptive pseudo labeling and informative active annotation. In: *Proceedings of the IEEE/CVF Conference on Computer Vision and Pattern Recognition;* 2022. pp. 20666–20676.

[57] Li D, Mersereau RM, Simske S. Blind image deconvolution through support vector regression. *IEEE Transactions on Neural Networks.* 2007;18(3): 931–935.

[58] Selvakumar J, Lakshmi A, Arivoli T. Brain tumor segmentation and its area calculation in brain MR images using K-mean clustering and Fuzzy C-mean algorithm. In: *IEEE-International Conference on Advances in Engineering, Science and Management (ICAESM-2012).* IEEE; 2012. pp. 186–190.

[59] Wu W, Li D, Du J, *et al.* An intelligent diagnosis method of brain MRI tumor segmentation using deep convolutional neural network and SVM algorithm. *Computational and Mathematical Methods in Medicine.* 2020;2020:Article ID 6789306.

[60] Khedher L, Ramírez J, Górriz JM, *et al.* Early diagnosis of Alzheimer's disease based on partial least squares, principal component analysis and support vector machine using segmented MRI images. *Neurocomputing.* 2015; 151:139–150.

[61] De Bruijne M. *Machine Learning Approaches in Medical Image Analysis: From Detection to Diagnosis.* New York, NY: Elsevier; 2016.

[62] Wang Y, Morrell G, Heibrun ME, *et al.* 3D multi-parametric breast MRI segmentation using hierarchical support vector machine with coil sensitivity correction. *Academic Radiology.* 2013;20(2):137–147.

[63] Liu J, Guo L. A new brain MRI image segmentation strategy based on K-means clustering and SVM. In: *2015 7th International Conference on Intelligent Human-Machine Systems and Cybernetics.* vol. 2. IEEE; 2015. pp. 270–273.

[64] Gitto S, Cuocolo R, Albano D, *et al.* MRI radiomics-based machine-learning classification of bone chondrosarcoma. *European Journal of Radiology.* 2020;128:109043.

[65] Aamani R, Vannala A, Murthy ASD, *et al.* Heart disease diagnosis process using Mri segmentation and lasso net classification ML. *Journal of Critical Reviews.* 2019;7(6):2020.

[66] Inbarani H H, Azar AT, Jothi G. Leukemia image segmentation using a hybrid histogram-based soft covering rough K-means clustering algorithm. *Electronics.* 2020;9(1):188.

[67] Dhawan AP, Chitre Y, Kaiser-Bonasso C. Analysis of mammographic micro-calcifications using gray-level image structure features. *IEEE Transactions on Medical Imaging*. 1996;15(3):246–259.

[68] Zaw HT, Maneerat N, Win KY. Brain tumor detection based on Naive Bayes Classification. In: *2019 5th International Conference on Engineering, Applied Sciences and Technology (ICEAST)*. IEEE; 2019. pp. 1–4.

[69] Brahimetaj R, Willekens I, Massart A, *et al*. Improved automated early detection of breast cancer based on high resolution 3D micro-CT microcalcification images. *BMC Cancer*. 2022;22(1):1–13.

[70] Padilla P, López M, Górriz JM, *et al*. NMF-SVM based CAD tool applied to functional brain images for the diagnosis of Alzheimer's disease. *IEEE Transactions on Medical Imaging*. 2011;31(2):207–216.

[71] Depeursinge A, Kurtz C, Beaulieu C, *et al*. Predicting visual semantic descriptive terms from radiological image data: preliminary results with liver lesions in CT. *IEEE Transactions on Medical Imaging*. 2014;33(8):1669–1676.

[72] Mahdy LN, Ezzat KA, Elmousalami HH, *et al*. Automatic X-ray COVID-19 lung image classification system based on multi-level thresholding and support vector machine. MedRxiv. 2020.

[73] Sable G, Bodhey H. Lung segmentation and tumor identification from CT scan images using SVM. *International Journal of Science and Research (IJSR)* ISSN (Online). 2014;64–69.

[74] Sarker P, Shuvo MMH, Hossain Z, *et al*. Segmentation and classification of lung tumor from 3D CT image using K-means clustering algorithm. In: *2017 4th International Conference on Advances in Electrical Engineering (ICAEE)*. IEEE; 2017. pp. 731–736.

[75] Huynh T, Gao Y, Kang J, *et al*. Estimating CT image from MRI data using structured random forest and auto-context model. *IEEE Transactions on Medical Imaging*. 2015;35(1):174–183.

[76] Polan DF, Brady SL, Kaufman RA. Tissue segmentation of computed tomography images using a Random Forest algorithm: a feasibility study. *Physics in Medicine & Biology*. 2016;61(17):6553.

[77] Tuncer SA, Alkan A. Spinal cord based kidney segmentation using connected component labeling and K-means clustering algorithm. *Traitement du Signal*. 2019;36(6):521–527.

[78] Parages FM, O'Connor JM, Pretorius PH, *et al*. A Naive-Bayes model observer for a human observer in detection, localization and assessment of perfusion defects in SPECT. In: *2013 IEEE Nuclear Science Symposium and Medical Imaging Conference (2013 NSS/MIC)*. IEEE; 2013. pp. 1–5.

[79] Darma PN, Takei M. High-speed and accurate meat composition imaging by mechanically-flexible electrical impedance tomography with K-nearest neighbor and fuzzy K-means machine learning approaches. *IEEE Access*. 2021;9:38792–38801.

[80] McDermott B, Elahi A, Santorelli A, *et al*. Multi-frequency symmetry difference electrical impedance tomography with machine learning for human stroke diagnosis. *Physiological Measurement*. 2020;41(7):075010.

[81] Darma PN, Ibrahim KA, Takei M. Super high-speed cross-sectional imaging of fat, muscle, and bone by machine learning and EIT. In: *2021 International Conference on Instrumentation, Control, and Automation (ICA)*. IEEE; 2021. pp. 4–8.

[82] Sami H, Sagheer M, Riaz K, *et al.* Machine learning-based approaches for breast cancer detection in microwave imaging. In: *2021 IEEE USNC-URSI Radio Science Meeting (Joint with AP-S Symposium)*. IEEE; 2021. pp. 72–73.

[83] Ojaroudi M, Bila S, Salimitorkamani M. A novel machine learning approach of hemorrhage stroke detection in differential microwave head imaging system. In: *2020 European Conference on Antennas and Propagation*; 2020.

[84] Rana SP, Dey M, Tiberi G, *et al.* Machine learning approaches for automated lesion detection in microwave breast imaging clinical data. *Scientific Reports*. 2019;9(1):1–12.

[85] Roohi M, Mazloum J, Pourmina MA, *et al.* Machine learning approaches for automated stroke detection, segmentation, and classification in microwave brain imaging systems. *Progress in Electromagnetics Research C*. 2021;116:193–205.

[86] LeCun Y, Bengio Y, Hinton G. Deep learning. *Nature*. 2015;521(7553):436–444.

[87] Minsky M. Steps toward artificial intelligence. *Proceedings of the IRE*. 1961;49(1):8–30.

[88] Zhang X, Zhao J, LeCun Y. Character-level convolutional networks for text classification. In: Cortes C, Lawrence N, Lee D, *et al.*, editors, *Advances in Neural Information Processing Systems*. vol. 28. Red Hook, NY: Curran Associates Inc.; 2015. Available from: https://proceedings.neurips.cc/paper/2015/file/250cf8b51c773f3f8dc8b4be867a9a02-Paper.pdf.

[89] Dürr O, Sick B, Murina E. *Probabilistic Deep Learning: With Python, Keras and TensorFlow Probability*. Shelter Island, NY: Manning Publications; 2020.

[90] Weidman S. *Deep Learning from Scratch: Building with Python from First Principles*. Sebastopol, CA: O'Reilly Media; 2019.

[91] Goodfellow I, Bengio Y, Courville A. *Deep Learning*. London: MIT Press; 2016.

[92] Bengio Y, Goodfellow I, Courville A. *Deep Learning*. vol. 1. Cambridge, MA: MIT Press; 2017.

[93] LeCun Y, Bottou L, Bengio Y, *et al.* Gradient-based learning applied to document recognition. *Proceedings of the IEEE*. 1998;86(11):2278–2324.

[94] Zhang A, Lipton ZC, Li M, *et al.* Dive into Deep Learning. arXiv preprint arXiv:210611342. 2021.

[95] Muhammad R. LeNet-5-A Classic CNN Architecture [homepage on the Internet]. Data Science Central; 2018 [updated 16/10/2018; cited 2022 Jun 9]. Available from: https://www.datasciencecentral.com/lenet-5-a-classic-cnn-architecture/.

[96] Kim J, Hong J, Park H. Prospects of deep learning for medical imaging. *Precision and Future Medicine*. 2018;2(2):37–52.

[97]   Goodfellow I, Pouget-Abadie J, Mirza M, *et al.* Generative adversarial nets. *Advances in Neural Information Processing Systems.* 2014;27:3422622.

[98]   Yi X, Walia E, Babyn P. Generative adversarial network in medical imaging: a review. *Medical Image Analysis.* 2019;58:101552.

[99]   Yang G, Lv J, Chen Y, *et al.* Generative Adversarial Networks (GAN) Powered Fast Magnetic Resonance Imaging–Mini Review, Comparison and Perspectives. arXiv preprint arXiv:210501800. 2021.

[100]   Xun S, Li D, Zhu H, *et al.* Generative adversarial networks in medical image segmentation: a review. *Computers in Biology and Medicine.* 2022; 140:105063.

[101]   Sheikhjafari A, Noga M, Punithakumar K, *et al.* Unsupervised deformable image registration with fully connected generative neural network. 1st Conference on Medical Imaging with Deep Learning (MIDL), Amsterdam, The Netherlands; 2018.

[102]   Punithakumar K, Boulanger P, Noga M. A GPU-accelerated deformable image registration algorithm with applications to right ventricular segmentation. *IEEE Access.* 2017;5:20374–20382.

[103]   Feng C, Kang K, Xing Y. Fully connected neural network for virtual monochromatic imaging in spectral computed tomography. *Journal of Medical Imaging.* 2018;6(1):011006.

[104]   Dachena C, Fedeli A, Fanti A, *et al.* Microwave imaging of the neck by means of artificial neural networks for tumor detection. *IEEE Open Journal of Antennas and Propagation.* 2021;2:1044–1056.

[105]   Roth HR, Shen C, Oda H, *et al.* Deep learning and its application to medical image segmentation. *Medical Imaging Technology.* 2018;36(2): 63–71.

[106]   Yamashita R, Nishio M, Do RKG, *et al.* Convolutional neural networks: an overview and application in radiology. *Insights into Imaging.* 2018;9(4): 611–629.

[107]   Yang R, Yu Y. Artificial convolutional neural network in object detection and semantic segmentation for medical imaging analysis. *Frontiers in Oncology.* 2021;11:573.

[108]   Samhan LF, Alfarra AH, Abu-Naser SS. Classification of Alzheimer's disease using convolutional neural networks. *International Journal of Academic Information Systems Research (IJAISR).* 2022;6(3):18–23.

[109]   Parisi L, Neagu D, Ma R, *et al.* Quantum ReLU activation for convolutional neural networks to improve diagnosis of Parkinson's disease and COVID-19. *Expert Systems with Applications.* 2022;187:115892.

[110]   Ahmad M, Qadri SF, Qadri S, *et al.* A lightweight convolutional neural network model for liver segmentation in medical diagnosis. *Computational Intelligence and Neuroscience.* 2022;2022:Article ID 7954333.

[111]   Moghbel M, Mashohor S, Mahmud R, *et al.* Automatic liver segmentation on computed tomography using random walkers for treatment planning. *EXCLI Journal.* 2016;15:500.

[112]   Ahmad M, Ai D, Xie G, *et al.* Deep belief network modeling for automatic liver segmentation. *IEEE Access.* 2019;7:20585–20595.

[113] Hatamizadeh A, Xu Z, Yang D, *et al*. UNetFormer: a unified vision transformer model and pre-training framework for 3D medical image segmentation. arXiv preprint arXiv:220400631. 2022.

[114] Biasi N, Gargano A, Arcarisi L, *et al*. Physics-based simulation and machine learning for the practical implementation of EIT-based tactile sensors. *IEEE Sensors Journal*. 2022;22:4186–4196.

[115] Capps M, Mueller JL. Reconstruction of organ boundaries With deep learning in the D-Bar method for electrical impedance tomography. *IEEE Transactions on Biomedical Engineering*. 2020;68(3):826–833.

[116] Qin Y, Ran P, Rodet T, *et al*. Breast imaging by convolutional neural networks from joint microwave and ultrasonic data. *IEEE Transactions on Antennas and Propagation*. 2022;70:6265–6276.

[117] Zhang X, Lee VC, Rong J, *et al*. Multi-channel convolutional neural network architectures for thyroid cancer detection. *PLoS One*. 2022;17(1): e0262128.

[118] Bayraktar Y, Ayan E. Diagnosis of interproximal caries lesions with deep convolutional neural network in digital bitewing radiographs. *Clinical Oral Investigations*. 2022;26(1):623–632.

[119] Chen X, Li L, Sharma A, *et al*. The application of convolutional neural network model in diagnosis and nursing of MR imaging in Alzheimer's disease. *Interdisciplinary Sciences: Computational Life Sciences*. 2022;14(1):34–44.

[120] Aaraji ZS, Abbas HH. Automatic classification of Alzheimer's disease using brain MRI data and deep convolutional neural networks. arXiv preprint arXiv:220400068. 2022.

[121] Ahmed AS, Salah HA. The IoT and registration of MRI brain diagnosis based on genetic algorithm and convolutional neural network. *Indonesian Journal of Electrical Engineering and Computer Science*. 2022;25(1):273–280.

[122] Ogawa T, Yoshii T, Oyama J, *et al*. Detecting ossification of the posterior longitudinal ligament on plain radiographs using a deep convolutional neural network: a pilot study. *The Spine Journal*. 2022;22(6):934–940.

[123] Kumar A, Tripathi AR, Satapathy SC, *et al*. SARS-Net: COVID-19 detection from chest x-rays by combining graph convolutional network and convolutional neural network. *Pattern Recognition*. 2022;122:108255.

[124] Sarki R, Ahmed K, Wang H, *et al*. Automated detection of COVID-19 through convolutional neural network using chest x-ray images. *PLoS One*. 2022;17(1):e0262052.

[125] Nasiri H, Hasani S. Automated detection of COVID-19 cases from chest X-ray images using deep neural network and XGBoost. *Radiography*. 2022;28(3):732–738.

[126] Gour M, Jain S. Uncertainty-aware convolutional neural network for COVID-19 X-ray images classification. *Computers in Biology and Medicine*. 2022;140:105047.

[127] Huang ML, Wu YZ. Semantic segmentation of pancreatic medical images by using convolutional neural network. *Biomedical Signal Processing and Control*. 2022;73:103458.

[128] Maity A, Nair TR, Mehta S, *et al*. Automatic lung parenchyma segmentation using a deep convolutional neural network from chest X-rays. *Biomedical Signal Processing and Control*. 2022;73:103398.

[129] Battalapalli D, Rao BP, Yogeeswari P, *et al*. An optimal brain tumor segmentation algorithm for clinical MRI dataset with low resolution and non-contiguous slices. *BMC Medical Imaging*. 2022;22(1):1–12.

[130] Khan TM, Robles-Kelly A, Naqvi SS. T-Net: a resource-constrained tiny convolutional neural network for medical image segmentation. In: *Proceedings of the IEEE/CVF Winter Conference on Applications of Computer Vision*; 2022. pp. 644–653.

[131] Cai Y, Wang Y. Ma-unet: an improved version of unet based on multi-scale and attention mechanism for medical image segmentation. In: *Third International Conference on Electronics and Communication; Network and Computer Technology (ECNCT 2021)*. vol. 12167. SPIE; 2022. pp. 205–211.

[132] Hoorali F, Khosravi H, Moradi B. IRUNet for medical image segmentation. *Expert Systems with Applications*. 2022;191:116399.

[133] Hatamizadeh A, Tang Y, Nath V, *et al*. Unetr: transformers for 3d medical image segmentation. In: *Proceedings of the IEEE/CVF Winter Conference on Applications of Computer Vision*; 2022. pp. 574–584.

[134] Shukla PK, Zakariah M, Hatamleh WA, *et al*. AI-DRIVEN novel approach for liver cancer screening and prediction using cascaded fully convolutional neural network. *Journal of Healthcare Engineering*. 2022;2022:4277436.

[135] Shaaf ZF, Jamil MMA, Ambar R, *et al*. Automatic left ventricle segmentation from short-axis cardiac MRI images based on fully convolutional neural network. *Diagnostics*. 2022;12(2):414.

[136] De RF, Hämäläinen E, Manninen E, *et al*. Convolutional neural networks enable robust automatic segmentation of the rat hippocampus in MRI after traumatic brain injury. *Frontiers in Neurology*. 2022;13:820267–820267.

[137] Russo C, Liu S, Di Ieva A. Spherical coordinates transformation preprocessing in Deep convolution neural networks for brain tumor segmentation in MRI. *Medical & Biological Engineering & Computing*. 2022;60(1): 121–134.

[138] Ranjbarzadeh R, Tataei Sarshar N, Jafarzadeh Ghoushchi S, *et al*. MRFE-CNN: multi-route feature extraction model for breast tumor segmentation in Mammograms using a convolutional neural network. *Annals of Operations Research*. 2022;1–22, https://doi.org/10.1007/s10479-022-04755-8.

[139] Mojabi P, Hughson M, Khoshdel V, *et al*. CNN for compressibility to permittivity mapping for combined ultrasound-microwave breast imaging. *IEEE Journal on Multiscale and Multiphysics Computational Techniques*. 2021;6:62–72.

[140] Autorino MM, Franceschini S, Ambrosanio M, *et al*. Deep learning strategies for quantitative biomedical microwave imaging. In: *2022 16th European Conference on Antennas and Propagation (EuCAP)*. New York, NY: IEEE; 2022. pp. 1–4.

[141] Hossain A, Islam MT, Islam MS, *et al.* A YOLOv3 deep neural network model to detect brain tumor in portable electromagnetic imaging system. *IEEE Access.* 2021;9:82647–82660.

[142] Shi Y, Tian Z, Wang M, *et al.* Residual convolutional neural network-based stroke classification with electrical impedance tomography. *IEEE Transactions on Instrumentation and Measurement.* 2022;71:1–11.

[143] Wu Y, Chen B, Liu K, *et al.* Shape reconstruction with multiphase conductivity for electrical impedance tomography using improved convolutional neural network method. *IEEE Sensors Journal.* 2021;21(7):9277–9287.

[144] Suzuki K. Overview of deep learning in medical imaging. *Radiological Physics and Technology.* 2017;10(3):257–273.

[145] Abdou MA. Literature review: efficient deep neural networks techniques for medical image analysis. *Neural Computing and Applications.* 2022;4: 1–22.

[146] Azizi S, Bayat S, Yan P, *et al.* Deep recurrent neural networks for prostate cancer detection: analysis of temporal enhanced ultrasound. *IEEE Transactions on Medical Imaging.* 2018;37(12):2695–2703.

[147] Monteiro M, Figueiredo MA, Oliveira AL. Conditional random fields as recurrent neural networks for 3d medical imaging segmentation. arXiv preprint arXiv:180707464. 2018.

[148] Antropova N, Huynh B, Giger M. Recurrent neural networks for breast lesion classification based on DCE-MRIs. In: *Medical Imaging 2018: Computer-Aided Diagnosis.* vol. 10575. SPIE; 2018. pp. 593–598.

[149] Hosseini SAH, Yaman B, Moeller S, *et al.* Dense recurrent neural networks for accelerated MRI: history-cognizant unrolling of optimization algorithms. *IEEE Journal of Selected Topics in Signal Processing.* 2020;14(6): 1280–1291.

[150] Wang H, Zhao S, Dong Q, *et al.* Recognizing brain states using deep sparse recurrent neural network. *IEEE Transactions on Medical Imaging.* 2018;38(4):1058–1068.

[151] Dvornek NC, Yang D, Ventola P, *et al.* Learning generalizable recurrent neural networks from small task-fmri datasets. In: *International Conference on Medical Image Computing and Computer-Assisted Intervention.* New York, NY: Springer; 2018. pp. 329–337.

[152] Dvornek NC, Li X, Zhuang J, *et al.* Jointly discriminative and generative recurrent neural networks for learning from fMRI. In: *International Workshop on Machine Learning in Medical Imaging.* New York, NY: Springer; 2019. pp. 382–390.

[153] Bhurwani MMS, Sommer KN, Ionita CN. Recovery of complete time density curves from incomplete angiographic data using recurrent neural networks. In: Gimi BS, Krol A, editors, *Medical Imaging 2022: Biomedical Applications in Molecular, Structural, and Functional Imaging.* vol. 12036. International Society for Optics and Photonics. SPIE; 2022. pp. 335–343. Available from: https://doi.org/10.1117/12.2611225.

[154]   Geng K, Yin G. Using deep learning in infrared images to enable human gesture recognition for autonomous vehicles. *IEEE Access*. 2020;8: 88227–88240.

[155]   Blasch E, Liu Z, Zheng Y. Advances in deep learning for infrared image processing and exploitation. In: *Infrared Technology and Applications XLVIII*. vol. 12107. SPIE; 2022. pp. 368–383.

[156]   Jiang M, Zhi M, Wei L, *et al*. FA-GAN: fused attentive generative adversarial networks for MRI image super-resolution. *Computerized Medical Imaging and Graphics*. 2021;92:101969.

[157]   Zhang K, Hu H, Philbrick K, *et al*. SOUP-GAN: super-resolution MRI using generative adversarial networks. *Tomography*. 2022;8(2):905–919.

[158]   Fernandes Jr FE, Yen GG. Pruning of generative adversarial neural networks for medical imaging diagnostics with evolution strategy. *Information Sciences*. 2021;558:91–102.

[159]   Amer Zakaria IJ, LoVetri J, Zakaria A. Full-vectorial parallel finite-element contrast source inversion method. *Progress in Electromagnetics Research*. 2013;142:463–483.

[160]   Edwards K, Krakalovich K, Kruk R, *et al*. The implementation of neural networks for phaseless parametric inversion. In: *2020 XXXIIIrd General Assembly and Scientific Symposium of the International Union of Radio Science*. New York, NY: IEEE; 2020. pp. 1–3.

Chapter 7

Deep learning techniques for direction of arrival estimation

Zhang-Meng Liu[1], Liuli Wu[2] and Philip S. Yu[3]

This chapter presents an overview of how deep learning (DL) techniques can be exploited to solve the problem of direction-of-arrival (DOA) estimation, and also provides a solution to this problem using a feasible and efficient hierarchical deep neural network (DNN). The chapter begins with a general introduction to existing DOA estimation and DL techniques in Section 1.1, then formulates the DOA estimation problem mathematically under different conditions in Section 1.2, and summarizes the most common DL frameworks that have been applied to DOA estimation in Section 1.3, including mainly their neural network configurations and the most widely used strategies for algorithm implementation. Section 1.4 presents a hierarchical DNN framework to solve the DOA estimation problem, and carries out simulations to demonstrate its predominance in generalization over previous machine learning (ML)-based methods, and in array-imperfection adaptation over conventional parametric methods. Finally, this chapter ends in Section 1.5 by providing some clues on several future research trends of this area.

7.1 Introduction

In the past several decades, direction-of-arrival (DOA) estimation has been a hot topic in widespread areas, such as radar, sonar, acoustics, wireless communications and astronomy [1–20]. Most of the research in this area focus on improving DOA estimation precision and super-resolution, and also enhancing adaptation to complex scenarios with low signal-to-noise ratio (SNR), limited snapshots, etc. [3]. Following this guideline, many methods have been proposed, e.g., beam-forming methods [4,6,7], subspace-based methods [8–10], maximum-likelihood methods [18–20] and sparsity-inducing methods [11–17]. These methods share a common feature, i.e., they are parametric methods that formulate a forward mapping from signal

[1] School of Electronic Science and Technology, National University of Defense Technology, China
[2] China Electronic Device System Engineering Corporation, Beijing, China
[3] Department of Computer Science, University of Illinois at Chicago, USA

directions to array outputs, and they also assume that the mapping is reversible. After that, DOA estimators are proposed by matching the array outputs to pre-formulated parametric mapping. Different methods differ from each other according to their matching criteria, e.g., beam-forming methods use a manifold correlation criterion [5–7], subspace-based methods use a superplane fitting criterion [8–10], maximum-likelihood methods use a fitting criterion on the raw array output [18–20] and sparsity-inducing methods reconstruct raw array outputs on overcomplete dictionaries [11–17]. Performances of these parametric methods depend heavily on the consistency between the two mappings, namely the forward mapping from signal directions to array outputs that works during data collection, and the inverse mapping from array outputs to signal directions that works during DOA estimation.

However, the forward mapping from signal directions to array outputs in practical systems is far more complicated than that used in parametric DOA estimation methods [21,22]. That is because various imperfections may exist in array systems, such as non-ideal sensor design, array installation, inter-sensor mutual interference, background radiation, etc. [23]. Some of these imperfections are so complicated that they can hardly be modelled precisely, and inaccurate modeling of them may significantly deteriorate DOA estimation performance [24,25]. Nevertheless, the effects of various imperfections are usually artificially simplified during array output formulation, so as to facilitate the implementation of auto-calibration methods and improve DOA estimation precision [26–32]. Most of the simplifications hold approximately only under some additional assumptions, such as inter-sensor independence of gain and phase errors [31,32], uniform linear or circular array geometries [26–28], and constrained sensor location errors within a particular line or plane [29,30]. In previous literatures, simulations have been carried out to prove the effectiveness of the auto-calibration DOA estimation methods [26–32]. However, in these simulations, the array outputs are generated based on the artificially simplified models with respect to some unknown variables only. These simplifications deviate actual array models with different degrees, and it is very difficult to clarify how the auto-calibration methods will behave in practical systems, especially when additional assumptions, such as linear/circular array geometries, do not hold. Moreover, if multiple kinds of imperfections coexist in the same array, the array output is much more difficult to formulate accurately, and high-precision DOA estimation becomes a very demanding task [33–35].

Due to the difficulties caused by unknown or mathematically unformulable array input–output mappings, researchers have resorted to Machine Learning (ML) techniques to solve DOA estimation problems [36–46]. The earliest work, as far as we known, dates back to the 1990s, when researchers considered the application of shallow artificial Neural Networks (NNs) to solve the DOA estimation problem [44,45]. After that, Support Vector Machine (SVM) is introduced in the 2000s to achieve satisfactory results in many fields including DOA estimation [36,37].

ML-based DOA estimation methods generally consist of two phases: the training phase and the testing phase. In the training phase, a function is learned from the training dataset containing input–output pairs using an ML technique, such as Support Vector Regression (SVR), SVM or Radial Basis Function (RBF). In the area of DOA estimation, the learned function stands for a surrogate mapping from array outputs to signal directions. In the testing phase, the surrogate function is exploited to estimate signal directions based on array outputs that are probably unseen in the training dataset. No assumption on array geometry or calibration status is required for ML-based DOA estimation methods, and they have been demonstrated via simulations to exceed subspace-based methods in computational complexity [36,46], and also perform comparably with them in DOA estimation precision in experiments [37].

Performances of most ML-based DOA estimation methods [36,37,46] rely heavily on the generalization characteristic of the ML techniques. High DOA estimation precision can be expected when the training data and test data have nearly identical distributions [47,48]. However, as too many unknown parameters are contained in the array observation model, such as signal number, signal directions, SNR, signal waveforms and noise samples, etc., it is very difficult to learn a surrogate DOA estimation function based on the training dataset that generalizes well in various testing scenarios.

In the past few years, Deep Learning (DL) techniques have experienced a rapid development and a widespread application in various areas, such as image classification, speech recognition, etc. [49–53]. Compared with shallow NN and other ML techniques, DNN has more layers and more units in each layer, and thus has a largely enhanced modeling capability of complex functions [54–59]. Some DL techniques, especially DNNs, have also been introduced to solve the DOA estimation problem. The foremost work in this field was proposed for sound source localization with microphone arrays [60–68]. After that, researchers considered more demanding scenarios with dynamic acoustic signals [60], wideband signals [64,66] and reverberant environments [62,63]. In these scenarios, the signal propagation models are too complicated to be formulated precisely, and the signal directions and locations are very difficult to be estimated with parametric methods. However, DL-based methods are data-driven and can learn the propagation models precisely based on a large training dataset, and they have been proved to perform satisfyingly in such direction and location estimation problems.

When introducing DNN techniques to estimate directions of acoustic signals, which generally last for seconds and contain redundant time–frequency features, the collected signal samples are usually transformed to the time–frequency domain first, and then inputted to DNNs to derive DOA estimates in a similar way as image recognition [60,64,66]. However, when general electro-magnetic (EM) signals are concerned, the snapshot number for DOA estimation is usually on scales of tens or hundreds, which is not large enough for time-frequency transformation as that used on acoustic signals. Therefore, in spite of the success of applying DL techniques to the field of acoustic signal processing [60,62–64,66], the methods can hardly be used directly to solve general DOA estimation problems of EM signals. Moreover, the settings of the DOA estimation problems for acoustic signals are quite simple, where only a single signal is considered [62,63], or the direction estimates are obtained on very coarse grids with inter-grid spacings of 5° [60,64] or even 10° [66]. Such simple settings deviate largely from most DOA estimation scenarios of EM signals, where super-resolution of temporally overlapped signals and high DOA estimation precisions of each signal are usually expected.

Table 7.1   Related notations

Notations	Definitions
X	matrix
x	vector
x	scalar
$\|x\|_2$	l_2-norm of vector x
$\|x\|$	the dimension of vector x
$\lceil x \rceil$	the smallest integer not smaller than x
I_M	$M \times M$ identity matrix
$0_{K \times M}$	$K \times M$ zero matrix
j	imaginary unit
$\mathbb{C}^{M \times N}$	$M \times N$ dimensional complex matrix set
$\mathbb{R}^{M \times N}$	$M \times N$ dimensional real matrix set
$Re\{\alpha\}$	the real part of a complex-valued variable α
$Im\{\alpha\}$	the imaginary part of a complex-valued variable α
$\angle \alpha$	the phase of a complex-valued variable α
$\mathbb{E}\{\cdot\}$	expectation operator
$(\cdot)^H$	conjugate transpose operator
$(\cdot)^T$	transpose operator
$\mathscr{CN}(m, C)$	the white circularly-symmetric Gaussian distribution with mean m and covariance C

Recently, many attempts are being devoted to applying DL techniques to solve the problem of DOA estimation of EM signals. They are believed to have potential in super-resolution and noise robustness [58,59,69–82]. For instance, [58] introduces a Deep Convolution Network (DCN) to learn the inverse transformation from array outputs to the DOA spectrum with a large training dataset. In [73], a DNN scheme for super-resolution DOA estimation in MIMO systems has been developed. In [75], Xiang *et al.* utilize an autoencoder to extract DOA-related features from radar signals in the presence of heavy multipath. In [76], a DNN with Fully Connected (FC) layers is presented for DOA classification of two targets using the covariance matrix. These works and the results therein show that DL is emerging as a powerful tool in solving demanding DOA estimation problems in complex scenarios with coherent signals [77,78], low SNR [79], near-field signals [80], etc.

The notations to be used in this chapter are listed in Table 7.1.

7.2   Problem formulation

7.2.1   Conventional observation model

Assume that K narrow-band uncorrelated far-field EM signals $s(t) = [s_1(t), \cdots, s_K(t)]^T$ impinge onto an M-element array from directions $\phi = [\phi_1, \cdots, \phi_K]$. A conventional model of the array output can be written as

$$x(t) = \sum_{k=1}^{K} a(\phi_k)s_k(t) + v(t) = A(\phi)s(t) + v(t) \qquad (7.1)$$

where $x(t) = [x_1(t), \cdots, x_M(t)]^T$ is the array output vector, $A(\phi) = [a(\phi_1), \cdots, a(\phi_K)]$ is the array steering matrix and $a(\phi_k)$ denotes the array steering vector corresponding to ϕ_k, $a(\phi_k)$ is assumed to be a unitary vector, i.e., $\|a(\phi_k)\|_2 = 1$. ϕ_k and $s_k(t)$ represent the incident direction and waveform of the kth signal, respectively. $v(t) = [v_1(t), \cdots, v_M(t)]^T$ is assumed to be additive zero-mean white Gaussian noise and is uncorrelated with the source signals, i.e. $v(t) \sim \mathcal{CN}(0, \sigma^2 I_M)$ with σ^2 denoting the power of white Gaussian noise, and $\mathbb{E}\{s(t)v^H(t)\} = 0_{K \times M}$.

The array output is usually sampled at N uniquely-spaced time instants $t_1, \cdots, t_N$ to obtain snapshot matrix X,

$$X = [x(t_1), \ldots, x(t_N)] = A(\phi)S + V \tag{7.2}$$

where $S = [s(t_1), \ldots, s(t_N)] \in \mathbb{C}^{K \times N}$ and $V = [v(t_1), \ldots, v(t_N)] \in \mathbb{C}^{M \times N}$ denote the signal waveform matrix and the noise matrix, respectively. Based on this observation model, DOA estimators are designed to estimate the unknown DOA vector ϕ from measurement matrix X.

7.2.2 Overcomplete formulation of array outputs

In some of the works, the DOA estimation problem is formulated as a classification task, and the signal directions are elected from a candidate angular grid set. This formulation also helps to highlight the spatial sparsity of the incident signals [11–14, 17,35,58,77]. Following this formulation, the array output $x(t)$ can be presented in the following overcomplete form,

$$x(t) = \sum_{i=1}^{I} a(\varphi_i)\bar{s}_i(t) + v(t) = A(\varphi)\bar{s}(t) + v(t) \tag{7.3}$$

where $\varphi = [\varphi_1, \varphi_2, \ldots, \varphi_I]$ is a discrete direction set sampled from the potential space of the incident signals, with $\Delta\varphi$ denoting the sampling interval. $\bar{s}(t)$ is the zero-padded extension of $s(t)$ from ϕ to φ, which satisfies

$$\bar{s}_i(t) = \begin{cases} s_k(t), if \ |\varphi_i - \phi_k| < \dfrac{\Delta\varphi}{2} \ or \ \varphi_i - \phi_k = \dfrac{\Delta\varphi}{2} & for \ i = 1, \cdots, I \\ 0, \qquad \qquad otherwise \end{cases} \tag{7.4}$$

Equation (7.3) is not an accurate extension of (7.1) due to the quantization errors in φ. However, previous literatures have shown that the model error in (7.3) can be overlooked given that $\Delta\varphi$ is adequately small. According to (7.4), $\bar{s}(t)$ only has K nonzero elements at or around the true source locations, i.e., $\bar{s}(t)$ is a sparse vector having much fewer nonzero elements than its dimension.

When N snapshots are collected, the observation matrix X in (7.2) can be rewritten as follows,

$$X = \overline{A}\,\overline{S} + V \tag{7.5}$$

where $\overline{S} = [\bar{s}(t_1), \ldots, \bar{s}(t_N)] \in \mathbb{C}^{I \times N}$ denotes the sparse signal matrix. $\overline{A} = A(\varphi) \in \mathbb{C}^{M \times I}$ denotes the known over-complete dictionary. The problem of DOA estimation can be solved by reconstructing $\overline{S}$ from x, and the locations of nonzero rows in the reconstructed $\hat{\overline{S}}$ indicate signal directions.

7.2.3   Array imperfections

Various kinds of array imperfections exist in almost all practical array systems, but they are usually overlooked in most academic works on array signal processing, and the array responding function $a(\phi)$ corresponding to direction ϕ in (7.1) is often supposed to be deterministic and known beforehand. Among the imperfections, gain and phase inconsistences, sensor position errors and inter-sensor mutual coupling are widely studied ones. These imperfections cause the actual array responding function to deviate from its imperfection-free counterpart $a(\phi)$, and the mapping from signal directions to array outputs in (7.1) does not hold any longer. Denote the imperfection parameters by e, the array outputs should be modified accordingly as follows,

$$x(t_n) = \sum_{k=1}^{K} a(\phi_k, e)s_k(t_n) + v(t_n), \quad for \ n = 1, \cdots, N. \tag{7.6}$$

Different kinds of array imperfections cast different influences on the array responding function, and it is still an open problem to figure out a precise formulation for $a(\phi, e)$ in practical arrays. Only after moderate simplifications can analytical mappings between (ϕ, e) and $a(\phi, e)$ be formulated approximately. However, such simplifications and approximations adapt only to particular array geometries and applications, the model and the corresponding auto-calibration methods can hardly be generalized to most practical arrays, where the simplifications and approximations do not hold.

7.3   Deep learning framework for DOA estimation

This section summarizes the DL architectures that have been applied in DOA estimation. Figure 7.1 presents a general framework of applying supervised learning techniques to solve the direction finding problem, which epitomizes the ideas of most existing research in the field of supervised DL-based DOA estimation.

Observation model in the figure refers to the mathematical formulation introduced in Section 7.2, which contains all the basic settings and corresponding parameters required in direction finding, such as array configuration, number of array elements, application scenarios, and signal environment.

Dotted arrows in the figure represent the data processing flow in the training phase, while solid arrows represent the flow in the testing or DOA estimation phase. The training phase learns a nonlinear function mapping the array output to signal directions based on a dataset of Q input–output training pairs $\mathscr{D} = \{[X_q, \phi_q]; q = 1, \ldots, Q\}$, with X_q being the array measurement matrix of the qth training sample and ϕ_q being the corresponding DOA vector. In general, the training samples in $\mathscr{D}$ are obtained in scenarios with signals impinging from known directions. Based on the mapping function learned in the training phase, the test phase estimates the DOA vector $\hat{\phi}$ for a particular input X.

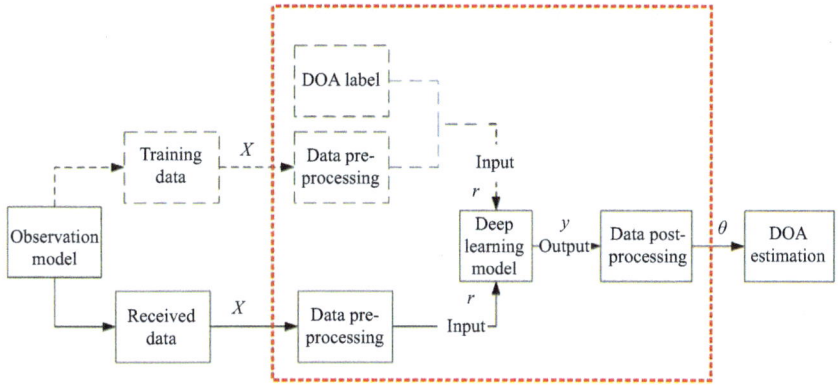

Figure 7.1   An overall framework for DL-based DOA estimation

The big red box in the figure highlights the three core components of this frame-work, including data pre-processing, DL-model building and post-processing. These three parts will be introduced in detail in the following subsections.

7.3.1   Data pre-processing

Array measurement matrix $X \in \mathbb{C}^{M \times N}$ contains necessary information for DOA esti-mation, but its dimension is usually too high to be inputted to a DL framework directly. Lots of redundant information irrelevant to the signal directions ϕ are also contained in x to enlarge its dimension, such as the signal waveforms. Inputting the raw mea-surements into a DL framework may cause significant negative effect on the DOA estimation performance due to the *curse of dimensionality*.

The purpose of data pre-processing is to make a proper transformation $r = \mathscr{F}(X)$ to reduce the dimension of the raw array output, while preserving the information related to ϕ as much as possible. This procedure is important in DL-based direction finding methods since the low dimension of r helps to reduce the training complexity of DL frameworks, and speed up the learning convergence. Different pre-processing schemes can be designed to adopt to different conditions, such as low SNR [79], near-field signals [80,83], coherent signals [77,78,84], complicated array configurations [83,85–87], and two-dimensional DOA estimation requirement [88,89], and also dif-ferent purposes like reducing training complexity or improving estimation accuracy.

In some works, pre-processing is also called *feature extraction*. Commonly used features or inputs in DL-based DOA estimation methods include: time difference [90–92], phase difference [61,93,94], power measurements [81,95], Generalized Cross-Correlation (GCC) matrix [62,63,67,96], Channel State Information (CSI) [97] and covariance matrix [58,59,73,75,76,82,98], etc. Among them, the covariance matrix is the most frequently used feature, which is given as

$$R = \mathbb{E}\{x(t)x^{\mathrm{H}}(t)\} = A(\phi)R_s A^{\mathrm{H}}(\phi) + \sigma^2 I_M \tag{7.7}$$

where $R_s = \mathbb{E}\{s(t)s^H(t)\}$ is the covariance matrix of the incident signals, which is diagonal when the signals are completely uncorrelated. It can be seen from (7.7) that the array covariance matrix $R \in \mathbb{C}^{M \times M}$ retains the information about signal directions ϕ, but its dimension has reduced to $M \times M$ from $M \times N$ of X, which is a very significant reduction given that N is usually much greater than M.

Most DL-based DOA estimation methods taking R as input only use the upper (or lower) triangle part of R by excluding the diagonal elements. The simplification does not introduce information loss because R is a conjugate symmetric matrix and its diagonal elements are independent with ϕ. The corresponding secondary array measurement is

$$\bar{r} = [R_{1,2}, R_{1,3}, \cdots, R_{1,M}, \cdots, R_{2,M}, \cdots, R_{M-1,M}]^T \tag{7.8}$$

where $R_{i,j}$ stands for the element of R in the ith row and jth column. Then a real-valued input vector is formulated as follows,

$$r = \frac{[Re\{\bar{r}\}, Im\{\bar{r}\}]}{\|\bar{r}\|_2} \tag{7.9}$$

References [79] and [99] adopt a different strategy by formulating an input r with a dimension of $M \times M \times 3$, whose third dimension relates to different *channels*. In particular, the three channels represent the real part $Re\{R\}$, the imaginary part $Im\{R\}$ and the phases $\angle R$ of R.

By considering the spatial sparsity of impinging signals, references [12,58,77, 86,87,100] preprocess the array outputs based on the over-complete formulation of the covariance matrix R according to (7.3),

$$R = \mathbb{E}\{x(t)x^H(t)\} = \sum_{i=1}^{I} \eta_i a(\varphi_i)a^H(\varphi_i) + \sigma^2 I_M \tag{7.10}$$

where $\eta_i = \mathbb{E}\{\bar{s}_i(t)\bar{s}_i^H(t)\}$. Then, a new measurement vector $z \in \mathbb{C}^{M^2 \times 1}$ can be obtained by stacking the columns of R one-after-another, i.e.,

$$z = \text{vec}(R) = \tilde{W}\eta + \sigma^2 \tilde{\xi} \tag{7.11}$$

where $\tilde{W} = [W_1; \ldots; W_M] \in \mathbb{C}^{M^2 \times I}$, $W_m = [a(\varphi_1)a^H(\varphi_1)\xi_m, \ldots, a(\varphi_I)a^H(\varphi_I)\xi_m]$, $\eta = [\eta_1, \eta_2, \ldots, \eta_I]^T \in \mathbb{R}^{I \times 1}$, $\tilde{\xi} = [\xi_1; \ldots; \xi_M] \in \mathbb{R}^{M^2 \times 1}$, ξ_m is an $M \times 1$ vector with the mth element being 1 and others being 0. The operator $[\bullet; \ldots; \bullet]$ stacks arrays or vectors in sequence vertically, $\text{vec}(\bullet)$ vectorizes a matrix by stacking its columns one by one. After these preprocessings, the input vector to the DL network in [12,58,77,100] is formulated as

$$r = \tilde{W}^H z \tag{7.12}$$

In practical applications, the matrix R in (7.7) is usually not available, it should be replaced by its estimate based on the raw array outputs, i.e.,

$$\hat{R} = \frac{1}{N} \sum_{n=1}^{N} x(t_n)x^H(t_n) \tag{7.13}$$

7.3.2   Deep learning model

In the context of DOA estimation in Figure 7.1, the DNN is used to build a surrogate model $y = \mathscr{G}(r)$ from a set of input–output pairs to approximate the DOA estimator. y represents the DNN output, which can be exploited to obtain DOA estimates $\hat{\phi}$ efficiently via a post-processing procedure.

7.3.2.1   Fundamentals of DL-based DOA estimation

So far, the problems addressed by DL techniques mainly divide into two categories: classification and regression. Accordingly, there are two kinds of DL-based DOA estimation methods.

(A) **DOA estimation via classification.** This kind of methods treats the DOA estimation problem as a classification task, where the potential angle space of incident signals is divided into I discrete *class*es by grid sampling. For example, the potential incident space $[-90°, 90°)$ can be divided into $I = 180$ directional intervals with a resolution of $1°$, and directions falling in the scope of each interval is defined as a *class*, as shown in Figure 7.2. Signals impinging from directions in the same angular interval belong to the same *class*, while those in different intervals belong to different *class*es. In this way, the DOA estimation problem is transformed to one of identifying the *class* of an unlabelled input r.

DL techniques have made remarkable achievements in solving classification problems, however, they may face some severe difficulties if used for DOA estimation straightforwardly. First, DOA estimates obtained from the classification framework are discrete and their precision relies heavily on the resolution of the angular grids, so unbiased estimates can hardly be obtained. Second, when multiple signals impinge onto the array simultaneously, the input r contains multiple components corresponding to different *class*es, making the problem a demanding *multi-label* labelling one. For example, if two signals arrive at the array from $20°$ and $40°$ respectively, the output of the DNN should fall in the $20°$ *class* and the $40°$ *class* at the same time. *Multi-label* labelling is quite different from the conventional classification problem, it is a fast-developing research field and a lot of key techniques still require to be addressed. Due to the above limits, the idea of solving the DOA estimation problem as a classification one has only been used in limited areas, such as indoor sound positioning, where

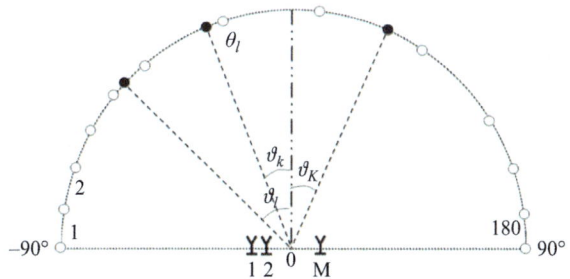

Figure 7.2   Space grid sampling

single-signal scenarios are usually considered and high DOA estimation precision is not predominantly required.

Many research efforts have been devoted to mitigate the disadvantages of existing classification techniques when used for DOA estimation, and one of the most widely adopted strategy is using a 0-1 vector to replace the *class* label as the output of DNN. As is illustrated in Figure 7.3, each direction interval is represented by an output node of DNN. During the model training stage, the value of the ith($i = 1, \ldots, I$) output node y_i is set to 1 if an incident signal is present in the corresponding angular interval; otherwise, y_i is set to 0. A well-trained DNN is expected to output an estimated *spatial spectrum $\hat{y}$*, which indicates the directions of the incident signals. Once the *spatial spectrum $\hat{y}$* is obtained, interpolation methods can be adopted to reduce angular quantization errors of the spatial grids and get DOA estimates with improved accuracies.

(B) DOA estimation via regression. In this kind of methods, the output of DNN, i.e., y, is formulated directly to be an estimate of ϕ, as is shown in Figure 7.4. The DNN is introduced to learn the nonlinear function $\phi = \mathscr{G}(r)$ to represent the mapping from r to ϕ.

When compared with classification-type DNN-based DOA estimation methods, a major advantage of a regression-type method is that, continuous DOA estimates with high precision can be achieved directly. While the disadvantage is also significant,

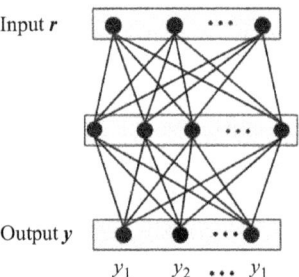

Figure 7.3   A variant of classification method

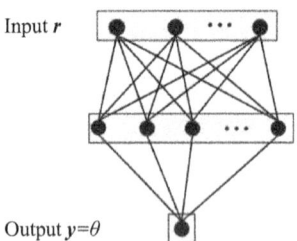

Figure 7.4   Regression method

i.e., the nonlinear mapping $\phi = \mathscr{G}(r)$ is usually very complicated and much difficult to approximate with both high precision and good generalization, as it relies on many factors such as signal number, signal directions, SNR, signal waveforms and noise samples, etc. Moreover, in scenarios of coexisting multiple signals, the number of signals should be estimated first to determine how many DNNs are needed, so that the directions of all the incident signals can be obtained via regression [101].

7.3.2.2 DNNs used for DOA estimation

Different DNNs have diverse capabilities in model learning, and they perform differently in various classification and regression tasks. Many DNN architectures, such as Recurrent Neural Network (RNN) [102], LeNet [103], Long Short-Term Memory (LSTM) [104], AlexNet [105], Residual Network (ResNet) [106], have been proposed to serve applications in various areas, and new DNN architectures are being proposed to adapt particular learning purposes and requirements. However, several factors may affect the performance of a certain DNN in a task, e.g., the number of hidden layers, the number of neurons in each layer, the activation function of each neuron, etc., and researchers in the area of DOA estimation have not agreed on the most appropriate DNN architecture. In this part, we make a brief introduction to the DNN architectures that have been applied to DOA estimation.

(A) **Fully connected neural network**. Fully Connected neural network (FCNN) is a type of straight feedforward neural network where neurons in two adjacent layers are fully connected, i.e., each neuron in one layer is connected to all the neurons in the next layer. FCNNs have been introduced to solve the DOA estimation and some other related problems in [59,73,75,77,81,95,107].

(B) **Convolutional neural network**. Convolutional Neural Network (CNN) is one of the most comprehensive NN frameworks that have been used in various fields. It employs a mathematical operation of convolution to replace general matrix multiplications in at least one of its layers. CNN is suitable for processing data that has a grid-like format and has achieved great successes in many applications including DOA estimation [58,79,84,97]. Advances in deep learning techniques can be invented to enrich the family of CNN framework, e.g., residual blocks [80,87,108] help to enhance the feasibility of CNNs having more layers.

(C) **Recurrent neural network**. Recurrent Neural Network (RNN) [102] is another kind of comprehensive NN besides CNN, it is designed for processing sequential data. Its neurons receive information not just from neurons in the previous layer, but also from the adjacent history state of themselves. Advances in RNN framework design, e.g., Long Short-Term Memory (LSTM) [104] and Gated Recurrent Unit (GRU) [109], have been made to overcome many shortcomings of the original RNN, such as vanishing or exploding backward derivatives. Some of the advanced RNNs have been successfully used to solve the DOA estimation problem. In [110], a sequence of covariance matrixes R is inputted to an LSTM-based RNN to estimate source directions.

(D) **Convolutional recurrent neural network**. Convolutional Recurrent Neural Network (CRNN) combines more than one of the above three architectures, i.e., FCNN, CNN and RNN, to exploit and synthesize the superiority of each NN. In the area of array signal processing, researchers have proposed to combine CNN and RNN to build different CRNNs to gain improved DOA estimation performance [78,101].

(E) Hybrid neural network. The mapping between array output X and source directions ϕ is influenced by many variables, such as signal number, signal directions, SNR, signal waveforms, and noise samples, and a single DNN can hardly generalize well in all conditions. Therefore, some researchers resort to hybrid neural networks to divide the DOA estimation task to smaller ones, and adopt several cascaded [82,83,89,100,107,108] or parallel [88,99,101] DNNs to reduce the influence of various variables on each individual subtask. More concretely, Ref.[100] presents a two-stage cascaded neural network for DOA estimation, it estimates DOAs coarsely with a CNN in the first stage, and refines the estimates by a tuning vector with an FCNN in the second stage. Reference [101] proposes a DOA estimation method based on a hybrid DNN, which consists of three parts: an autoencoder for noise filtering, an FCNN for signal number detection, and a series of parallel directed acyclic graph (DAG) networks for DOA estimation. Reference [107] first inputs array measurements to a SNR grading network to evaluate the SNR of the array output, and then calls a particular DOA estimation DNN module corresponding to this SNR from a series of DNN candidates to address the DOA estimation problem. The cascaded DNN scheme is specially designed to enhance the generality of the DL-based DOA estimation technique under different SNR conditions. Besides the cascaded and parallel DNNs, some other schemes have also been designed to combine conventional DNNs to propose new DOA estimation methods [84,86,111,112]. For example, Ref. [111] proposes to solve the DOA estimation problem by combining a CNN-based initialization step and a model-aware gradient step on the stochastic maximum-likelihood function.

(F) Complex-valued neural network. In most practical DOA estimation applications, receivers sample incident signals with I/Q channels in parallel, thus array outputs in EM DOA estimation problems have both real and imaginary parts and they form complex-valued inputs to DOA estimating RNNs. As a result, some researchers have made great efforts to replace conventionally-used real-valued DNNs with Complex-Valued Neural Networks (CVNN) to better serve the task of processing complex-valued array outputs. Compared with real-valued DNNs, CVNNs contain complex-valued weight parameters and performs complex arithmetics to extract DOA-related information from array outputs, and have been demonstrated to improve DOA estimation performance in some conditions [80,113].

(G) Transfer learning and unsupervised training. In practical direction finding systems, it is usually very difficult to collect a large enough dataset with DOA labels. If only a limited amount of labelled data is available, DNNs with satisfying precision can hardly be derived via supervised learning to approach the surrogate mapping from array outputs to signal directions. As an alternative, other DL techniques such as unsupervised pre-training [59,114] (such as autoencoders), Semi-Supervised Learning (SSL) [115–117] and Transfer Learning (TL) [89,118–120] have been employed to improve the efficiency of unlabelled data and better solve various problems.

Unsupervised pre-training extracts underlying features from raw array outputs in an unsupervised way, and provides initial weights for feature-extracting DNN modules in practical classification and regression tasks. SSL exploits both labelled data and unlabelled data to improve model learning performance. TL reduces the dependence of model learning on the amount of training data by transferring the knowledge shared in different but related scenarios and datasets. These techniques have been partially applied in the area of DOA estimation. Reference [115] proposes an SSL-based localization approach via deep generative modeling with Variational AutoEncoders (VAEs). Ref. [119] utilizes simulated data to train a basic CNN model and then fine-tunes its parameters using TL techniques to better adapt practical applications. Similarly, Ref. [89] first trains a CNN model in imperfection-free scenarios, and then transfers it to solve the DOA estimation problem in the presence of array imperfections.

(H) Deep unfolded network. The above DNNs learn DOA estimation models and solve DOA estimation problems in the form of black boxes, they do not exploit prior knowledge about the signal structure and optimize the neural network parameters by trial and error. Deep unfolded network [85,121] has recently been developed by treating the arithmetics of DNNs as unfolded optimization iterations, it enhances the interpretability of conventional DNNs. Related DOA estimation methods first unfold a well-understood iterative recovery algorithm to obtain a signal-flow graph with trainable variables, e.g., Ref. [85] unfolds Sparse Bayesian Learning (SBL) algorithm and Ref. [121] unfolds Fixed Point Continuation (FPC) algorithm, then tune these variables with supervised learning techniques, such as stochastic gradient descent algorithms based on back-propagation, to obtain more precise models for DOA estimation.

7.3.3 Post-processing for DOA refinement

DOA estimates gained from DNN outputs are generally discretized to predefined angular grids. Therefore, a data post-processing procedure is necessary to reduce the angular discretization errors and obtain high-precision DOA estimates $\hat{\phi}$ from the DNN output y. Different post-processing strategies should be designed according to the DNN type used for DOA estimation, and generally no post-processing is needed for DNNs belonging to the regression category, as the DNNs output refined DOA estimates directly.

DOA estimation DNNs belonging to the classification category outputs class labels or spatial spectrum on prefined angular grids. In cases when class labels are outputted, grid directions corresponding to the class labels are taken as DOA estimates. Otherwise, if a spatial spectrum y is outputted, peak searching and linear interpolation can be adopted to improve DOA estimation accuracy. In a DOA estimation scenario with K incident signals, y may contain K peak clusters with each containing several adjacent spectrum lines having significant magnitudes, and each peak cluster corresponds to one of the K incident signals. Take the kth cluster as an example, suppose that it contains two adjacent spectrum lines with significant magnitudes of $y_{k,1}$ and $y_{k,2}$, and the corresponding directions are $\phi_{k,1}$ and $\phi_{k,2}$, then a refined DOA estimate of the kth signal can be calculated via the following linear interpolation,

$$\hat{\phi}_k = \frac{\sqrt{y_{k,1}}}{\sqrt{y_{k,1}} + \sqrt{y_{k,2}}}\phi_{k,1} + \frac{\sqrt{y_{k,2}}}{\sqrt{y_{k,1}} + \sqrt{y_{k,2}}}\phi_{k,2}; \quad k = 1,\ldots,K \tag{7.14}$$

7.4   A hybrid DNN architecture for DOA estimation

In this section, a hybrid DNN framework is presented to solve the DOA estimation problem. This framework consists of a multi-task autoencoder and multiple parallel multi-layer classifiers. First, we introduce the DNN structure and interpret how it extracts DOA-related information from array outputs, and also how it fits DOA estimation requirements in Section 7.4.1. Then we clarify the training strategies of the DNN framework and highlight its behavior in array imperfection adaptation in Section 7.4.2. Section 7.4.3 carries out simulations to show the predominance of this DL-based method to its counterparts, which do not use DL techniques.

7.4.1   The hierarchical DNN structure

The hybrid DNN framework built in this subsection consists of two parts, a multi-task autoencoder for spatial filtering and a group of parallel multi-layer classifiers for spatial spectral reconstruction. A schematic diagram of the DNN structure is shown in Figure 7.5.

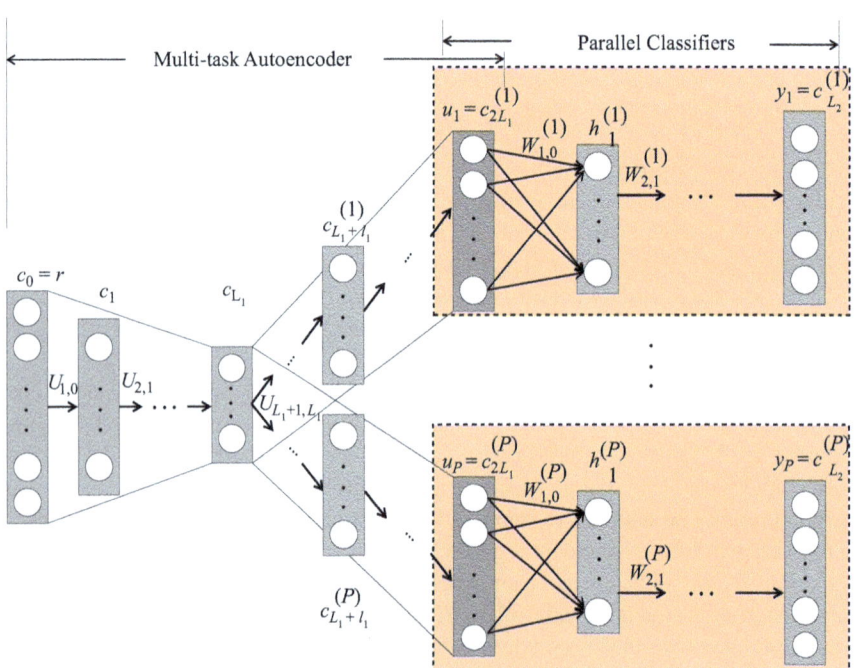

Figure 7.5   Structure of the proposed hybrid DNN for DOA estimation. The neural network consists of two parts, one is a multi-task autoencoder for spatial filtering, the other is a group of fully-connected multi-layer neural networks for spatial spectrum estimation.

The multi-task autoencoder is added before the parallel classifiers to denoise the input of DNN, and decompose it into multiple components in P different spatial subregions. If the input of DNN contains a signal in the pth subregion (possibly together with some other signals located in the other $P - 1$ subregions), the output of the pth decoder is expected to be equal to the DNN input when the other signals are absent. If there is no signal present in this subregion, the output of the pth decoder should be equal to zero. That is to say, the autoencoders act as filters that help to separate spatially apart but temporally overlapped signals.

Afterwards, a series of parallel fully-connected multi-layer neural networks are designed for signals in each subregion. Each of them behaves as a multi-class classifier that determines whether a signal impinges from the preset directional grids within the corresponding subregion. The output of a DNN node will be nonzero if a signal is located near the corresponding grid, and the output value indicates how much the signal location departs from this direction grid.

By adding a spatial filtering module in front of the DNN classifiers, the signals inputted to each classifier has a much smaller angular spread range and thus has more concentrated distributions than the raw input data. The classifiers realize DOA estimation in each subregion, and they do not need to consider the signal components located in other subregions, which helps a lot to facilitate the DNN training and strengthen its generalization to previously unseen scenarios.

7.4.1.1   Spatial filtering autoencoder

As shown in Figure 7.5, the autoencoder consists of an encoder and a group of decoders, which first compresses the input vector into a lower-dimensional one to extract the principal components in the original input through encoding, and then restores it to the original dimension through multi-task decoding with the components belonging to different subregions recovered by different decoders. The encoding-decoding process helps to reduce the influence of noise and interference of other signals in the DNN input [54].

It is assumed that each of the encoder and decoders has L_1 layers respectively, the vectors c in the $(L_1 - l_1)$th and the $(L_1 + l_1)$th layers have the same dimensions for $0 < l_1 \leq L_1$, and generally $|c_{l_1}^{(p)}| < |c_{l_1-1}^{(p)}|$, where $|c|$ represents the dimension of vector c. Neighbor layers of the autoencoder are fully-connected with feedforward computations, i.e.,

$$
\begin{aligned}
o_{l_1}^{(p)} &= U_{l_1,l_1-1}^{(p)} c_{l_1-1}^{(p)} + b_{l_1}^{(p)} \\
c_{l_1}^{(p)} &= f_{l_1}(o_{l_1}^{(p)}), \\
p &= \begin{cases} 1 & \text{for } l_1 = 1, \cdots, L_1, \\ 1, \cdots, P & \text{for } l_1 = L_1 + 1, \cdots, 2L_1. \end{cases}
\end{aligned}
\tag{7.15}
$$

In the above equations, P denotes the number of spatial subregions, subscripts $(\bullet)_{l_1}$ and $(\bullet)_{l_1-1}$ denote the indexes of autoencoder layer, superscript $(\bullet)^{(p)}$ denotes variables corresponding to the pth subregion and the pth autoencoder task, and $c_{l_1}^{(p)}$ stands for the output of l_1th layer in the pth autoencoder. The superscript $(\bullet)^{(p)}$ can be ignored when $l_1 \leq L_1$ as the P autoencoders share the same encoder. We also define

$c_0 = r$ as the input of the autoencoder. $U_{l_1, l_1-1}^{(p)} \in \mathbb{R}^{|c_{l_1}^{(p)}| \times |c_{l_1-1}^{(p)}|}$ is the weight matrix from the $(l_1 - 1)$-th layer to the l_1th layer of the pth task, and $b_{l_1}^{(p)} \in \mathbb{R}^{|c_{l_1}^{(p)}| \times 1}$ is the additive bias vector in the l_1th layer of the pth task. $f_{l_1}(\bullet)$ represents the element-wise activation function in the l_1th layer.

The multi-task autoencoder aims at decomposing the inputs $c_0 = r$ into P components with respect to incident signals belonging to P spatial subregions. A straightforward strategy to define the subregions is choosing $P + 1$ particular directions $\phi^{(0)} < \phi^{(1)} < \cdots < \phi^{(P)}$, which satisfy $\phi^{(1)} - \phi^{(0)} = \phi^{(2)} - \phi^{(1)} = \cdots = \phi^{(P)} - \phi^{(P-1)}$ and the interval $[\phi^{(0)}, \phi^{(P)})$ covers the potential scope of the incident signals. If a signal impinge from the pth subregion to the autoencoder, the output of the pth decoder, which is also denoted as $u_p = c_{2L_1}^{(p)}$ is expected to be equal to the input r, while outputs of the other decoders equal to zero. A way of enhancing the denoising ability of the autoencoder is using the noise-free counterpart of the input as the decoder output. However, it can hardly be implemented as it is not an easy task to collect noise-free counterparts of the training dataset in real systems.

The expected autoencoder outputs are $F^{(p)}(r) = r$ if the signal direction $\phi \in [\phi^{(p-1)}, \phi^{(p)})$, and $F^{(p)}(r) = 0$ otherwise, where $F^{(p)}(\bullet)$ is the over-all function of the pth autoencoder task. In addition, in DOA estimation applications, the autoencoder should have a linearity characteristic, i.e.,

$$F^{(p)}(r_1 + r_2) = F^{(p)}(r_1) + F^{(p)}(r_2) \qquad (7.16)$$

The additive characteristic is required since the autoencoder is used to decompose the input vector into different decoder outputs according to signals impinging from different subregions. In order to satisfy (7.16), the activation functions $f_{l_1}^{(p)}[\bullet]$ should be linear, therefore, we use a unit function for activation, i.e., $c_{l_1}^{(p)} = o_{l_1}^{(p)}$. Then there is no longer any nonlinear transformation in the hidden layers of the autoencoder, and the encoding and decoding process of the multi-layer neural network can be simplified to a single-layer implementation, i.e., $L_1 = 1$, and the autoencoder can be rewritten as

$$\begin{aligned} c_1 &= U_{1,0}r + b_1, \\ u_p &= U_{2,1}^{(p)}c_1 + b_2^{(p)}, \quad p = 1, \cdots, P. \end{aligned} \qquad (7.17)$$

7.4.1.2   Spectrum estimation with parallel multi-layer classifiers

Existing ML methods based on RBF [46] and SVR [36,37] treat the DOA estimation problem as a regression task, they assume the number of incident signals K is known beforehand, and then fix the number of output nodes in the ML model accordingly. If the number of incident signals changes, the trained model does not work anymore. Therefore, in order to exploit these methods to well address the DOA estimation problem, one need to train a series of models in case of different K's. Even though, such models can hardly be integrated effectively to deal with DOA estimation problems when K is not known beforehand.

A more flexible way to enhance the generalization of ML models to situations with unknown K is to use a series of one-vs-all classifiers instead. Each output node of the classifier corresponds to a preset directional grid, and the final output value of the node represents the signal power locating on the corresponding grid. DOAs of signals impinging from off-grid directions can be estimated through interpolation between two adjacent grids.

As shown in Figure 7.5, there are P parallel multi-layer classifiers in total, with the pth classifier taking the output of the pth decoder as input. Then, the pth classifier analyzes the components of the input within the pth spatial subregion to reconstruct the corresponding DOA spectrum. All the P classifiers have the same structure and there are no mutual connections between them. Neighbor layers of each classifiers are fully-connected according to feedforward computations, i.e.,

$$o_{l_2}^{(p)} = W_{l_2,l_2-1}^{(p)} h_{l_2-1}^{(p)} + q_{l_2}^{(p)}$$
$$h_{l_2}^{(p)} = g_{l_2}[o_{l_2}^{(p)}],$$
$$p = 1, \cdots, P; \quad l_2 = 1, \cdots, L_2$$
(7.18)

where $h_{l_2}^{(p)}$ is the output vector in the l_2th layer of the pth classifier, with $h_0^{(p)} = u_p$ and $h_{L_2}^{(p)} = y_p$; $W_{l_2,l_2-1}^{(p)} \in \mathbb{R}^{|h_{l_2}^{(p)}| \times |h_{l_2-1}^{(p)}|}$ is the fully-connected feed-forward weight matrix between the $(l_2 - 1)$th layer and the l_2th layer, $q_{l_2}^{(p)}$ is the additive bias vector on the l_2th layer of the pth classifier; $g_{l_2}[\bullet]$ is an element-wise activation function for the inputs of the l_2th layer.

After obtaining the outputs $[y_1, y_2, \cdots, y_P]$ of the P parallel classifiers based on the P decoder outputs, the spatial spectrum associated with DNN input r can be reconstructed by concatenating the P outputs in order, i.e.,

$$y = [y_1^T, \cdots, y_P^T]^T.$$
(7.19)

There are totally $|y|$ one-vs-all classifiers in this part of the DNN, and the output vector y represents the discrete spatial spectrum estimate associated with DNN input r. y only takes positive values on the grid nodes close to the true signal directions, while all the others have zero values.

7.4.2 Training strategy of the hybrid DNN model

Besides the DNN framework, the design of the training dataset and the model training strategy are two other factors that greatly affect the performance of DNN-based DOA estimation methods. Since the autoencoder and parallel classifiers in the hybrid DNN framework perform different functions during DOA estimation, and training the entire network as a whole increases the risk of getting trapped in undesirable local minima [122], we propose to train the two parts of the DNN in separate procedures.

In order to reduce the variability of the DNN input, which is influenced significantly by factors such as signal waveforms, we follow the pre-processing strategy in Section 7.3.1 to compute the array covariance matrix, and reformulate the off-diagonal upper right matrix elements as the input vector to the DNN according to (7.13), (7.8), and (7.9), i.e.,

$$\bar{r} = [R_{1,2}, R_{1,3}, \cdots, R_{1,M}, R_{2,3}, \cdots, R_{2,M}, \cdots, R_{M-1,M}]^{\mathrm{T}},$$

$$r = \frac{[\operatorname{Re}\{\bar{r}\}, \operatorname{Im}\{\bar{r}\}]}{\|\bar{r}\|_2} \tag{7.20}$$

7.4.2.1 Training strategy for autoencoder

Since the autoencoder is designed to be linear and satisfy the linearity property, its performance of spatial filtering can be guaranteed if it performs well in single-signal scenarios. Therefore, we construct the training dataset for autoencoder using r corresponding to a series of single-signal scenarios, where the signal direction ϕ is sampled within the range $[\phi^{(0)}, \phi^{(P)})$. One straightforward selection of the signal directions is the equally spaced grids corresponding to the classifier outputs, which are denoted as $\varphi_1, \varphi_2, \cdots, \varphi_I$. It is supposed that I is dividable by P with $\frac{I}{P} = I_0$ being an integer.

If the covariance vector $r(\varphi_i)$ corresponding to a signal from direction φ_i is inputted to the autoencoder, the output of the p_ith ($p_i = \lceil i/I_0 \rceil$) decoder is expected to be $r(\varphi_i)$, while the outputs of the other $P - 1$ decoders expected to be $0_{\beta \times 1}$ where $\beta = |r|$. By concatenating the outputs of all the P decoders, the expected output of the whole autoencoder can be written as

$$u = [u_1^{\mathrm{T}}, \cdots, u_P^{\mathrm{T}}]^{\mathrm{T}} = \left[\underbrace{0_{\beta \times 1}^{\mathrm{T}}, \cdots, 0_{\beta \times 1}^{\mathrm{T}}}_{p-1}, r^{\mathrm{T}}(\varphi_i), \underbrace{0_{\beta \times 1}^{\mathrm{T}}, \cdots, 0_{\beta \times 1}^{\mathrm{T}}}_{P-p} \right]^{\mathrm{T}}. \tag{7.21}$$

When φ_i varies from $\phi^{(0)}$ to $\phi^{(P)}$, the p_i's corresponding to $i = 1, \cdots, I$ are $\underbrace{1, \cdots, 1,}_{I_0}$

$\underbrace{2, \cdots, 2,}_{I_0} \cdots, \underbrace{P, \cdots, P}_{I_0}.$

Denote the autoencoder output corresponding to input vector $r(\varphi_i)$ by $u(\varphi_i)$, and the input dataset for autoencoder training by

$$\Xi^{(AE)} = [r(\varphi_1), \cdots, r(\varphi_I)]. \tag{7.22}$$

Moreover, the column-wise output label-set associated with the input dataset $\Xi^{(AE)}$ is

$$\psi^{(AE)} = [u(\varphi_1), u(\varphi_2), \cdots, u(\varphi_I)] = \begin{bmatrix} \Upsilon_1 & 0_{\beta \times I_0} & 0_{\beta \times I_0} & 0_{\beta \times I_0} \\ 0_{\beta \times I_0} & \Upsilon_2 & 0_{\beta \times I_0} & 0_{\beta \times I_0} \\ 0_{\beta \times I_0} & 0_{\beta \times I_0} & \ddots & 0_{\beta \times I_0} \\ 0_{\beta \times I_0} & 0_{\beta \times I_0} & 0_{\beta \times I_0} & \Upsilon_P \end{bmatrix} \tag{7.23}$$

where superscript $(\bullet)^{(AE)}$ is used for variables related to the autoencoder, and superscript $(\bullet)^{(CF)}$ will be used for the classifiers, and

$$\Upsilon_p = [r(\varphi_{(p-1)I_0+1}), \cdots, r(\varphi_{pI_0})], \quad for\ p = 1, \cdots, P. \tag{7.24}$$

The data-label pair of ($\Xi^{(AE)}, \psi^{(AE)}$) are then used as input and expected output to train the autoencoder. The squared l_2-norm distance between the expected output and the actual one is used as the loss function, i.e.,

$$\varepsilon^{(AE)}(\varphi_i) = \frac{1}{2}\|\tilde{\boldsymbol{u}}(\varphi_i)\|_2^2, \tag{7.25}$$

where $\tilde{\boldsymbol{u}}(\varphi_i) = \boldsymbol{u}(\varphi_i) - \hat{\boldsymbol{u}}(\varphi_i)$ and $\hat{\boldsymbol{u}}(\varphi_i)$ is the actual output of the autoencoder when $r(\varphi_i)$ is inputted.

The weight matrices $\boldsymbol{U}_{1,0}$ and $\boldsymbol{U}_{2,1}$, together with bias vectors $\boldsymbol{b}_1$ and $\boldsymbol{b}_2$, are then updated based on the back-propagated gradients of the loss function $\varepsilon^{(AE)}(\varphi_i)$ with respect to the variables. The gradients can be computed via straightforward mathematical derivations and are listed below,

$$\frac{\partial \varepsilon^{(AE)}(\varphi_i)}{\partial [\boldsymbol{U}_{2,1}]_{i_1,i_2}} = [\tilde{\boldsymbol{u}}(\varphi_i)]_{i_1}[\boldsymbol{U}_{1,0}r(\varphi_i) + \boldsymbol{b}_1]_{i_2}, \tag{7.26}$$

$$\frac{\partial \varepsilon^{(AE)}(\varphi_i)}{\partial [\boldsymbol{U}_{1,0}]_{i_1,i_2}} = \tilde{\boldsymbol{u}}^T(\varphi_i)[\boldsymbol{U}_{2,1}]_{:,i_1}[r(\varphi_i)]_{i_2}, \tag{7.27}$$

$$\frac{\partial \varepsilon^{(AE)}(\varphi_i)}{\partial [\boldsymbol{b}_1]_l} = \tilde{\boldsymbol{u}}^T(\varphi_i)[\boldsymbol{U}_{2,1}]_{:,l}, \tag{7.28}$$

$$\frac{\partial \varepsilon^{(AE)}(\varphi_i)}{\partial [\boldsymbol{b}_2]_l} = [\tilde{\boldsymbol{u}}(\varphi_i)]_l, \tag{7.29}$$

where $[\boldsymbol{\alpha}]_l$ represents the lth element of vector $\boldsymbol{\alpha}$, $[\boldsymbol{A}]_{i_1,i_2}$ represents the (i_1,i_2)th element of matrix $\boldsymbol{A}$. The variants are then updated iteratively as

$$\alpha_{\text{new}} = \alpha_{\text{old}} + \mu_1 \frac{\partial \varepsilon^{(AE)}(\varphi_i)}{\partial \alpha}, \tag{7.30}$$

where α can be any element in matrices $\boldsymbol{U}_{1,0}$, $\boldsymbol{U}_{2,1}$ or vectors $\boldsymbol{b}_1$, $\boldsymbol{b}_2$, μ_1 is the learning rate, α_{old} and α_{new} denote the values of the variables before and after the current update, respectively.

Figure 7.6 shows the spatial responses when $P = 6$ spatial filtering decoders are trained in the spatial scope of $[-60°, 60°)$ according to the above strategy. Detailed descriptions of the simulation settings can be found in Section 7.4.3. Figure 7.6(a) shows the spatial gains of the filters, which is defined as,

$$g_a^{(p)} = |\bar{\boldsymbol{r}}^H(\varphi_i)\bar{\boldsymbol{u}}_p|, \ \text{for } p = 1, \cdots, P; \ i = 1, \cdots, I, \tag{7.31}$$

where $\bar{\boldsymbol{u}}_p$ is the complex-valued variant of $\boldsymbol{u}_p$ by taking the first half of the vector as the real part and the second half as imaginary part. Figure 7.6(b) shows the phase responses of the filters, i.e.,

$$g_b^{(p)} = \frac{|\bar{\boldsymbol{r}}^H(\varphi_i)\bar{\boldsymbol{u}}_p|}{\|\bar{\boldsymbol{r}}(\varphi_i)\|_2\|\bar{\boldsymbol{u}}_p\|_2}, \ \text{for } p = 1, \cdots, P; \ i = 1, \cdots, I. \tag{7.32}$$

$g_b^{(p)}$ describes how much the phase shifts between the elements of $\bar{\boldsymbol{r}}(\varphi_i)$ are kept unchanged after being filtered by the autoencoder, and $g_a^{(p)}$ combines the effect of

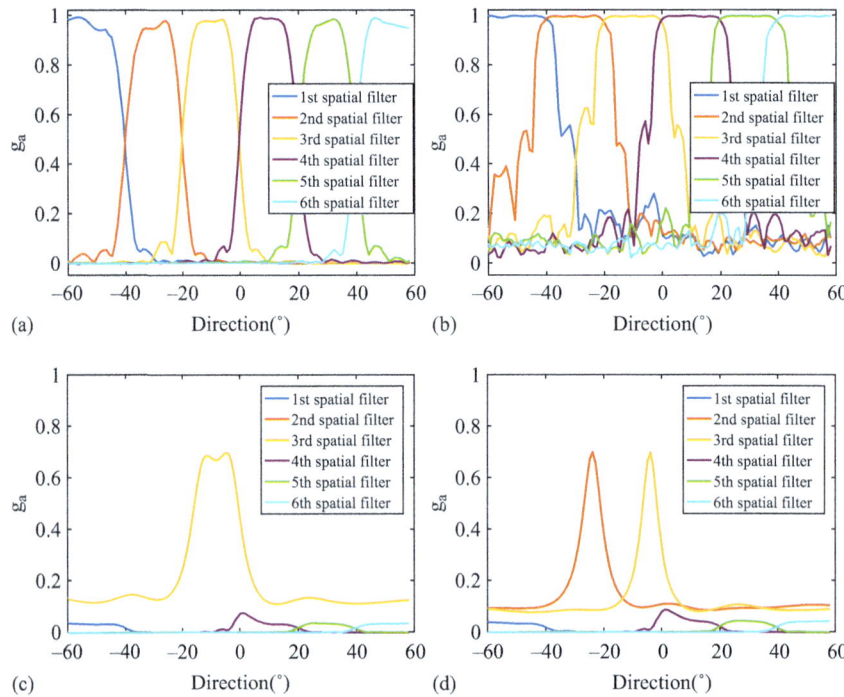

Figure 7.6 *Performance of the multi-task autoencoder for spatial filtering, (a) gain responses; (b) phase responses; (c) filter outputs of two signals in the same subregion ($\phi_1 = -13°$, $\phi_2 = -3°$); (d) filter outputs of two signals in different subregions ($\phi_1 = -24°$, $\phi_2 = -4°$).*

amplitude attenuation of different filters. Figure 7.6(a) and (b) show that the phase consistency between the autoencoder input and its expected output is maintained within the divided subregions, and the gain decays rapidly at the edge of each subregion. The outputs of the decoders other than the corresponding one have small amplitudes and weak phase consistency.

In order to test the linearity property of the autoencoder, a vector r corresponding to two-signal scenarios is inputted to the autoencoder. First, two signals located in the same subregion of $[-20°, 0°)$ impinge onto the array from directions of $\phi_1 = -13°$ and $\phi_2 = -3°$, respectively. The gain response $g_a^{(p)}$ of the six decoder outputs are shown in Figure 7.6(c). The spatial gain response of the corresponding decoder is similar to that of the beamformers, while the gains of the decoders in the other subregions are very small. Then repeat the simulation with two signals from directions $\phi_1 = -24°$ and $\phi_2 = -4°$, which locate in two adjacent subregions. The $g_a^{(p)}$ of the six decoder outputs in this scenario are shown in Figure 7.6(d). As can be seen from the figure, the two signal components are well separated by the corresponding filters, and the output amplitudes of the other filters are negligible.

7.4.2.2   Training strategy for parallel classifiers

The P parallel classifiers take the output of the corresponding decoder as inputs, and estimate the spatial spectrum in the corresponding subregions separately. When compared with r, each of the autoencoder outputs u_p $(p = 1, \cdots, P)$ contains signals impinging from a much smaller subregion. Since spatially closer signals generally have more similar steering vectors, u_p's should have much more concentrated distributions than r. In the parallel classifiers, we use multiple hidden layers and nonlinear activation functions to enhance expressivity of the network, so as to realize refined DOA estimation. In order to retain the polarity of the inputs at each layer of the classifiers, an element-wise hyperbolic tangent function is used as the activation function, i.e.,

$$
\begin{aligned}
\tanh(\boldsymbol{\alpha}) &= [\tanh(\alpha_1), \tanh(\alpha_2), \cdots, \tanh(\alpha_{-1})]^{\mathrm{T}}, \\
\tanh(\alpha) &= \frac{e^\alpha - e^{-\alpha}}{e^\alpha + e^{-\alpha}}
\end{aligned}
\tag{7.33}
$$

where α_{-1} denotes the last element of $\boldsymbol{\alpha}$.

When the training of the autoencoder has been completed, its weights and biases are kept fixed. Then a new end-to-end neural network framework is formed between the input vector r and the reconstructed spectrum y. The weights and biases of the classification neural networks should be trained to estimate the directions of incident signals in different subregions. To achieve this goal, another training dataset is constructed in scenarios with two simultaneous signals.

We choose several inter-signal angle intervals $\boldsymbol{\Delta} = \{\Delta_j\}_{j=1}^J$. For each interval, form the input vectors $r(\phi, \Delta_j)$ corresponding to two incident signals from directions ϕ and $\phi + \Delta_j$, where $\phi^{(0)} \leq \phi < \phi^{(P)} - \Delta_j$ and $j = 1, \cdots, J$. The expected classifier output corresponding to input $r(\phi, \Delta_j)$ is denoted by $y(\phi, \Delta_j)$ as follows,

$$
[y(\phi, \Delta_j)]_l = \begin{cases} \dfrac{\overline{\phi} - \varphi_{l-1}}{\varphi_l - \varphi_{l-1}}, & \varphi_{l-1} \leq \overline{\phi} < \varphi_l, \overline{\phi} \in \{\phi, \phi + \Delta_j\}, \\ \dfrac{\varphi_{l+1} - \overline{\phi}}{\varphi_{l+1} - \varphi_l}, & \varphi_l \leq \overline{\phi} < \varphi_{l+1}, \overline{\phi} \in \{\phi, \phi + \Delta_j\}, \\ 0, & \text{otherwise.} \end{cases}
\tag{7.34}
$$

The above equation shows that the reconstructed spatial spectrum is expected to have non-zero positive values only on the grids adjacent to the true signal direction, and the direction of each signal can be estimated accurately by linear amplitude interpolation between the two adjacent grids. The training dataset of the classifiers can be written as

$$
\Xi^{(CF)} = \left[\Xi_1^{(CF)}, \cdots, \Xi_J^{(CF)} \right],
\tag{7.35}
$$

where $\Xi_j^{(CF)} = \left[r(\varphi_1, \Delta_j), \cdots, r(\varphi_l - \Delta_j, \Delta_j) \right]$, and the associated label-set is

$$
\psi^{(CF)} = \left[\psi_1^{(CF)}, \cdots, \psi_J^{(CF)} \right],
\tag{7.36}
$$

where $\psi_j^{(CF)} = \left[y(\varphi_1, \Delta_j), \cdots, y(\varphi_l - \Delta_j, \Delta_j) \right]$.

During the training process, the reconstruction error of the spatial spectrum is calculated and back-propagated to optimize the parameters of the parallel classifiers. Denote the expected and actual classifier outputs corresponding to $r(\phi, \Delta)$ as $y(\phi, \Delta)$ and $\hat{y}(\phi, \Delta)$, respectively, and the reconstruction error can be expressed as

$$\tilde{y}(\phi, \Delta) = \hat{y}(\phi, \Delta) - y(\phi, \Delta). \tag{7.37}$$

The loss function of each classifier is the squared l_2-norm of the spatial spectral reconstruction error, i.e.

$$\varepsilon^{(CF)}(\phi, \Delta) = \frac{1}{2} \|\tilde{y}(\phi, \Delta)\|_2^2, \tag{7.38}$$

The gradients of the loss function with respect to the variables of the classifier can be derived by straightforward mathematical differentiation. Details of the derivation are omitted here, and interested readers can refer to the previous relevant literature such as [57]. Most deep learning platforms, such as TensorFlow [123], also provide callable instructions for calculating the gradients automatically.

Subsequently, the elements of the weight matrices and bias vectors are then optimized using their gradients as follows,

$$\alpha_{\text{new}} = \alpha_{\text{old}} + \mu_2 \frac{\partial \varepsilon^{(CF)}(\phi, \Delta)}{\partial \alpha}, \tag{7.39}$$

where μ_2 represents the learning rate.

After training the classifiers with settings that will be detailed in Section 7.4.3, the corresponding array covariance vectors $r(\phi = -13°, \Delta = -3°)$ and $r(\phi = -24°, \Delta = -4°)$ associated with Figure 7.6(c) and (d) are re-inputted to the entire DNN to obtain reconstructed spatial spectra, and the results are shown in Figure 7.7(a) and (b). The simulation results indicate that the established DNN framework can separate two simultaneously incident signals satisfyingly, no matter they impinge from the same or different spatial subregions, and there are only slight residuals on directions without incident signals. Refined signal directions can finally be estimated via linear interpolation within the spectrum peaks.

7.4.2.3   Adaptation to array imperfections

As has been discussed in Sections 7.2 and 7.3, ML-based DOA estimation methods adopt a data-driven implementation and are expected to have built-in adaptability to various array imperfections. In this subsection, we further analyze and validate such a property of the presented method.

Suppose that the array responding function is perturbed by a specific type or a combination of imperfections with parameters e, and the mapping from signal direction to covariance vector is denoted as $\phi \xmapsto{e} r_e(\phi)$. It is also assumed that there is no prior information about the array imperfections. When the perturbed vector $r_e(\varphi_i)$

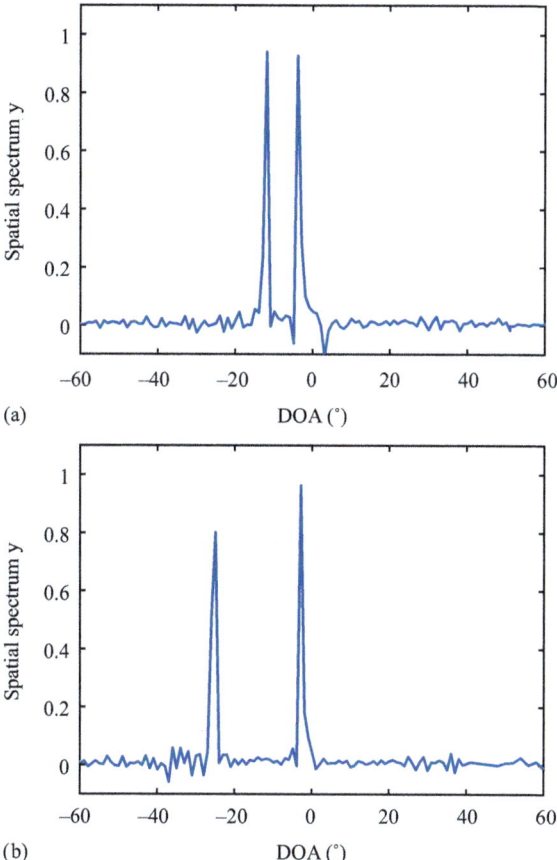

(a)

(b)

Figure 7.7   Reconstructed spatial spectrum of two signals, (a) $\phi_1 = -13°$,
$\phi_2 = -3°$; (b) $\phi_1 = -24°$, $\phi_2 = -4°$

with $\lceil i/I_0 \rceil = p$ is inputted to the autoencoder, the corresponding label vector can be represented as

$$u = \left[u_1^T, \cdots, u_P^T\right]^T = \left[\underbrace{0_{\beta\times1}^T, \cdots, 0_{\beta\times1}^T}_{p-1}, r_e^T(\varphi_i), \underbrace{0_{\beta\times1}^T, \cdots, 0_{\beta\times1}^T}_{P-p} \right]^T. \tag{7.40}$$

That is to say, the input vector $r_e^T(\varphi_i)$ will be filtered into the pth decoder even in the presence of array imperfections.

Subsequently, the output of the decoder is fed into the parallel classifiers. Since the signal component from direction φ_i is embedded in the output of the pth decoder, it will be processed by the pth classifier. The corresponding spatial spectrum contains a spectral peak on one or two adjacent grids closest to φ_i, which can be further interpolated to obtain a DOA estimate of φ_i. Thus, the well-trained hybrid DNN (the autoencoder together with the parallel classifiers) actually reconstructs the inverse mapping relationship of $r_e(\phi) \mapsto^e \phi$, no matter which kinds of imperfections are present and how they perturbs the array responding function. The reconstructed inverse mapping embeds the influence of array imperfections in it, so it can also adapt to any test data in case of the same imperfections, and is expected to obtain unbiased DOA estimates in despite of array imperfections.

7.4.3 Simulations and analyses

This subsection carries out simulations to verify the advantages of the presented DL-based DOA estimation method over some other ML-based methods [36,37] in generalization, and also its predominance over the most widely cited parametric method of MUSIC [8] in imperfection adaptation. The simulations are implemented on the deep learning platform TensorFlow [123], and the gradients are calculated using its embedded tools directly. The more recently proposed DL-based methods [60,62–64,66] are not chosen as baselines, because some of them adapt only to single-source scenarios [62,63], while others take time-frequency representations of incident signals as inputs [60,64,66]. Therefore, they do not adapt to the considered multi-signal direction finding scenarios with only a few hundreds of snapshots. Auto-calibration techniques [26–32] are not considered, since no prior information of the array imperfections is assumed to be known beforehand. Such settings help to show the robustness of different methods to un-calibrated arrays and make fairer performance comparisons.

7.4.3.1 Simulation settings

In the following simulations, we use a 10-element uniform linear array (ULA) to estimate directions of signals impinging from the spatial scope of $[-60°, 60°)$, i.e., $M = 10$, $\phi^{(0)} = -60°$, $\phi^{(P)} = 60°$. The inter-element spacing of the ULA is half-wavelength, and the potential space is divided equally into $P = 6$ subregions and sampled with a grid interval of $1°$, resulting in $I = 120$ grids in total with $\varphi_1 = -60°, \varphi_2 = -59°, \cdots, \varphi_I = 59°$, and each spatial subregion has $I_0 = 20$ grids. The covariance vectors r in the training datasets of both the autoencoder and the classifiers, and also in the test datasets, are obtained with $N = 256$ snapshots.

During the training of the autoencoder, the $[-60°, 60°)$ space is also sampled with an interval of $1°$ to obtain a direction set with $\phi_1 = -60°, \phi_2 = -59°, \ldots, \phi_I = 59°$. The covariance vectors and associated labels are computed according to (7.22) and (7.23). For each $\varphi_i, i = 1, \ldots, 120$, only one group of snapshots is collected to calculate the covariance vector r, with the signal-to-noise ratio (SNR) being 5dB. The mini-batch training strategy [124] is used with a batch size of 32 and learning rate of $\mu_1 = 0.001$, and 1,000 epochs are taken for the training with the dataset shuffled in each epoch. The dimension of the input layer is $\beta = M(M - 1)/2 = 45$, and that of the hidden and output layers are set to $\lfloor 45/2 \rfloor = 22$ and $\beta P = 45 \times 6$, respectively.

The parameters of the autoencoder remain unchanged after the training process, and another dataset according to two-signal scenarios is collected to train the parallel classifiers. The covariance vectors and associated labels are computed according to (7.35) and (7.36). The angle Δ between the two signals is sampled from the set of $\{1°, 2°, \ldots, 39°, 40°\}$, which covers scenarios from very close signals to signals separated by twice the width of a subregion. Then the direction of the first signal (denoted by ϕ) is traversed with an interval of $1°$ from $-60°$ to $60° - \Delta$, and the direction of the second signal is $\phi + \Delta$. The SNR of both signals is 5 dB, and 10 groups of snapshots are generated by adding different random noises to calculate r in each direction setting. Finally, a total of $(119 + 118 + \cdots + 80) \times 10 = 39,800$ covariance vectors are collected in the dataset.

The data-label pairs are used for training the classifiers with a mini-batch size of 32 and a learning rate of $\mu_2 = 0.001$, and the order of the vectors is shuffled during each of the 300 training epochs. The number of hidden layers is chosen to be $L_2 - 1 = 2$ as a trade-off between the expressivity power (which improves with deeper networks [55]) and under-training risk (which aggravates with more network parameters [122]) of the classifiers, and the sizes of the hidden and output layers in each classifier are $\lfloor 2/3 \times \beta \rfloor = 30$, $\lfloor 4/9 \times \beta \rfloor = 20$ and $I_0 = 20$, respectively. All the weights and biases of the DNN are randomly initialized according to a uniform distribution in the range $[-0.1, 0.1]$.

Three typical kinds of array imperfections are considered in the simulations, including gain and phase inconsistence, sensor position error and inter-sensor mutual coupling. The imperfections in practical arrays may be very complicated to formulate mathematically, so we use simplified models to facilitate simulations. However, the simplification does not cause any loss of generality to the results.

The gain biases of the array sensors are set as

$$e_{\text{gain}} = \rho \times [0, \underbrace{0.2, \ldots, 0.2}_{5}, \underbrace{-0.2, \ldots, -0.2}_{4}]^{\text{T}}, \tag{7.41}$$

where the parameter $\rho \in [0, 1]$ is introduced to control the strength of the imperfections. The phase biases are

$$e_{\text{phase}} = \rho \times [0, \underbrace{-30°, \ldots, -30°}_{5}, \underbrace{30°, \ldots, 30°}_{4}]^{\text{T}}. \tag{7.42}$$

The position biases are

$$e_{\text{pos}} = \rho \times [0, \underbrace{-0.2, \ldots, -0.2}_{5}, \underbrace{0.2, \ldots, 0.2}_{4}]^{\text{T}} \times d, \tag{7.43}$$

where d is the inter-sensor spacing of the ULA. And the mutual coupling coefficient vector is

$$e_{\text{mc}} = \rho \times [0, \gamma^1, \cdots, \gamma^{M-1}]^{\text{T}}, \tag{7.44}$$

where $\gamma = 0.29e^{j58°}$ is the mutual coupling coefficient between adjacent sensors.

By specializing ρ, the array imperfections will be determined, and the perturbed array responding function is rewritten as follows,

$$a(\phi, e) = (I_M + \delta_{mc} E_{mc}) \times (I_M + \text{Diag}(\delta_{gain} e_{gain}))$$
$$\times \text{Diag}(\exp(j\delta_{phase} e_{phase})) \times a(\phi, \delta_{pos} e_{pos}), \tag{7.45}$$

where impulse function $\delta(\bullet)$ is used to indicate the existence of a certain kind of imperfection, $\text{Diag}(\bullet)$ forms diagonal matrices with the given vector on the diagonal, E_{mc} is a toeplitz matrix generated with the parameter vector e_{mc} [26], $a(\phi, \delta_{pos} e_{pos})$ indicates the actual array responding vector corresponding to the signal from direction ϕ when position error e_{pos} is included in the array geometry.

The array responding function given in (7.45) has been greatly simplified when compared with its counterpart in actual applications. Although the actual array response function can be measured more precisely with computational electromagnetic methods, such as [125–127] and the array imperfection formulations in (7.41)–(7.44) can also be modelled more accurately following previous existing literatures, such as [128–131] for mutual coupling, we use the simplified formulations mainly to facilitate simulation, and we believe that these simplifications are reasonable for performance comparison. That is because the proposed DL-based method does not utilize any prior information about the array imperfections and steering vectors, and the simplifications do not affect the adaptability of the direction finding method to various kinds of array imperfections. The proposed end-to-end training and testing strategies can be generalized straightforwardly to other array geometries and imperfections, no matter how the antennas are fed and how much the array steering vector has been perturbed by imperfections.

7.4.3.2   Generalization to unseen scenarios

In this subsection, we compare the performance of the presented DNN-based DOA estimation method with the SVR-based DOA estimator [36,37] to show how they generalize to scenarios not included in the training dataset. Array imperfections are not considered in this subsection.

First, two signals with an angular distance of 13.5° and SNR=5dB are assumed to impinge onto the array simultaneously, and the direction of the first signal varies from −60° to 50°. This angular distance is not contained in the training set Δ, and the direction of the second signal deviates from the preset training directions and the output spectrum grids. The final DOA estimates are obtained via amplitude interpolation within the two most significant peaks of the reconstructed spectra according to (7.14). When the direction of the first signal increases from −60° to 50° with a step of 1°, the estimated directions and the estimation errors of the two signals are shown in Figure 7.8(a) and (b), respectively. The DOA estimates of the presented method well match their true values, and most of the estimation errors are smaller than 2°. In Figure 7.8(c) and (d), we plot the results of the SVR-based DOA estimation results in the same scenarios. In Figure 7.8(c) and (d), the SVR's are trained with the same training dataset as the DNN classifiers, except that the training dataset in Figure 7.8(c) is noise-free. The results indicate that the SVR's also perform well when the training

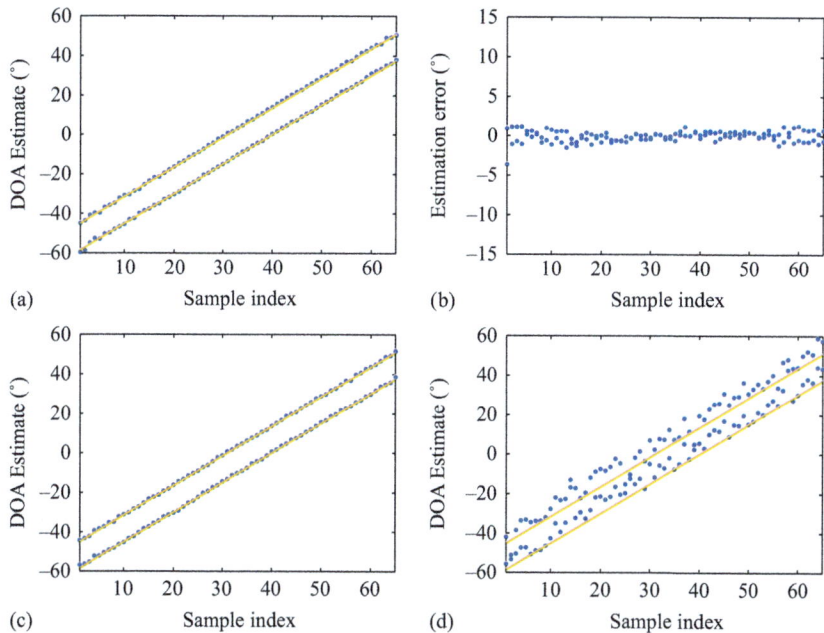

Figure 7.8   *DOA estimation performance of off-grid signals, (a) DNN-based DOA estimates; (b) DNN-based DOA estimation errors; (c) SVR-based DOA estimates with noise-free training data; (d) SVR-based DOA estimates with training data of SNR=5dB*

data are noise-free, but their performance aggravates significantly when there are perturbations in the training data. As it is very difficult or even impossible to collect noise-free training data, the presented method is believed to behave better than the SVR-based method in practice.

We then keep the SNR of the two signals fixed at 5dB, and enlarge their angular distance to 50.4°, 60.1° and 70.7°, respectively, which deviates from the Δ's in the training set largely. When the first signal direction varies from $-60°$ to $60° - \Delta$, the DOA estimates of the DNN-based method and the SVR-based method are shown in Figure 7.9. The DNN-based method again shows much better adaptation to these scenarios that are unseen in the training dataset, while the SVR-based method fails to obtain valid DOA estimates for the signals.

Finally, we show how the DNN-based method behaves when the testing data contains different numbers of signals as the training data. The DNN and SVR models are trained with data in two-signal scenarios, but tested in two much different scenarios, with one containing a single signal with a DOA of $-14.7°$ and the other containing three signals with DOA's of $-34.7°$, 5.3°, and 20.3°. The SVR-based method forms two regression machines for processing test data and outputs two DOA estimates for each given data [36,37]. If the input covariance vector contains more or fewer signal

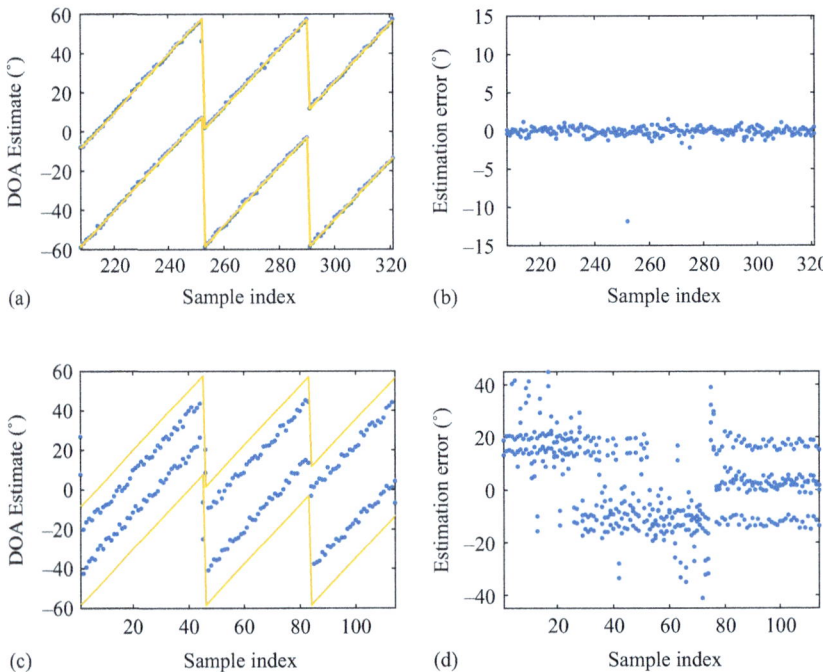

Figure 7.9 *DOA estimation results of two signals separated by 50.4°, 60.1°, 70.7°, which are much larger than the separations in the training dataset, (a) DNN-based DOA estimates; (b) DNN-based DOA estimation errors;(c) SVR-based DOA estimates; (d) SVR-based DOA estimation errors*

components, the SVR outputs make no sense. However, one can conclude from the results in Figure 7.10 that the DNN-based method still performs satisfyingly in the one-signal and three-signal scenarios.

7.4.3.3 Adaptation to array imperfections

In this subsection, we carry out simulations to verify the adaptability of DNN-based DOA estimation method to various kinds of array imperfections, and compare it with the most widely cited parametric DOA estimation method MUSIC [8] in DOA estimation precision. Although many newly proposed parametric methods perform better than MUSIC in demanding scenarios, such as the sparsity-inducing ones [11–15], the DOA estimation performance of different parametric methods is similar in the presence of significant array imperfections. MUSIC is chosen as a baseline since it is widely known among existing parametric DOA estimation methods. The SVR-based estimation method is also excluded here, because the results in Figure 7.8 show that it lacks robustness to noisy training datasets, and SVR models trained with noise-free datasets make the comparisons unfair.

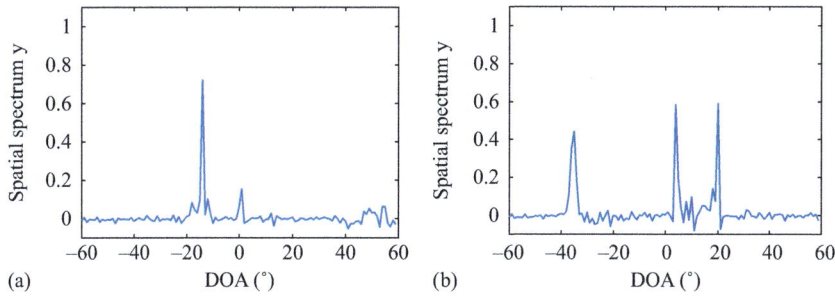

(a)   (b)

Figure 7.10 DOA estimation results in one-signal and three-signal scenarios with DNN models trained in two-signal scenarios, (a) one-signal scenario; (b) three-signal scenario

Assume that two signals with the same SNR of 10dB impinge onto the array simultaneously from directions of 32.5° and 43.4°, which both deviate from the direction grids of the training dataset and the reconstructed spatial spectrum, and different types of array imperfections are considered by adjusting parameter ρ from 0 to 1 in (7.41)–(7.44). When $\rho = 0$, no imperfection is contained in the array responding functions. Four cases with different array imperfections are considered by setting the $\delta(\bullet)$'s in (7.45) to different values.

First, set $\delta_{gain} = \delta_{phase} = 1$ and $\delta_{pos} = \delta_{mc} = 0$, i.e., only gain and phase imperfections are added to the array. The estimation root-mean-square-errors (RMSE's) of the DNN-based method and the MUSIC method in 100 Monte-Carlo simulations with ρ varying from 0 to 1 are shown in Figure 7.11(a). Then set $\delta_{gain} = \delta_{phase} = \delta_{mc} = 0$ and $\delta_{pos} = 1$ to consider the sensor position error only, and the DOA estimation RMSE's are shown in Figure 7.11(b). After that, set $\delta_{gain} = \delta_{phase} = \delta_{pos} = 0$ and $\delta_{mc} = 1$ to retain the mutual coupling effect only, and the corresponding DOA estimation RMSE's are shown in Figure 7.11(c). Finally, set $\delta_{gain} = \delta_{phase} = \delta_{pos} = \delta_{mc} = 1$ to consider a combination of all the three kinds of array imperfections at the same time, and the DOA estimation RMSE's are shown in Figure 7.11(d).

When $\rho = 0$, the array responding function is unbiased and consistent with its formulation in parametric methods, so MUSIC obtains DOA estimates with very high precisions. However, as array imperfections become more and more significant, the DOA estimation error of MUSIC increases almost linearly, indicating that the method is weakly adaptable to unknown array imperfections, and it should be modified with various auto-calibration methods to improve DOA estimation precisions. On the contrary, the DNN-based method performs slightly worse than MUSIC when no imperfections are present. But it is very robust to different or even combined array imperfections, and its DOA estimation precision seldom deteriorates with the increase of the imperfections strength. That is because it does not rely on any pre-assumption about the array geometry or array responding function. When ρ is as small as 0.1–0.3, the DNN-based method performs comparably as MUSIC, and when ρ

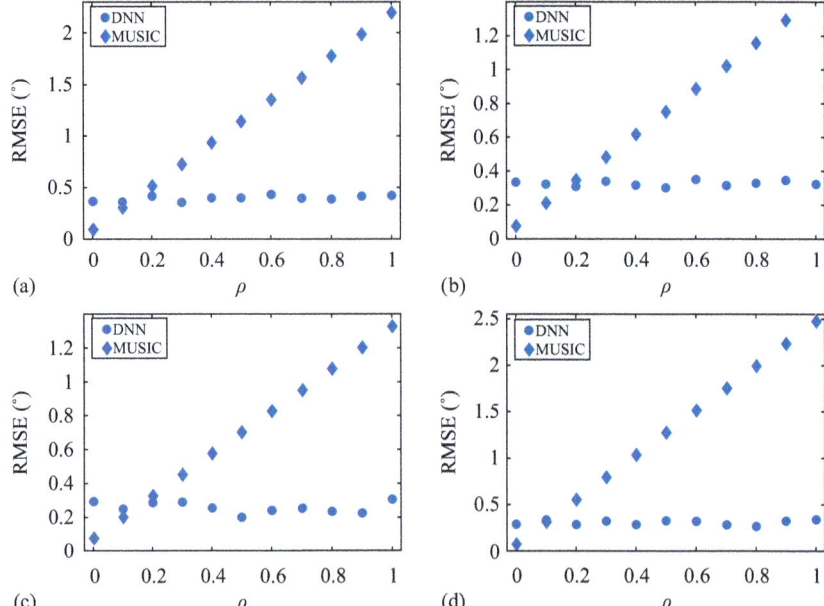

Figure 7.11 *DOA estimation RMSE's of DNN-based and MUSIC methods when ρ*
increases from 0 to 1 in the presence of different array imperfections,
(a) gain and phase inconsistence; (b) sensor position error; (c) mutual
coupling; (d) combination of three kinds of imperfections

becomes larger and the array responding function deviates more significantly away
its imperfection-free counterpart, the DNN-based method performs much better than
MUSIC.

In order to illustrate the contribution of the multi-task autoencoder in DOA estimation, the next set of simulation experiments shows how the number of spatial filters
(or the decoders) affects the accuracy of DOA estimation.

The numbers of filters are set to be 3, 6 and 12, and the corresponding DNN
models are trained and tested in scenarios with different kinds of array imperfections
and different ρ's. Parameters of the training and testing datasets are the same as that
in the simulations corresponding to Figure 7.11. The DOA estimation RMSE with
respect to ρ in the case of different numbers of spatial filters are shown in Figure
7.12. The results show that, when the number of filters is as small as 3, each of the
filter covers a wide spatial subregion and the output vector of the filter has a relatively
dispersed distribution. As a result, the trained DNN model does not work well in some
of the testing scenarios and the DOA estimation RMSE's have very large variances.

When the number of filters is increased to 6, the DOA estimation RMSE reduces
significantly and keeps stable in different scenarios. If we further increase the number

of filters to 12, the RMSE does not decrease largely anymore. It can be concluded from this group of simulation results that, increasing the number of decoders leads to improved DOA estimation performance in the presented DNN framework, but when the number of decoders is greater than a certain threshold, significant performance improvement can hardly be gained anymore. Therefore, we have empirically set the number of decoders to be 6 in previous simulations. Another special value of the decoder number is 1, which is equivalent to removing the autoencoder module from the DNN framework shown in Figure 7.5. In this case, a single classifier will be trained for DOA estimation directly. From the results in Figure 7.12, it is somewhat straightforward to infer that, the DOA estimation performance of the DNN framework with a single classifier will be much worse than that of the DNN containing an additional 6-task autoencoder.

This group of simulation results provides a good illustration for the role of the multi-task autoencoder in the DNN framework. It spatially filters the input of the whole model to concentrate its distribution, and helps to significantly reduce the generalization burden of the subsequent classifiers, and finally improves DOA estimation performance.

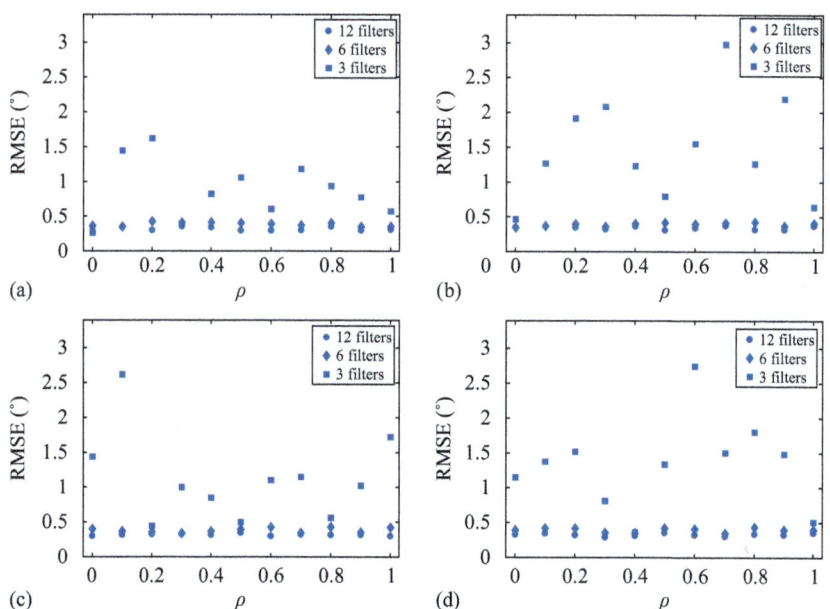

Figure 7.12   *DOA estimation RMSE of DNN-based method using different numbers of spatial filters in the case of different array imperfections, (a) gain and phase inconsistence; (b) sensor position error; (c) mutual coupling; (d) combined imperfections*

7.5   Concluding remarks and future trends

In this chapter, we present an overview and a concrete example of state-of-the-art DL-based DOA estimation research. We start with a brief introduction to existing methods in this area, and then make a survey of the DL frameworks presented in publications to summarize how different DL techniques have been exploited to solve diverse DOA estimation problems. Finally, a hierarchical DNN framework is presented as an example to show the implementation of DL-based DOA estimation methods, and demonstrate partially their predominance over existing methods not using DL techniques.

DL-based methods provide a better solution to DOA estimation problems from many perspectives, such as reducing computational complexity in the testing phase, improving DOA estimation precision in demanding scenarios when analytic modeling of array outputs is either impossible or very challenging. Despite the fruitful research of DL-based DOA estimation methods in recent years, many issues should be further clarified before these research results can be applied in comprehensive DOA estimation applications.

1. DL-based methods require a big labelled dataset to train a DOA estimator with satisfying performance. However, as the requirement is usually very difficult to meet in practical applications, researchers should make efforts to optimize the methods to relax the requirement on big dataset. Some attempts have been carried out following ideas of unsupervised training and transfer learning, but there is still a gap between the available results and practical demands.

2. In some DOA estimation systems, labelled data are accumulated gradually during usage, and thus evolvable models should be built to improve with time. Based on this goal, it might be necessary to introduce reinforcement and incremental learning techniques to this area to propose new methods, which may be quite different from the current predominant research on batch learning methods.

3. Most of existing DL-based DOA estimation methods are purely data-driven, they are unable to exploit prior knowledge about the array manifold or signal features, such as completely or partially known array configuration, and cyclostationarity or modulation information of incident signals, etc. Further investigations are needed to work out a way to make better use of these information during the pre-processing, post-processing or DNN building procedures, so as to improve DOA estimation performance.

4. In the area of DOA estimation, many factors may affect the mapping from signal directions to array outputs, e.g., number of incident signals, signal amplitudes and waveforms, it is usually impossible to enumerate all scenarios and collect enough data in each scenario to train a universal DNN model. Therefore, more efforts should be devoted to enhance the generalization of the DL-based DOA estimation methods, so that they will be able to perform satisfyingly in unseen scenarios.

References

[1] Kovaly JJ. Statistical description of nonlinear signal processing in array antenna systems. *IEEE Transactions on Military Electronics.* 1965;9(3): 237–246.

[2] Johnson DH, Dudgeon DE. *Array Signal Processing: Concepts and Techniques.* New York, NY: Simon & Schuster; 1992.

[3] Krim H, Viberg M. Two decades of array signal processing research: the parametric approach. *IEEE Signal Processing Magazine.* 1996;13(4):67–94.

[4] Zhang W, Wang J, Wu S. Robust Capon beamforming against large DOA mismatch. *Signal Processing.* 2013;93(4):804–810. Available from: https://search-ebscohost-com-s.nudtproxy.yitlink.com:443/login. aspx?direct=true&db=aci&AN=84477631&lang=zh-cn&site=ehost-live.

[5] Van Veen BD, Buckley KM. Beamforming: a versatile approach to spatial filtering. *IEEE ASSP Magazine.* 1988;5(2):4–24.

[6] Litva J, Lo TK. *Digital Beamforming in Wireless Communications.* London: Artech House Inc.; 1996.

[7] Li J, Stoica P. *Robust Adaptive Beamforming*, vol. 88. New York, NY: John Wiley & Sons; 2005.

[8] Schmidt R. Multiple emitter location and signal parameter estimation. *IEEE Transactions on Antennas and Propagation.* 1986;34(3):276–280.

[9] Roy R, Kailath T. ESPRIT-estimation of signal parameters via rotational invariance techniques. *IEEE Transactions on Acoustics, Speech, and Signal Processing.* 1989;37(7):984–995.

[10] Gonen E, Mendel JM. Subspace-based direction finding methods. In: Madisetti VK and Williams DB, editors, *The Digital Signal Processing Handbook*, Boca Raton, FL: CRC Press LLC. chapitre. 1999. p. 62.

[11] Malioutov D, Cetin M, Willsky AS. A sparse signal reconstruction perspective for source localization with sensor arrays. *IEEE Transactions on Signal Processing.* 2005;53(8):3010–3022.

[12] Liu ZM, Huang ZT, Zhou YY. Direction-of-arrival estimation of wideband signals via covariance matrix sparse representation. *IEEE Transactions on Signal Processing.* 2011;59(9):4256–4270.

[13] Liu ZM, Huang ZT, Zhou YY. An efficient maximum likelihood method for direction-of-arrival estimation via sparse Bayesian learning. *IEEE Transactions on Wireless Communications.* 2012;11(10):1–11.

[14] Yang Z, Xie L, Zhang C. Off-grid direction of arrival estimation using sparse Bayesian inference. *IEEE Transactions on Signal Processing.* 2013;61(1): 38–43.

[15] Liu ZM, Guo FC. Azimuth and elevation estimation with rotating long-baseline interferometers. *IEEE Transactions on Signal Processing.* 2015;63(9):2405–2419.

[16] Wu LL, Liu ZM, Huang ZT. Wideband DOA estimation for SOBFN. *Iet Radar Sonar and Navigation.* 2017;11(6):972–977.

[17]   Wu LL, Liu ZM, Jiang WL. A direction finding method for spatial optical beam-forming network based on sparse Bayesian learning. *Signal, Image and Video Processing*. 2016;11(2):1–7.

[18]   Jaffer AG. Maximum likelihood direction finding of stochastic sources: a separable solution. In: *1988 International Conference on Acoustics, Speech, and Signal Processing*, 1988. ICASSP-88, New York, NY: IEEE; 1988. pp. 2893–2896.

[19]   Stoica P, Nehorai A. MUSIC, maximum likelihood, and Cramer-Rao bound. *IEEE Transactions on Acoustics, Speech, and Signal Processing*. 1989;37(5):720–741.

[20]   Miller MI, Fuhrmann DR. Maximum-likelihood narrow-band direction finding and the EM algorithm. *IEEE Transactions on Acoustics, Speech, and Signal Processing*. 1990;38(9):1560–1577.

[21]   Porat B, Friedlander B. Accuracy requirements in off-line array calibration. *IEEE Transactions on Aerospace and Electronic Systems*. 1997;33(2):545–556.

[22]   Hopkinson GR, Goodman TM, Prince SR. *A Guide to the Use and Calibration of Detector Array Equipment*. vol. 142. SPIE Press; 2004.

[23]   Allen B, Ghavami M. *Adaptive Array Systems: Fundamentals and Applications*. New York, NY: John Wiley and Sons; 2006.

[24]   Viberg M, Swindlehurst AL. Analysis of the combined effects of finite samples and model errors on array processing performance. *IEEE Transactions on Signal Processing*. 1994;42(11):3073–3083.

[25]   Liu Z, Huang Z, Zhou Y. Bias analysis of MUSIC in the presence of mutual coupling. *IET Signal Processing*. 2009;3(1):74–84.

[26]   Friedlander B, Weiss AJ. Direction finding in the presence of mutual coupling. *IEEE Transactions on Antennas and Propagation*. 1991;39(3):273–284.

[27]   Svantesson T. Modeling and estimation of mutual coupling in a uniform linear array of dipoles. In: *1999 IEEE International Conference on Acoustics, Speech, and Signal Processing*, 1999, vol. 5. New York, NY: IEEE; 1999. pp. 2961–2964.

[28]   Lin M, Yang L. Blind calibration and DOA estimation with uniform circular arrays in the presence of mutual coupling. *IEEE Antennas and Wireless Propagation Letters*. 2006;5(1):315–318.

[29]   Weiss AJ, Friedlander B. Array shape calibration using eigenstructure methods. Signal Processing. 1991;22(3):251–258.

[30]   Flanagan BP, Bell KL. Array self-calibration with large sensor position errors. *Signal Processing*. 2001;81(10):2201–2214.

[31]   Paulraj A, Kailath T. Direction of arrival estimation by eigenstructure methods with unknown sensor gain and phase. In: *IEEE International Conference on Acoustics, Speech, and Signal Processing*, ICASSP'85, vol. 10. IEEE; 1985. pp. 640–643.

[32]   Li Y, Er M. Theoretical analyses of gain and phase error calibration with optimal implementation for linear equispaced array. *IEEE Transactions on Signal Processing*. 2006;54(2):712–723.

[33] Ng BC, See CMS. Sensor-array calibration using a maximum-likelihood approach. *IEEE Transactions on Antennas and Propagation*. 1996;44(6): 827–835.

[34] Stavropoulos KV, Manikas A. Array calibration in the presence of unknown sensor characteristics and mutual coupling. In: *2000 10th European Signal Processing Conference*, IEEE; 2000. pp. 1–4.

[35] Liu ZM, Zhou YY. A unified framework and sparse Bayesian perspective for direction-of-arrival estimation in the presence of array imperfections. *IEEE Transactions on Signal Processing*. 2013;61(15):3786–3798.

[36] Pastorino M, Randazzo A. A smart antenna system for direction of arrival estimation based on a support vector regression. *IEEE Transactions on Antennas and Propagation*. 2005;53(7):2161–2168.

[37] Randazzo A, Abou-Khousa M, Pastorino M, *et al.* Direction of arrival estimation based on support vector regression: experimental validation and comparison with MUSIC. *IEEE Antennas and Wireless Propagation Letters*. 2007;6:379–382.

[38] Rawat A, Yadav RN, Shrivastava SC. Neural network applications in smart antenna arrays: a review. *AEUE – International Journal of Electronics and Communications*. 2012;66(11):903–912.

[39] Terabayashi K, Natsuaki R, Hirose A. Ultrawideband direction-of-arrival estimation using complex-valued spatiotemporal neural networks. *IEEE Transactions on Neural Networks and Learning Systems*. 2014;25(9):1727–1732.

[40] Gao Y, Hu D, Chen Y, *et al.* Gridless 1-b DOA estimation exploiting SVM approach. *IEEE Communications Letters*. 2017;21(10): 2210–2213.

[41] Huang ZT, Wu LL, Liu ZM. Toward wide-frequency-range direction finding with support vector regression. *IEEE Communications Letters*. 2019;23(6):1029–1032.

[42] Wu LL, Huang ZT. Coherent SVR learning for wideband direction-of-arrival estimation. *IEEE Signal Processing Letters*. 2019;26(4):642–646.

[43] Massa A, Oliveri G, Salucci M, *et al.* Learning-by-examples techniques as applied to electromagnetics. *Journal of Electromagnetic Waves and Applications*. 2017 11;32:1–26.

[44] Jha S, Durrani T. Direction of arrival estimation using artificial neural networks. *IEEE Transactions on Systems, Man, and Cybernetics*. 1991;21(5):1192–1201.

[45] Southall HL, Simmers JA, O'Donnell TH. Direction finding in phased arrays with a neural network beamformer. *IEEE Transactions on Antennas and Propagation*. 1995;43(12):1369–1374.

[46] El Zooghby AH, Christodoulou CG, Georgiopoulos M. A neural network-based smart antenna for multiple source tracking. *IEEE Transactions on Antennas and Propagation*. 2000;48(5):768–776.

[47] Quionero-Candela J, Sugiyama M, Schwaighofer A, *et al. Dataset Shift in Machine Learning*. New York, NY: The MIT Press; 2009.

[48] Sugiyama M, Kawanabe M. *Machine Learning in Non-stationary Environments: Introduction to Covariate Shift Adaptation*. New York, NY: MIT Press; 2012.

[49] He K, Zhang X, Ren S, *et al.* Deep Residual Learning for Image Recognition. In: *2016 IEEE Conference on Computer Vision and Pattern Recognition (CVPR)*; 2016. pp. 770–778.

[50] Sze V, Chen YH, Yang TJ, *et al.* Efficient processing of deep neural networks: a tutorial and survey. *Proceedings of the IEEE*. 2017;105(12):2295–2329.

[51] Chen S, Wang H, Xu F, *et al.* Target classification using the deep convolutional networks for SAR images. *IEEE Transactions on Geoscience and Remote Sensing*. 2016;54(8):4806–4817.

[52] Deng L, Hinton G, Kingsbury B. New types of deep neural network learning for speech recognition and related applications: an overview. In: 2013 *IEEE International Conference on Acoustics, Speech and Signal Processing*; 2013. pp. 8599–8603.

[53] Goodfellow I, Bengio Y, Courville A. *Deep Learning*. New York, NY: MIT Press; 2016. http://www.deeplearningbook.org.

[54] Hinton GE, Salakhutdinov RR. Reducing the dimensionality of data with neural networks. *Science*. 2006;313(5786):504–507.

[55] LeCun Y, Bengio Y, Hinton G. Deep learning. *Nature*. 2015;521(7553):436–444.

[56] Schmidhuber J. Deep learning in neural networks: an overview. *Neural Networks*. 2015;61:85–117.

[57] Nielsen M. *Neural Networks and Deep Learning*. Determination Press; 2015.

[58] Wu LL, Liu ZM, Huang ZT. Deep convolution network for direction of arrival estimation with sparse prior. *IEEE Signal Processing Letters*. 2019;26(11):1688–1692.

[59] Wu LL, Liu ZM, Huang ZT, *et al.* Deep Neural Network for DOA estimation with unsupervised pretraining. In: *2019 IEEE International Conference on Signal, Information and Data Processing (ICSIDP)*; 2019. pp. 1–5.

[60] Takeda R, Komatani K. Discriminative multiple sound source localization based on deep neural networks using independent location model. In: *Spoken Language Technology Workshop (SLT), 2016 IEEE*. IEEE; 2016. pp. 603–609.

[61] Takeda R, Komatani K. Sound source localization based on deep neural networks with directional activate function exploiting phase information. In: *2016 IEEE International Conference on Acoustics, Speech and Signal Processing (ICASSP)*; 2016.

[62] Xiao X, Zhao S, Zhong X, *et al.* A learning-based approach to direction of arrival estimation in noisy and reverberant environments. In: *2015 IEEE International Conference on Acoustics, Speech and Signal Processing (ICASSP)*, IEEE; 2015. pp. 2814–2818.

[63] Vesperini F, Vecchiotti P, Principi E, *et al.* A neural network based algorithm for speaker localization in a multi-room environment. In: *2016 IEEE 26th International Workshop on Machine Learning for Signal Processing (MLSP)*, IEEE; 2016. pp. 1–6.

[64] Chakrabarty S, Habets EAP. Broadband doa estimation using convolutional neural networks trained with noise signals. In: *2017 IEEE Workshop on Applications of Signal Processing to Audio and Acoustics (WASPAA)*; 2017. pp. 136–140.

[65] Chakrabarty S, Habets EAP. Multi-speaker DOA estimation using deep convolutional networks trained with noise signals. *IEEE Journal of Selected Topics in Signal Processing*. 2019;13(1):8–21.

[66] Adavanne S, Politis A, Virtanen T. Direction of arrival estimation for multiple sound sources using convolutional recurrent neural network. arXiv preprint arXiv:171010059. 2017.

[67] He W, Motlicek P, Odobez JM. Deep neural networks for multiple speaker detection and localization. In: *2018 IEEE International Conference on Robotics and Automation (ICRA)*; 2018. pp. 74–79.

[68] Ahmad M, Muaz M, Adeel M. A survey of deep neural network in acoustic direction finding. In: *2021 International Conference on Digital Futures and Transformative Technologies (ICoDT2)*; 2021. pp. 1–6.

[69] Erricolo D, Chen PY, Rozhkova A, *et al*. Machine learning in electromagnetics: a review and some perspectives for future research. In: *2019 International Conference on Electromagnetics in Advanced Applications (ICEAA)*; 2019. pp. 1377–1380.

[70] You MY, Lu AN, Ye YX, *et al*. A review on machine learning-based radio direction finding. *Mathematical Problems in Engineering*. 2020; pp. 1–9.

[71] Zardi F, Nayeri P, Rocca P, *et al*. Artificial intelligence for adaptive and reconfigurable antenna arrays: a review. *IEEE Antennas and Propagation Magazine*. 2021;63(3):28–38.

[72] Massa A, Marcantonio D, Chen X, *et al*. DNNs as applied to electromagnetics, antennas, and propagation—: a review. *IEEE Antennas and Wireless Propagation Letters*. 2019;18(11):2225–2229.

[73] Huang H, Yang J, Huang H, *et al*. Deep learning for super-resolution channel estimation and DOA estimation based massive MIMO system. *IEEE Transactions on Vehicular Technology*. 2018;67(9):8549–8560.

[74] Shengguo Ge SNBMR Kuo Li. Deep learning approach in DOA estimation: a systematic literature review. *Mobile Information Systems*. 2021;2021:1139–1151.

[75] Xiang H, Chen B, Yang M, *et al*. Altitude measurement based on characteristics reversal by deep neural network for VHF radar. *Radar Sonar and Navigation Iet*. 2019;13(1):98–103.

[76] Kase Y, Nishimura T, Ohgane T, *et al*. DOA estimation of two targets with deep learning. In: *2018 15th Workshop on Positioning, Navigation and Communications (WPNC)*; 2018. pp. 1–5.

[77] Xiang H, Chen B, Yang M, *et al*. Angle separation learning for coherent DOA estimation with deep sparse prior. *IEEE Communications Letters*. 2020;2020(99):1–1.

[78] Yao Y, Lei H, He W. A-CRNN-based method for coherent doa estimation with unknown source number. *Sensors (Basel, Switzerland)*. 2020;20(8):2296.

[79] Papageorgiou GK, Sellathurai M, Eldar YC. Deep networks for direction-of-arrival estimation in low SNR. *IEEE Transactions on Signal Processing.* 2021;69:3714–3729.

[80] Cao Y, Lv T, Lin Z, *et al.* Complex ResNet aided DoA estimation for near-field MIMO systems. *IEEE Transactions on Vehicular Technology.* 2020;69(10):11139–11151.

[81] Kim D, Kim SH, Cha SG, *et al.* DNN-based direction finding by time modulation. In: *2020 IEEE International Symposium on Antennas and Propagation and North American Radio Science Meeting;* 2020. pp. 439–440.

[82] Chen M, Gong Y, Mao X. Deep neural network for estimation of direction of arrival with antenna array. *IEEE Access.* 2020;8:140688–140698.

[83] Su X, Hu P, Liu Z, *et al.* Mixed near-field and far-field source localization based on convolution neural networks via symmetric nested array. *IEEE Transactions on Vehicular Technology.* 2021;70(8):7908–7920.

[84] A Massa MSNA G Oliveri, Rocca P. Learning-by-examples techniques as applied to electromagnetics. *Journal of Electromagnetic Waves and Applications.* 2018;32(4):516–541.

[85] Wan L, Sun Y, Sun L, *et al.* Deep learning based autonomous vehicle super resolution DOA estimation for safety driving. *IEEE Transactions on Intelligent Transportation Systems.* 2021;22(7):4301–4315.

[86] Papageorgiou GK, Sellathurai M. Fast direction-of-arrival estimation of multiple targets using deep learning and sparse Arrays. In: *ICASSP 2020 – 2020 IEEE International Conference on Acoustics, Speech and Signal Processing (ICASSP);* 2020. pp. 4632–4636.

[87] Chen Y, Xiong KL, Huang ZT. Robust direction-of-arrival estimation via sparse representation and deep residual convolutional network for co-prime arrays. In: *2020 IEEE 3rd International Conference on Electronic Information and Communication Technology (ICEICT);* 2020. pp. 514–519.

[88] Zhu W, Zhang M, Li P, *et al.* Two-dimensional DOA estimation via deep ensemble learning. *IEEE Access.* 2020;8:124544–124552.

[89] Chen D, Joo Y. Novel approach to 2D DOA estimation for uniform circular arrays using convolutional neural networks. *International Journal of Antennas and Propagation.* 2021;2021:Article ID 5516798.

[90] Chen H, Ser W. Acoustic source localization using LS-SVMs without calibration of microphone arrays. In: *International Symposium on Circuits and Systems (ISCAS 2009), 24-17 May 2009*, Taipei, Taiwan, 2009. pp. 1863–1866.

[91] Tan X, Hu H, Cheng R, *et al.* Direction of arrival estimation using self-organizing map. *Mathematical Problems in Engineering.* 2015;2015(8):1–8.

[92] Youssef K, Argentieri S, Zarader JL. A learning-based approach to robust binaural sound localization. In: *IEEE/RSJ International Conference on Intelligent Robots and Systems;* 2014.

[93] Shieh CS, Lin CT. Direction of arrival estimation based on phase differences using neural fuzzy network. *IEEE Transactions on Antennas Propagation.* 2000;48(7):1115–1124.

[94] Xiang H, Chen B, Yang M, *et al.* A novel phase enhancement method for low-angle estimation based on supervised DNN learning. *IEEE Access.* 2019;2019(99):1–1.

[95] Abeywickrama S, Jayasinghe L, Fu H, *et al.* RF-based direction finding of UAVs using DNN. In: *2018 IEEE International Conference on Communication Systems (ICCS)*; 2018. pp. 157–161.

[96] Sun Y, Chen J, Yuen C, *et al.* Indoor sound source localization with probabilistic neural network. *IEEE Transactions on Industrial Electronics.* 2017;68: 1–1.

[97] Zhu Y, Cai Y, Zhu H, *et al.* DeepAoA: online vehicular direction finding based on a deep learning method. In: *2019 IEEE 25th International Conference on Parallel and Distributed Systems (ICPADS)*; 2019. pp. 782–789.

[98] Chen J, Yi H, Zhou X. Direction of arrival estimation method with eigenvector-based radial basis function neural network. *Journal of Shanghai Jiaotong University.* 2003;37(003):373–375.

[99] Elbir AM. DeepMUSIC: multiple signal classification via deep learning. *IEEE Sensors Letters.* 2020;4(4):1–4.

[100] Chung H, Seo H, Joo J, *et al.* Off-grid DoA estimation via two-stage cascaded neural network. *Energies.* 2021;14(1):1–11.

[101] Cong J, Wang X, Huang M, *et al.* Robust DOA estimation method for MIMO radar via deep neural networks. *IEEE Sensors Journal.* 2021;21(6): 7498–7507.

[102] Elman JL. Finding structure in time. *Cognitive Science.* 1990;14(2): 179–211.

[103] Lecun Y, Bottou L, Bengio Y, *et al.* Gradient-based learning applied to document recognition. *Proceedings of the IEEE.* 1998;86(11):2278–2324.

[104] Hochreiter S, Schmidhuber J. Long short-term memory. *Neural Computation.* 1997;9(8):1735–1780.

[105] Krizhevsky A, Sutskever I, Hinton G. ImageNet classification with deep convolutional neural networks. In: *Advances in Neural Information Processing Systems*, vol. 25, 2nd ed., New York, NY: MIT Press; 2012.

[106] He K, Zhang X, Ren S, *et al.* Deep residual learning for image recognition. In: *2016 IEEE Conference on Computer Vision and Pattern Recognition (CVPR)*; 2016. pp. 770–778.

[107] Guo Y, Zhang Z, Huang Y, *et al.* DOA estimation method based on cascaded neural network for two closely spaced sources. *IEEE Signal Processing Letters.* 2020;27:570–574.

[108] Xu Y, Guo H, Fan R, *et al.* An intelligent direction finding method with deep neural network. In: *2020 IEEE 2nd International Conference on Civil Aviation Safety and Information Technology (ICCASIT)*; 2020. pp. 1094–1098.

[109] Chung J, Gulcehre C, Cho KH, *et al.* Empirical Evaluation of Gated Recurrent Neural Networks on Sequence Modeling. Eprint Arxiv. 2014.

[110] Xiang H, Chen B, Yang M, *et al.* Improved direction-of-arrival estimation method based on LSTM neural networks with robustness to array imperfections. *Applied Intelligence.* 2021;51:1–14.

[111] Barthelme A, Utschick W. A machine learning approach to DoA estimation and model order selection for antenna arrays with subarray sampling. *IEEE Transactions on Signal Processing*. 2021;69:3075–3087.

[112] Barthelme A, Utschick W. DoA estimation using neural network-based covariance matrix reconstruction. *IEEE Signal Processing Letters*. 2021;28: 783–787.

[113] Shi B, Ma X, Zhang W, *et al*. Complex-valued convolutional neural networks design and its application on UAV DOA estimation in urban environments. *Journal of Communications and Information Networks*. 2020;5(2):130–137.

[114] Yuan Y, Wu S, Wu M, *et al*. Unsupervised learning strategy for direction-of-arrival estimation network. *IEEE Signal Processing Letters*. 2021;28: 1450–1454.

[115] Bianco MJ, Gannot S, Fernandez-Grande E, *et al*. Semi-supervised source localization in reverberant environments with deep generative modeling. *IEEE Access*. 2021;9:84956–84970.

[116] Varanasi V, Hegde R. Robust online direction of arrival estimation using low dimensional spherical harmonic features. In: *2017 IEEE International Conference on Acoustics, Speech and Signal Processing (ICASSP)*; 2017. pp. 511–515.

[117] Laufer-Goldshtein B, Talmon R, Gannot S. Semi-supervised sound source localization based on manifold regularization. *IEEE/ACM Transactions on Audio, Speech, and Language Processing*. 2016;24(8):1393–1407.

[118] Elbir AM, Mishra KV. Sparse array selection across arbitrary sensor geometries with deep transfer learning. *IEEE Transactions on Cognitive Communications and Networking*. 2021;7(1):255–264.

[119] Cao H, Wang W, Su L, *et al*. Deep transfer learning for underwater direction of arrival using one vector sensor. *The Journal of the Acoustical Society of America*. 2021;149(3):1699–1711.

[120] Tan Q, Zhu L. A GFK-based SNR adaptive DOA estimation method. In: *2019 IEEE 19th International Conference on Communication Technology (ICCT)*; 2019. pp. 233–238.

[121] Xiao P, Liao B, Deligiannis N. DeepFPC: a deep unfolded network for sparse signal recovery from 1-Bit measurements with application to DOA estimation. *Signal Processing*. 2020;176:n.pag. Available from: https://search-ebscohost-com-s.nudtproxy.yitlink.com:443/login. aspx?direct=true&db=a9h&AN=145408621&lang=zh-cn&site=ehost-live.

[122] Ioffe S, Szegedy C. Batch normalization: accelerating deep network training by reducing internal covariate shift. In: *International Conference on Machine Learning*; 2015. pp. 448–456.

[123] Abadi M, Agarwal A, Barham P, *et al*. Tensorflow: large-scale machine learning on heterogeneous distributed systems. arXiv preprint arXiv:160304467. 2016.

[124] Cotter A, Shamir O, Srebro N, *et al*. Better mini-batch algorithms via accelerated gradient methods. In: *Advances in Neural Information Processing Systems*, New York, NY: MIT Press; 2011. pp. 1647–1655.

[125]   Kindt RW, Sertel K, Volakis JL. A review of finite array modeling via finite-element and integral-equation-based decomposition methods. *The Radio Science Bulletin*. 2011;(336):12–22.

[126]   Hu J, Lu W, Shao H, *et al*. Electromagnetic analysis of large scale periodic arrays using a two-level CBFs method accelerated with FMM-FFT. *IEEE Transactions on Antennas and Propagation*. 2012;60(12):5709–5716.

[127]   Ludick DJ, Botha MM, Maaskant R, *et al*. The CBFM-enhanced Jacobi method for efficient finite antenna array analysis. *IEEE Antennas and Wireless Propagation Letters*. 2017;16:2700–2703.

[128]   Pasala KM, Friel EM. Mutual coupling effects and their reduction in wideband direction of arrival estimation. *IEEE Transactions on Aerospace and Electronic Systems*. 1994;30(4):1116–1122.

[129]   Adve RS, Sarkar TK. Compensation for the effects of mutual coupling on direct data domain adaptive algorithms. *IEEE Transactions on Antennas and Propagation*. 2000;48(1):86–94.

[130]   Edwin Lau CK, Adve RS, Sarkar TK. Minimum norm mutual coupling compensation with applications in direction of arrival estimation. *IEEE Transactions on Antennas and Propagation*. 2004;52(8):2034–2041.

[131]   Lui HS, Hui HT. Direction-of-arrival estimation: measurement using compact antenna arrays under the influence of mutual coupling. *IEEE Antennas and Propagation Magazine*. 2015;57(6):62–68.

Chapter 8

Deep learning techniques for remote sensing

Qian Song[1] and Feng Xu[2]

Due to the wide swath and acceptable cost, remote sensing (RS) techniques have been widely applied in extensive applications, such as land cover land use (LCLU), flood detection, urbanization monitoring. A number of airborne and space-borne missions are conducted to acquire remote sensing data—ALOS, TerraSAR-X, Sentinel-1, Sentinel-2, GEDI, UAVSAR, Landsat, to name a few. They carried different types of sensors that differ from each other in terms of resolution, penetration ability, and imaging mechanism, thus are suitable for different applications scenarios. Accordingly, it calls for specifically designed models for different types of data.

With the accumulation of years of the vast amount of data, how to effectively use them especially in an automatic manner to serve for practical applications becomes a challenge. Deep learning (DL), which has achieved great success in other tasks in the computer vision field, is employed as a powerful tool for dealing with remote sensing data [1–6]. Previous studies reviewed the basic deep learning models and their applications in remote sensing data regarding either the data types [2,4] or the task types [1,3,5,6]. They mainly align the remote sensing tasks with the computer vision tasks. In this chapter, however, we revisit several hot topics that come from the fields of target recognition, land cover and land use (LCLU), weather forecasting, and forest monitoring, introduce how various deep learning models are employed and fitted into these specific tasks.

8.1 Target recognition

8.1.1 Ship detection

Traditionally, detection ships in synthetic aperture radar (SAR) images relies on constant false alarm rate (CFAR) algorithms [7]. It assumes that ships are "brighter" than the background clutter, and its main effort is paid to determine a threshold to discriminate the current pixel. It uses the probability density function (PDF) of pixel values in a guarding window to fit a statistical model. When sets a constant false

[1]German Aerospace Center, Oberpfaffenhofen, Germany
[2]School of Information and Technology, Fudan University, Shanghai, China

alarm rate p_f, the threshold is calculated as the point when the cumulative density function equals $1 - p_f$. But in practical scenarios, CFAR which ignores ships' spatial features, obtains either a low detection accuracy or a high false alarm rate, especially in inshore ranges. Thus many deep learning models are developed in recent years for ship detection.

In [8], Jiao *et al.* used a faster RCNN framework which consists of a region proposal network and a detection network. Then detection network predicts whether the input proposal contains a ship or not, and outputs the anchor box of the detected ship. The focal loss is adopted to replace the original cross-entropy loss, which is defined as

$$\mathcal{L} = -(1 - p_t)^\gamma \log p_t, \tag{8.1}$$

$$p_t = \begin{cases} p, & y = 1, \\ 1 - p, & y = 0 \end{cases} \tag{8.2}$$

where p and y denote the network's output and ground truth. It is known that small ships are prone to be omitted by the deep networks trained with multi-scale samples. In order to deal with this problem, a dense attention Pyramid network is proposed in [9]. It stacks the feature maps extracted from different convolutional layers, which allows for grasping the intrinsic features of multi-scale targets. A similar framework, as shown in Figure 8.1, is applied in [10], where the backbone feature extractor is redesigned and trained from scratch. Experiments demonstrate that the proposed model outperforms those transfer learning-based baseline models in terms of metrics such as F1-score, mAP.

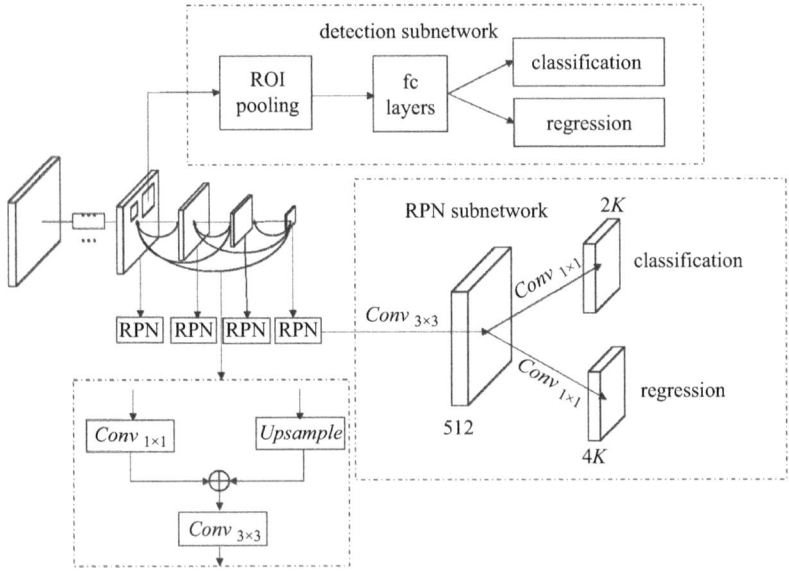

Figure 8.1   Architecture of the Faster RCNN-based ship detector proposed in [8]

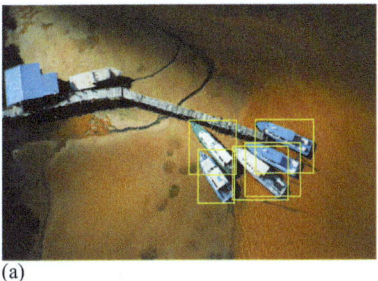

(a)   (b)

Figure 8.2   Comparison of horizontal bounding box (a) and rotated bounding box (b) for ship localization

On the other hand, ships near the shores are docked near each other, thus their horizontally placed bounding boxes overlap, as shown in Figure 8.2(a). When post-processing the candidate bounding boxes with non-maximum suppression (NMS) which merges the overlapped boxes into a big one, multiple small ships will be wrongly detected as one big ship. To deal with that, rotated bounding boxes are used accordingly, as shown in Figure 8.2(b). Then the location, size, and orientation need to be estimated simultaneously.

In [11], Wang *et al.* modified the original Single Shot Detector (SSD) network to allow for orientation angle prediction. Then the angle-related IoU (ArIoU) defined in (8.3) is employed for quantitatively comparing the proposed bounding box with the ground truth bounding box:

$$ArIoU(A, B) = \frac{A \bigcap B}{A \bigcup B} |\cos(\theta_A - \theta_B)| \tag{8.3}$$

In [12], An *et al.* proposed an improved rotatable bounding box-based ship detector named DRBox-v2 to boost the performance of DRBox-v1. It modifies the bounding box encoding method and the backbone feature extraction network with a pyramid network. Besides, the hard negative mining loss is combined with the focal loss in (8.1). Yang *et al.* introduced a task-wise feature pyramid network that calibrates the extracted feature maps with two separated channels for the two sub-tasks [13], as shown in Figure 8.3. The threshold of IoU is adaptively set according to the averaged IoU to incrementally increase the difficulty of the task.

8.1.2   *Aircraft recognition*

Aircraft recognition models can be categorized by the types of data used. In SAR images, aircraft have higher intensities compared with ground, thus SAR data are widely-used for aircraft detection. However, the aircraft in SAR images consists of discrete scattering clusters, and it increases the difficulty of detecting the aircraft as a whole, which becomes the main obstacle.

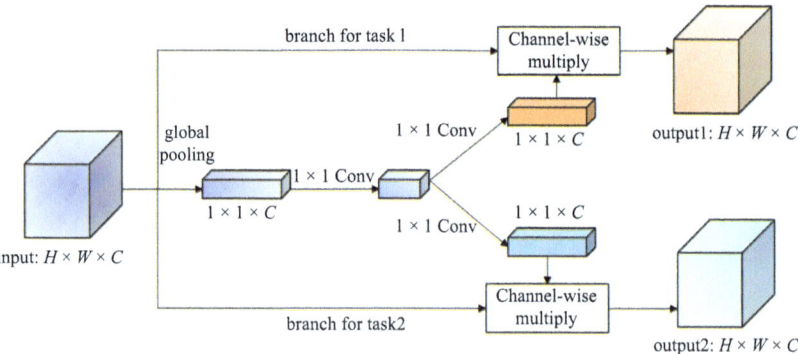

Figure 8.3   Illustration of the task-wise attention module proposed in [10]

In [14], convolutional neural network is applied for high-resolution SAR aircraft detection. A three-look network framework is proposed in [15] to detect the airport, aircraft, and the runway respectively using ResFaster R-CNN. To deal with the discontinuity of the strong scattering points, Zhao *et al.* proposed pyramid attention dilated network [16]. Its main building block, dilated attention block, is composed of a dilated convolution module and an attention module. The dilated convolution, as shown in Figure 8.4, skips several points according to the dilation rate in input images while applying convolution operation, which can increase the receptive field without introducing more parameters. Thus using dilated convolution helps integrate the discrete parts of aircraft. The pyramid network is also adopted in [17], as shown in Figure 8.5. The proposed algorithm firstly extracts the airport area, and then detects and integrates the strong scattering points, and the pyramid network extracts multi-scale feature maps and makes predictions.

High-resolution optical images allow for higher detection accuracy and aircraft recognition applications. In [18], Wu *et al.* use the binarized normed gradients algorithm to extract region proposals, and then a convolutional neural network is adopted to remove the false alarms. In [19], a classification framework that separately predicts the segmentation mask and the keypoints of aircraft, and then compares the rotated aircraft mask with the templates is developed. Thus the classification task is decomposed into two sub-tasks, segmentation and regression. Considering the size differences of different types of aircrafts, the objective of the regression network is defined as the normalized distances between predicted and labeled keypoints $\mathbf{p}_i$ and $\mathbf{p}'_i$ by the aircraft length l, i.e.

$$\mathcal{L} = \frac{1}{2} \sum_{i=1}^{N} \frac{1}{l_i} \parallel \mathbf{p}_i - \mathbf{p}'_i \parallel_2^2 . \tag{8.4}$$

In [20], Jia *et al.* also suggest rotating the aircraft slices to the same direction, and the rotation parameters are automatically estimated from the convolutional neural

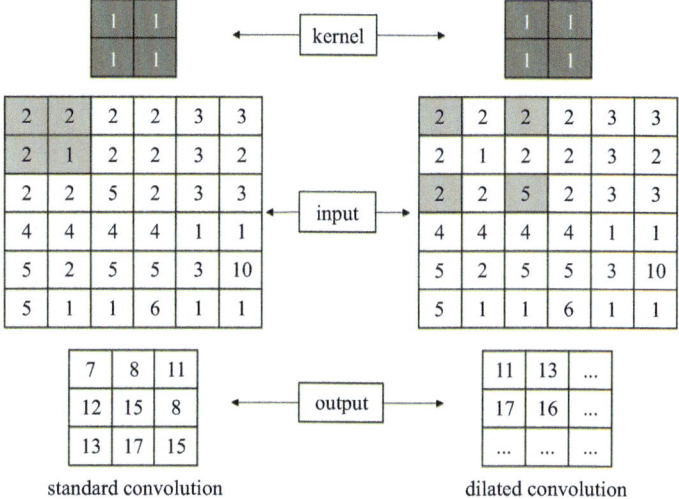

*Figure 8.4   Comparison of standard convolutional and dilated convolutional
operations, where the dilated convolution skips some points*

network. A component-based discrimination strategy using the rotated images is adopted to boost the classification results predicted by the front-stage rotatable boundingbox-based pyramid network. Experiments with Gaofen dataset demonstrate that the proposed network achieves over 70% accuracy in a ten-category recognition task.

8.1.3   Footprint extraction

According to the output of the networks, the existing footprint generation methods can be divided into three types: binary classification, classification, and regression. The first two types assign each pixel in the region of interest (ROI) with one of the two or more categories. In this regard, many segmentation-based deep learning models were applied directly or indirectly for extraction building footprints. In [21], a fully convolutional network was used to replace the patched-based architecture to make the borders of different patches more continuous. In [22], the classical encoder-decoder convolutional neural networks (CNN) was used to extract the multi-scale spatial features from the input data for building boundary delineation (as shown in Figure 8.6), and a follow-up CNN to classify these boundary images into footprint maps. Ji *et al.* proposed a Siamese U-Net (SiU-Net) for building extraction from high-resolution remote sensing images [23]. The original input and their down-sampled images are fed into the network simultaneously, and the discrepancy between their outputs and the corresponding ground truth as well as the down-sampled labels is calculated as the loss function. In [24], a multi-task framework that uses two U-Net networks to predict the building footprint and the road network respectively is proposed, as shown in

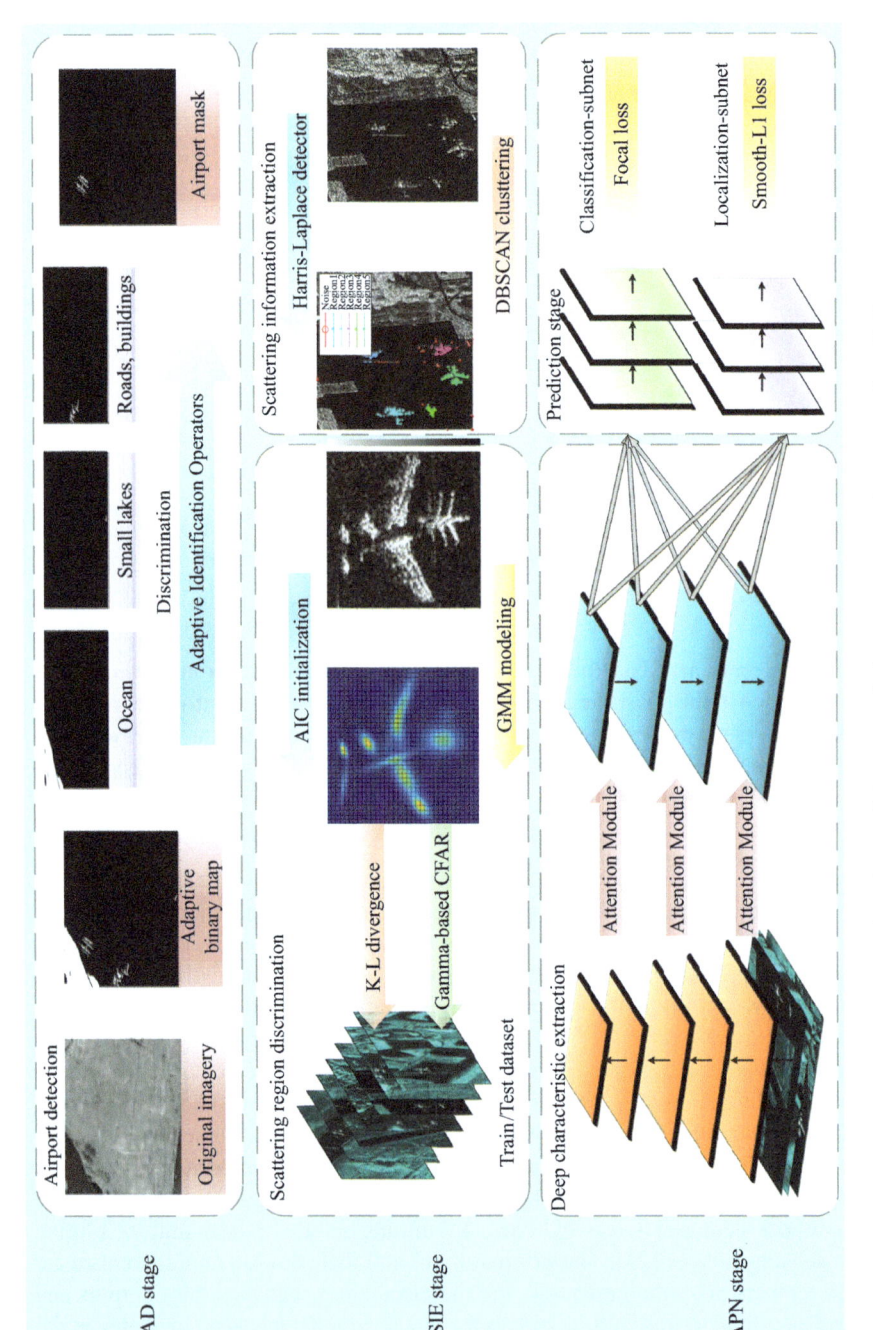

Figure 8.5   Flowchart of the aircraft detection network proposed in [17]

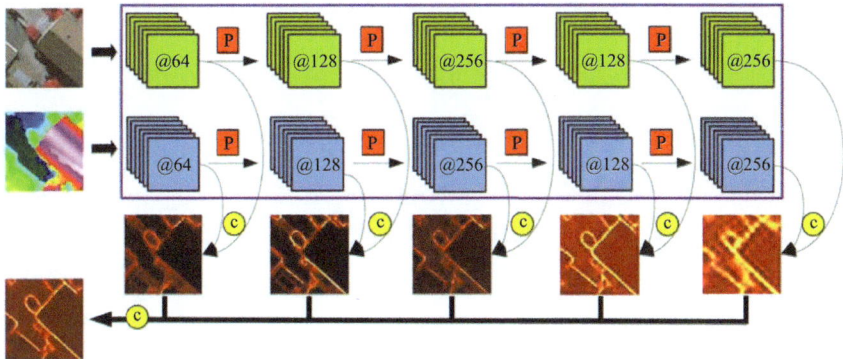

Figure 8.6   The used class-boundary network module for footprint extraction network in [22]

Figure 8.7. With the help of the vector data of OpenStreetMap, the networks are able to yield sub-pixel levels (2.5 m resolution) of building and road detection results using satellite images (10 m resolution). In [25], Li *et al.* proposed to concatenate two sub-networks to predict the attraction field representation (AFM) from input images and extract the footprint from AFM. AFM is calculated as the horizontal and vertical distances between the current pixel with the nearest pixel on the boundaries. Results showed that using AFM representation can yield sharper boundaries and reduce the false-alarms.

In [26], Yuan *et al.* proposed a network to predict the signed distance between the current pixel with the nearest boundary pixel, instead of a binary label. The distance was further divided into 128 classes, and the first 63 classes correspond to the non-building class. Thus, the building footprint generation task is regarded as a multi-class classification problem. Furthermore, Yang *et al.* adopted the same idea [27]. Two signed distance prediction networks are trained with different sources of data (RGB images and near IR-G-B images) respectively, and the averaged output from softmax layers is used as the final output.

In [28], the active contours model (ACM) was used to add geometric constraints, such as continuity, smooth edges to the extracted boundaries. The model is trained to learn the four parameters in the ACM, thus building footprint boundary generation was treated as a regression task. Although the predicted boundaries were not highly aligned with the ground truth, ACM did output more smooth and accurate edges as compared with the baseline model. In [29], a network named PolyMapper which combines convolutional neural networks with a recurrent neural network was proposed to directly extract the building boundaries and the roads polygons. In [30], Girard *et al.* proposed to use the deep learning models to predict the frame field and the segmentation map in a multi-task framework. The frame field can be converted into a polygon through a polygonization method, which allows the model to improve the building extraction performance.

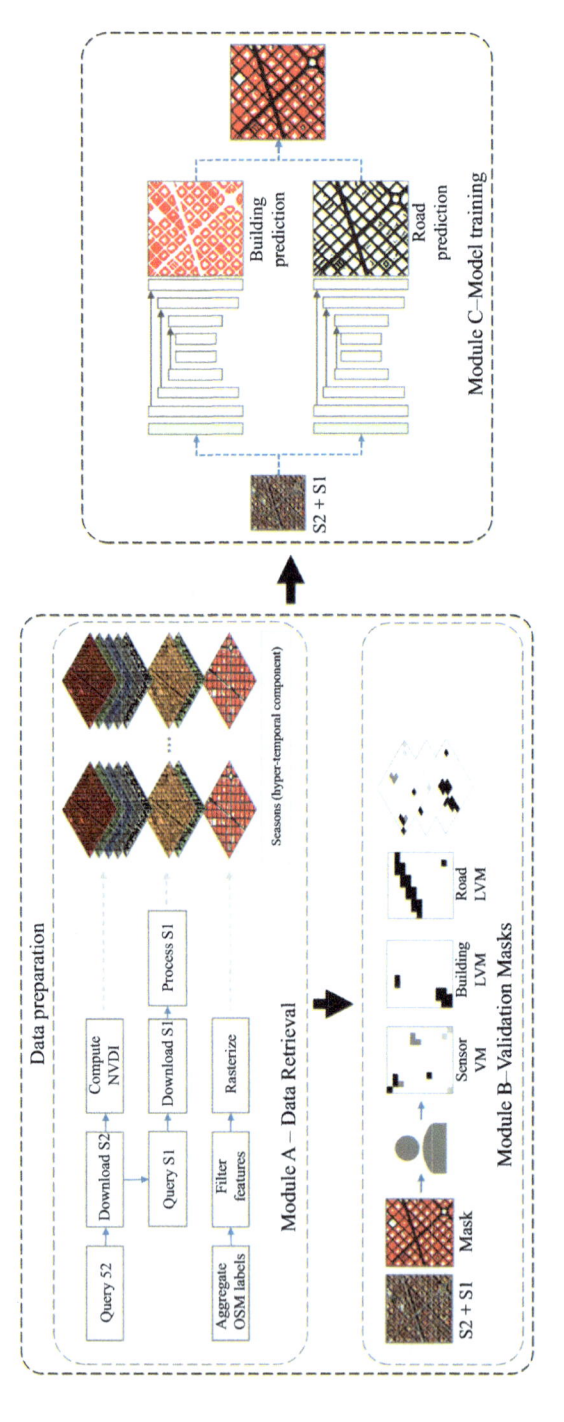

Figure 8.7   Flowchart of the proposed enhanced building footprint and road detection method in [24]

8.1.4   Few-shot recognition of SAR targets

The back-scattering intensities of SAR signals over man-made targets such as ships, buildings, vehicles are significantly stronger than that over other targets such as sea surface, road, and trees. Besides, SAR signals can penetrate rain, cloud, snow, tree canopy, and even the ground, thus are able to detect the targets under the cover. Thus SAR is widely-used for target recognition. On the other hand, however, due to its incoherent imaging mechanism, targets in synthetic aperture radar images vary with observation angles, and the cost of acquiring SAR samples is expensive, which limits the number of training samples. Thus few-shot learning (FSL) of target recognition in SAR images is essential.

Existing methods can be divided into two types: using generative models to augment the dataset and transfer learning from other data sources. The generative models take several parameters (such as category label, observation angle) as input, and output a pseudo-SAR image. In this case, a continuous vector corresponds to a unique SAR image. By varying the input vector, a vast number of SAR images can be generated. In [31], Song and Xu proposed a deep neural network for zero-shot learning, as shown in Figure 8.8. The learned mapping relationship is then used to inverse the feature representations from SAR images via the interpreter DNN. The unseen targets without available training samples can be recognized by comparing

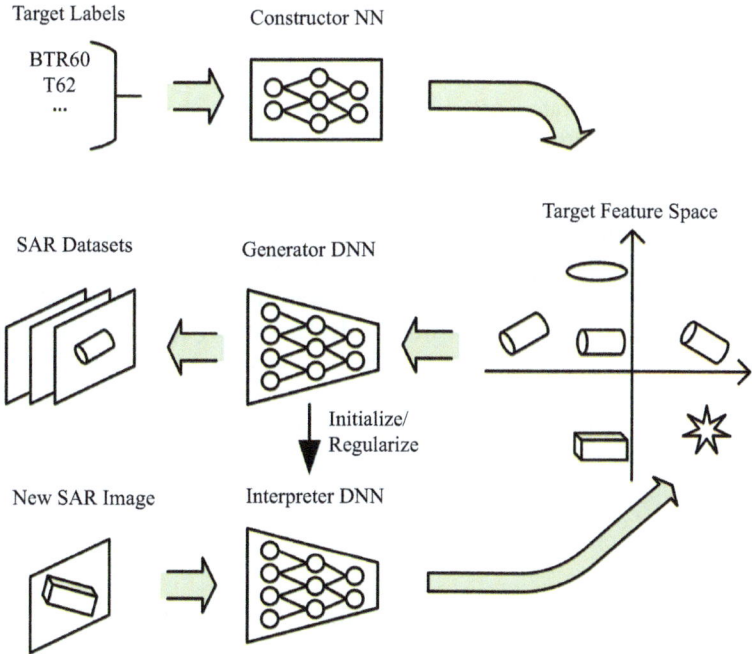

Figure 8.8   The proposed generative neural network for zero-shot target recognition in [31]

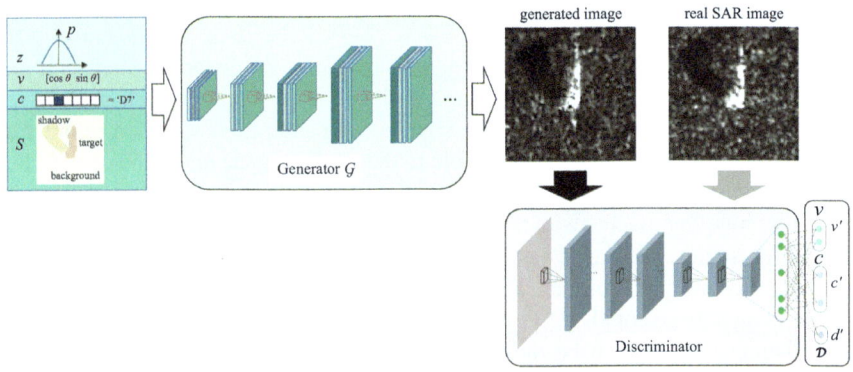

Figure 8.9   Framework of adversarial autoencoder proposed in [35]

the similarity/distances between the mapped features of the know targets and theirs. Interestingly, Gui *et al.* [32] used the feature representations learned from Word2Vec generative model [33] which is trained with texts. However, for more general classification tasks, such as ship classification, aircraft classification, it is not always possible to obtain reliable representations through such semantic embedding models. Since it was proposed in 2014 [34], the generative adversarial nets (GAN) has gained great popularity due to its novel design of adversarial structure which allows an increase of the fidelity of generated images. Following this idea, an adversarial auto-encoder that consists of a generator (decoder) and discriminator (encoder) as illustrated in Figure 8.9 is proposed to generate SAR-alike images with a few-shot training set in [35]. The generated images are then used to substitute part of real images for training, and the results show that the generated images can boost the test accuracy by 5.77%.

Transfer learning either directly applies the simulated SAR images as training data, or fine-tune the model that is trained with other data sources. In [36], Song *et al.* proposed a three-step pre-processing method to reduce the gap between the simulated and real data, and the processed images are used to train a truncated VGG16 network. The average margin index which calculates the averaged distance between the classification boundaries and the extracted features of simulated data is proposed to predict the test accuracy level, and determines when to stop the training process. In [37], Huang *et al.* used nearly 50,000 TerraSAR-X patches as a source dataset to train an auto-encoder which is then be used as a feature extractor of the target dataset. To avoid training the entire network from scratch, transfer learning can effectively reduce the risk of overfitting.

8.2   Land use and land classification

8.2.1   *Local climate zone classification*

The concept of local climate zone (LCZ) which was proposed for urban heat island (UHI) studies, is now also used for classifying the urban areas, and for urbanization

1.)	2.)	3.)	4.)	5.)	6.)	7.)	8.)	9.)
Compact high-rise	Compact midrise	Compact low-rise	Open high-rise	Open midrise	Open low-rise	Lightweight low-rise	Large low-rise	Sparsely built

10.)	A.)	B.)	C.)	D.)	E.)	F.)	G.)
Heavy industry	Dense trees	Scattered trees	Bush, scrub	Low plants	Bare rock or paved	Bare soil or land	Water

Figure 8.10   Examples and explanation of the 17 types local climate zones included in [38]

monitoring [38]. It is defined as different climate-relevant urban structures in terms of the height, density, and type of the surface covers and anthropogenic parameters. Figure 8.10 shows examples of the 17 types of local climate zones.

The World Urban Database and Portal Tool (WUDAPT), which was developed by Bechtel *et al.*, provided the LCZ community with a large dataset that covers nearly 100 cities with a resolution of 100 m [39]. In [38], Zhu *et al.* build a larger and finer LCZ dataset named "So2Sat LCZ42," which consists of 400,673 pairs of Sentinel-1, Sentinel-2 patches with their corresponding LCZ maps and covers 42 + 10 cities around the world.

LCZ classification is a challenging task, due to that: (1) *imbalanced classes:* the numbers of different categories pixels are not equal. For example, there are much more low plants zones than compact high-rise zones, open high-rise zones, or scattered trees zones in Paris. Then if trained with Paris data, the model would pay more attention to the low plants class, which decreases the classification accuracy of the others. Besides, the distributions of the 17 categories vary with locations, which hinders improving the model's transferability. For example, the amount of samples of water zones in Zurich is significantly more than that in Rome. (2) *noise label:* according to [38], the overall confidence of the So2Sat LCZ42 dataset labels is 85%, and the quality of WUDAPT is worse. The noise label not only influences the training of deep models but also limits the maximum classification accuracy.

Existing deep learning models mainly focus on fusion methods of different types of data for LCZ classification. In [40], Qiu *et al.* analyzed the importance of different sources of data for LCZ classification using the residual convolutional neural network (ResNet). The results showed that introduce of other data sources such as Open Street Map (OSM) and Nighttime Light data (NTL) can boost the overall and weighted accuracy. Then, they proposed to use the multi-spectra data acquired in four seasons to further extract the temporal features [41]. The input Sentinel-2 data obtained in each season are fed into four residual blocks to extract the spectral features respectively, and the extracted features are the input for the follow-up long short-term memory

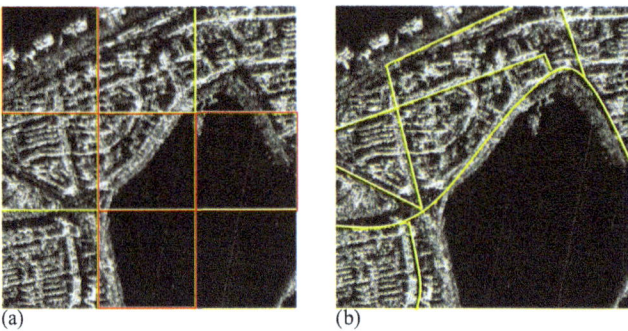

*Figure 8.11   Illustration of concepts of regular segments (a) with segments by the
road network (b) of remote sensing images proposed in [44]*

(LSTM) network for LCZ mapping. To reduce the gap between different sources of
data, Elshamli *et al*. proposed a domain adaption method [42]. Suppose there are N
sources of data, the loss is defined as the weighted average of the N losses calculated
when using different data, i.e.

$$L = \sum_{i=1}^{N} \lambda_i L_i, \ \lambda_i = 1/a_i, \tag{8.5}$$

where a_i is the validation accuracy tested on the ith source of data. In the LCZ
classification task, building areas are classified into several sub-classes according
to the density and height of the buildings. In [43], Yoo fused the building footprints,
building heights, and the number of stories with Sentinel-2 data to improve the overall
accuracy of the CNN model for urban LCZ classes. Due to the complicated urban
structures, the input patch may contain several LCZ types, as shown in Figure 8.11(a).
In [44], the road network data is used to segment the Sentinel-2 images, thus each
segment includes only a few similar patterns, as shown in Figure 8.11(b). Then the
segments are post-processed to have equal sizes, and fed into a CNN network for
classification.

   Besides, other models put their attention on the design of network architectures.
In [45], Liu *et al*. proposed to treat the LCZ mapping as a scene classification task, and
the results show a patch size of $320 \times 320 \ \text{m}^2$–$640 \times 640 \ \text{m}^2$ is suitable. A multitask
learning framework was proposed to make the LCZ maps and estimate the human
settlement extent (HSE) simultaneously in [46]. In [47], Ma *et al*. summarized the
state-of-the-art LCZ classification methods from different aspects, and proposed an
object-based image analysis method.

8.2.2   *Crop-type classification*

Previous studies showed that the agriculture industry has a significant effect on the
total gross domestic product. In Europe, there are nearly 10 million farms and over

20 million people working in this industry [48]. Crop classification on a large scale is of great potential in benefiting policymakers. With the launch of Sentinel-1 and Sentinel-2 satellites, the number of available synthetic aperture radar (SAR) and multi-spectral data is rising, which enables the automatic monitoring of crop growth.

Traditional methods often use handcrafted features, such as the normalized difference vegetation index (NVDI) and other spectral features, together with classifiers such as random forest, support vector machine (SVM). Different crops have slight differences in their spectral characteristic due to their structural diversities. In addition, due to the different growth cycles of different crops, temporal features are also important for crop classification. Thanks to the short revisit time (5 days with 2 satellites) of Sentinel-2 and Sentinel-1 (6 days with the two-satellite constellation), analysis of the time-series remote sensing data for crop classification becomes possible, and gained a lot of attention in recent years. Usually, each sample is a $M \times N \times C \times T$ data cube, also named by parcel, where $M \times N$ is the patch size, C denotes the numbers of spectral bands, and T denotes the length in the temporal dimension. Thus the feature representation learning for crops is a challenging but primary task, and using the handcrafted features is not sufficient for mining the information from parcels.

With the development of deep learning, the essential features for crop type classification purpose are automatically extracted by the networks. More specifically, CNN are used to extract the spatial and spectral features, and recurrent neural networks (RNN) are applied to learn the temporal attributes. Hybrid networks of CNN and RNN are employed on the parcels to explore their hyper-dimensional features.

There are several typical network architectures for crop classification: (1) MLP: use multi-layer perceptron (MLP) network to directly classify the input data; (2) CNN: use CNN to extract the spatial features for each date, and concatenate the extracted features, and these features are the inputs for a follow-up classifier; (3) RNN: calculate a vector for each time date, and the time-series vectors are fed to a standard RNN; (4) RCNN: the standard recurrent module in RNN is modified to a CNN structure instead of MLP to handle tensor data; (5) 3D-CNN: take advantage of three-dimensional convolutional operation instead of two-dimensional one to extract the spatial and temporal features simultaneously.

In 2015, Kussul *et al.* proposed an MLP method to classify the multi-temporal space-borne images for crop type classification [49]. The results showed that the parcel-based approach is superior to the pixel-based approach, and verified the importance of taking both spatial and temporal features into consideration. And then the convolutional neural network was used as a feature extractor with a reduced number of model parameters [50]. Current deep learning models are trained end-to-end, so that the networks can yield the classification results directly. In [51], the effectiveness of using RNN (long short-term memory networks, LSTM) is validated as compared with the SVM classifier. In [52], Ji *et al.* proposed a 3D-CNN in which a convolutional operation is applied to a 3D input instead of 2D data, as shown in Figure 8.12. Thus, the convolutional layers extract not only the spatial features, but also the temporal features. The classification results show that compared with 2D CNN, spatio-temporal 3D CNN can obtain higher accuracy by 2%.

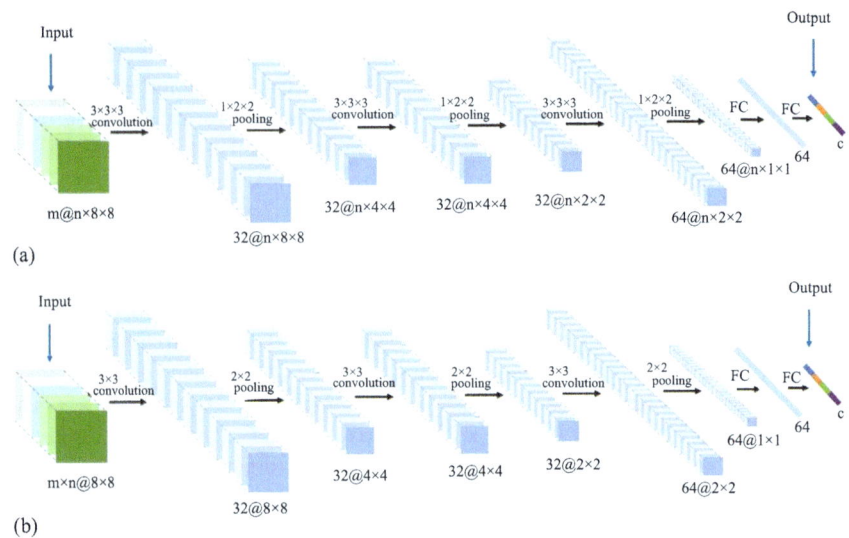

Figure 8.12   Comparison of 3D-CNN (a) and 2D-CNN (b) for crop classification. The figures are adopted from [52]

In 2018, Rußwurm *et al.* adopted a hybrid deep learning network architecture that combines convolutional neural layers with recurrent networks [53]. One main challenge in multi-spectral parcel data classification is the coverage of clouds that hide the crops. The results showed a great potential of applying RCNN for cloud-induced temporal noise.

In [54], Garnot *et al.* compared the classification performances of CNN, RNN, RCNN models which extract spatial, temporal, and spatio-temporal features respectively. The used data are obtained by Sentinel-2 over a region of $12{,}100 \text{ km}^2$ in southern France, and consists of 199,464 parcels of $32 \times 32 \times 10 \times 24$. The best performance was achieved by the RCNN model, when a majority of the parameters are related to the temporal feature extraction. It implies that when using Sentinel-2 data for crop type classification, temporal features are essential.

In [55], Rußwurm *et al.* compared the self-attention networks transformer with LSTM-RNN, and CNNs in pixel-based crop type classification. The self-attention layer can be formulated as

$$\mathbf{h} = softmax(\tan(\Theta_A^T \mathbf{X})\Theta_K)^T \mathbf{X}, \tag{8.6}$$

where the $\mathbf{X}$ is the input, and Θ_A^T and Θ_K are the parameters to be determined. The term of $softmax(\tan(\Theta_A^T \mathbf{X})\Theta_K)$ is regarded as the attention scores (as visualized in Figure 8.13) and each element is normalized to [0, 1]. As tested on a Sentinel-2 dataset over a large area for over 10 category crop type classification using both the raw and the preprocessed data, results show that all the models yield equally high accuracy on

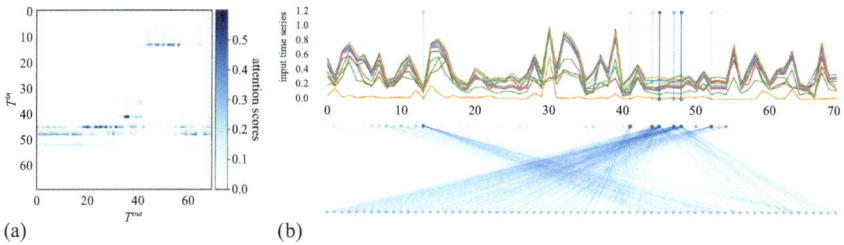

(a)   (b)

Figure 8.13   Visualization of the learned attention score matrix in [55]

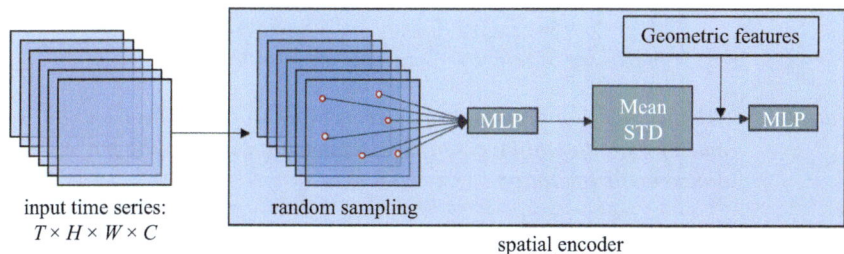

Figure 8.14   Encoder for extracting the spatial features proposed in [48]. The randomly sampled pixels instead of the whole image are fed into the network.

the preprocessed data, and the transformer and LSTM-RNN outperform other models on the raw multi-spectral data.

In [48], unlike convolutional neural networks which explore the spatial structures in the images, the proposed deep learning model takes the input coarser-resolution images as unordered sets of pixels. For each data, several pixels are randomly sampled from the input multi-spectral data several times, and are fed into an MLP layer, as shown in Figure 8.14. Then the mean and standard deviation of the output vectors are concatenated and fed into the follow-up modules. The proposed network was compared with several classical deep learning models, yielding the highest classification accuracy with less time and memory usage.

8.2.3   SAR-optical fusion for land segmentation

The fusion of cloud-free SAR data and high-resolution optical data is of great value in rapid large-scale land monitoring. For example, floods are usually accompanied by heavy rainfall, when a large area of the hyperspectral images is occupied by cloud pixels. In this case, the fusion of SAR and optical images provides a solution for high-accuracy disaster assessment.

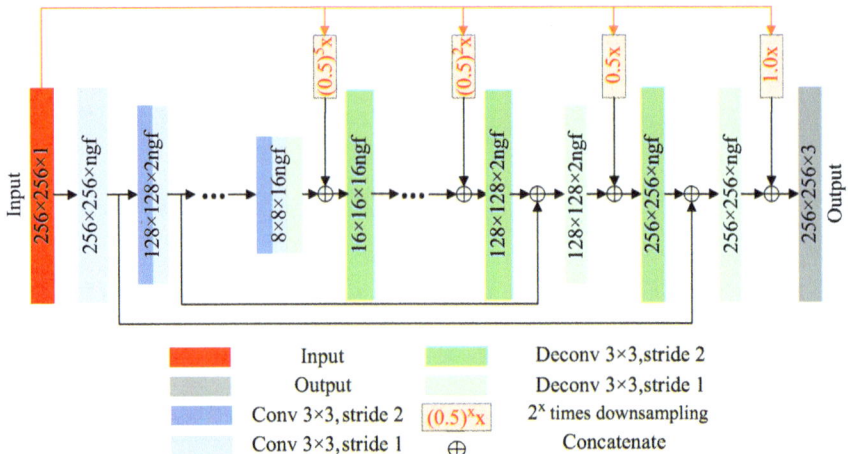

Figure 8.15   Architecture of proposed translator network in [57], where feature maps from different layers in the decoder are cascaded by the concatenate operation

In [56], a dataset that consists of 282,384 SAR and optical images pairs acquired from space-borne Sentinel-1 and Sentinel-2 data are collected. The SAR and optical images are co-registered, and images with large cloud coverage ($>1\%$) and distorted colors are excluded. It has global geographic coverage and includes regions of various climate types, which is of potential for fostering future SAR-optical fusion models development.

An intuitive way of fusion SAR and optical images is to learn a mapping from SAR and optical pairs. In [57], Fu *et al.* proposed a SAR-optical translation network using cascaded-residual adversarial networks under the CycleGAN framework. It comprises two translators (generators) and two discriminators: the translators, as shown in Figure 8.15, are leveraged to translate SAR/optical images to optical/SAR data; the discriminators are used to distinguish whether the input image is the real image or the synthesized one. Experiments conducted with 12,854 pairs of images demonstrate that the proposed network is applicable to the data obtained over various terrain types.

A large land segmentation dataset is proposed in [58], which divides the image into five categories: water, road, building, vegetation, and others. It includes 610 pairs of single polarimetric SAR images acquired by Gaofen-3 mission and optical images, as well as the corresponding equal-size ground-truth labels. Each patch has a size of $1,024 \times 1,024$. A deep model based on encoder-decoder architecture, as shown in Figure 8.16, is proposed for the land segmentation task. The results show that the proposed network achieved an overall accuracy of 75.84%, and outperforms the classical deep models such as U-Net, SegNet. Since it only uses SAR images as inputs, fusion with optical images may further boost segmentation accuracy.

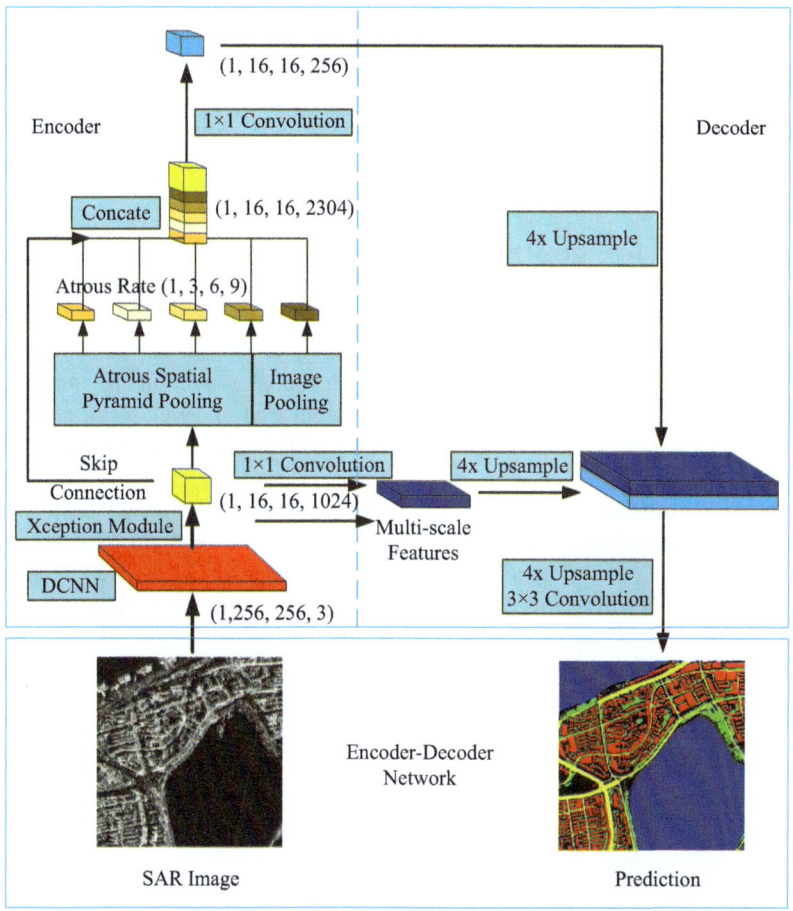

Figure 8.16 An encoder–decoder network designed for land segmentation (the image is adapted from [58])

8.3 Disaster monitoring

According to the previous studies, with climate change, the frequency of extreme weather events, such as storms, floods, typhoons, and droughts, increases. These extreme weather events would have a significant influence on the electric power supply, agricultural yield, and human activities, and threaten personal safety, and social security. For example, a winter storm in North America in January 2022 left thousands of homes without power. In this section, we focus on the application of deep learning models to disaster monitoring, and briefly introduce the development of flood detection, storms, and lighting nowcasting models.

8.3.1 Flood detection

Severe floods not only damage buildings, infrastructures, crops but also cause deaths. To reduce the losses caused by the lack of accurate and timely flood information, flood detection has gained increasing popularity in recent years. A number of datasets and deep learning models are proposed dedicated to this task. Here we list three public datasets:

FloodNet [59]: It is collected between August 30 and September 04 (after the hurricane) in 2017 over Ford Bend County in Texas and other areas directly impacted by Hurricane Harvey with drones. It consists of 3,200 high-resolution images which are annotated with 9 classes, and about 11,000 question-image pairs. The dataset is collected for three levels of post-disaster damage assessment: image classification (Flooded/Non-Flooded), image segmentation (9 classes), and visual question answering (VQA). Some examples of the annotated data are shown in Figure 8.17 as in [59].

Water Segmentation Open Collection (WSOC) [60]: It is collected from Twitter posts and the existing image segmentation datasets that contain water category, such as COCO, the Microsoft Research in Cambridge v2 (MSRC v2). It includes 120,061 optical images as well as pixel-level segmentation annotations, which is available at https://zenodo.org/record/3642406.

UNOSAT Flood Dataset [61]: The dataset consists of Sentinel-1 data with a resolution of 10 m × 10 m, and their corresponding flood extent boundary maps. It

Figure 8.17   Three tasks (i.e. image classification, semantic segmentation and visual question answering (VQA)) included in the FloodNet dataset in [59]

includes 58,128 samples of 256 × 256 pixels obtained over multiple regions (8 cities/countries) from 2015 to 2020.

Various sources of data and multiple deep learning networks that are used for image segmentation tasks in the computer vision field are utilized for flood detection. In 2017, Lopez-Fuentes *et al.* used social media posts (both images and text) for flood classification [62]. The famous InceptionV3 network which was pre-trained on the ImageNet dataset was used to extract the features in images, and the bidirectional Long Short-Term Memory network was used for text mining. And it achieved an 81.6% average precision on the test set. In [59], the authors tested three networks on the FloodNet dataset, and the results showed that PSPNet [63] yield significantly higher accuracy than the other two. In [64], Akiva *et al.* proposed a self-supervised learning method named as H2O-Net for flood detection in satellites images. It trains a generative adversarial network (GAN) to generate SWIR images from input RGB images, and used the synthetic SWIR and the RGB image for segmentation.

Due to its ability to work all-day and all-weather, space-borne SAR sensors are able to acquire images through rain and cloud on a large scale, which is beneficial for flood detection. In [65], a fully convolutional network was used to derive the binary segmentation maps with equal size as the input images from PolSAR data. The generalization ability of the network is further explored by using a dataset obtained from multiple countries and regions in [61]. In [66], Li *et al.* applied a convolutional neural network on multi-temporal TerraSAR-X data, and used an active learning strategy to make use of the unlabeled data during training. In SAR images, the scattering intensities in both the flood and shadow areas are low. In [67], the proposed multi-resolution dense encoder-decoder network (MRDED) segments the input SAR images into three categories: background, flood, and shadow. As shown in Figure 8.18, the network encodes and decodes the different scales of features from different convolution and deconvolution layers. Results showed lower accuracy of water (80.12%) and shadow (88%) compared with background class (95.16%), which implies that water pixels in SAR images may be confused with shadow class.

8.3.2 Storm nowcasting

Unlike forecasts, storm nowcasting means the short-term prediction of the storm activity for up to 3 hours. Usually, the distribution map of hydrometeors in the atmosphere can be plotted by the weather radars. Then the storm nowcasting is done by using the radar echoes from a past period to predict the future radar echoes. Traditionally, it is taken as a task of extrapolation in the temporal dimension. It can also be regarded as an image sequence to image sequence translation task, which is similar to the video prediction task in the computer vision field [68].

Traditionally, storm nowcasting methods rely on optical flow, fuzzy logic, and/or numerical models. The optical flow calculates the movement between two images at different times. More specifically, to assume that

$$I(x + \Delta x, y + \Delta y, t + \Delta t) = I(x, y, t) + \frac{\partial I}{\partial x}\Delta x + \frac{\partial I}{\partial y}\Delta y + \frac{\partial I}{\partial t}\Delta t + \delta, (8.7)$$

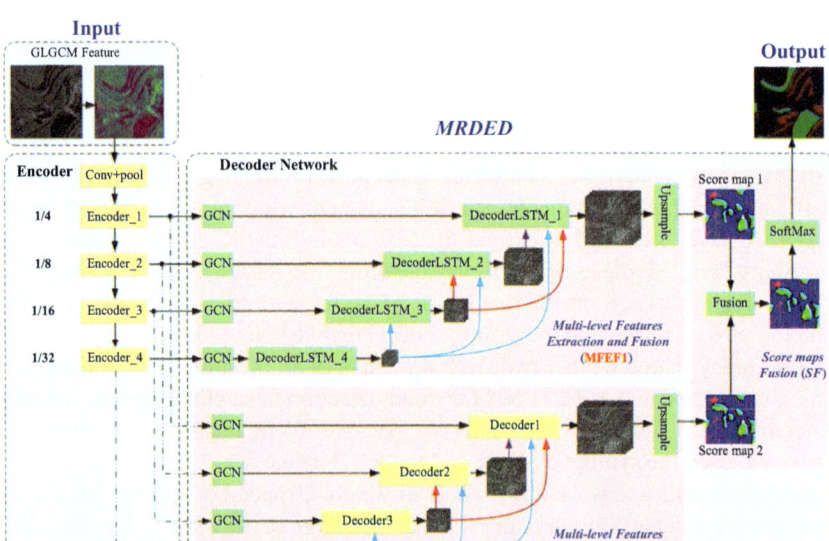

Figure 8.18   Flowchart of MRDED proposed in [67]

where the $I(x, y, t)$ is the value at the location (x, y) at time t, and δ is the higher-order terms of Taylor series. Some successful storm nowcasting systems are DARTS [69], STEPS [70], CO-TREC [71], and SWIRLS [72].

The advance of deep learning models, especially those that come from the video processing field, has motivated the application of deep learning to storm nowcasting. Their ability to learn relevant latent features which is not capturable by classical physical-based models has attracted a lot of attention [73]. A convolutional long short-term memory network is proposed in [73], and the results show a better performance of the proposed network over the optical flow based ROVER algorithms. And it also has the advantage of coping with the cases in which a sudden appearance of cloud agglomeration at the boundary. Figure 8.19 shows an example of a convolutional long-term memory module. In [74], Sato *et al.* proposed a skip-connected PredNet (SDPredNet) which has a different encoder-decoder architecture. After being trained on a 3-year real dataset, the test results showed a comparable prediction accuracy of SDPredNet as compared with the baseline model in terms of (CSI) and Heidke Skill Score (HSS). In [75], the proposed deep learning model that combines CNN and recurrent structure was verified to outperform the classical CO-TREC model [71].

Other classical deep learning models are also applied. In [76], the widely-used segmentation network U-Net is used for short-term precipitation nowcasting. Chen et al. proposed a GAN network to increase the resolution of the radar echo images [77]. In [78], Jing proposed an adversarial extrapolation neural network (AENN) for radar echo extrapolation. The time-series radar echo maps from the last four hours are fed to the generator as partial inputs, and there are two discriminators, i.e. echo-frame

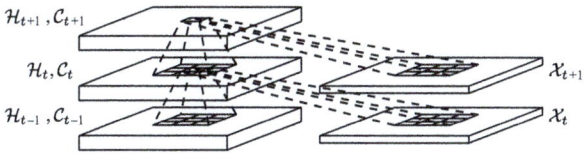

Figure 8.19   The proposed ConvLSTM module in [73]

discriminator and echo-sequence discriminator. The experiments were carried out using the data from 2016 to 2017 that acquired over five cities in China. The results showed that AENN outperforms the ConvLSTM [73], ROVER, and TREC for most predictions. In [79], the convolutional gated recurrent unit (ConvGRU) are trained with an adversarial strategy to increase the sharpness of extrapolated images.

In [80], Prudden *et al.* reviewed the current state-of-the-art precipitation prediction methods. In [68], Cuomo *et al.* compared three different deep learning models from several aspects: used multiple loss functions including the non-convex ones; analyzed the results when making use of the different numbers of input radar echo images; compared the models' performances when different outputs types are applied; extensively compared different model using 12 years of data from 2005 to 2017 for training, and data from 2018 and 2019 for validation and test.

8.3.3   Lightning nowcasting

Although lightning is a common natural phenomenon, it can also threaten human safety and cause damage to man-made facilities. Every year, thousands of people die from lightning. Thus, lighting monitoring is of great importance. For example, the National Lightning Detection Network (NLDN) of China records the lightning occurrence frequency from 394 ground-based sensors. And satellite data and radar data have been explored for lightning nowcasting [81].

The lightning nowcasting can be formulated as

$$Y_t = F(X), \ Y_t \in \mathbb{R}^{M \times N}, \ X \in \mathbb{R}^{M \times N \times C \times T}, \tag{8.8}$$

where the input is T-hour data, and $X(m, n, c, t)$ corresponds to the acquired data at location (m, n) in the tth time interval, and the model outputs the lightning frequency in the interested region at the $T + t$ interval [82]. The input data can be the simulated data by the weather research and forecasting (WRF) model, and/or accumulated number of lightning occurrences per hour [82], the satellite data, and/or radar reflectivity data [81].

Traditional lightning forecast methods are based on numerical weather prediction (NMP) systems. The NWP systems calculate the meteorological parameters by solving the partial differential equations derived by physical modeling. Although the future lightning can not be directly forecasted in this way, the meteorological parameters are related to it. Additional lightning parameterization schemes are used to estimate the lightning density from the simulated meteorological parameters. Here we list two models for example. The lightning parameterization scheme PR92 relates the lightning flash rate to the vertical windspeed w [83] with a power law, which is formulated as

$$F = 5.7 \times 10^{-6} w_{max}^{4.5}, \tag{8.9}$$

Another widely-used lightning parameterization meteorology scheme is Threat F2 which takes the air density ρ, ice, snow, and graupel mixing ratios i, s, g as inputs [84]. It is formulated as

$$F_2 = 0.02 \int \rho(i_z + s_z + g_z)dz, \tag{8.10}$$

where the subscript z denotes the height.

The deep learning models are introduced to explore the potential value of combining the physical-aware numerical results and the actual historic lightning observations, and/or the satellite and radar data [81,82,85,86]. It is natural since the history observations are equally important for lightning nowcasting, and the deep learning models' ability in extracting spatio-temporal features is famous in other fields. In [82], an attention-based dual-source spatio-temporal neural network (ADSNet) is proposed to fuse the numerically simulated results and the historic lightning observations acquired from ground-based stations. Its overall architecture is shown in Figure 8.20. A channel-wise convolutional operation is adopted to learn the importance of different input simulated data. The normal and channel-wise convolutional layers can be formulated as follows:

$$Y_{m, n, c'} = \sum_{k, l, c} X(m + k, n + l, c) \times W(k, l, c, c'), \tag{8.11}$$

$$Y_{m, n, c} = \sum_{k, l} X(m + k, n + l, c) \times W(k, l, c). \tag{8.12}$$

In (8.11) and (8.12), X, W, Y are the input multi-channel data (different simulated meteorological parameters), the weights, and the layer output respectively. Additionally, the ConvLSTM module is applied as did in precipitation nowcasting [73].

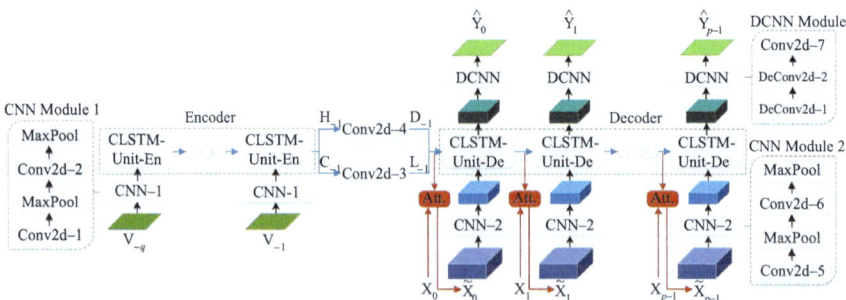

Figure 8.20   Illustration of overall architecture of the attention-based dual-source spatio-temporal neural network (ADSNet) proposed in [82]

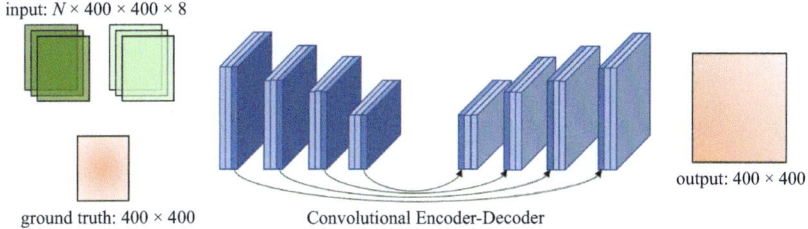

input: $N \times 400 \times 400 \times 8$

ground truth: 400×400

output: 400×400

Convolutional Encoder-Decoder

Figure 8.21   Illustration of the architecture of LightningNet in [81]. It is a typical auto-encoder with four groups of convolution layers in encoder and decoder respectively.

As compared with numerical simulation methods, the deep learning models are superior, and the proposed ADSNet performs best based on the experiments.

In [85], LightNet which consists of two encoder modules, one fusion module and one decoder module was proposed. The two encoders extract the spatio-temporal features for the past and simulated future meteorological data, and the lightning observations, respectively, and then the extracted are concatenated. Results suggested that LightNet is capable of both short- (first-three-hour) and long-term (second-three-hour) lightning forecasts. The follow-up network LightNet+ [86] has a similar architecture as LightNet [85], and is introduced to connect lightning nowcasting with automatic weather station (AWS) data as well. Although when using only AWS data, the network performs worst, the results showed that when fuses the AWS data with lightning observations and simulated data, the model yield higher accuracy compared with LightNet.

On the other hand, the obtained data from meteorological satellites and radar also are highly correlated with lightning prediction. For example, the received echos reflected by large amounts of graupel and hail particles is large, and the study in [87] showed that the high radar reflectivity and low temperature is a sign of lightning. In [81], Zhou *et al.* proposed to use stacked 6-band spectral images, one radar reflectivity image and one observation data to nowcast the lightning occurrence. The proposed deep network LightningNet as shown in Figure 8.21, consists of one encoder and decoder, which is a modified version of SegNet [88]. LightningNet was tested on different types of data, and the results showed that fusing the lightning data with radar and/or satellite data is essential, and significantly improved the prediction accuracy.

8.4   Forest applications

As shown in Figure 8.22, forests covered about one-third of the land on earth. According to the Food and Agriculture Organization of the United Nations (FAO)'s report, there are 4,058.93 million ha forests over the globe.* Tropical (dark green),

*https://fra-data.fao.org/.

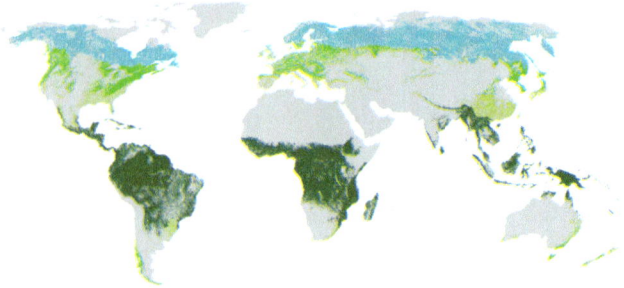

Figure 8.22   The geographic distribution of forest covered on earth. The picture is adapted from https://fra-data.fao.org/.

boreal(blue), temperate (green), and sub-tropical (yellow) forests contribute 45%, 27%, 16%, and 11% proportion of the world's forests.

Forests, which stock carbon and produce oxygen, and are home to various animals, are essential for climate change and biodiversity preservation. Previous research has shown the high cost of climate change if greenhouse gas emission keeps the same. Thus regular monitoring of forests is necessary. Using remote sensing techniques and deep learning models enables fast, automatic, large-scale forests applications. In this section, we introduced the recent developments of deep learning models that applied in tree species classification, deforestation mapping, and fire monitoring.

8.4.1   Tree species classification

Species classification in forests identifies the species of either an individual tree or a forest stand, which can help improve the forest biomass calculation accuracy. Currently, forest-stand-level species classification has been not addressed using deep learning models yet. Thus, this section focuses only on individual-level methods.

In this regard, LiDAR data are able to facilitate the species classification a lot [89–91] in (1) detecting the position of trees using the LiDAR-derived heights and local maximum algorithms; (2) segmenting the remote sensing images for individual trees according to the trees' positions; (3) extracting the 3D structure information from the LiDAR point cloud for discriminating different species. In [92], the listed 4×6 LiDAR samples of four types of trees reveal that different species vary in the canopy structures, leave types, etc., and the 3D features are one of the important clues for species classification. Besides, high-resolution multi-spectral images are also used for extracting spatial and spectral features.

In [92], Zou *et al.* isolated different trees, and constructed a voxel for each tree using terrestrial LiDAR data. Then several projected 2D images from the voxel data are fed into a deep belief network (DBN). Sun *et al.* segmented a number of Worldview-2 patches according to the LiDAR-derived tree heights, and compared the usage of AlexNet, VGG16, ResNet50, DenseNet, RF for classifying the segments into species [89]. Results suggest that ResNet50 is superior to DenseNet and RF

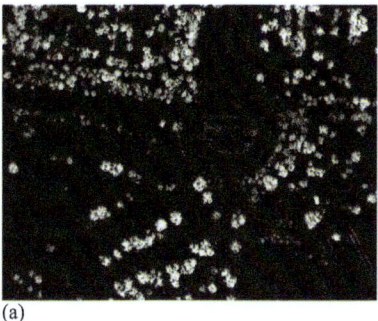 

(a) (b)

Figure 8.23   Examples of LiDAR-derived relative height model (a) and LiDAR
intensity return image (b). The figures are adopted from paper [93].

on seven-type-species classification task, and VGG16 is better than ResNet50 and AlexNet on the fine species classification task (18 tree species). A similar workflow is adopted in [93], where DenseNet has achieved over 80% accuracy on the 8-species tree classification task, while SVM and RF can only achieve about 60% overall accuracy. And results also suggest that fusing VNIR, SWIR and LiDAR data (such as relative height model and intensity return image as shown in Figure 8.23) can boost the performances of the classifiers. In [90], a simple but effective convolutional neural network that combines leaf-off and leaf-on remote sensing images is proposed to classify the leaf type of the trees.

Schiefer *et al.* cast the species classification task as a segmentation problem that learns a mapping that

$$f : X \in \mathbb{R}^{M \times N \times C} \Rightarrow Y \in \mathbb{R}^{M \times N}. \tag{8.13}$$

The classical segmentation network U-Net is utilized to process the very high-resolution (<2 cm) RGB imagery, and obtained a high classification accuracy of 0.89 with an F1-score of 0.73.

In [94], the point cloud $\{(x, y, z)\}$ acquired by LiDAR is directly used for tree species classification. Two issues behind this are (1) how to handle the disordered points set and (2) the numbers of points vary among species and trees. In dealing with the first problem, the famous PointNet++ [95] which is good at set abstraction is adopted. To solve the second problem, a threshold-based preprocessing step is applied first: it randomly discards some points when the number of points $n > \theta_1$; and when $\theta_2 =< n < \theta_1$, it randomly copies $\theta_1 - n$ points.

To make the best of both the LiDAR data and multi-spectral images, a dual CNN network named Silvi-Net that extracts features from two sources of data respectively is proposed in [91]. As illustrated in Figure 8.24, the cropped multi-spectral images and the projected LiDAR 2D images are fed into two ResNet-18, and the outputs features are combined for final classification prediction. Besides, the explainability analysis via class activation mapping (CAM) shows that the trained network can put

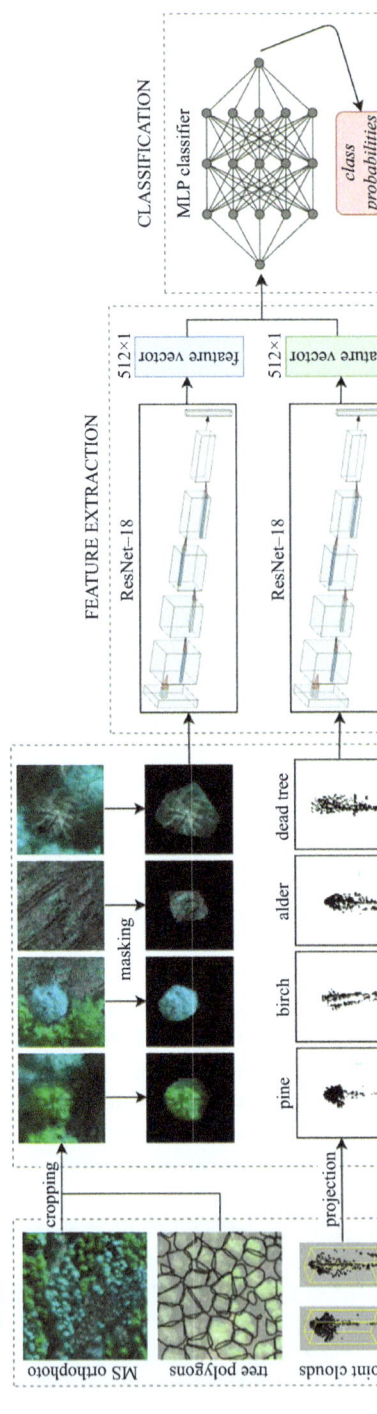

Figure 8.24   Flowchart of the dual-CNN network Silvi-Net that designed for species classification which is proposed in [91]

its attention on the image center (the tree), and the incorrect classification results are captured when the network misplaced its attention.

In summary, various deep learning models have been developed for individual-level tree species classification using LiDAR and multi-spectral data. These models are able to extract effective features from disordered point cloud data. However, existing methods are trained and tested with the data obtained over a small region (a park, or a forest region), thus are only applicable to a small region. Furthermore, temporal features which are useful in distinguishing conifers from deciduous trees have not been explored yet.

8.4.2 Deforestation mapping

From 2000 to 2020, the forest area has decreased from 4,158 million ha to 4,059 million ha.[†] As a result, the total carbon stock has reduced by about 1.17 Gigaton. The disturbance of forests, either caused by natural hazards or human activities, is one of the main sources of greenhouse gas emissions. Thus, regularly monitoring and mapping deforestation is crucial for forest and biodiversity preservation, and climate change mitigation.

A convolutional neural network is used to extract the spatial features of each of the time-series input images respectively, and the outputs after the first fully-connection layer are fused by a max-pooling operation along the temporal domain in [96]. Experiments reveal that the deep CNN model is significantly better than the traditional SVM and RF algorithms.

U-Net and its variations have dominated the deep deforestation mapping methods [97–101]. In [97], a 93-layer ResUnet is used and compared with RF, MLP, SharpMask, and U-Net models. The 7-band Landsat data acquired over two dates are stacked and fed into the deep networks, which is also named as early fusion (EF) method. Test results (as shown in Figure 8.25) over a different site show that deep models are superior to the classical methods, among which the ResUnet achieved the best accuracy in terms of F1 score, Kappa coefficient, mIoU, and recall metrics. The EF U-Net is compared with Siamese Network (SN) and convolutional SVM in [102]. On the other hand, a late fusion (LF) U-net which concatenates the feature maps of the input images at the same scale in U-Net is proposed in [98]. And the results demonstrate a slight superiority of LF U-Net over the baseline U-Net and EF U-Net models.

In [99], seven variations of U-Net are compared using Sentinel-2 data over Ukraine, where the UNet-diff and UNet-CH which both include the difference images in the inputs achieved higher F1-score over the others. In order to deal with the unbalanced classes, the weighted sum of binary cross-entropy loss and Dice loss are the final objective of the deep network, which is defined as

$$\mathcal{L} = w_{BCE} \times \mathcal{L}_{BCE} + w_{dice} \times \mathcal{L}_{dice}, \tag{8.14}$$

[†]https://fra-data.fao.org/WO/fra2020/home/.

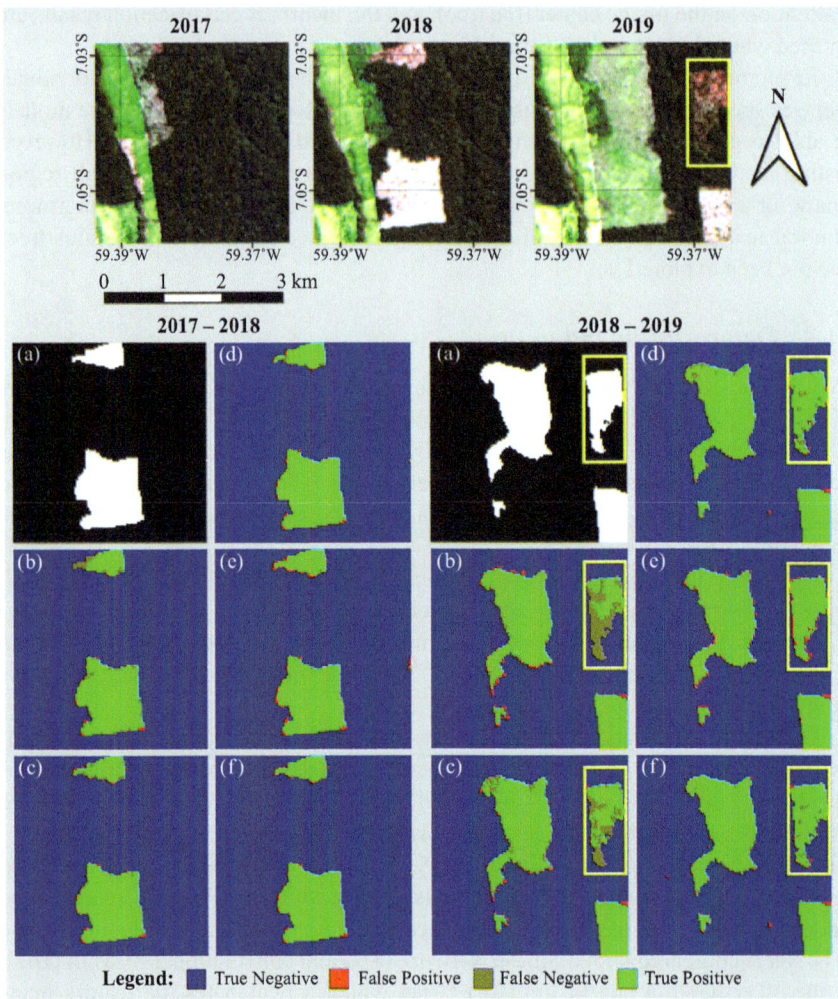

*Figure 8.25   Comparison of (a) ground truth deforestation maps with
            classifications made by the (b) RF, (c) MLP, (d) SharpMask, (e) U-Net,
            and (f) ResUnet models in 2017–2018 and 2018–2019 reported
            in [97].*

$$\mathcal{L}_{BCE} = -\frac{1}{m} \sum_{i=1}^{m} (y_i \times \log \hat{y}_i) + (1 - y_i) \times \log (1 - \hat{y}_i), \tag{8.15}$$

$$\mathcal{L}_{dice} = 1 - \frac{2 \times \sum_{i=1}^{m} (y_i \times \hat{y}_i) + \varepsilon}{\sum_{i=1}^{m} y_i^2 + \sum_{i=1}^{m} \hat{y}_i^2 + \varepsilon}. \tag{8.16}$$

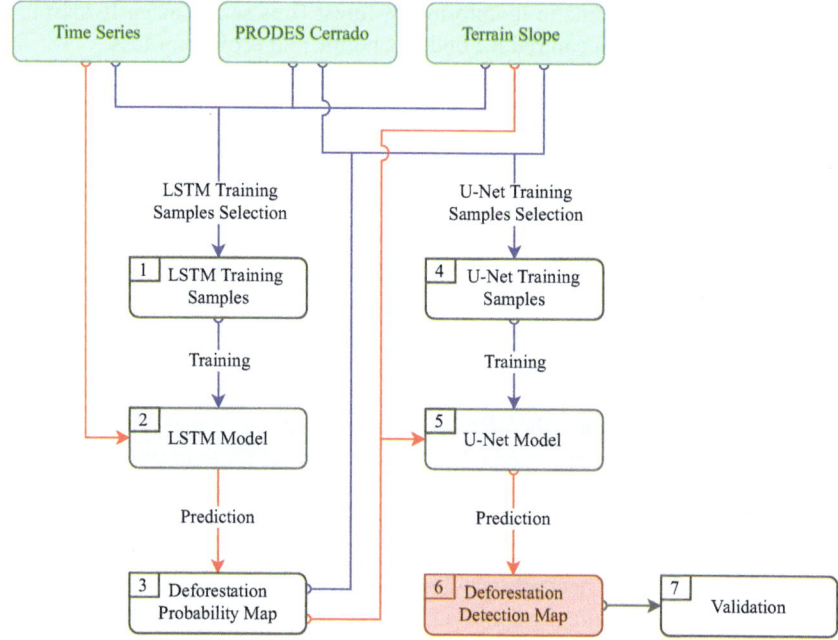

Figure 8.26   Flowchart of the LSTM and U-Net based deforestation detection method (adapted from [101])

Attention modules are added into the U-Net model in [100]. The attention module learns an attention score matrix from two neighboring scales of feature maps in the encoder, and multiplies it with the decoded features before fused with feature maps in the decoder. LSTM is the most commonly used model to handle time-series data. A novel framework that uses the LSTM predicted deforestation probability map with PRODES and slope data as the input of U-Net for deforestation type classification is proposed in [101], as shown in Figure 8.26. As high as 99% overall accuracy and over 0.7 of F1-score are obtained as tested with Landsat and Sentinel data over Mato Grosso and Bahia.

8.4.3   Fire monitoring

Forest fire happens more frequently in recent years due to climate change and forest degradation, and its impact on the ecological environments is also getting worse. According to the World Wild Fund's report,[‡] from 2019 to 2020, the Australian bushfire has killed about three billion animals, and burned up over 12 million hectares of

---

[‡]https://www.wwf.org.au/what-we-do/bushfires.

forest and bushland. Automatic monitoring of forest fires serves as early alarms to the decision-maker, which can reduce both economic and ecological loss.

In [103], Naderpour *et al.* applied a multilayer perceptron (MLP) to predict the forest fire vulnerability based on 12 factors including wind rainfall, temperature, normalized difference vegetation index (NDVI), etc. The outputs are then combined with the calculated physical and social susceptibility indexes based on the analytic hierarchy process (AHP) model, to generate the forest fire risk map.

Besides the effort of pre-event fire estimation, existing studies mainly focus on locating the fire in the remote sensing images. It can be formulated as outputting a binary matrix $Y \in \{0, 1\}^{M \times N}$ given a C-channel images $X \in \mathbb{R}^{M \times N \times C}$ using the trained models. The binary values in Y denote either fire or non-fire class. In [104], a deep neural network is used to transfer the MODIS data into the fire map. Considering that there are many more non-fire class pixels in the censused data, an early stopping method is adopted to prevent the network from over-fitting. Equal numbers of fire and non-fire classes data are spared for validation, and the training process will stop when the validation is not further improved. A similar framework is proposed in [105] to classify the MODIS data into five types (burned area, vegetation, cloud, bare soil, and cloud shadow).

Convolutional neural networks are also used to extract the representative spatial features from remote sensing images. In [106], Ban *et al.* use a CNN to classify the Sentinel-1 SAR time-series images, as shown in Figure 8.27. The differences between pre-fire and ongoing time-series images which represent the rapid change caused by wildfire are beneficial. The time-series images are stacked and concatenated with the digital elevation model (DEM) products, which are fed into the CNN model. Experiments conducted with three wildfires demonstrate that the CNN-based method can achieve at least 0.11 higher accuracy in Kappa than the log-ratio-based algorithm.

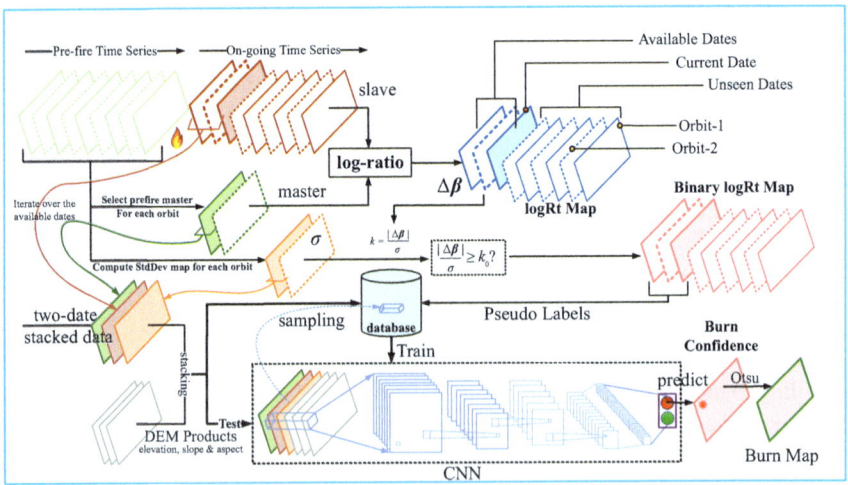

Figure 8.27   Flowchart of the online learning fire detection framework using time-series data based on CNN which is proposed in [106]

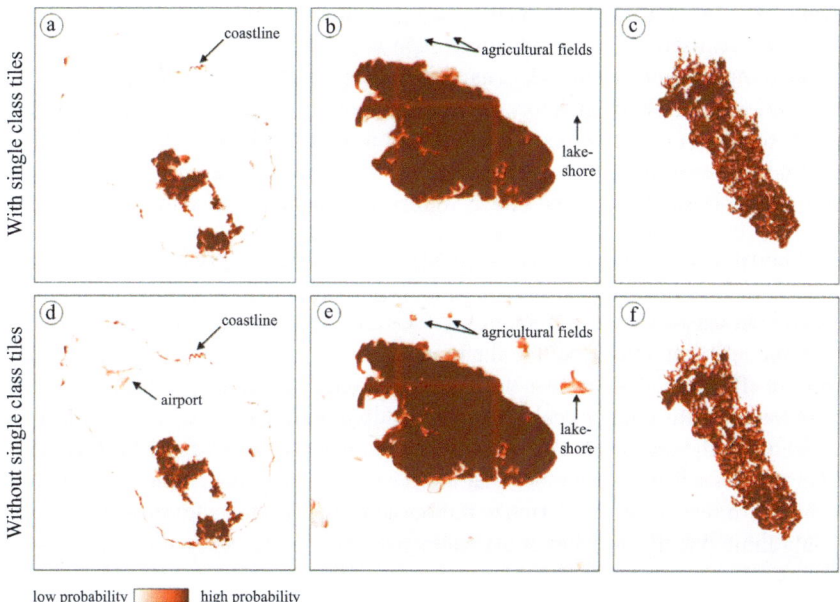

low probability [   ] high probability

Figure 8.28   Comparison of U-Net's predictions of probability of being burned
when trained with and without single class tiles. The image is adapted
from [107].

As a famous and successful image transformation network, U-Net is also adopted
for burned area mapping [107–109]. In [107], Knopp *et al.* applied the standard
U-Net for burned area segmentation in Sentinel-2 images. Results show that by adding
the single-class samples into the training set, U-Net can better handle the unburned
class, and cast fewer false alarms, as shown in Figure 8.28. In [108], Pinto *et al.* used
time-series images for large-scale burned area mapping with U-Net. Four sources of
satellite data, which are required from Sentinel-1, Sentinel-2, Sentinel-3, and MODIS,
are fused and compared using U-Net as a benchmark model in [109]. Since the dataset
is severely unbalanced with significantly more unburned pixels, the combination
of binary cross-entropy and dice losses that is defined in (8.14)–(8.16) are used in
this case. Results suggest fusing Sentinel-2 and Sentinel-3 images during the clear
conditions, and using Sentinel-1 data on cloudy days.

8.5   Conclusions

In this chapter, we summarized the state-of-the-art deep learning models that are
developed for remote sensing applications. Specifically, we introduced the models
which are designed for 13 tasks related to target recognition, LCLU, disaster nowcasting, and forest monitoring. CNN, as a major feature extraction tool, is popular in nearly

all tasks. Besides, since most tasks can be regarded as an image translation problem, U-Net is widely used in these applications such as deforestation and wildfire mapping, building footprint generation. When handling tasks where the outputs are related with a time-series before current states, LSTM is often applied as one of the important modules to extract temporal features. In addition, specifically tailored loss functions are employed for dealing with problems such as unbalanced training sets, and target bounding box estimation. In the future, to further boost the prediction accuracy,

- explainable artificial intelligence (XAI) tools can be explored to understand the functional mechanisms of the applied deep models. For example, the gradient-weighted class activation mapping (Grad-CAM) method can be used to analyze the model's "attention" paid to the inputs;
- future efforts need to increase the applicable scope of the models. Current models are trained with and applicable to data acquired within a/several regions. To cope with it, on the one hand, a large scale of benchmark datasets should be censused. On the other hand, techniques that are developed for transfer learning and few-shot learning can be transferred to remote sensing tasks to handle scenarios where only limited annotated data is available;
- physical-aware models can be combined with current deep neural networks to increase the models' robustness and generalization ability. Current models directly applied the networks which are proposed in the computer vision field, and these models may not work well for microwave vision tasks. But, how to combine them is still an open issue.

References

[1]   Zhu XX, Tuia D, Mou L, *et al.* Deep learning in remote sensing: a comprehensive review and list of resources. *IEEE Geoscience and Remote Sensing Magazine.* 2017;5(4):8–36.

[2]   Zhang L, Zhang L, Du B. Deep learning for remote sensing data: a technical tutorial on the state of the art. *IEEE Geoscience and Remote Sensing Magazine.* 2016;4(2):22–40.

[3]   Ma L, Liu Y, Zhang X, *et al.* Deep learning in remote sensing applications: a meta-analysis and review. *ISPRS Journal of Photogrammetry and Remote Sensing.* 2019;152:166–177.

[4]   Zhu X, Montazeri S, Ali M, *et al.* Deep learning meets SAR: concepts, models, pitfalls, and perspectives. *IEEE Geoscience and Remote Sensing Magazine.* 2021;pp. 0–0.

[5]   Yuan Q, Shen H, Li T, *et al.* Deep learning in environmental remote sensing: achievements and challenges. *Remote Sensing of Environment.* 2020;241:111716.

[6]   Ball JE, Anderson DT, Sr CSC. Comprehensive survey of deep learning in remote sensing: theories, tools, and challenges for the community. *Journal of Applied Remote Sensing.* 2017;11(4):1–54.

[7]   Hou X, Ao W, Song Q, *et al.* FUSAR-Ship: building a high-resolution SAR-AIS matchup dataset of Gaofen-3 for ship detection and recognition. *Science China Information Sciences*. 2020;63(4):1–19.

[8]   Jiao J, Zhang Y, Sun H, *et al.* A densely connected end-to-end neural network for multiscale and multiscene SAR ship detection. *IEEE Access*. 2018;6:20881–20892.

[9]   Cui Z, Li Q, Cao Z, *et al.* Dense attention pyramid networks for multi-scale ship detection in SAR images. *IEEE Transactions on Geoscience and Remote Sensing*. 2019;57(11):8983–8997.

[10] Deng Z, Sun H, Zhou S, *et al.* Learning deep ship detector in SAR images from scratch. *IEEE Transactions on Geoscience and Remote Sensing*. 2019;57(6):4021–4039.

[11] Wang J, Lu C, Jiang W. Simultaneous ship detection and orientation estimation in SAR images based on attention module and angle regression. *Sensors*. 2018;18(9):2851.

[12] An Q, Pan Z, Liu L, *et al.* DRBox-v2: an improved detector with rotatable boxes for target detection in SAR images. *IEEE Transactions on Geoscience and Remote Sensing*. 2019;57(11):8333–8349.

[13] Yang R, Pan Z, Jia X, *et al.* A novel CNN-based detector for ship detection based on rotatable bounding box in SAR images. *IEEE Journal of Selected Topics in Applied Earth Observations and Remote Sensing*. 2021;14:1938–1958.

[14] Siyu W, Xin G, Hao S, *et al.* An aircraft detection method based on convolutional neural networks in high-resolution SAR images. *Journal of Radars*. 2017;6(2):195–203.

[15] Zhang L, Li C, Zhao L, *et al.* A cascaded three-look network for aircraft detection in SAR images. *Remote Sensing Letters*. 2020;11(1):57–65.

[16] Zhao Y, Zhao L, Li C, *et al.* Pyramid attention dilated network for aircraft detection in SAR images. *IEEE Geoscience and Remote Sensing Letters*. 2020;18(4):662–666.

[17] Guo Q, Wang H, Xu F. Scattering enhanced attention pyramid network for aircraft detection in SAR images. *IEEE Transactions on Geoscience and Remote Sensing*. 2021;59(9):7570–7587.

[18] Wu H, Zhang H, Zhang J, *et al.* Fast aircraft detection in satellite images based on convolutional neural networks. In: *2015 IEEE International Conference on Image Processing (ICIP)*. IEEE; 2015. pp. 4210–4214.

[19] Zuo J, Xu G, Fu K, *et al.* Aircraft type recognition based on segmentation with deep convolutional neural networks. *IEEE Geoscience and Remote Sensing Letters*. 2018;15(2):282–286.

[20] Jia H, Guo Q, Chen J, *et al.* Adaptive component discrimination network for airplane detection in remote sensing images. *IEEE Journal of Selected Topics in Applied Earth Observations and Remote Sensing*. 2021;14:7699–7713.

[21] Maggiori E, Tarabalka Y, Charpiat G, *et al.* Convolutional neural networks for large-scale remote-sensing image classification. *IEEE Transactions on Geoscience and Remote Sensing*. 2016;55(2):645–657.

[22] Marmanis D, Schindler K, Wegner JD, *et al.* Classification with an edge: Improving semantic image segmentation with boundary detection. *ISPRS Journal of Photogrammetry and Remote Sensing.* 2018;135: 158–172.

[23] Ji S, Wei S, Lu M. Fully convolutional networks for multisource building extraction from an open aerial and satellite imagery data set. *IEEE Transactions on Geoscience and Remote Sensing.* 2018;57(1):574–586.

[24] Ayala C, Sesma R, Aranda C, *et al.* A deep learning approach to an enhanced building footprint and road detection in high-resolution satellite imagery. *Remote Sensing.* 2021;13(16):3135.

[25] Li Q, Mou L, Hua Y, *et al.* Building footprint generation through convolutional neural networks with attraction field representation. *IEEE Transactions on Geoscience and Remote Sensing.* 2021;60:3109844.

[26] Yuan J. Learning building extraction in aerial scenes with convolutional networks. *IEEE Transactions on Pattern Analysis and Machine Intelligence.* 2017;40(11):2793–2798.

[27] Yang HL, Yuan J, Lunga D, *et al.* Building extraction at scale using convolutional neural network: Mapping of the united states. *IEEE Journal of Selected Topics in Applied Earth Observations and Remote Sensing.* 2018;11(8):2600–2614.

[28] Marcos D, Tuia D, Kellenberger B, *et al.* Learning deep structured active contours end-to-end. In: *Proceedings of the IEEE Conference on Computer Vision and Pattern Recognition*; 2018. pp. 8877–8885.

[29] Li Z, Wegner JD, Lucchi A. Topological map extraction from overhead images. In: *Proceedings of the IEEE/CVF International Conference on Computer Vision*; 2019. pp. 1715–1724.

[30] Girard N, Smirnov D, Solomon J, *et al.* Polygonal Building Segmentation by Frame Field Learning. arXiv preprint arXiv:200414875. 2020.

[31] Song Q, Xu F. Zero-shot learning of SAR target feature space with deep generative neural networks. *IEEE Geoscience and Remote Sensing Letters.* 2017;14(12):2245–2249.

[32] Gui R, Xu X, Wang L, *et al.* A generalized zero-shot learning framework for PolSAR land cover classification. *Remote Sensing.* 2018;10(8):1307.

[33] Mikolov T, Sutskever I, Chen K, *et al.* Distributed representations of words and phrases and their compositionality. In: *Advances in Neural Information Processing Systems.* 2013. p. 26.

[34] Goodfellow I, Pouget-Abadie J, Mirza M, *et al.* Generative adversarial networks. In: *Advances in Neural Information Processing Systems.* 2014. p. 27.

[35] Song Q, Xu F, Zhu XX, *et al.* Learning to generate SAR images with adversarial autoencoder. *IEEE Transactions on Geoscience and Remote Sensing.* 2021;60:5210015.

[36] Song Q, Chen H, Xu F, *et al.* EM simulation-aided zero-shot learning for SAR automatic target recognition. *IEEE Geoscience and Remote Sensing Letters.* 2019;17(6):1092–1096.

[37] Huang Z, Pan Z, Lei B. Transfer learning with deep convolutional neural network for SAR target classification with limited labeled data. *Remote Sensing*. 2017;9(9):907.

[38] Zhu XX, Hu J, Qiu C, *et al*. So2Sat LCZ42: a benchmark data set for the classification of global local climate zones [Software and Data Sets]. *IEEE Geoscience and Remote Sensing Magazine*. 2020;8(3):76–89.

[39] Bechtel B, Alexander PJ, Böhner J, *et al*. Mapping local climate zones for a worldwide database of the form and function of cities. *ISPRS International Journal of Geo-Information*. 2015;4(1):199–219.

[40] Qiu C, Schmitt M, Mou L, *et al*. Feature importance analysis for local climate zone classification using a residual convolutional neural network with multi-source datasets. *Remote Sensing*. 2018;10(10):1572.

[41] Qiu C, Mou L, Schmitt M, *et al*. Local climate zone-based urban land cover classification from multi-seasonal Sentinel-2 images with a recurrent residual network. *ISPRS Journal of Photogrammetry and Remote Sensing*. 2019;154:151–162.

[42] Elshamli A, Taylor GW, Areibi S. Multisource domain adaptation for remote sensing using deep neural networks. *IEEE Transactions on Geoscience and Remote Sensing*. 2019;58(5):3328–3340.

[43] Yoo C, Lee Y, Cho D, *et al*. Improving local climate zone classification using incomplete building data and Sentinel 2 images based on convolutional neural networks. *Remote Sensing*. 2020;12(21):3552.

[44] Zhou Y, Wei T, Zhu X, *et al*. A parcel-based deep-learning classification to map local climate zones from Sentinel-2 images. *IEEE Journal of Selected Topics in Applied Earth Observations and Remote Sensing*. 2021;14:4194–4204.

[45] Liu S, Shi Q. Local climate zone mapping as remote sensing scene classification using deep learning: a case study of metropolitan China. *ISPRS Journal of Photogrammetry and Remote Sensing*. 2020;164:229–242.

[46] Qiu C, Liebel L, Hughes LH, *et al*. Multitask learning for human settlement extent regression and local climate zone classification. *IEEE Geoscience and Remote Sensing Letters*. 2020;19:1–5.

[47] Ma L, Zhu X, Qiu C, *et al*. Advances of local climate zone mapping and its practice using object-based image analysis. *Atmosphere*. 2021;12(9):1146.

[48] Garnot VSF, Landrieu L, Giordano S, *et al*. Satellite image time series classification with pixel-set encoders and temporal self-attention. In: *Proceedings of the IEEE/CVF Conference on Computer Vision and Pattern Recognition*; 2020. pp. 12325–12334.

[49] Kussul N, Lemoine G, Gallego J, *et al*. Parcel based classification for agricultural mapping and monitoring using multi-temporal satellite image sequences. In: *2015 IEEE International Geoscience and Remote Sensing Symposium (IGARSS)*. IEEE; 2015. pp. 165–168.

[50] Nijhawan R, Sharma H, Sahni H, *et al*. A deep learning hybrid CNN framework approach for vegetation cover mapping using deep features. In: *2017 13th International Conference on Signal-Image Technology & Internet-Based Systems (SITIS)*. IEEE; 2017. pp. 192–196.

[51]   Rußwurm M, Korner M. Temporal vegetation modelling using long short-term memory networks for crop identification from medium-resolution multi-spectral satellite images. In: *Proceedings of the IEEE Conference on Computer Vision and Pattern Recognition Workshops*; 2017. pp. 11–19.

[52]   Ji S, Zhang C, Xu A, *et al.* 3D convolutional neural networks for crop classification with multi-temporal remote sensing images. *Remote Sensing*. 2018;10(1):75.

[53]   Rußwurm M, Körner M. Convolutional LSTMs for Cloud-Robust Segmentation of Remote Sensing Imagery. arXiv preprint arXiv:181102471. 2018.

[54]   Garnot VSF, Landrieu L, Giordano S, *et al.* Time–space tradeoff in deep learning models for crop classification on satellite multi-spectral image time series. In: *IGARSS 2019-2019 IEEE International Geoscience and Remote Sensing Symposium*. IEEE; 2019. pp. 6247–6250.

[55]   Rußwurm M, Körner M. Self-attention for raw optical satellite time series classification. *ISPRS Journal of Photogrammetry and Remote Sensing*. 2020;169:421–435.

[56]   Schmitt M, Hughes LH, Zhu XX. The SEN1-2 Dataset for Deep Learning in SAR-Optical Data Fusion. arXiv preprint arXiv:180701569. 2018.

[57]   Fu S, Xu F, Jin YQ. Reciprocal translation between SAR and optical remote sensing images with cascaded-residual adversarial networks. *Science China Information Sciences*. 2021;64(2):1–15.

[58]   Shi X, Fu S, Chen J, *et al.* Object-level semantic segmentation on the high-resolution Gaofen-3 FUSAR-map dataset. *IEEE Journal of Selected Topics in Applied Earth Observations and Remote Sensing*. 2021;14:3107–3119.

[59]   Rahnemoonfar M, Chowdhury T, Sarkar A, *et al.* Floodnet: A High Resolution Aerial Imagery Dataset for Post Flood Scene Understanding. arXiv preprint arXiv:201202951. 2020.

[60]   Zaffaroni M, Rossi C. Water segmentation with deep learning models for flood detection and monitoring. In: *Proceedings of the 17th ISCRAM Conference*, Blacksburg, VA, USA; 2020. pp. 24–27.

[61]   Nemni E, Bullock J, Belabbes S, *et al.* Fully convolutional neural network for rapid flood segmentation in synthetic aperture radar imagery. *Remote Sensing*. 2020;12(16):2532.

[62]   Lopez-Fuentes L, van de Weijer J, Bolanos M, *et al.* Multi-modal deep learning approach for flood detection. MediaEval. 2017;17:13–15.

[63]   Zhao H, Shi J, Qi X, *et al.* Pyramid scene parsing network. In: *Proceedings of the IEEE Conference on Computer Vision and Pattern Recognition*; 2017. pp. 2881–2890.

[64]   Akiva P, Purri M, Dana K, *et al.* H2O-Net: self-supervised flood segmentation via adversarial domain adaptation and label refinement. In: *Proceedings of the IEEE/CVF Winter Conference on Applications of Computer Vision*; 2021. pp. 111–122.

[65]   Kang W, Xiang Y, Wang F, *et al.* Flood detection in gaofen-3 SAR images via fully convolutional networks. *Sensors*. 2018;18(9):2915.

[66] Li Y, Martinis S, Wieland M. Urban flood mapping with an active self-learning convolutional neural network based on TerraSAR-X intensity and interfero-metric coherence. *ISPRS Journal of Photogrammetry and Remote Sensing.* 2019;152:178–191.

[67] Zhang P, Chen L, Li Z, *et al.* Automatic extraction of water and shadow from SAR images based on a multi-resolution dense encoder and decoder network. *Sensors.* 2019;19(16):3576.

[68] Cuomo J, Chandrasekar V. Developing deep learning models for storm now-casting. *IEEE Transactions on Geoscience and Remote Sensing.* 2021;60: 4103713.

[69] Ruzanski E, Chandrasekar V, Wang Y. The CASA nowcasting sys-tem. *Journal of Atmospheric and Oceanic Technology.* 2011;28(5): 640–655.

[70] Bowler NE, Pierce CE, Seed AW. STEPS: a probabilistic precipitation fore-casting scheme which merges an extrapolation nowcast with downscaled NWP. *Quarterly Journal of the Royal Meteorological Society: A Journal of the Atmospheric Sciences, Applied Meteorology and Physical Oceanography.* 2006;132(620):2127–2155.

[71] Li L, Schmid W, Joss J. Nowcasting of motion and growth of precipitation with radar over a complex orography. *Journal of Applied Meteorology and Climatology.* 1995;34(6):1286–1300.

[72] Li P, Lai ES. Short-range quantitative precipitation forecasting in Hong Kong. *Journal of Hydrology.* 2004;288(1–2):189–209.

[73] Xingjian S, Chen Z, Wang H, *et al.* Convolutional LSTM network: a machine learning approach for precipitation nowcasting. In: *Advances in Neural Information Processing Systems*; 2015. pp. 802–810.

[74] Sato R, Kashima H, Yamamoto T. Short-term precipitation prediction with skip-connected prednet. In: *International Conference on Artificial Neural Networks.* New York, NY: Springer; 2018. pp. 373–382.

[75] Shi E, Li Q, Gu D, *et al.* A method of weather radar echo extrapolation based on convolutional neural networks. In: *International Conference on Multimedia Modeling.* New York, NY: Springer; 2018. pp. 16–28.

[76] Agrawal S, Barrington L, Bromberg C, *et al.* Machine Learning for Pre-cipitation Nowcasting from Radar Images. arXiv preprint arXiv:191212132. 2019.

[77] Chen H, Zhang X, Liu Y, *et al.* Generative adversarial networks capabilities for super-resolution reconstruction of weather radar echo images. *Atmosphere.* 2019;10(9):555.

[78] Jing J, Li Q, Ding X, *et al.* Aenn: a generative adversarial neural network for weather radar echo extrapolation. *The International Archives of Pho-togrammetry, Remote Sensing and Spatial Information Sciences.* 2019;42: 89–94.

[79] Tian L, Li X, Ye Y, *et al.* A generative adversarial gated recurrent unit model for precipitation nowcasting. *IEEE Geoscience and Remote Sensing Letters.* 2019;17(4):601–605.

[80]   Prudden R, Adams S, Kangin D, *et al*. A Review of Radar-Based Nowcasting of Precipitation and Applicable Machine Learning Techniques. arXiv preprint arXiv:200504988. 2020.

[81]   Zhou K, Zheng Y, Dong W, *et al*. A deep learning network for cloud-to-ground lightning nowcasting with multisource data. *Journal of Atmospheric and Oceanic Technology*. 2020;37(5):927–942.

[82]   Lin T, Li Q, Geng YA, *et al*. Attention-based dual-source spatiotemporal neural network for lightning forecast. *IEEE Access*. 2019;7:158296–158307.

[83]   Price C, Rind D. A simple lightning parameterization for calculating global lightning distributions. *Journal of Geophysical Research: Atmospheres*. 1992;97(D9):9919–9933.

[84]   McCaul Jr EW, Goodman SJ, LaCasse KM, *et al*. Forecasting lightning threat using cloud-resolving model simulations. *Weather and Forecasting*. 2009;24(3):709–729.

[85]   Geng Ya, Li Q, Lin T, *et al*. Lightnet: a dual spatiotemporal encoder network model for lightning prediction. In: *Proceedings of the 25th ACM SIGKDD International Conference on Knowledge Discovery & Data Mining*; 2019. pp. 2439–2447.

[86]   Geng Ya, Li Q, Lin T, *et al*. A deep learning framework for lightning forecasting with multi-source spatiotemporal data. *Quarterly Journal of the Royal Meteorological Society*. 2021;147(741):4048–4062.

[87]   Mosier RM, Schumacher C, Orville RE, *et al*. Radar nowcasting of cloud-to-ground lightning over Houston, Texas. *Weather and Forecasting*. 2011;26(2):199–212.

[88]   Badrinarayanan V, Kendall A, Cipolla R. Segnet: a deep convolutional encoder-decoder architecture for image segmentation. *IEEE Transactions on Pattern Analysis and Machine Intelligence*. 2017;39(12):2481–2495.

[89]   Sun Y, Xin Q, Huang J, *et al*. Characterizing tree species of a tropical wetland in southern china at the individual tree level based on convolutional neural network. *IEEE Journal of Selected Topics in Applied Earth Observations and Remote Sensing*. 2019;12(11):4415–4425.

[90]   Hamraz H, Jacobs NB, Contreras MA, *et al*. Deep learning for conifer/deciduous classification of airborne LiDAR 3D point clouds representing individual trees. *ISPRS Journal of Photogrammetry and Remote Sensing*. 2019;158:219–230.

[91]   Briechle S, Krzystek P, Vosselman G. Silvi-Net—a dual-CNN approach for combined classification of tree species and standing dead trees from remote sensing data. *International Journal of Applied Earth Observation and Geoinformation*. 2021;98:102292.

[92]   Zou X, Cheng M, Wang C, *et al*. Tree classification in complex forest point clouds based on deep learning. *IEEE Geoscience and Remote Sensing Letters*. 2017;14(12):2360–2364.

[93]   Hartling S, Sagan V, Sidike P, *et al*. Urban tree species classification using a WorldView-2/3 and LiDAR data fusion approach and deep learning. *Sensors*. 2019;19(6):1284.

[94] Briechle S, Krzystek P, Vosselman G. Classification of tree species and stand-ing dead trees by fusing UAV-based lidar data and multispectral imagery in the 3D deep neural network PointNet++. *ISPRS Annals of the Photogrammetry, Remote Sensing and Spatial Information Sciences.* 2020;2:203–210.

[95] Qi CR, Yi L, Su H, *et al.* Pointnet++: Deep hierarchical feature learning on point sets in a metric space. In: *Advances in Neural Information Processing Systems.* 2017. p. 30.

[96] Khan SH, He X, Porikli F, *et al.* Forest change detection in incomplete satellite images with deep neural networks. *IEEE Transactions on Geoscience and Remote Sensing.* 2017;55(9):5407–5423.

[97] De Bem PP, de Carvalho Junior OA, Fontes Guimarães R, *et al.* Change detection of deforestation in the Brazilian Amazon using landsat data and convolutional neural networks. *Remote Sensing.* 2020;12(6):901.

[98] Maretto RV, Fonseca LM, Jacobs N, *et al.* Spatio-temporal deep learning approach to map deforestation in amazon rainforest. *IEEE Geoscience and Remote Sensing Letters.* 2020;18(5):771–775.

[99] Isaienkov K, Yushchuk M, Khramtsov V, *et al.* Deep learning for regular change detection in Ukrainian forest ecosystem with Sentinel-2. *IEEE Journal of Selected Topics in Applied Earth Observations and Remote Sensing.* 2020;14:364–376.

[100] John D, Zhang C. An attention-based U-Net for detecting deforestation within satellite sensor imagery. *International Journal of Applied Earth Observation and Geoinformation.* 2022;107:102685.

[101] Matosak BM, Fonseca LMG, Taquary EC, *et al.* Mapping deforestation in Cerrado based on hybrid deep learning architecture and medium spatial resolution satellite time series. *Remote Sensing.* 2022;14(1):209.

[102] Ortega Adarme M, Queiroz Feitosa R, Nigri Happ P, *et al.* Evaluation of deep learning techniques for deforestation detection in the Brazilian Amazon and cerrado biomes from remote sensing imagery. *Remote Sensing.* 2020;12(6):910.

[103] Naderpour M, Rizeei HM, Ramezani F. Forest fire risk prediction: a spatial deep neural network-based framework. *Remote Sensing.* 2021;13(13):2513.

[104] Langford Z, Kumar J, Hoffman F. Wildfire mapping in Interior Alaska using deep neural networks on imbalanced datasets. In: *2018 IEEE International Conference on Data Mining Workshops (ICDMW).* IEEE; 2018. pp. 770–778.

[105] Ba R, Song W, Li X, *et al.* Integration of multiple spectral indices and a neural network for burned area mapping based on MODIS data. *Remote Sensing.* 2019;11(3):326.

[106] Ban Y, Zhang P, Nascetti A, *et al.* Near real-time wildfire progression monitoring with Sentinel-1 SAR time series and deep learning. *Scientific Reports.* 2020;10(1):1–15.

[107] Knopp L, Wieland M, Rättich M, *et al.* A deep learning approach for burned area segmentation with Sentinel-2 data. *Remote Sensing.* 2020;12(15):2422.

[108] Pinto MM, Libonati R, Trigo RM, *et al*. A deep learning approach for mapping and dating burned areas using temporal sequences of satellite images. *ISPRS Journal of Photogrammetry and Remote Sensing*. 2020;160:260–274.

[109] Rashkovetsky D, Mauracher F, Langer M, *et al*. Wildfire detection from multisensor satellite imagery using deep semantic segmentation. *IEEE Journal of Selected Topics in Applied Earth Observations and Remote Sensing*. 2021;14:7001–7016.

Chapter 9

Deep learning techniques for digital satellite communications

Federico Garbuglia[1], Tom Dhaene[1] and Domenico Spina[1]

9.1 Introduction

Satellite communication (SatCom) has been in rapid evolution during the last decades. The rapid growth of space industry is making satellite technologies available to more companies and private customers. However, despite the technology improvements, the capacity of SatCom systems is still subject to the limitations presented by physical communication channels. In fact, signals travelling from the earth surface to near-space artificial satellites are strongly attenuated by long propagation distances, and deteriorated by atmospheric and extraterrestrial noise. At the same time, compensating for noise and attenuation is only possible at the cost of increasing the power consumption or reducing the data rate of the transmitting devices. Hence, SatCom systems need intelligent strategies for the allocation of power and bandwidth resources [1].

Thankfully, machine learning (ML) can be employed to automate resource allocation, as described in Section 9.2. In particular, deep learning (DL) techniques, which exploit artificial neural networks, are suitable to perform such complex tasks in sophisticated SatCom systems. For a proper resource allocation, the receiver in a communication system has to be informed about characteristics of the channel and the operating conditions of transmitting devices, such as propagation losses and operating point of high-power amplifiers. However, carrying such information reduces the serviceable capacity of the channel, as described in Section 9.2. Instead, a DL model can be employed to directly extract this information from the signal samples incoming at the SatCom receiver, thus preserving communication link capacity.

In particular, this chapter specifically focuses on the characterisation of noise and nonlinear distortion in digital satellite communication links, which is of paramount importance to ensure the desired performance of modern SatCom systems, as described in Sections 9.3 and 9.4. The noise is due to the propagation of the transmitted signals in the channel, while the distortion is mainly caused by nonlinear high-power amplifiers used at the transmitter. The goal is to *efficiently* estimate these quantities

[1]IDLab, Ghent University – imec, Belgium

directly at the receiver of a SatCom system via suitable DL techniques. Two independent DL strategies are proposed for the estimation of noise and distortion from received signals. First, a convolutional autoencoder is presented in Section 9.5 for noise estimation. Next, a deep convolutional classifier is described in Section 9.6 to evaluate the distortion introduced by high-power amplifiers. Both strategies achieve high accuracy when tested on suitable application examples of SatCom systems. Furthermore, results indicate that the trained DL models can be used simultaneously on signals that are affected by channel noise and amplifier distortion.

More generally, this chapter illustrates how DL models can be used to extract the value of system parameters or transmitting conditions from samples of an electric signal. Hence, the use of these strategies can be extended to other domains, such as electromagnetic compatibility or signal integrity, where it is valuable to estimate unknown system conditions from detected signals.

9.2   Machine learning for SatCom

SatCom has become an omnipresent technology in contemporary times. Artificial satellites are employed for every communication service that requires a world-wide coverage, such as radio, video broadcasting, internet access, aeronautical and maritime communications.

Recently, SatCom has become increasingly prominent in cellular communication networks. In fact, the latest mobile communication standards like LTE and 5G put heavy traffic loads on the cell backhaul systems. Thanks to the latest technological improvements, such as high-throughput satellites, data prefetching, and high-order adaptive modulation schemes, satellite communication provides a viable option to re-design the cell backhauls [2]. In fact, these innovations allow to increase the available bandwidth while reducing the delay introduced by SatCom systems. In addition, SatCom has become essential to extend the cellular network and the internet access to remote areas, where building the infrastructure for terrestrial communications is infeasible. Moreover, the cost of deploying communication satellites is progressively reduced by the advancements in orbital launch systems. Both governments and private companies can now deploy large constellations of satellites and provide world-wide broadband services [3].

A consequence of these innovations is that SatCom systems are getting more complex, increasing the need for automated control strategies. In fact, it is desirable to make control strategies adjustable to numerous factors, including user traffic and weather. From the software side, ML algorithms can help fulfil the need for automation. Indeed, ML has natural application in the processing of the information that is gathered from – or carried through – artificial satellites. For example, it has been employed in weather prediction, earth observation, navigation and positioning systems [4–6]. Additionally, ML has been successfully used to solve issues related to the design and implementation of communication systems. In fact, it can be employed for typical SatCom problems such as interference mitigation, allocation of spectrum and power resources, optimisation of multi-input–multi-output (MIMO) communications [7], or optimisation of network architecture [8].

Furthermore, the evolution of SatCom is headed towards the inclusion of advanced functionalities like network virtualisation, active antennas and mega constellations. Thankfully, ML techniques like digital twins enable the seamless integration of these increasingly complex functionalities. In fact, digital twins can virtually replicate the functioning of an entire SatCom system, from the physical layer to the user application layer. New designs and functionalities can be tested on digital twins, before the actual implementation. Moreover, a digital twin can be interfaced with the real system to automate monitoring and control policies. Thus, digital twins can significantly reduce the cost of performance optimisation for complex systems [9].

Within this framework, this chapter focuses on the use ML for the estimation of power resources. In SatCom systems, the operating conditions of the transmitter and the characteristics of the satellite channel are in continuous mutation, due to numerous factor, such as weather, users traffic, or even the battery status of the transmitting device. Thus, in order to limit the power consumption of a transmitter, it is necessary to control the transmission power by taking variable factors into account.

The first step to design a power control strategy is to estimate the status of the received signal. Specifically, the amount of noise and distortion on the received signal has to be measured. Noise and distortion are commonly measured at the receiver by checking a sequence of incoming known sequences, called pilot symbols. In fact, noise and distortion affect the reconstruction of the received pilot symbols. However, to implement this solution, part of the SatCom link capacity has to be reserved for the transmission of pilot symbols. Instead, ML techniques can be applied to estimate noise and distortion directly from any signal pulse arriving at the receiver, as described in the following. Once this estimation is executed, the signal can be corrected at the receiver using techniques like symbol-based equalisation. Furthermore, if the transmitter is informed about the estimation, corrective actions such as predistortion can be applied [10].

9.2.1 Deep learning

Depending on the available data and the interaction with the system, three main classes of ML algorithms can be distinguished: supervised, unsupervised and reinforcement learning. In *supervised learning*, a set of data samples are provided together with labels. Such labels represent the target values for the ML model. Then, the task of the ML model is to produce a mapping between the input data and the corresponding labels. If the ML model is well trained, it will be able to predict the correct labels for previously unseen data samples. However, labelled dataset are typically expensive to collect, since they require an external 'oracle', such as a human or a simulator, to provide the true label for each sample. A common example of supervised learning is the classification of images: after collecting and labelling a wide collection of input images, a ML model can be trained until it assigns the correct label to all the possible inputs. The training process is usually driven by an error metric between the correct label and the model prediction: the model modifies its internal parameters such that the output error is reduced.

In cases when only unlabelled data is available, *unsupervised learning* techniques can be applied. In unsupervised learning, the task of the ML model is to provide a mapping of the input data in which common patterns can be identified. For example, unsupervised learning can be applied to image classification when the image labels are missing. In this case, the model usually provides an internal representation of the input images that are similar among inputs of the same class. Thus, clusters of images with similar representation can be identified and labelled. Additionally, in unsupervised learning, the model typically provides an inverse mapping from the internal representation to a reconstructed input data. Thus, the model training is driven by an error metric between original input and the reconstructed input: the lower the error, the more accurate is the internal representation. Additionally, the inverse mapping grants to the model the ability to generate new data, by choosing an internal representation that does not correspond to any of the training data samples.

The last class of ML algorithm is *reinforcement learning*. In reinforcement learning, the ML model is capable of extracting the data by performing an action in a specific environment. In this case, data is typically extracted iteratively: the ML model selects the data observation that maximises a specified reward function. Thus, the ML model performs a decision based on the data collected from the environment during previous iterations. Differently from supervised learning, the model is not trained on the desired output (i.e. the labels), but on a reward function built on the response of the environment. For examples, in robot kinematics, a model can be trained to move the robot to a specific position. Then, reward function can be computed on one or multiple movements and used to progressively update the model. The performance of reinforcement learning techniques strongly depends on the definition of the reward function. In fact, the definition has still to be provided by a human interpreter, who evaluates the actions performed by the model.

In this chapter, supervised ML techniques will be adopted: more specifically, deep learning techniques based on artificial neural networks (ANN) [11]. The strength of neural networks derives from their ability to model extremely complex functions, due to their high number of learning parameters. Additionally, they possess a versatile architecture: if higher complexity is needed, multiple layers of neurons can be stacked, creating a *deep* structure. The downside of such modelling capability is that neural networks need a large dataset of examples (i.e. large amount of training data) to learn a specific task. Thus, high capacity storage and computation power are typically required to train DL models. Another advantage of DL models is their ability to operate on high-dimensional data vectors. Consequently, in the last decade, countless DL techniques have been designed to operate on complex data such as text, images, audio and telecommunication signals.

In this chapter, a specific use case of DL for satellite communication is presented: Section 9.3 gives an overview of the general architecture of digital satellite communication systems, while the specific SatCom model studied is described in Section 9.4.

9.3   Digital satellite communication systems

In general, a SatCom system can be divided into two ground segments and one space segment. The first ground segment is the *uplink*, which is responsible for transmitting the user signal to a satellite in orbit around the earth. The radio components of the artificial satellite constitute the *space segment*. Finally, the signal is relayed by the satellite to the receiving ground segment, that recovers the original signal sent by the user. This last segment is called *downlink*.

9.3.1   Uplink segment

Typically, the uplink segment consists of the components shown in Figure 9.1. Here, the digital signal at the user terminal is encoded and modulated according to a suitable digital modulation scheme. Typical schemes include Amplitude Phase-Shift Keying (APSK), Quadrature Amplitude Modulation (QAM) or Quadrature Phase-Shift Keying (QPSK). The choice of the modulation scheme is strongly dependent on the bandwidth and power resources available in the radio channel. In fact, increasing the order of the modulation, i.e. the number of encoding symbols, allows to carry more information in the same bandwidth. However, high-order schemes are more sensitive to noise and distortion, since they present a reduced separation among symbols.

After modulation, the resulting analog signal is allocated in the transmission band by the upconverter. Depending on the type of the satellite service, several transmission bands can be reserved, ranging from 1–2 GHz (L-band) to 26.5–40 GHz (KA-band). Due to the high propagation loss in the atmosphere and in free space, the upconverted signal needs to be amplified by a high-power amplifier, so that it can reach the satellite transponder at a sufficient signal-to-noise ratio (SNR). The SNR is the ratio between the signal power and the background noise power. If its values is too low, the noise can compromise the reconstruction of the original message at the receiver. Finally, the transmitter antenna, which is the last component of the uplink segment, adapts the signal to the open space propagation. For earth station uplinks, wide parabolic antennas are usually used, since they provide high gains. In fact, higher gains correspond to higher signal power, thus reducing the amplification needed at the satellite receiver.

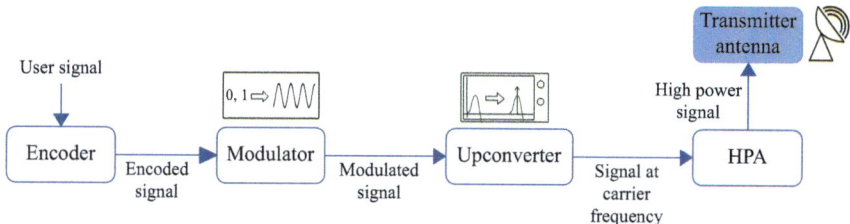

Figure 9.1   Schematic of the uplink ground station

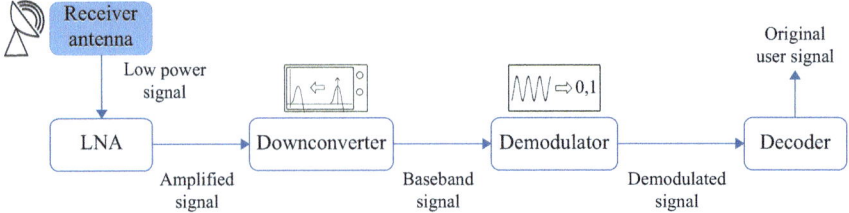

Figure 9.2   Schematic of the downlink ground station

9.3.2   Space segment

In a SatCom system, the function of an artificial satellite is to re-amplify and re-transmit the signal from the user terminal to the receiving earth station that constitutes the downlink segment. For this purpose, each artificial satellite incorporates one or multiple transponders that operate at different carrier frequencies. The re-amplification capability of artificial satellites is limited by the maximum power consumption of the spacecraft bus, which include a power generator and an energy storage system. Currently, the trend in the satellite industry is to deploy high numbers of small, energy-efficient satellites. Although this choice reduces the serviceable power for each satellite, it allows for the creation of mega-constellations that can provide high connectivity services on a global scale. However, a high number of satellites require a precise subdivision of the communication bandwidth and the adjustment of the antennas towards the ground areas that need to be covered. Therefore, the bandwidth and power resources are more and more stringent in modern SatCom systems.

9.3.3   Downlink segment

This segment consists of components that are dual to the uplink system, as shown in Figure 9.2. First, the signal reaches the receiving antenna at low power level, due to the high channel losses and the low re-amplification at the satellite. Therefore, a low-noise amplifier (LNA) subsequently restores the necessary signal power. Next, the downconverter returns the baseband signal. Then, the demodulator converts the incoming analog signal to the encoded signal. Finally, the decoder recovers the original user signal.

9.4   SatCom systems modelling

A DL model necessitates of a dataset of examples in order to execute the requested task. In fact, the dataset has to be sufficiently large to allow a proper training of the ML model. The training can be considered successful if the model can operate accurately on data that was not provided during training. In particular, for the estimation of noise and distortion in SatCom systems, many signal instances have to be measured in different transmitting conditions and with different information content. However, the cost of performing this operation on real, deployed systems or experimental setups can be prohibitively high, due to the following reasons. First, signals at intermediate stages of the SatCom segments can be hard to access, because of the high level of integration between the physical components. Second, the collection signals at different transmitting conditions may require multiple measurement runs, thus increasing the time cost of the acquisition. Therefore, a preliminary investigation is conducted in this chapter using simulated signals, in order to assess the applicability of DL techniques and to estimate their performance. For this purpose, a simulation model of the SatCom link described in Section 9.3 is introduced. This model is able to rapidly generate communicated signals at any stage of the system and for any transmitting condition desired.

The simulation model of the SatCom return link is implemented in MATLAB®* according to the simplified scheme in Figure 9.3. The return link is constituted by a transmitter, corresponding to an uplink ground station, and a receiver, corresponding to a downlink ground station. In its entirety, the return link represents the travelling of a signal from the user terminal to the central gateway.

This MATLAB model is able to simulate signals that are affected by distortion caused by the high-power amplifier in the transmitter and by the noise in the communication channel [12]. Specifically, the signal message from the user terminal is initiated as a sequence of random bits in the transmitter model. Then, the bit sequence is modulated such that a baseband signal is obtained in the form of complex-valued symbols. The modulation is executed according to a specific modulation scheme,

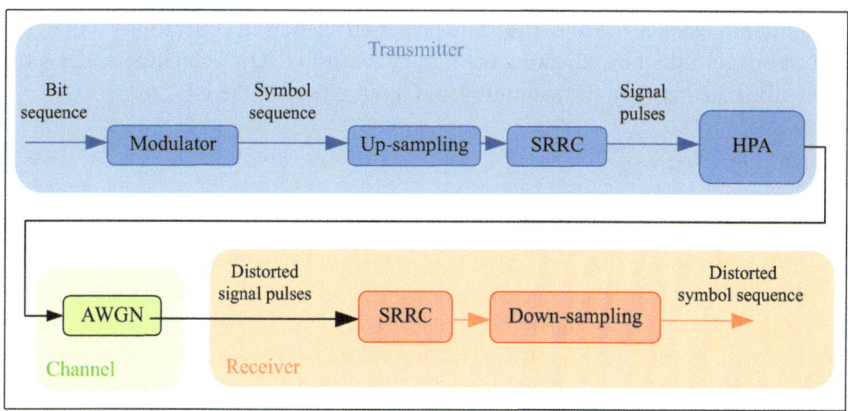

Figure 9.3 *MATLAB model for the SatCom return link. The high-power amplifier is indicated with the acronym HPA.*

---

*The Mathworks Inc., Natick, MA, USA.

which can be adapted to the channel characteristics and bandwidth resources. Next, the signal is up-sampled and shaped into pulses by a square-root raised cosine (SRRC) filter. The last two operations correspond to the upconverter in the uplink ground station (Figure 9.1). Before transmission, the signal is sent to an high-power amplifier. Note that such amplifier introduces a nonlinear distortion to the transmitted signal, as discussed in detail in the next section.

An additive white Gaussian noise (AWGN) block simulates the noise introduced by the communication channel. As the name indicates, this block adds a Gaussian noise sequence to both the real and the imaginary part of the signal corresponding to a specified SNR. Finally, the signal is provided to the receiver model, where it is filtered via a second SRRC, that is matched to the one used in the transmitter, and then down-sampled at the same rate.

9.4.1   High-power amplifier modelling

The ideal functioning of a high-power amplifier is to increase the amplitude of the signal to prepare it for atmospheric and space transmission. In fact, the amplification gain has to compensate for the attenuation and the noise introduced by the communication channel towards the satellite transponder and the receiving ground station. If the amplification is not sufficient, then the low SNR at the receiver can cause excessive errors that can not be corrected by the decoder.

However, the amplification can not be set arbitrarily high, but it is limited by the maximum power consumption of the uplink device. Moreover, solid-state amplifiers commonly operate near the saturation region, where their efficiency is higher. However, the input–output characteristic of HPAs becomes highly non-linear near saturation, thus introducing significant distortion in the output signal.

In order to balance the distortion and power efficiency, the operating point of the HPA can be controlled by adjusting the input back-off (IBO). The IBO indicates the ratio in dB scale between the saturation and input power of the HPA as:

$$\text{IBO} = 10 \log_{10} \frac{P_{sat,in}}{P_{in}} \tag{9.1}$$

where $P_{sat,in}$ and P_{in} are the saturation power and the average power of the input signal, respectively. Figure 9.4 illustrates a typical input–output power characteristic of a HPA. Note that the IBO is zero when the power of the input signal coincides with the saturation power of the HPA. Conversely, increasing the IBO means to move the operating point of the HPA towards its linear region, thus reducing the distortion.

As the HPA distortion is mainly amplitude-dependent, two other characteristics are important: the Amplitude-to-Amplitude (AM–AM) and the Amplitude-to-Phase (AM–PM) curves, which represent the magnitude and the phase, respectively, of the complex gain for any operating point of the amplifier [13]. An example of measured gain curves is shown in Figure 9.5. Thus, at any IBO, the behaviour of an HPA can be simulated by simply multiplying the amplitude and phase of the input signal by the corresponding gains on the AM–AM and AM–PM curves. Ultimately, the described SatCom model allows one to collect a dataset of received complex symbols sequences, with SNR and IBO specified by the user. These two quantities respectively measure the noise and the distortion on the signal.

The techniques discussed in the following sections aim at estimating the SNR and IBO from the received signals. In particular, the SNR estimation strategy introduced in [12] is explained in Section 9.5. This technique is designed to work on samples of the analog signals, incoming at the input of the receiver. Subsequently, a new IBO estimation strategy is developed in Section 9.6. Rather than working with analog signals, this technique works on sequences of modulated symbols. Since modulated symbols can be measured more easily, the IBO strategy is better suited to be implemented in real receiver systems.

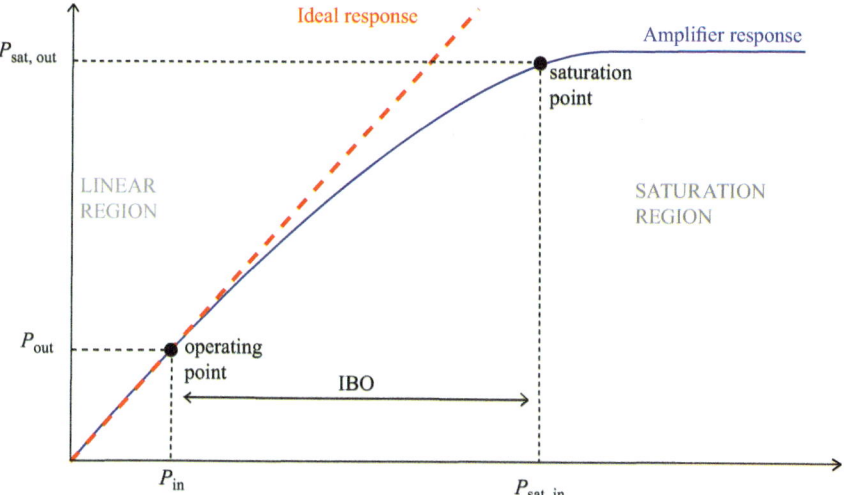

Figure 9.4   Example of input–output power characteristics of a HPA

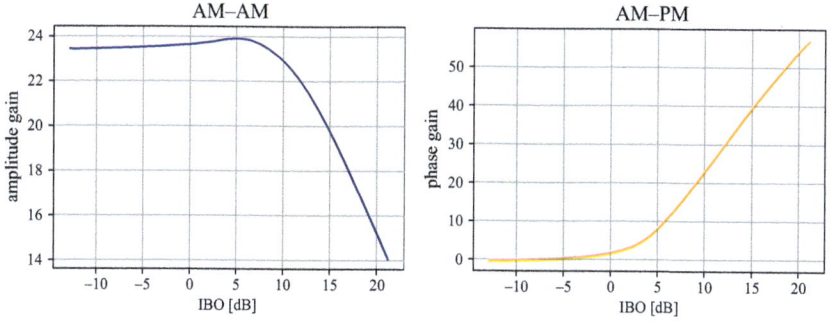

Figure 9.5   Example of input–output gain characteristics of a HPA

9.5   SNR estimation

As anticipated in Section 9.2, informing the receiver about the channel characteristics is crucial for the application of noise reduction techniques. However, acquiring such information requires the transmission of known signal (pilot symbols), which causes overhead and reduces the serviceable data rate [10]. Instead, a methodology is introduced in [12] to estimate the SNR at the receiver directly from any incoming signal, without the need of pilot symbols. This methodology is designed to operate in a SatCom return link: the signal is transmitted by the user terminal to the ground station through a satellite transponder. The core of the technique is a ML model called autoencoder (AE), which is described in Section 9.5.1. Then, the SNR estimation procedure is discussed in detail in Section 9.5.2.

9.5.1   Autoencoders

AEs are DL models able to encode input data into a lower-dimensional space, which is known as latent space. At the same time, an AE can reconstruct the input data with negligible loss of information starting from its latent space representation. Thanks to these properties, AEs have been successfully employed for a wide variety of problems, ranging from anomaly detection [14,15], design optimisation [16] and dimensionality reduction [17].

More specifically, an AE is a particular type of ANN presenting the architecture shown in Figure 9.6. Here, two main structures can be identified: the encoder and the decoder. The task of the encoder is to convert the data in the input layer into the latent space. Conversely, the decoder converts the latent space representation into the output layer, thus recovering the information at the input layer. The key property of this architecture is that input and output layers have the same dimensionality, while the latent space is lower-dimensional. Note that encoder and decoder networks may consist of one or multiple neural layers. Furthermore, due to the non-linear activation function at the output of each neuron, both the latent space and the output layer are non-linear representations of the input layer [18].

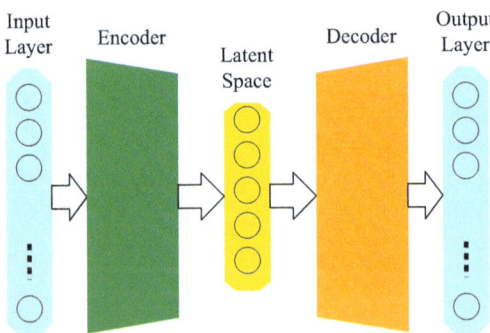

Figure 9.6   General architecture of an autoencoder

Similarly to other neural networks, AEs can be trained via back-propagation [11]. First, the output of the network is produced for one or multiple input instances. Second, a suitable loss function is computed based on the AE's output. Next, the parameters of each node in the network are updated such that the loss function is minimised. This process is repeated for all the available training instances until the loss function is sufficiently low.

By defining the loss function as a suitable error metric between the output and the input layer, an AE learns to reproduce the input instances that are used for training. In particular, the decoder network is trained to exactly reconstruct the original data instance from its latent space representation.

In order to increase the AE's reconstruction accuracy, one or multiple neural layers can be replaced with convolution operations. The resulting architecture is called convolutional autoencoder (CAE). Convolutional layers are useful to detect and extract features of the input data that are shift-invariant, while keeping a low model complexity [19]. Hence, the convolution is best suited for time sequences data like received SatCom signals, which presents time-invariant characteristics, such as noise and HPA distortion level. The architecture of the CAE used for SNR identification is described in detail in Section 9.5.4.

9.5.2 SNR estimation methodology

The proposed noise estimation technique consists of two steps: the first is the training phase, which follows the scheme illustrated in Figure 9.7. First, a dataset of signals is generated via the SatCom model described in Section 9.4. For this phase, the signal instances are generated by selecting a *high value* of SNR and IBO, so that low noise and distortion is introduced. Next, incoming signals are pre-processed in order to be analysed via an AE. The pre-processing comprises standardisation and frame splitting: the signal samples are suitably scaled such that their mean is zero and their standard deviation is equal to one; then, the signal is split in frames of constant size. Now, a CAE can be trained to reproduce the signal frames. The goal of training is to estimate

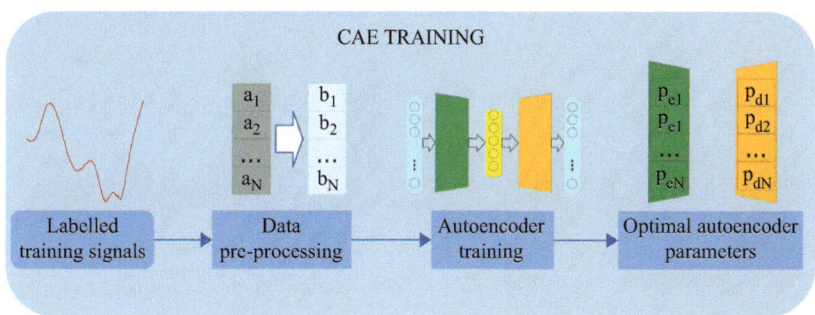

Figure 9.7   Training phase of the proposed approach for quantifying the SNR in received signals

the value of the CAE parameters producing the best input reconstruction. Note that the size of the CAE input and output layer must match the number of samples in each frame.

The second step is the testing phase, shown in Figure 9.8. Similar to the training phase, new test signals are generated, pre-processed and then split in frames of suitable length. However, *variable* values of SNR are now used when producing the signals. Next, these signal frames are fed to the previously trained CAE, that tries to to reconstruct the input from its latent space encoding. The idea behind this approach is that noisy signals will produce a different latent space encoding than the training instances. More specifically, signals with lower SNR will correspond to anomalous points in the latent space. As a consequence, the reconstruction error at the CAE output will be sensibly higher than during the training phase. Therefore, several metrics are defined in [12] to estimate the SNR from the latent space or from the output reconstruction of the CAE. These metrics are discussed in detail in Section 9.5.3.

However, noise is not the only cause of signal deterioration. As discussed in Section 9.4.1, the HPA introduces distortion due to its non-ideal characteristics. Hence, it is important to have an accurate SNR estimation even in presence of different levels of distortion. Figure 9.9 shows the real and imaginary part of a signal containing the same information content, but with different noise and distortion levels, simulated at the input of the SatCom receiver. It is apparent that noise and distortion produce different effects on the analog pulses. Hence, in order to make the proposed algorithm independent from the HPA distortion, the CAE is trained on a dataset of signals with high SNR, but *varying* IBO. In such manner, the encoding and the reconstruction error in the testing phase will be robust to variations in the distortion level.

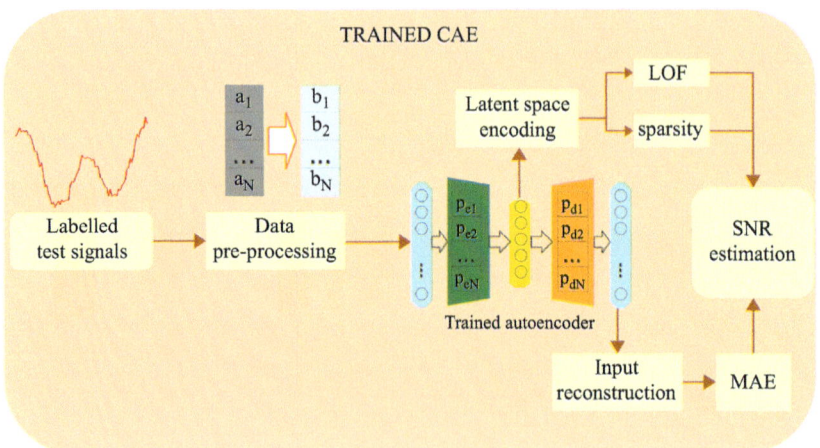

Figure 9.8   Testing phase of the proposed approach for quantifying the SNR in received signals

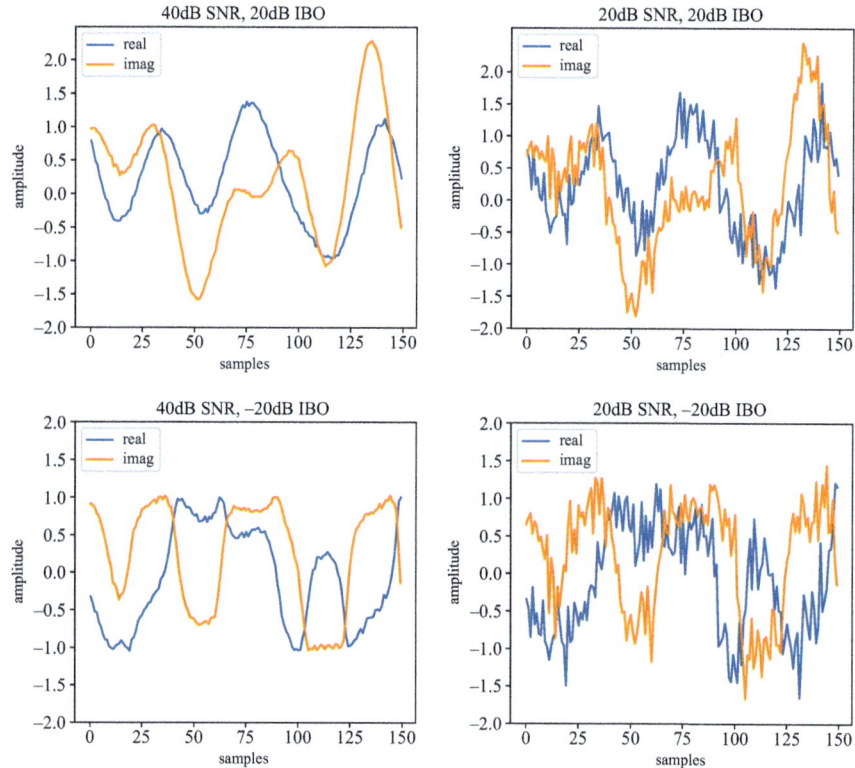

*Figure 9.9   Real and imaginary part of signals containing the same information
content, but with different distortion levels, simulated at the input
of a SatCom receiver*

Hence, this SNR estimation technique can work on any received signal and
requires only a sufficient amount of 'clean' signals that are transmitted by the SatCom
system to train the CAE.

9.5.3   *Metrics*

In this section, three different metrics to estimate the SNR via a suitable CAE architecture are presented. The first metric is the Local Outlier Factor (LOF). The LOF is
applied to the encoding of the signals in the latent space, where incoming signals are
represented as point vectors. The LOF provides a comparison between the density of
each point in the latent space and the local density of the k-nearest points [20] as:

$$LOF_k(X) = \frac{\sum_{Y \in N_k(X)} \frac{rd_k(Y)}{rd_k(X)}}{|N_k(X)|} \tag{9.2}$$

where X represents a point in the latent space, and rd is called reachability distance and is defined as:

$$rd(X) = \left(\frac{\sum_{Y \in N_k(X)} \max\{k_d(X), d(X, Y)\}}{|N_k(X)|} \right)^{-1} \tag{9.3}$$

In (9.3), k_d is the distance of X from its kth nearest point Y, $d(X, Y)$ is the absolute distance between X and Y, while $|N_k(X)|$ represents the number of nearest neighbours. The term $rd(Y)$ in (9.2) can be defined in a similar way.

This metric is designed for anomaly detection via AEs [16]. An anomaly, or outlier, is any data item that presents inconsistency with the rest of the data, which may indicate a different origin. The LOF works under the assumption that the anomalies occur in areas of the latent space with lower density of points, compared to normal instances. Hence, points with high value of LOF corresponds to points that are outliers in the latent space, and thus an anomaly in the input signal. However, our problem setting is different than typical anomaly detection tasks: the noise is spread over the entire duration of the signal, such that it does not constitute a sporadic anomaly, but a continuous effect. Therefore, a single anomaly score has to be assigned to all the CAE input frames that originate from the same signal at specific SNR and IBO. For this purpose, the SNR is estimated using the standard deviation of the LOF computed on all the available frames, rather than using directly the LOF score.

The sparsity is an alternative anomaly metric on the latent space. In fact, rather than comparing local densities, it employs only the absolute distances between each point and its k-nearest neighbours in the latent space. Then, the distances are averaged over all the encoded points, in order to obtain a single value metric. The sparsity can be defined by the following expression [12]:

$$\text{Sparsity} = \frac{\sum_{X=1}^{N} \frac{\sum_Y |X-Y|}{|\mathbb{H}_Y|} \Big|_{Y \in \mathbb{H}_Y}}{N}, \tag{9.4}$$

where X and Y are two points in the latent space, $\mathbb{H}_Y$ is the set of k-nearest neighbours of Y and N is the total number of encoded points. As a result, high sparsity values corresponds to greater average distances between points, which indicates high noise level in the input signal.

Finally, the SNR can be estimated by computing the mean absolute error (MAE) between the CAE's input and output layers for each signal frame. The MAE can be simply defined as:

$$MAE = \frac{\sum_{i=1}^{n} |g(x_i) - x_i|}{n} \tag{9.5}$$

where x_i is the ith component of the input vector x and $g(x_i)$ is the corresponding CAE output. Unlike the previous two metrics, the MAE considers only the CAE reconstruction rather than the corresponding latent space.

Each one the presented metric can be employed for the SNR estimation, as it is illustrated in Figure 9.8. A comparison of the performance of these metrics is presented in the following.

9.5.4 Application example

The proposed SNR estimation technique is applied on a SatCom system model following the scheme in Figure 9.3, featuring a 16-APSK modulation scheme. For the HPA model described in Section 9.4.1, the AM-AM and AM-PM characteristics are sampled and tabulated from a solid state HPA draining 3W and operating in the Ka-band. Using the MATLAB model, signals at specific values of SNR and IBO are generated starting from a random bit sequence. Next, the modulator encodes the random bits in a sequence of analog samples. An oversampling rate of 16 is chosen for the modulator, such that each modulation symbol corresponds to 16 samples in the analog signal. Then, the signals arriving at the receiver are obtained by setting a specific value for the IBO, ranging from -20 dB to 15 dB. Conversely, the signals SNR is fixed at 40 dB, which corresponds to a negligible amount of noise. The simulated analog signals constitute the training set for the following CAE.

Next, the signals are pre-processed as indicated in Section 9.5.2. Thus, the signals are normalised and split into vectors of 2^{16} input frames of 16 complex-valued samples. The imaginary part of the samples is discarded, while the real part is kept for training. In fact, the noise degrades equally the real and imaginary part, such that their information about the SNR is redundant. The size of the input and the output layer of the CAE can be halved by discarding the imaginary part, thus reducing the complexity and the training time of the neural network, without compromising its accuracy.

In the training phase, the CAE learns to reproduce the signals independently from the distortion introduced by the HPA. Therefore, after training, the reconstruction accuracy of the CAE will degrade if signals with lower SNR are provided at the input. In order to obtain the best accuracy on the training set, the CAE is carefully tuned, until the architecture in Figure 9.10 is obtained. It is formed by four convolutional

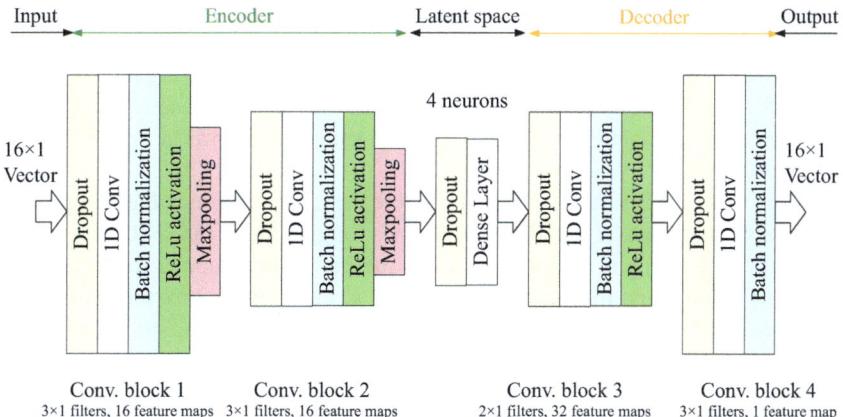

Figure 9.10   Convolutional autoencoder architecture for the SNR quantification

layers and one intermediate dense layer that represents the latent space. Note that the size of the input and the output layers matches the duration of a single frame, which corresponds to one modulation symbol. In each convolutional block, dropout operations are added to improve regularisation [21], while batch normalization is introduced to improve the training speed and stability [22].

The CAE model parameters are updated during the training phase by executing the Gradient Descent algorithm [23] to minimise the loss function. For this purpose, the contractive loss is chosen to increase the autoencoder robustness to small variations in the training set. In fact, this function improves over the mean square error (MSE) loss by adding a term for the contraction of the latent space representation [14]. Furthermore, the CAE is trained for 120 epochs, setting a learning rate of $8 \cdot 10^{-4}$ and batch size of 64 for the Gradient Descent algorithm.

During the testing phase, new signals are simulated with the same procedure discussed above. However, the simulation is performed with SNR ranging from 0 dB to 40 dB, in steps of 5 dB. At the same time, the range for the IBO is kept from -20 dB to 15 dB. Moreover, different signals are obtained for 10 different bit sequences, which are generated by changing the seed of a random bit generator. This procedure allows one to verify the robustness of the CAE to variations in the information content of the signals. Thus, the complete test set for the CAE consists of 720 analog sample sequences, one for each the possible combinations of 9 values of SNR, 8 values of IBO and 10 different random generator seeds.

Next, the test sequences are pre-processed and fed to the trained CAE. The CAE returns the latent space representation and its reconstruction at the output, for each frame of all the sequences in the test set. Finally, the three SNR metrics discussed in Section 9.5.3 are computed from the latent and the output space.

9.5.5   *Metrics tuning and consistency analysis*

When computing the LOF standard deviation and the sparsity metric, the number of k-nearest neighbours has to be tuned for a better estimation of the SNR. Therefore, the computation of both LOF and sparsity is repeated on the test set for different numbers of k. In this section, an example of metrics tuning is presented, fixing the IBO value at 20 dB for simplicity. After training the CAE, the metrics to be tuned are computed, and then plotted in function of the chosen SNR values, as shown in Figure 9.11 for the LOF and in Figure 9.12 for the sparsity.

Let us focus on the LOF first. From Figure 9.11, it can be seen that the relation between SNR and the LOF standard deviation is roughly linear for SNR values lower than 30 dB. It is apparent that $k = 1,600$ is the best choice among the considered numbers of neighbours. Indeed, for this value of k, the metric approximates a monotonic function of the SNR. This property allows one to define a one-to-one mapping from the LOF to the SNR, through linear interpolation.

Regarding the sparsity, Figure 9.12 shows that the relation between sparsity and SNR is strongly linear. Therefore, the sparsity can effectively be used to estimate SNR

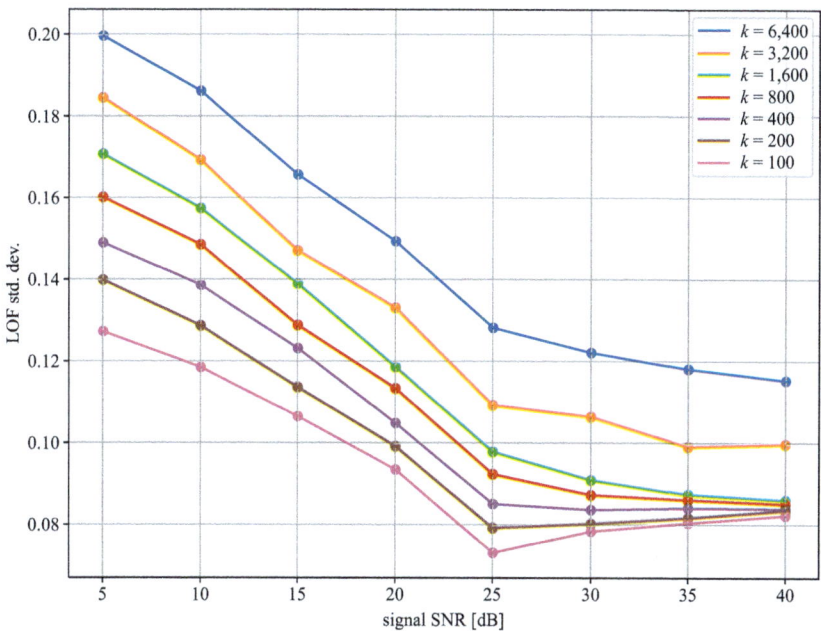

Figure 9.11   LOF standard deviation metric as a function of the test signal SNR for varying k-nearest points

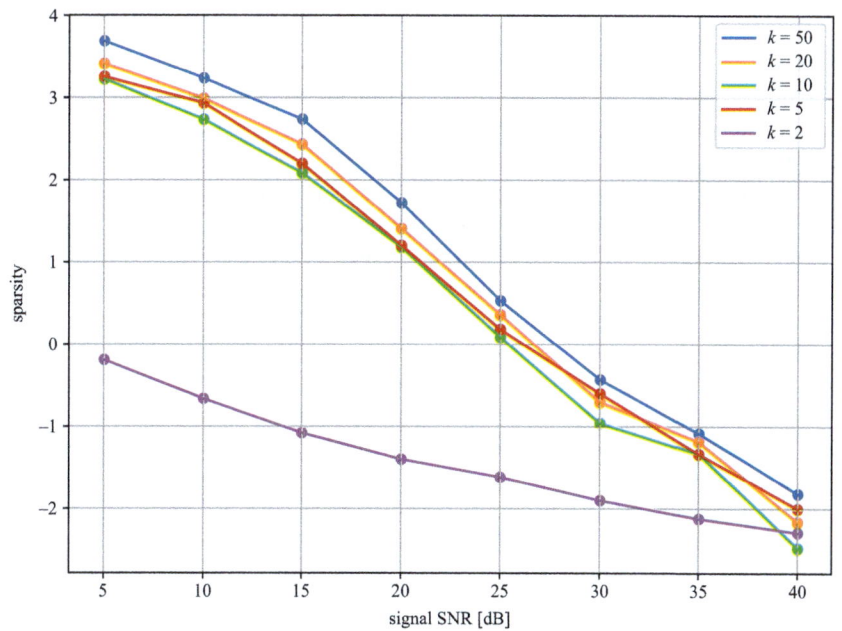

Figure 9.12   Sparsity metric as a function of the test signal SNR for varying k-nearest points

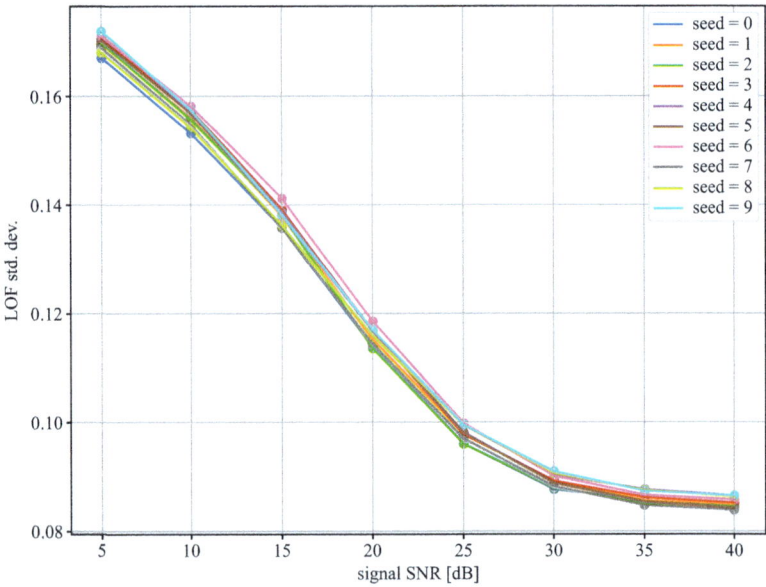

Figure 9.13   LOF standard deviation metric as a function of the test signal SNR for varying random seeds, for k = 1,600

values. However, a sufficient number of neighbours should be chosen: for $k < 5$, the differentiation among the sparsity values is lower and the precision of the mapping is reduced. Then, $k = 5$ is selected for the sparsity, since it higher values require higher computational time (see (9.4)).

Furthermore, it is crucial to assess the generalisation capability of the CAE network: the SNR estimation has to be consistent and independent from the information content of the signal. For this purpose, the test phase is repeated with signals generated from 10 different bit sequences, by varying the seed of a random bit generator. The relation between metrics and SNR for different random seeds are recorded in Figures 9.13–9.15, for the LOF standard deviation, the sparsity and the MAE, respectively. Note that, the optimal value of k for the LOF and sparsity is chosen for this analysis. In these figures, it can be observed that the LOF metric presents the highest variability across seeds. On the other hand, the sparsity and the MAE possess very low variability. It can be noticed in Figure 9.15 that the relation between MAE and SNR is also highly linear, for any value of SNR. Therefore, the sparsity and the MAE are more suitable metrics for the SNR. However, the MAE offers an important advantage: it does not require any parameter tuning.

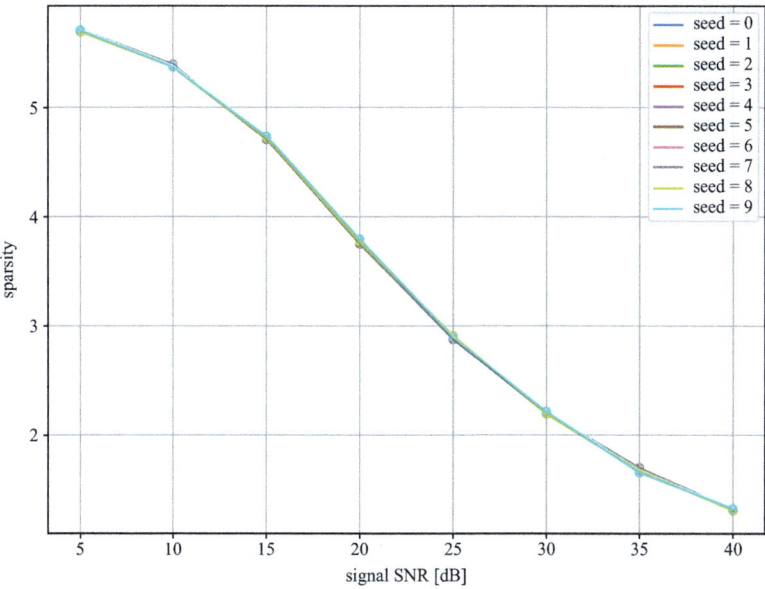

Figure 9.14   Sparsity metric as a function of the test signal SNR for varying random seeds, for k = 5

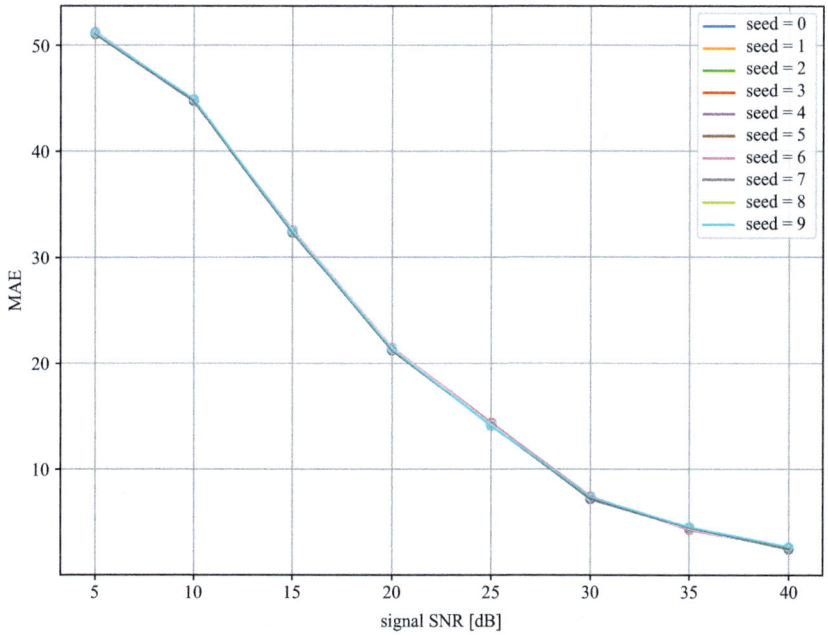

Figure 9.15   MAE metric as a function of the test signal SNR for varying random seeds

9.5.6   Results and discussion

From the previous section, the MAE emerges as the best metric for the SNR quantification when the signal distortion is negligible. Subsequently, the metrics are re-tested on signals with different levels of IBO, to verify the robustness of the CAE to the distortion introduced by the HPA. In this case, the test signals from one single random seed are employed. Next, the metrics for each IBO value are computed in function of the SNR. For example, the resulting MAE values are reported in Figure 9.16. Interestingly, the variation of the MAE with respect to the SNR remains linear for any value of IBO. Therefore, it is still possible to estimate intermediate values of SNR by interpolating this curve at the corresponding MAE provided by the CAE. However, for specific value of SNR, the different levels of IBO introduces a small variation of the MAE. Similar plots can be produced for the LOF and the sparsity. For these two metrics, the k parameter need to be re-tuned on test signals with different values of IBO, as explained in the previous section. The resulting optimal value of k is 800 for the LOF standard deviation and 200 for the sparsity.

Then, the variability of all the metrics with respect to the IBO is shown in Table 9.1. For clarity, Table 9.1 shows the variability caused by the IBO across four different ranges of SNR values, called SNR regions. It can be seen that the MAE presents the lower variability from 10 to 40 dB all regions. On the other hand, the sparsity is more accurate in the 0–10 dB region. Note that in the region between 30 dB and 40 dB, the behaviour of the LOF standard deviation is not monotonic with respect to the SNR. Therefore, it is not possible to estimate the SNR values that fall in this range by using the LOF. Looking at Table 9.1, the MAE appears to be the best metric even for different levels of HPA distortion on the signal.

Finally, the results of the SNR estimation using MAE are summarised in Table 9.2. In this table, the first column represents the ratio of the MAE deviation due to seeds variability, for the considered SNR regions. The second column shows the error caused by such MAE deviation on the SNR value. Lastly, the third column shows the overall error in SNR estimation due to the seed variability and the variation in distortion levels. Interestingly, the accuracy of the technique is higher in between 10

Table 9.1   Accuracy of SNR estimation for the considered metrics

Metric	Robustness to signal randomness	IBO variability per SNR region			
LOF (st.dev.)	Average	2.6 dB	2.3 dB	2.3 dB	N.A.
Sparsity	Good	2.4 dB	1.8 dB	1.8 dB	3.1 dB
MAE	Good	2.5 dB	1.1 dB	1.6 dB	2.7 dB
		SNR region			
		0–10 dB	10–20 dB	20–30 dB	30–40 dB

Table 9.2   Impact of varying seeds on the SNR estimation

SNR region	Deviation in MAE across seeds	SNR estimation error across seeds	SNR estimation error
0–10 dB	0.34	1.0 dB	±2.5 dB
10–20 dB	0.17	0.3 dB	±1.1 dB
20–30 dB	0.22	0.6 dB	±1.6 dB
30–40 dB	0.30	1.5 dB	±2.7 dB

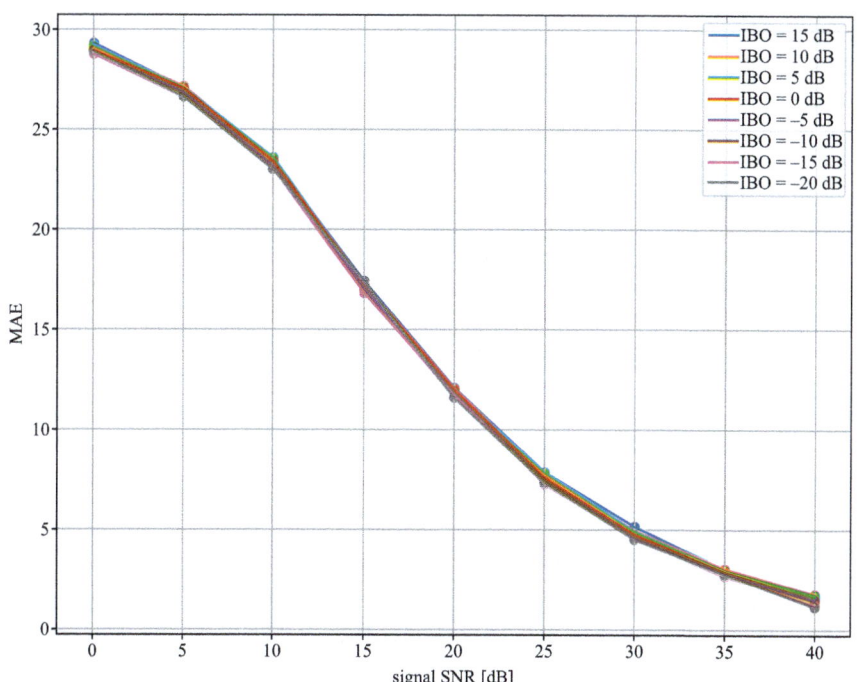

Figure 9.16   MAE as a function of the test signal's SNR for varying IBOs

and 30 dB of SNR. In fact, the MAE curve in Figure 9.16 appears more flat in the 0–10 dB and the 30–40 dB SNR regions. A possible explanation for this result is that the CAE is less able to distinguish among high levels or low levels of SNR. Thus, the reconstruction error may be similar among signals with very low or very high noise. These results are consistent for test signals generated with different random seeds, indicating that the neural network model is able to provide an accurate SNR estimation, independently from the information content at the input.

9.6   Input back-off estimation

The propagation loss in the communication channel is one of the main causes of performance degradation in a SatCom system: the signal power strongly decays while propagating in the atmosphere and in near-earth space. Thus, the signal may reach the receiver at a low signal-to-noise ratio, causing high error rates when decoding. Therefore, in order to maintain high data-rates, the signal transmitted from a user terminal has to reach the satellite and the ground receiver at a sufficient power level. For this reason, channel fading and noise often need to be compensated by rising the transmit power. However, the non-linearity of the HPA increases with transmit power and can introduce excessive distortion of the signal. Thus, informing the receiver about the operating point of the amplifier is essential to control the transmit power. Several techniques have been proposed to estimate the HPA operating point using channel state information (CSI) [13]. More recently, DL techniques have been successfully used to both improve CSI acquisition [24,25] and implement efficient power control schemes [26]. Nonetheless, both classical and recent DL approaches require pilot symbols to be transmitted, causing overhead. In this chapter, a different DL strategy is proposed to estimate the input back-off (IBO) of the amplifier: a neural network classifier predicts the IBO given any distorted symbols sequence at the receiver. In fact, the predicted IBO defines the operating point of the HPA on its input–output characteristic, as illustrated in Figure 9.4. The advantages of this method are that no pilot symbols are required and that the classifier can be trained for different modulation schemes and different HPA models. Moreover, the classifier is designed to operate at a highly variable level of signal-to-noise ratio at the receiver. Specifically, IBO estimation is applied to the return link system described in Section 9.4. Here, the received signal is corrupted by HPA distortion at the user terminal, while the noise in the link is modelled as AWGN.

9.6.1   Deep learning model for IBO estimation

A DL model can be trained to estimate the IBO of an HPA from a limited portion of the received symbol sequence. Indeed, the IBO estimation can be addressed as a supervised learning problem, since the SatCom model described in Section 9.4 allows one to specify the desired value of IBO, which can be used as a training label. For this purpose, a deep convolutional neural network (DCNN) is proposed as a classifier [27]. The task of the DCNN classifier is to identify the amplifier's IBO among a set of discrete values. An alternative approach would be using a DCNN regressor to select the IBO prediction from a continuous range of values. This would yield a higher resolution for the measure of distortion. Nonetheless, classification requires a simpler neural architecture and a significantly lower amount of training data, compared to regression, thus representing a good trade-off between computational complexity and accuracy. This is particularly useful when dealing with complex data, where both the real and imaginary parts are informative for the learner, since they are both affected by the distortion of the HPA.

The proposed DCNN classifier structure is represented in Figure 9.17. The input layer of the network consists of 512 values. In fact, a sequence of 256 consecutive

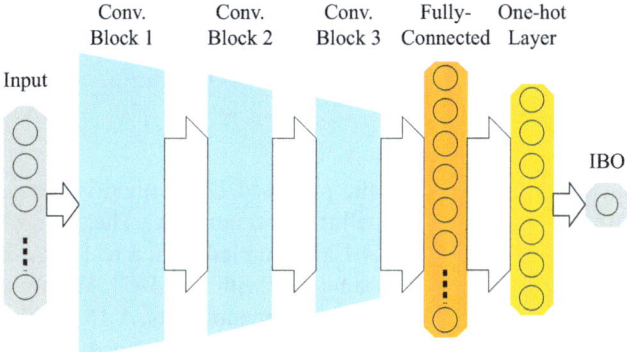

Figure 9.17 DCNN architecture used for IBO classification

symbols are fed to the network by alternating the real and imaginary parts of each symbol. Three convolutional blocks are sequentially applied to the input in order to detect time-shift invariant features of the received signal. In each block, a 1D convolution is followed by batch normalisation and a max pooling layer. These blocks performs analogous functions in the CAE for the SNR estimation (Section 9.5.4). The convolutional and max pooling operations enable an automatic extraction of the most relevant features from raw symbols, without pre-processing. In addition, batch normalisation improves the training speed and stability of the network. Moreover, a dropout operation is added at the end of each block to prevent overfitting via regularisation. Next, a single fully connected layer is inserted. Finally, the network is terminated by a fully connected layer with one-hot encoding for class selection. Hence, the final layer is formed by a neuron for each possible value of IBO. The predicted class is then selected by a softmax operation on the last layer. Once trained, the DCNN is able to associate an IBO value to any sequence of 256 consecutive symbols collected from the receiver. The network is trained by adopting the Adam Optimisation algorithm [28] to minimise a suitable loss function. The categorical cross-entropy [29] is chosen, which is the standard loss for multi-class classifiers. The training is performed for 50 epochs on batches of 1024 instances, using a learning rate of $5 \cdot 10^{-4}$ for the Adam Optimisation.

9.6.2 Performance metric

An estimation of the classifier accuracy is provided by the F1 score [30]. This metric is defined as the harmonic mean between precision and recall for each class of IBO:

$$\text{precision} = \frac{\text{true positives}}{\text{true positives} + \text{false positives}} \tag{9.6a}$$

$$\text{recall} = \frac{\text{true positives}}{\text{true positives} + \text{false negatives}} \tag{9.6b}$$

where '*true positives*' are the instances in the class that are correctly classified, while '*false*' are classification errors, which can be either '*positives*' (if predicted in the class) or '*negatives*' (if predicted outside of the class).

9.6.3   *Data generation*

In this section, the performance of the proposed DL method is demonstrated on symbols that are distorted by a real satellite power amplifier. The same AM–AM and AM–PM characteristics of Section 9.4.1 are sampled from a real HPA drawing 3 W at 29 GHz . Received signals are then simulated with the MATLAB model described in Section 9.4, starting from a sequence of 2^{22} random bits. A 16 Amplitude-Phase-Shift-Keying (APSK) modulation scheme and an oversampling rate of 16 are applied in the return link system. The IBO and SNR values used to generate distorted symbols are selected from the following discrete sets:

$$\text{IBO} = [-15, -10, -5, 0, 5, 10, 15] \text{ dB}, \tag{9.7a}$$

$$\text{SNR} = [5, 10, 15, 20, 25, 30] \text{ dB}. \tag{9.7b}$$

Hence, a total of seven classes is chosen for the IBO estimation, with constant step size of 5 dB. This choice provides a good compromise between resolution and computation cost. Indeed, resolution can be increased by generating a larger training dataset, which requires extra computational time or resources.

For each possible combination of IBO and SNR, the binary sequence is transformed into a distorted and noisy symbol sequence at the receiver. Next, all the obtained sequences are split into frames of 256 complex symbols. Each frame constitutes one input vector for the DCNN, i.e. one training instance. Consequently, a dataset of about 7 million instances is collected and fed into the neural network for training. Additionally, ten validation sets are produced by following the same procedure as described above, but starting from different sequences of 2^{16} random bits. The validation sets are used to test the capability of the network to classify previously unseen symbol sequences.

Figure 9.18 illustrates four examples of training instances, plotted in the complex plane. The instances are generated from the same symbol sequence with different values of IBO and SNR. In particular, Figure 9.18(a) and (b) illustrates the effect of a change in IBO value, while the SNR is kept constant: as the IBO decreases, the instance values get farther from the original 16-APSK modulation symbols (red circles). In fact, the distortion causes a dispersion and a slight rotation of the clusters of symbols in the complex plane. On the other hand, increasing the noise in the channel (i.e. decreasing the SNR) produces a spread of the received symbols in all directions, until the clusters become indistinct, as indicated in Figures 9.18(a), (c), and (d).

Compared to the SNR estimation technique proposed in Section 9.5, an additional challenge must be solved: rather than estimating the IBO from the signal pulses at the input of the receiver, the distorted sequence of modulated symbols is employed. In fact, in a real implementation, estimating the IBO from analog signal pulses would

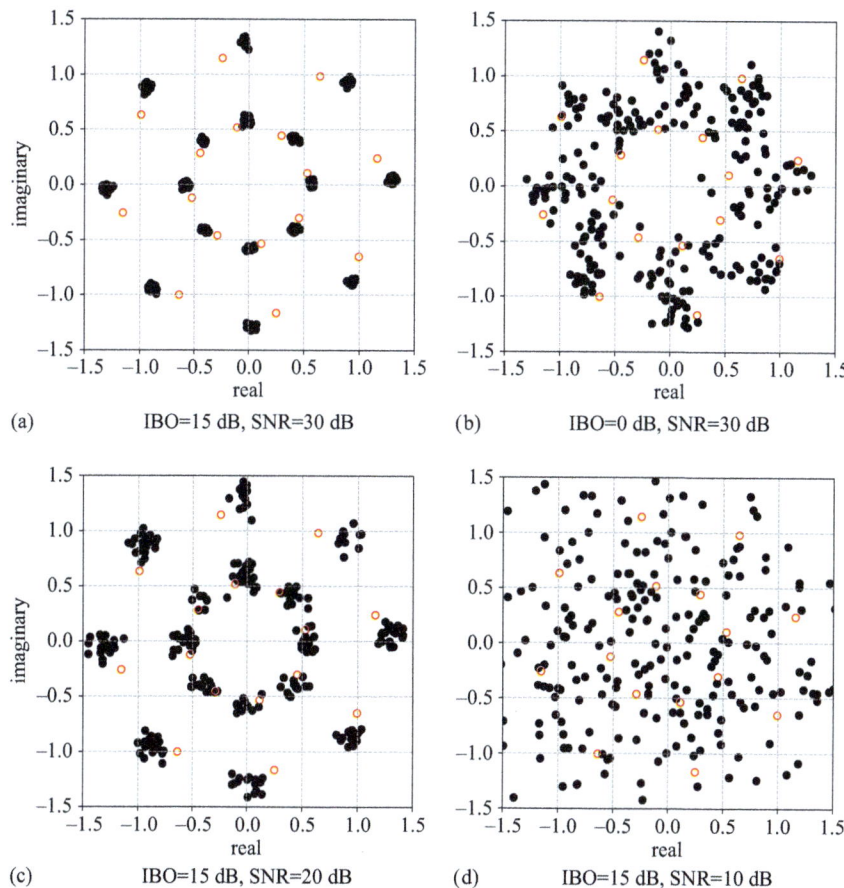

Figure 9.18   Example of values of four training instances in the complex plane for different IBO and SNR. All instances are generated from the same bit sequence. Red circles indicate the original symbols of the 16-APSK modulation scheme, corrected for the delays of the SRRC filters.

require measurements right after the receiving antenna and the low-noise amplifier. On the contrary, the reconstructed symbols sequence is more easily accessible, since it can be recorded at the input of the receiver demodulator. However, when the symbols are used, the user's message is encoded in a sequence of complex values which is much shorter than the corresponding signal pulses. Indeed, due to the oversampling rate, each symbol corresponds to 16 complex values in the sequence of signal pulses. Therefore, assuming that the DCNN requires a minimum number of input values for an accurate estimation, the length of training frames has to be increased by a factor equal to the oversampling rate. Moreover, unlike the additive noise, the HPA distortion unequally affects real and imaginary parts of the symbols: the imaginary part of the training sequence cannot be discarded, and the input size of the DCNN is doubled. For example, a minimum input size of 16 values can be assumed for the estimation from signal pulses. Note that this is the same size used for the SNR estimation network (Section 9.5.4). Then, 512 values ($16 \times 16 \times 2$) is an adequate size for the IBO estimation from modulated symbols.

The first consequence of the increased size of the input layer is a greater complexity of the DCNN. Second, since each signal frame is longer, longer signals need to be simulated in order to obtain the same number of training instances. Thus, estimating the IBO from received symbols, rather than the less accessible signal pulses, represents a harder task for the DL model.

9.6.4 Results and discussion

After training, the classifier reaches an average F1 score of 86% across all the validations sets. This demonstrates that the DCNN successfully predicts IBO values and that is able to generalise to previously unseen symbol sequences. As an example, the confusion matrix for a single validation set is presented in Figure 9.19. Note that one validation set contains 10,710 sample sequences, 1,785 for each value of SNR. The

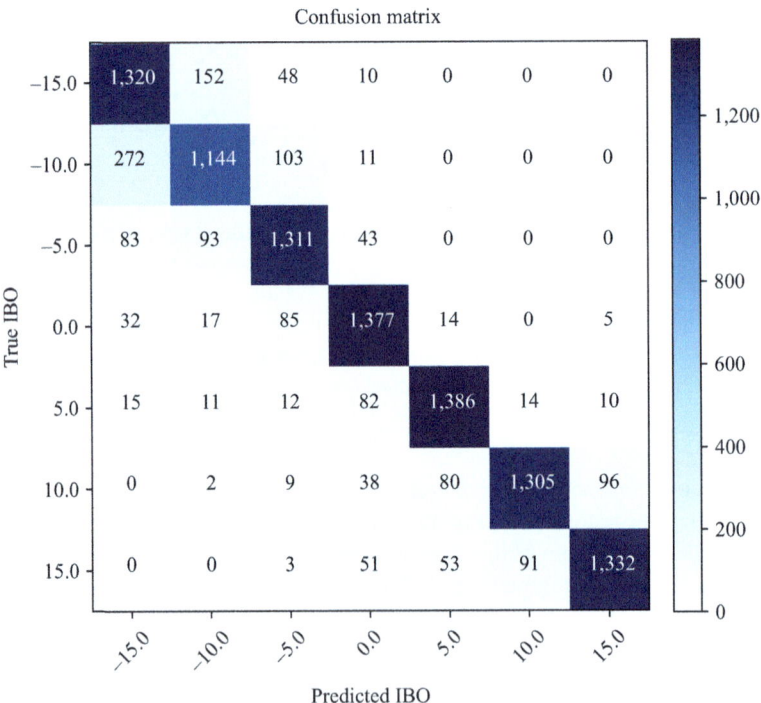

Figure 9.19 Confusion matrix of the DCNN classifier for a single validation set

Table 9.3 *F1 score of the classifier for each value of IBO, averaged among 10 validation sets*

IBO [dB]	−15	−10	−5	0	5	10	15
F1 score	0.82	0.79	0.86	0.89	0.91	0.89	0.90

confusion matrix shows that the large majority of symbols sequences are correctly classified, whereas most of the classification errors occur between adjacent values of IBO. This suggests that the resolution of the DCNN can be further increased by choosing more classes of IBO and providing additional training data. Additionally, the F1 score, averaged over all SNR values, is computed separately for each class of IBO and reported in Table 9.3. The obtained results indicate that the classifier accuracy is not uniform, but it decreases for negative IBO values, corresponding to higher values of distortion.

The impact of noise on the classification is analysed in Figure 9.20, where the average number of classification errors is recorded for each value of SNR in the validation sets. What stands out is that the number of errors falls sharply for SNR ≥ 15 dB. This denotes that large noise powers are the main cause of misclassification: for sufficiently high SNR, the DCNN is able to recognize any amount of distortion. In fact,

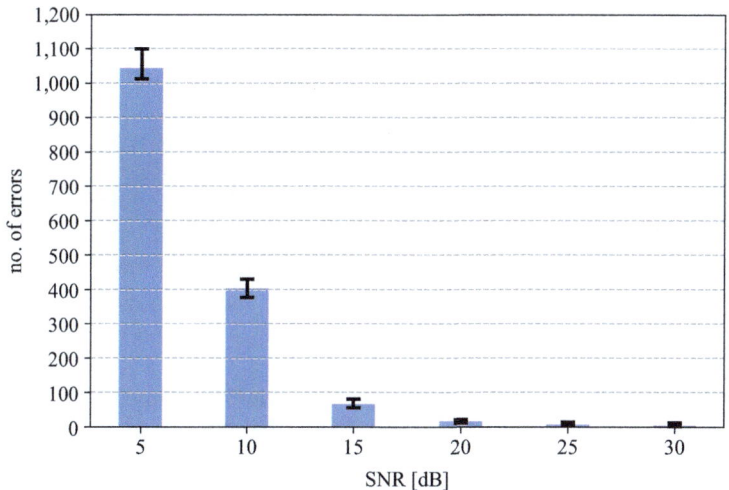

Figure 9.20 *Number of classification errors for different values of SNR, averaged across 10 validation sets. Error bars represent the variation between minimum and maximum number of errors among the sets and over all SNR values.*

discerning the effect of noise and distortion on training instances is intuitively easier at low noise powers (see Figure 9.18). Furthermore, high distortion levels, corresponding to negative values of IBO, become soon indistinct when the noise is increased, thus explaining the results in Table 9.3. Finally, there is only a small variation between the minimum and the maximum number of errors due to the SNR among the different validation sets: the results in Figure 9.20 prove that the DCNN performance is robust to the transmitted symbol sequence.

9.7 Conclusion

In this contribution, deep learning (DL) is applied in a satellite communication (Sat-Com) return link. The aim is to estimate the characteristics of the channel and the transmitter directly from received signals. In particular, two techniques are presented: one for the signal-to-noise ratio (SNR) estimation, and the other to evaluate the operating point of a transmitting high-power amplifier, measured via the input back-off (IBO). Obtaining a measure of these quantities at the receiver is essential to operate a transmit power control. Common approaches to measure SNR and IBO employ the transmission of pilot symbols, which causes overhead and thus reduces the actual data rate of SatCom links. Conversely, the presented DL techniques preserve the link capacity by performing the estimation on any incoming signal at the receiver without the need for pilot symbols.

In order to train DL architectures, a dataset of signals is collected by using a simplified SatCom model. Such a model allows one to simulate received signals at desired level of SNR and IBO, with any information content. Then, a convolutional autoencoder is proposed to extract the SNR from analog signal samples. The autoencoder is able to encode signal frames in a lower dimensional space and then reconstruct the input signal. After training on low noise signals, the SNR can be estimated for any signal frame by computing a suitable metric on the encoded representation or on the signal reconstruction.

Next, a deep convolutional classifier is employed to recognize the IBO of a received signal frame among a finite set of values. Instead of using analog signal samples, the classifier is designed to operate on a sequence of received modulation symbols. This simplifies the future acquisition of real signals, by allowing the collection the symbols right at the input of the receiver demodulator, rather than measuring the incoming analog pulses.

The two techniques here described are able to operate independently from each other on signals that have variable levels of both SNR and IBO. Both the SNR and IBO estimation reach high levels of accuracy on any received signal. Specifically, the SNR estimation reaches the highest accuracy in the range between 10 and 30 dB of SNR, for any considered level of amplifier distortion. Instead, the IBO technique is more accurate from -5 to 15 dB of IBO, while it is especially robust to the signal noise when the SNR is higher than 15 dB.

Moreover, the presented DL models can be trained on different modulation schemes and amplifier characteristics. Results suggest that their usage can be extended to more complex models of satellite communication systems. In addition, a dataset of real-world symbol sequences should be collected to re-train or optimise the networks for practical usage.

References

[1]   Kodheli O, Lagunas E, Maturo N, *et al*. Satellite communications in the new space era: a survey and future challenges. *IEEE Communications Surveys Tutorials*. 2021;23(1):70–109.

[2]   Zeydan E, Turk Y. On the impact of satellite communications over mobile networks: an experimental analysis. *IEEE Transactions on Vehicular Technology*. 2019;68(11):11146–11157.

[3]   Del Portillo I, Cameron BG, Crawley E. A technical comparison of three low earth orbit satellite constellation systems to provide global broadband. *Acta Astronautica*. 2019;159:123–135.

[4]   Gujanatti RB, Vijapur N, Jadhav SS, *et al*. Machine learning approaches used for weather attributes forecasting. In: *2021 2nd International Conference for Emerging Technology (INCET)*; 2021. pp. 1–5.

[5]   Thamaraikannan N, Manju S. Review on image classification techniques in machine learning for satellite imagery. In: *2021 International Conference on Artificial Intelligence and Smart Systems (ICAIS)*; 2021. pp. 144–149.

[6]   Xu H, Angrisano A, Gaglione S, *et al*. Machine learning based LOS/NLOS classifier and robust estimator for GNSS shadow matching. *Satellite Navigation*. 2020;1:1–15.

[7]   Liu Z, Zhang L, Ding Z. Overcoming the channel estimation barrier in massive MIMO communication via deep learning. *IEEE Wireless Communications*. 2020;27(5):104–111.

[8]   Hwang K, Chen M, Gharavi H, *et al*. Artificial intelligence for cognitive wireless communications. *IEEE Wireless Communications*. 2019;26(3): 10–11.

[9]   Glaessgen E, Stargel D. The digital twin paradigm for future NASA and U.S. air force vehicles. In: *53rd AIAA/ASME/ASCE/AHS/ASC Structures, Structural Dynamics and Materials Conference*; 2012.

[10]   Deleu T, Dervin M, Kasai K, *et al*. Iterative predistortion of the nonlinear satellite channel. *IEEE Transactions on Communications*. 2014;62(8):2916–2926.

[11]   Goodfellow I, Bengio Y, Courville A. *Deep Learning*. London: MIT Press; 2016. http://www.deeplearningbook.org.

[12]   Dhuyvetters B, Delaruelle D, Rogier H, *et al*. Machine learning-based characterization of SNR in digital satellite communication links. In: Dusseldorf GO, editor. *2021 15th European Conference on Antennas and Propagation (EuCAP)*; 2021. pp. 1–5.

[13]   Ghannouchi FM, Hammi O, Helaoui M. Characterization of wireless transmitter distortions. In: *Behavioral Modeling and Predistortion of Wideband Wireless Transmitters*, 3rd ed. New York, NY: John Wiley & Sons; 2015.

[14] Kiran BR, Thomas DM, Parakkal R. An overview of deep learning based methods for unsupervised and semi-supervised anomaly detection in videos. *Journal of Imaging.* 2018;4:2.

[15] Medico R, Spina D, Ginste DV, *et al.* Autoencoding density-based anomaly detection for signal integrity applications. In: *Proceedings of IEEE 27th Conference on Electrical Performance of Electronic Packaging and Systems (EPEPS)*, San Jose, CA; 2018. pp. 14–17.

[16] Medico R, Spina D, Ginste DV, *et al.* Machine-learning-based error detection and design optimization in signal integrity applications. *IEEE Transactions on Components, Packaging and Manufacturing Technology.* 2019;9(9): 1712–1720.

[17] Janakiramaiah N, Kalyani G, Narayana S, *et al. Smart Intelligent Computing and Applications.* Singapore: Springer; 2020.

[18] Bengio Y, Courville AC, Vincent P. Representation learning: a review and new perspectives. *IEEE Transactions on Pattern Analysis and Machine Intelligence.* 2013;35(8):1798–1828.

[19] O'Shea TJ, Corgan J, Clancy TC. Unsupervised representation learning of structured radio communication signals. In: *Proceedings of First International Workshop on Sensing. Aalborg, Denmark: Processing and Learning for Intelligent Machines (SPLINE)*; 2016. pp. 6–8.

[20] Breunig MM, Kriegel H, Ng RT, *et al.* LOF: identifying density-based local outliers. In: *Proceedings of the 2000 ACM SIGMOD International Conference on Management of Data*, Dallas, TX; 2000. pp. 16–18.

[21] Wan L, Zeiler M, Zhang S, *et al.* Regularization of neural networks using DropConnect. In: Dasgupta S, McAllester D, editors. *Proceedings of the 30th International Conference on Machine Learning*, vol. 28 of Proceedings of Machine Learning Research, Atlanta, GA: PMLR; 2013. pp. 1058–1066.

[22] Ioffe S, Szegedy C. Batch normalization: accelerating deep network training by reducing internal covariate shift. In: *CoRR.* 2015 August;abs/1502.03167. Available from: http://arxiv.org/abs/1502.03167.

[23] Ruder S. An Overview of Gradient Descent Optimization Algorithms. arXiv preprint:1609.04747; 2016.

[24] Ibnkahla M, Cao Y. A pilot-aided neural network for modeling and identification of nonlinear satellite mobile channels. In: *Canadian Conference on Electrical and Computer Engineering*, Niagara Falls, ON, Canada; 2008. pp. 1539–1542.

[25] Soltani M, Pourahmadi V, Mirzaei A, *et al.* Deep learning-based channel estimation. *IEEE Communications Letters.* 2019;23(4):652–655.

[26] Lee W, Kim M, Cho D. Deep power control: transmit power control scheme based on convolutional neural network. *IEEE Communications Letters.* 2018 June;22(7):1276–1279.

[27] Kiranyaz S, Ince T, Abdeljaber O, *et al.* 1-D convolutional neural networks for signal processing applications. In: Brighton UK, editor. *Proceedings of IEEE International Conference on Acoustics, Speech and Signal Processing (ICASSP)*; 2019. pp. 8360–8364.

[28] Kingma D, Ba J. Adam: A Method for Stochastic Optimization; 2014. ArXiv preprint.

[29] Golik P, Doetsch P, Ney H. Cross-entropy vs. squared error training: a theoretical and experimental comparison. In: *Proc. Interspeech*, vol. 13, Aug. 2013, pp. 1756–1760.

[30] Tharwat A. Classification assessment methods. *Applied Computing and Informatics*. 2021;17(1):168–192.

Chapter 10

Deep learning techniques for imaging and gesture recognition

Hongrui Zhang[1] and Lianlin Li[1]

10.1 Introduction

Nowadays, intelligent electromagnetic (EM) sensing, as a powerful all-weather all-day examination technique [1–6], is ever-increasingly demanded to probe people in daily lives in a way not to infringe on visual privacy, in particular, to recognize where people are, how their physiological states, what they are doing, what they want to express by their body signs, etc. We here mean by intelligence that sensing systems can adaptively organize the task-oriented sensing pipeline (data acquisition plus processing) without the human intervene. Although three kinds of EM sensing schemes of real-aperture [10–13], synthetic-aperture [2], and coding-aperture [7–9] have been proposed by now, they are hindered from many practical utilizations because of trading-off many critical factors effecting the cost-performance-index, especially when dealing with the so-called data crisis. To tackle this formidable challenge, recently, we have proposed the concept of hybrid-computing-based intelligent sensing by synergizing artificial materials (AMs; specifically, reprogrammable metasurfaces [14–20]) for flexible wave manipulation thereby analogy data compression on physical level with artificial intelligences (AMs; specifically, deep learning strategies [21–24]) for powerful data manipulation on digital level [25–29]. In this chapter, we discuss three recent progress: intelligent metasurface imager [31], variational-autoencoder (VAE)-based intelligent integrated metasurface sensor [32], and free-energy (FE)-based intelligent integrated metasurface sensor [33]. We mean by the integration that for a scene of interest and given hardware constraints, the settings of data acquisition and processing are simultaneously learned as a unique whole entity.

10.2 Design of reprogrammable metasurface

The intelligent metasurface sensor, as explicitly implied by its name, relies heavily on the utilization of the reprogrammable metasurface. The aforementioned three sensing systems are all based on a reflection-type one-bit reprogrammable metasurface

[1] State Key Laboratory of Advanced Optical Communication Systems and Networks, School of Electronics, Peking University, China

working at around 2.4 GHz. The reprogrammable metasurface is composed of a two-dimensional array of engineered structure elements (called meta-atoms), and each meta-atom with the size of 54×54 mm^2 is optimized to a sandwich structure integrated with a SMP1345-079LF PIN diode, as shown in Figure 10.1(a) and (b). Here, the one-bit meta-atom has binary EM status: the reflection phases of 0° and 180°, denoted as the digits "0" and "1," respectively. Figure 10.1(d) shows the numerical and experimental results of EM reflection responses of the meta-atom. Figure 10.1(c) reports the waveguide-based measurement setup, where a standard waveguide to coaxial adapter A-INFO 430WCAS is used. As for the simulations, the CST Microwave Transient Simulation Package 2017 was used. In our simulations, the PIN diode has been modeled as a series lumped-parameter circuit: it is represented by a 0.7-nH inductor in series with a 2Ω resistor when the PIN at ON, while a 1.8 pF capacitor in series with a 0.7 nH inductor when the PIN at OFF. From above results, one can see that when the state of the meta-atom is switched from "1" to "0" (or from "0" to "1"), the reflection phases of meta-atom are flipped approximately by 180° in the frequency range 2.41–2.48 GHz More details about the meta-atom are provided as follows: the meta-atom is composed of two substrate layers: the bottom substrate is FR4 and the top substrate is F4B with the relative permittivity of 2.55 and a loss

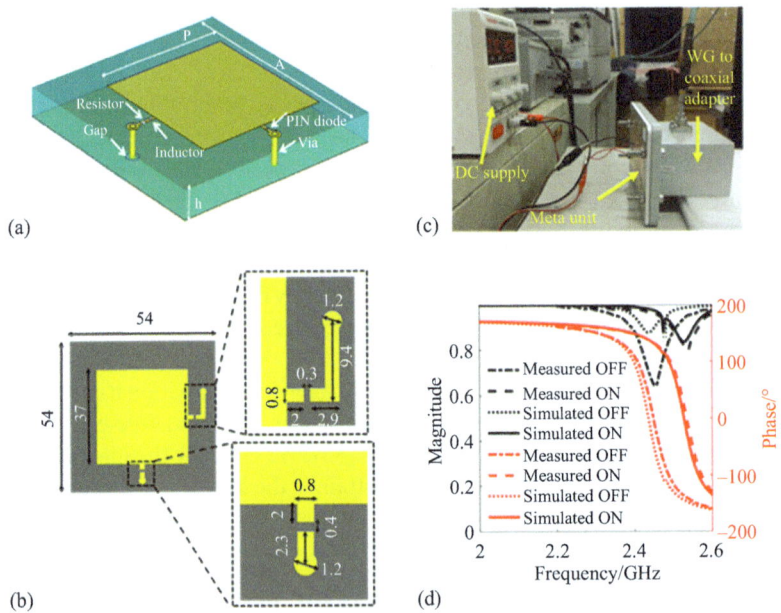

Figure 10.1   *Design and test of electronically controllable meta-atom [31]. (a) The sketched map of designed meta-atom. (b) The details of geometrical parameters of designed meta-atom, where the unit is mm. (c) The waveguide-based experimental setup of the designed meta-atom. (d) Experimental and simulated results of magnitude-frequency and phase-frequency responses of the designed meta-atom.*

tangent of 0.0019. The top square patch is embedded with a PIN diode, and a TDK chip inductor (MLK1005S33NJT000) is used to suppress the AC coupling to ground.

Here, we take the metasurface adopted in [31] to illustrate the design and operational principle of one-bit reprogrammable metasurface, as shown in Figure 10.2(a)

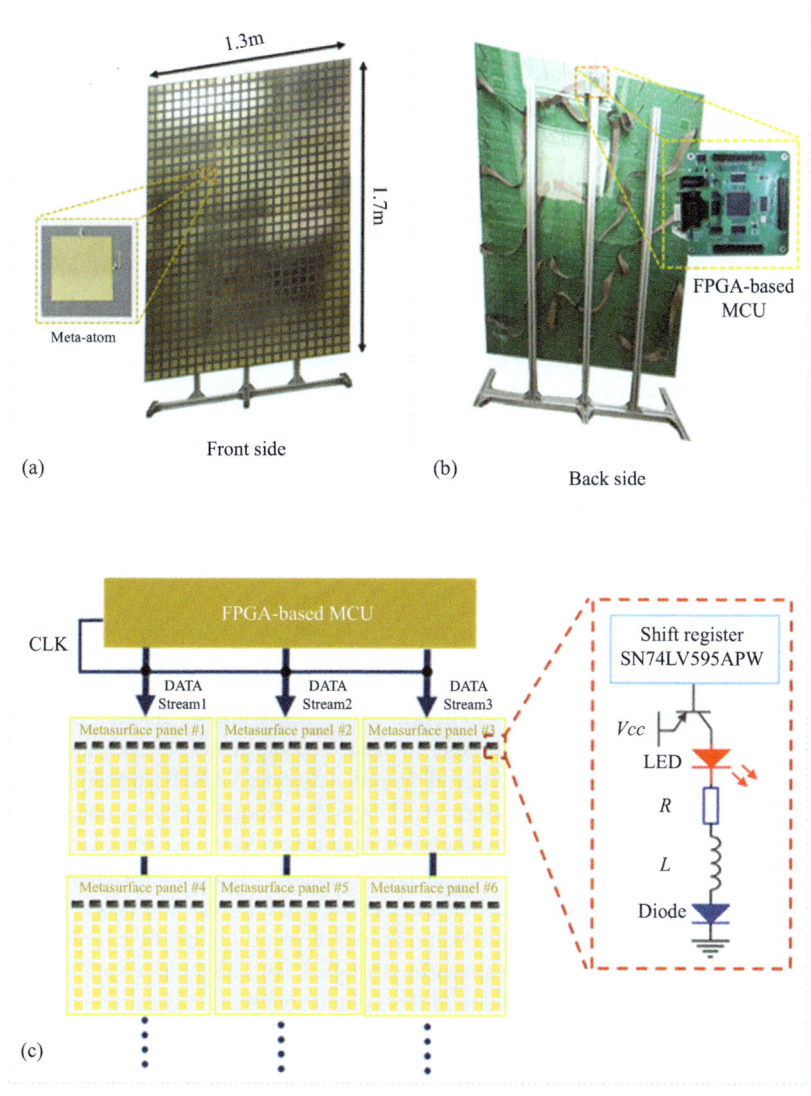

Figure 10.2 Designed large-aperture programmable metasurface and its controlling scheme [31]. (a) and (b) The pictures of the designed programmable metasurface with size of 1.3 × 1.7 m². (c) The control architecture of the FPGA-based MCU and zoomed version of logical circuit on the metasurface panel.

and b for its front and back views, respectively. Such reprogrammable metasurface is composed of independently controllable 32×24 meta-atoms, and has the size of 1.7×1.3 m^2 in total. For the sake of fabrication, the whole large-aperture metasurface was divided into 3×4 identical panels, and each metasurface panel consists of 8×8 meta-atoms. The whole metasurface is controlled with a FPGA-based Micro-Control-Unit (MCU), as shown in the inset of Figure 10.2(b), where the FPGA chip was used to distribute all commands to 768 PIN diodes. To achieve the real-time and flexible controls of 768 PIN diodes, the MCU with a size of 90×90 mm^2 was assembled on the upper rear of the metasurface. The MCU is responsible for dispatching all commands sent from a master computer subject to one common clock (CLK) signal. In this work, the adopted CLK is 50 MHz, and the switching time of PIN diode was about 10 μs each cycle.

Each metasurface panel is equipped with eight 8-bit shift registers (SN74L V595APW), and every 8 PIN diodes share the same shift register, as shown in Figure 10.2(c). With the use of shift registers, 8 PIN diodes are sequentially controlled. Then MCU will send commands over 24 independent channels, leading to almost real-time manipulations of all PIN diodes. In addition, 768 red-color LEDs are soldiered to indicate the status of the associated PIN diodes, in particular, to indicate clearly whether the PIN diode works well or not. It is remarked that the proposed control strategy can be readily extended for accommodating more PIN diodes by concatenating more metasurface panels, allowing adjustable rearrangement of metasurface panels to meet various needs.

10.3 Intelligent metasurface imager

Considering the unprecedent successes of deep learning techniques in data mining and knowledge discovery, Li *et al.* have proposed the concept of intelligent metasurface EM imager based on the hybrid computing scheme: analogy data compression with a one-bit reprogrammable metasurface on physical level, and digital data postprocessing with deep artificial neural networks (ANNs) on digital level [31]. A proof-of-principle demo system working at the frequency of around 2.4 GHz (exactly commodity Wi-Fi frequency band) has been developed. Li *et al.* have experimentally demonstrated that such imager is intelligent in sense that it is capable of adaptively accomplishing a series of successive high-quality sensing tasks including in situ imaging, body sign recognition and human respiration identification of multiple non-cooperative persons in real-world settings. In this chapter, we are restricted ourselves into the imaging setting due to the limited space.

10.3.1 System configuration

The schematic configuration and concept of an intelligent metasurface system is shown in Figure 10.3(a). As the hardware core of the whole system, a large-aperture one-bit reprogrammable metasurface was designed to control EM wavefields adaptively by manipulating its control coding sequences. In principle, it acts as an *analogy wave computer* for high-dimensional data compression on physical level. In addition, with the aid of deep ANNs on a *digital electronic computer*, the analogy wave computer is capable of processing data flow in smart and real-time way. With reference to Figure 10.3(a), the sensing system has two operational modes: active and passive modes. We here are limited to the active mode due to the limited space. In the active

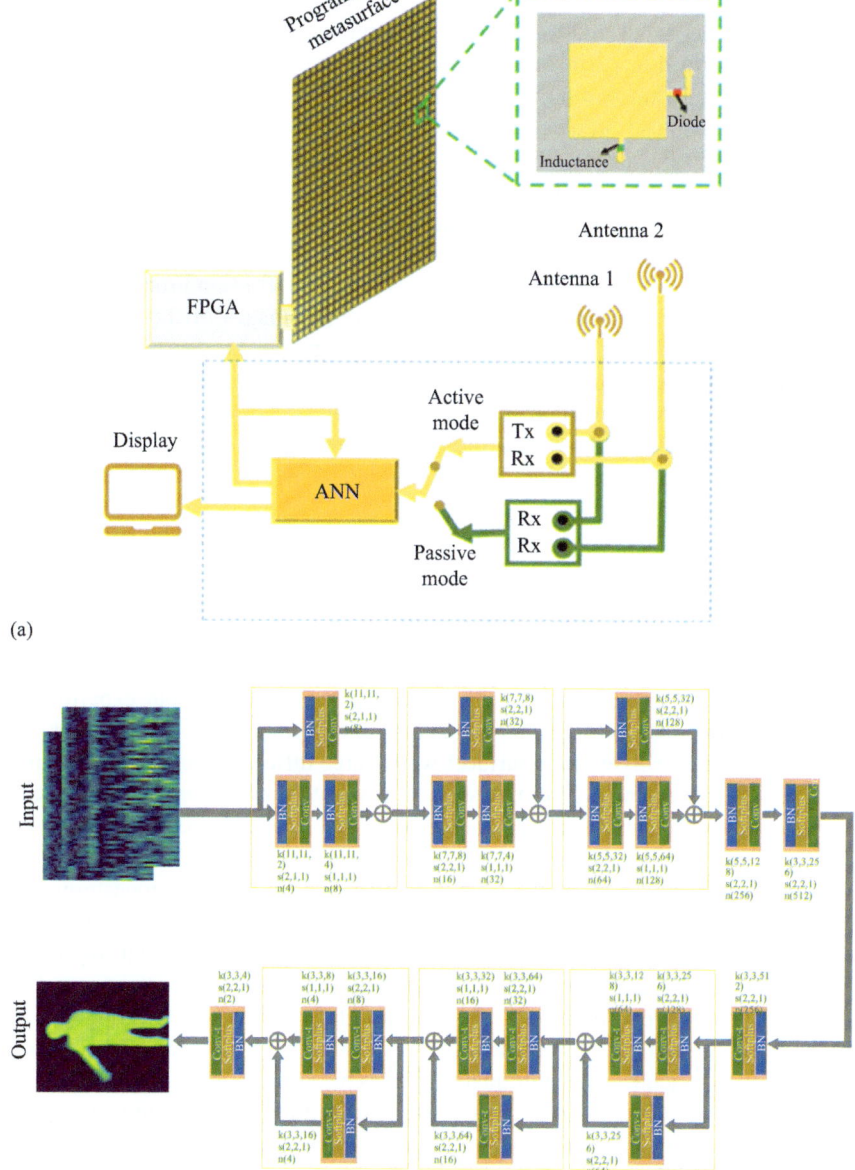

(a)

(b)

Figure 10.3   *(a) The configuration of intelligent metasurface system. (b) The architectures of proposed IM-CNN [31].*

mode, the intelligent metasurface sensing system is composed of a reprogrammable metasurface, ANNs [34], a transmitting antenna, a receiving antenna, and a vector network analyzer. For the imaging purpose, a deep convolutional network (called IM-CNN) has been developed, as shown in Figure 10.3(b). In order to train the metasurface sensing system in a supervised manner, a commercial 4-megapixel digital optical camera has been deployed and synchronized with the whole system, which was utilized to collect the labeled samples to train the deep ANNs.

The IM-CNN is an end-to-end nonlinear mapping from the complex-valued microwave data to desired images. In order to improve the efficiency of network optimization meanwhile avoid the gradient exploding and vanish, the IM-CNN was designed to be composed of a cascade of residual CNNs, as shown in Figure 10.3(b). In this figure, the BN denotes the batch normalization, Softmax denotes the softmax nonlinear activation function, $k\,(a,b,c)$ denotes the convolutional kernel with a size of $a \times b \times c$, and $n\,(a)$ denotes the number of convolutional kernels to be a. During the training stage of IM-CNN, the labeled human-body images captured by the optical camera after background removal and binarization processing can be approximately regarded as the microwave reflectivity images of the human body, because the microwave reflection of the human body can be approximated to be homogenous over the frequencies from 2.4 to 2.5 GHz. The training was done using the ADAM optimization method, with mini-batches size of 32, and epoch setting as 50. The complex-valued weights and biases were initialized by random weights with zero-mean Gaussian distribution and standard deviation of 10^{-3}, respectively. The computations were performed with AMD Ryzen Threadripper 1950X 16-Core processor, NVIDIA GeForce GTX 1080Ti, and 128 GB access memory.

10.3.2 Results

Here, we provide a set of experimental results selected from Ref. [31] to evaluate the performance of the intelligent metasurface sensing system. In our experiments, we collected more than 10^5 pairs of labeled data of two volunteers with different gestures to train the intelligent metasurface; while three different persons were invited to test it. The trained intelligent metasurface is able to produce the high-resolution images of the test persons, from which their body gestures can be readily recognized. A series of imaging results are presented in Figure 10.4. The first row shows the optical images of specimen, which include single person with different gestures, two persons with different gestures, and two persons behind a 5-cm-thick wooden wall. The corresponding imaging results by the metasurface sensing system and amplitudes of microwave data are, respectively, illustrated in the second and bottom row. Particularly, the "see-through-the-wall" ability is validated by clearly detecting notable movements of the test persons behind a 5 cm-thick wooden wall. All the results show that it is enough to achieve the high-quality images by using 53 coding patterns, where 101 frequency points from 2.4 to 2.5 GHz are utilized for each coding pattern. The switch time of coding patterns is around 10 μs, implying that the time in data acquisition is less than 0.7 ms in total even if 63 coding patterns are used. Consequently, it can be safely concluded that the intelligent metasurface integrated with IM-CNN can

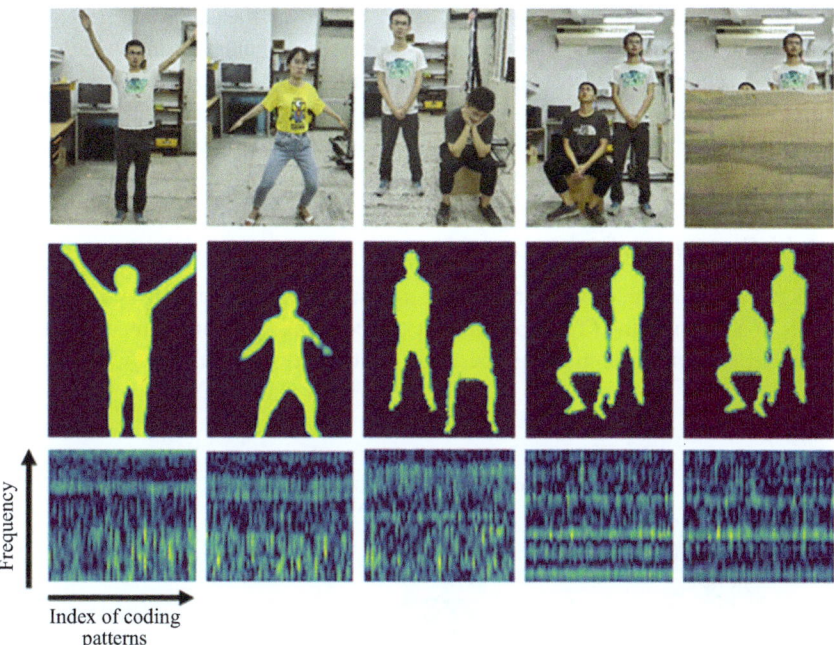

Frequency

Index of coding
patterns

*Figure 10.4   In situ imaging results using the intelligent metasurface with active
microwave [31]*

instantly produce high-quality images of multiple persons in real world, even when
they are behind obstacles.

10.4   VAE-based intelligent integrated metasurface sensor

Here, we remark that above sensing strategy remains not intelligent enough yet, in
the sense that it indiscriminately acquires all information, ignoring available knowledge about scene, sensing task and hardware constraints. To fully reap the benefits
of intelligence, Li *et al.* [32] and Wang *et al.* [33] proposed two frameworks of
intelligent integrated sensing, which could enable us to joint learn the optimal measurement and processing settings given the hardware and task in the frameworks
of variational-autoencoder (VAE) and free-energy (FE) framework, respectively, as
discussed in Sections 10.4 and 10.5. They experimentally demonstrated that using
the integrated sensing scheme, the performance improvements could be particularly
large especially when the number of measurements is limited. In this section, we here
discuss the hybrid-computing-based intelligent integrated sensing based on the variational autoencoder In particular a variational autoencoder (VAE) framework [35,36]
is explored to achieve data-driven learnable data acquisition by integrating it into a

data-driven learnable data-processing pipeline. Thereby, a measurement strategy can be learned jointly with a matching data post-processing scheme, optimally tailored to the specific sensing hardware, task, and scene, allowing us to perform high-quality imaging and high-accuracy recognition with a remarkably reduced number of measurements. This strategy drastically helps us to improve many critical metrics, such as speed, processing burden and energy consumption.

10.4.1   System configuration

Figure 10.5(a) shows the configuration of proposed integrated sensing system, which consists of a transmitting (TX) horn antenna, a receiving (RX) horn antenna, a large-aperture reprogrammable metasurface, and a vector network analyzer (VNA, Agilent

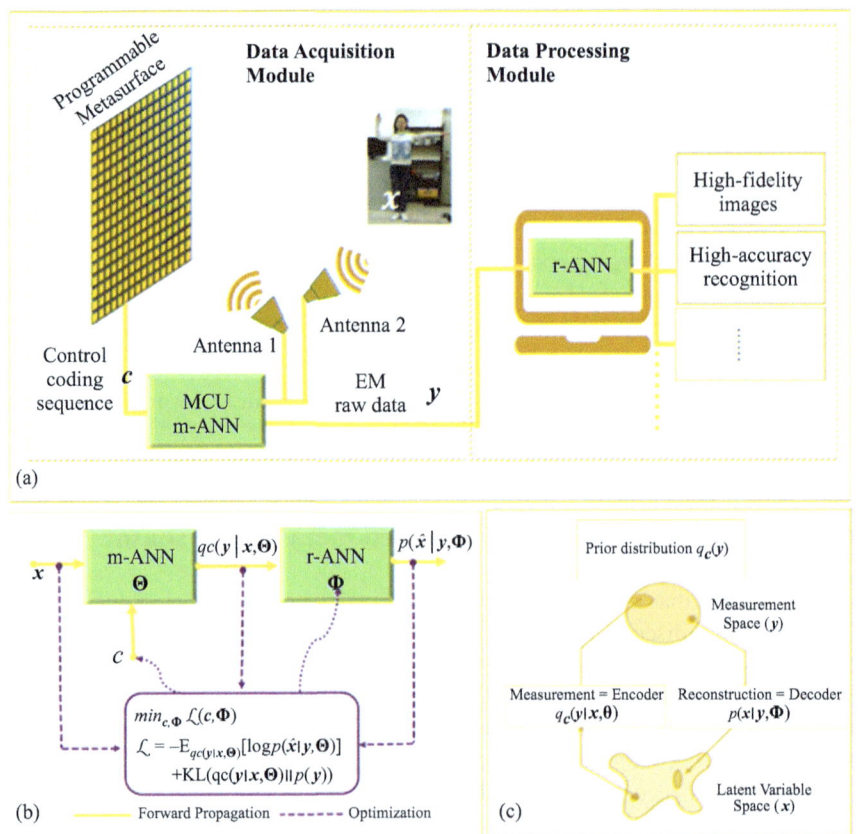

Figure 10.5   *System setup and working principle of the intelligent integrated metasurface sensor [32]. (a) The intelligent sensing system consists of two data-driven learnable modules. (b) Propagation and optimization process. (c) Interpretation of the entire sensing process in the VAE framework.*

E5071C). The two horn antennas are connected to two ports of the VNA via two 4 m-long 50-Ω coaxial cables, and the VNA is used to acquire the response data by measuring transmission coefficients. With reference to Figure 10.5(a) and (b), the sensing system consists of two data-driven learnable modules: the m-ANN-driven data acquisition module and the r-ANN-driven data processing module, and the m-ANN models the measurement process involving a pair of horn antennas and a coding metasurface reprogrammed with the MCU. The operational procedure of the sensing system is described as following. Antenna-1 connected to port-1 of VNA is used to emit periodically radio signals, which are shaped by the m-ANN-driven reprogrammable metasurface. After being scattered by the subject of interest, the wavefields are received by Antenna-2 connected to port-2 of VNA. Finally, the received microwave raw data are instantly processed by the r-ANN, producing the desired imaging or recognition results.

10.4.2 Variational auto-encoder (VAE) principle

Intelligent integrated metasurface sensor integrates the reconfigurable measurement process with data processing as a whole pipeline, enabling to jointly optimize the learnable physical and digital weights. To this end, the whole sensing pipeline was treated in the framework of VAE [35,36]. Figure 10.5(c) shows the interpretation of the entire sensing process in the VAE framework: the latent variable space, x, is encoded by the analog measurements in a measurement space, y and then the digital reconstruction decodes the measurements to return to the latent variable space. In a nutshell, we can view the entire sensing process (data acquisition and processing) as a user-controlled end-to-end process. Given a scene x, a set of complex-valued measurements y is generated by sampling from the $\mathcal{C}$-controllable conditional distribution $y \sim q_{\mathcal{C}}(y|x, \Theta)$. In other words, the latent scene variable of interest, x, is *encoded* in a measurement space y via a distribution controlled by the metasurface configuration $\mathcal{C}$. The goal of the processing is then to find an estimator that retrieves the relevant scene information x from the measurements y. This estimator inverts the action of the measurement process, in other words it *decodes* the information of interest to return from the measurement space to the latent variable space. Using the VAE framework, the digital decoder can be modeled as sampling the measurement space with a conditional distribution $x \sim p_{\Phi}(x|y)$, controlled by its digital weights Φ, to generate estimates of the latent variable of interest. The decoding is implemented with a deep ANN, called r-ANN, whose trainable weights can hence be identified as Φ.

To jointly learn optimal analog and digital weights, i.e. the metasurface control coding pattern $\mathcal{C}$ and the r-ANN weights Φ, respectively, we minimize the following objective function [34]:

$$\mathcal{L}(\mathcal{C}, \Phi) = -\mathbb{E}_{q_{\mathcal{C}}(y|x,\Theta)}[\log p_{\Phi}(x|y)] + \mathrm{KL}(q_{\mathcal{C}}(y|x, \Theta)|p(y)). \qquad (10.1)$$

The term $-\mathbb{E}_{q_{\mathcal{C}}(y|x,\Theta)}[\log p_{\Phi}(x|y)]$ can be interpreted as the "reconstruction error" of the VAE function: it is the log-likelihood of the true latent data given the inferred latent data. The term $\mathrm{KL}(q_{\mathcal{C}}(y|x, \Theta)|p(y))$ acts as a regularizer and encourages the distribution of the decoder to be close to a chosen prior distribution $p(y)$. Both analog encoder $q_C(y|x, \Theta)$ and digital decoder $p_\Phi(x|y)$ are treated with deep ANNs, namely m-ANN and r-ANN.

Note that in minimizing (10.1), there are two sets of different optimization variables, i.e., the continuously adjustable weights Φ in the r-ANN and Θ in the m-ANN, and the binary controllable variables C in the m-ANN. Starting with some initializations of C and Φ, we calculate C (resp. Φ) for Φ (resp. C) updated in the last iteration step, followed by calculating Φ (resp. C) based on the obtained C (resp. Φ). This procedure is repeated until a stopping criterion is fulfilled. Apparently, the optimization with respect to Φ and Θ can be efficiently realized with the well-known back propagation algorithm [38], which can be accomplished with well-developed optimizers in TensorFlow. However, it is really challenging to minimize (10.1) with respect to the binary control coding sequences C since it involves a NP-hard combinatorial optimization problem. To address this difficulty, the so-called randomized simultaneous perturbation stochastic approximation (r-SPSA) [39], originally developed for the problem of optimal well place and control in petroleum engineering, was modified for this problem. This heuristic optimization approach relies on two randomized descent strategies. First, as done by stochastic gradient descent approach, at each iteration, a fraction of training samples is randomly selected to determine a descent direction. Consequently, the concept of batch size is also applicable. Second, at each iteration, as done by the so-called randomized coordinate descent method, only a fraction of optimization components chosen to be updated. Here, partial coding meta-atoms are randomly selected, and their binary status is changed to their opposites correspondingly. If the change leads to the improvement on the objective function defined over randomly selected training samples, we save such change and go into next iteration. Otherwise, we need to randomly re-select some of coding meta-atoms and perform above operations. We repeat such procedure until some stop criterion is arrived and finally the whole VAE network is effectively converged.

10.4.3 Results

The sensing system is applied to the task of in situ high-resolution imaging of a human body in our laboratory environment. As outlined previously, *m-ANN* for data acquisition and r-ANN for smart data processing are integrated into a unique data-driven learnable sensing chain. To that end, we jointly optimized the coding patterns C of the m-ANN together with the weights Φ of r-ANN for the specific task of human body imaging. The integrated ANN is composed of a sequence of nonlinear convolution layers, which can be trained with a standard supervised learning procedure in TensorFlow. Following [37], in order to illustrate the significant improvement of the proposed learned sensing strategy on the image quality over conventional learning-based sensing methods, the training procedure is divided into two stages. During the first stage, the coding patterns of the m-ANN and the digital weights of the r-ANN are optimized separately as in [30]. The coding patterns of m-ANN are assigned following the two most common state-of-the-art approaches that correspond to using random or PCA-based scene illuminations. During the second stage, m-ANN and r-ANN are jointly trained to achieve the overall optimal sensing performance. In this two-stage training, the benefit reaped by the proposed sensing strategy over the conventional methods can be clearly demonstrated. Several people called as training person in short are used to train this intelligent microwave sensing system, and different persons called as test person are invited to test it. In addition, there are 1,000 random codes and 1,000 PCA-based codes (200 standard PCA-based codes and their 800 perturbations) are used as raining samples for training Θ. The details of the training people are provided by Li *et al.* [31]. In addition, the ground truth is defined as the binarized optical images of the scene.

In general, according to the difficulty of the sensing task and the signal-to-noise ratio [28], a measurement with a single coding pattern cannot be expected to obtain sufficient relevant information. Figure 10.6 displays the cross-validation errors over the course of the training iterations for different numbers M of coding patterns of the metasurface (3, 9, 15, and 20), from which the two stages of the

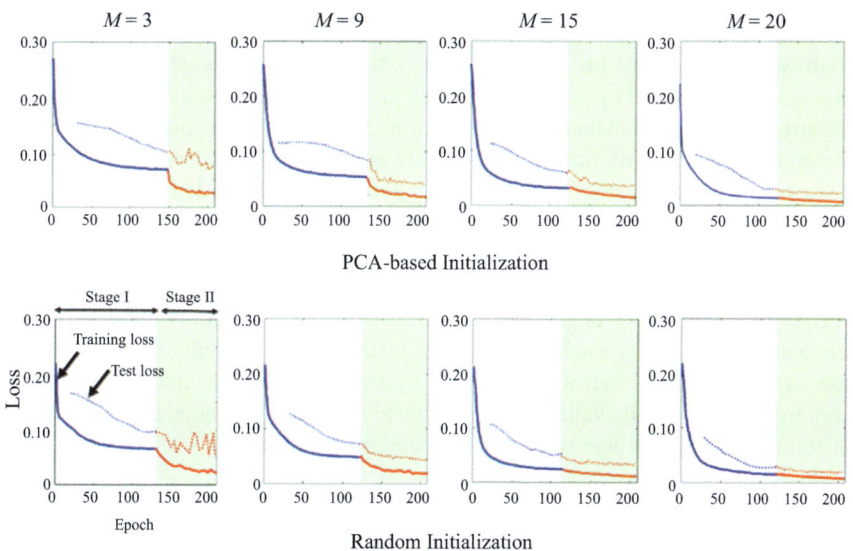

Figure 10.6   *Training dynamics for learned EM sensing applied to an imaging task [32]. The dependence of the training and test loss functions on the progress of iterative epochs is shown for different numbers of coding patterns M, i.e., M=3, 9, 15, and 20. The continuous lines indicate the training loss, and the dashed lines indicate the test loss. The control coding patterns of the metasurface are initialized randomly (top) or PCA-based (bottom). During stage I, only the digital decoder weights Φ are optimized. Then, during stage II, both the physical weights C and the digital weights Φ are jointly optimized.*

aforementioned training protocol can be clearly distinguished. In this figure, the continuous lines indicate the training loss and the dashed lines indicate the test loss and the control coding patterns of the metasurface are initialized randomly (top) or PCA-based (bottom). During stage I, only the digital decoder weights Φ are optimized. Then, during stage II, both the physical weights C and the digital weights Φ are jointly optimized. The presented results show a remarkable improvement of the image quality achieved by using the joint optimization of C and Φ during stage II, compared to that based on solely optimizing Φ (i.e. the end of stage I). The effect is especially striking when the number of measurements is very limited. Since the trainable physical (C) and digital (Φ) parameters are initialized randomly before training, except PCA-based C, we can conduct about 500 realizations in order to remove any sensitivity to the choices of random initializations made for m-ANN and r-ANN.

Figure 10.7 reports several selected image reconstruction results of the test person with different body gestures using the proposed integrated metasurface sensor with the corresponding coding patterns of the metasurface displayed in Figure 10.8. We display images of three different poses reconstructed with different numbers of coding patterns of the metasurface, M, for the case of only optimizing Φ (first row) or jointly optimizing C and Φ (second row). In this set of experiments, the random initialization is used. In line with [27], we observe that the sensing quality (here image quality) achieved by jointly optimizing physical (C) and digital (Φ) parameters is significantly better than that if only Φ is optimized. This may be intuitively expected since more trainable parameters are available and all a priori knowledge is used in the learned sensing scheme. These above experimental results demonstrate, in line with [27], that simultaneous learning of measurement and reconstruction settings is remarkably superior to the conventional sensing strategies in which measurement and/or reconstruction are optimized separately (if optimized at all). The benefits of integrated sensing are strong when the number of measurements is highly limited such that learned sensing enables a remarkable dimensionality reduction. Ultimately, these superior characteristics are enabled by training a unique integrated sensing chain, making use of all available prior knowledge about the probed scene, task, and constraints on measurement setting and processing pipeline.

10.5 Free-energy-based intelligent integrated metasurface sensor

Here, we discuss the intelligent integrated metasurface sensor in context of free energy minimization, developed by Wang *et al.* [33]. Guided by the free energy minimization principle, the metasurface sensing system can work in an intelligence way, in the sense that it can be trained such that the measurements can be adaptively collected on physical level and that the target can be recognized on digital level, similar to above. The system is the first effort with the reprogrammable metasurface to realize the high-frame-rate imaging in real-world settings. We here mean by the real-world setting that the target is in a really complicated indoor physical environment, and acquired signals are seriously disturbed by unknown co-channel interferences.

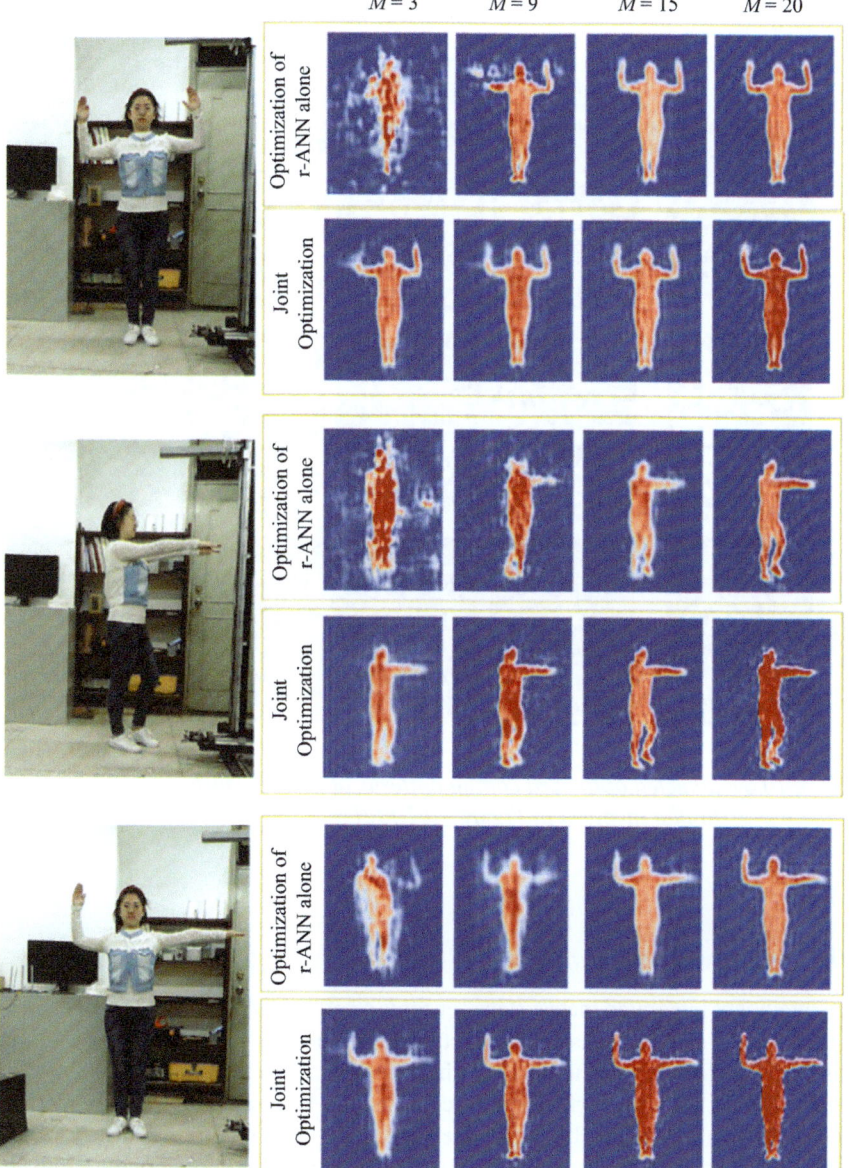

Figure 10.7   Experimental results for learned EM sensing applied to an imaging task [32]. We display images of three different poses reconstructed with different numbers of coding patterns of the metasurface, M, for the case of only optimizing Φ (first row) or jointly optimizing 𝒞 and Φ (second row).

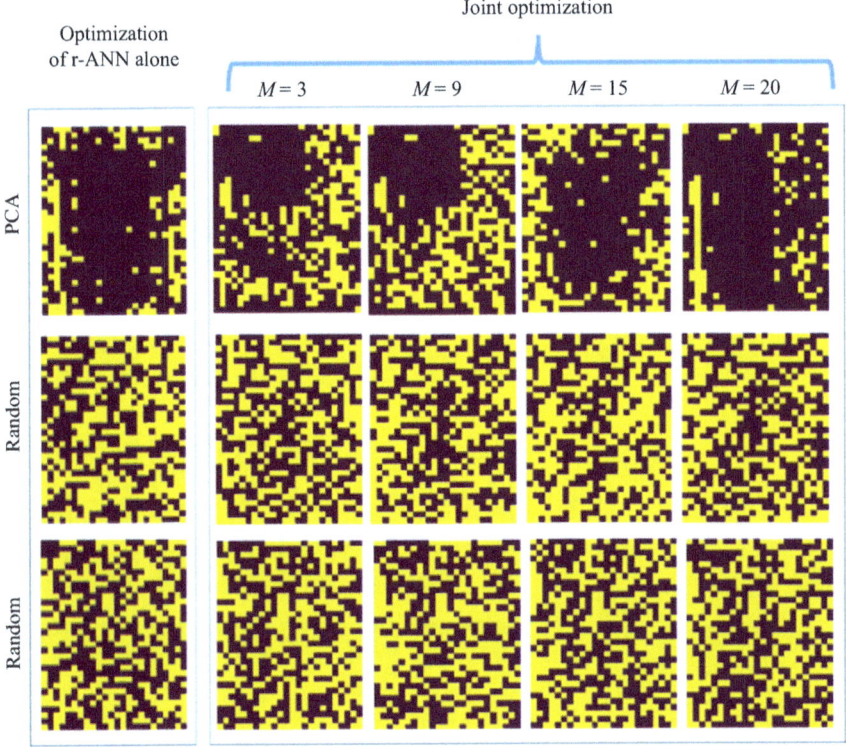

Figure 10.8   *Selected optimized coding patterns of the metasurface, corresponding to that involved in Figure 10.7 [32]*

10.5.1   *System configuration*

The intelligent integrated metasurface sensor is a software-defined system in favor of the high-frame-rate EM sensing. The metasurface sensor, working at around 2.4 GHz, was designed for monitoring human behaviors in indoor environment. As shown in Figure 10.9(a), the metasurface sensor consists of a large-aperture reprogrammable metasurface, a commercial software-defined radio device (Ettus USRP X310), a transmitting antenna, a three-antenna receiver and a personal computer. Both the USRP and metasurface are communicated with the host computer via the Ethernet under the transmission control protocol (TCP) and the USRP has I/O series communication with the metasurface. Figure 10.9(b) shows its operational procedure of data acquisition: The host computer is responsible of calculating the control patterns and sending these patterns to the metasurface through FPGA module; at the same time, it sends a command signal to the USRP for synchronizing its transmitting and receiving channels. Note that this initialization process takes about 10 ms. To trade-off the imaging quality with efficiency, 18 patterns for compressive microwave measurement per

(a)

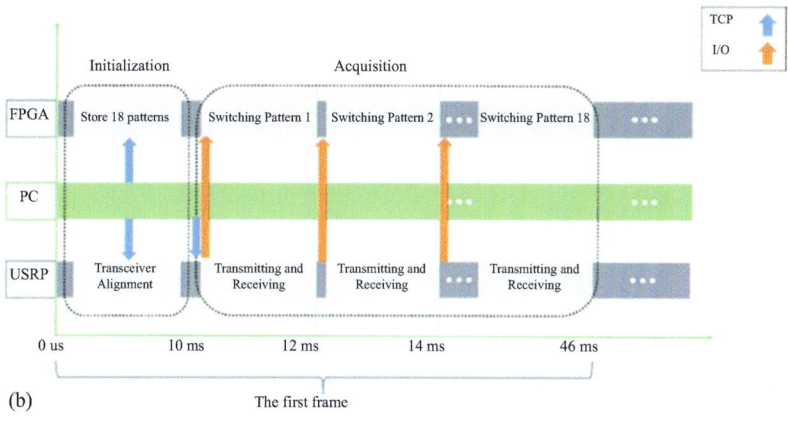

(b)

Figure 10.9   System configuration of the proposed intelligent integrated metasurface sensor [33]. (a) The experimental setup in indoor environment. (b) The operational procedure of data acquisition, where a 10 ms-length system initialization procedure is marked before data acquisition.

image are explored in this work. For each control pattern, the USRP under control of the host computer generates the Chirp radio signal, radiating it into the investigation domain through the transmitting antenna, and receive the echoes reflected from the target. It took about 36 ms to produce a microwave image, implying the frame rate achievable is about 27 Hz. As pointed out in Ref. [33], if the USRP is updated with more specialized transceiver devices, the frame rate achievable could be optimized to be in order of tens of kHz in principle, which means its performance has a lot of room to improve in the future. Afterwards, the acquired echoes are processed by ANNs, which is directly responsible for the object imaging and recognition. In this work, the chip signal waveform transmitted by the USRP reads:

$$s(t) = \exp\left(j\left(2\pi f_c t + \pi K t^2\right)\right) \quad 0 \le t \le T \tag{10.2}$$

where $j = \sqrt{-1}$, $f_c = 2.424$ GHz is the carrier frequency, $K = B/T$ denotes the sweep rate of the chirp, $B = 50$ MHz is the frequency bandwidth, and $T = 10$ µs is the Chirp pulse duration.

The sketch map of the proposed intelligent metasurface sensor is shown in Figure 10.10(a), which is composed of a large-aperture reprogrammable metasurface, a USRP X310, a transmitting antenna (TA), a three-antenna (RA1, RA2, RA3) receiver and a personal computer (PC). Here, the programmable metasurface controlled with artificial neural networks is utilized for two major purposes: (i) manipulating adaptively the EM wavefields towards the target, suppressing the

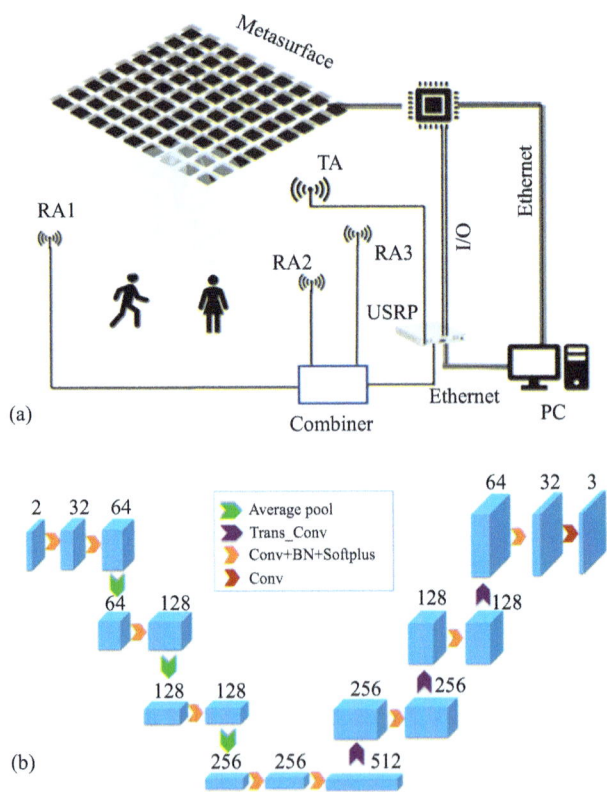

Figure 10.10 System structure and network diagram [33]. (a) The sketch map of proposed intelligent metasurface sensor. (b) The proposed U-net network along with necessary parameters.

unwanted disturbances from surrounding environment like walls, furnishings, and so on; (ii) serving as an electronically controllable coding aperture in compressive-sensing manner. Being different from the coding patterns explored in conventional compressive sensing strategies, the deep-learning-driven metasurface considered here is capable of generating the measurements which are consistent with those needed by the digital reconstruction, as detailed next section. Owing to the large-view field nature by the reprogrammable metasurface, the target's information can be fully captured. Therefore, the reconstruction of target's information can be easily achieved from the compressed measurements by a deep ANN, i.e., a U-net network. As shown in Figure 10.10(b), the designed U-net network has double-channel input: one channel is from the real-part of preprocessed microwave signal and another is from the imaginary part. Features of microwave signals were extracted layer by layer, gradually approaching to the labeled IUV three-channel images [46]. To improve the network performance, the residual structure is adopted at each layer of the U-net. Each residual network module is composed of three convolutional network layers, where a SoftPlus nonlinear activation operation and batch normalization follow after each convolutional layer. The training was performed over a GPU computer with a single Nvidia GTX2080Ti, and the training setup was made as follows: the optimizer was Adam [42], the learning rate was 10^{-3}, the weight decay rate was 5×10^{-5}, and the batch-size was 128. In order to speed-up the digital computation, Wang *et al.* proposed window Adam: the values of the pixels outside the target are enforced to be zero, each iteration during the training procedure. By using this simple scheme, the training efficiency can be considerably improved.

10.5.2 *Free-energy minimization principle*

The Bayesian principle says that a self-organizing system that is at equilibrium with its supporting environment must minimize its free energy [40]. For this problem, assuming the target's state s_t at time t, the metasurface sensor aims at organizing the measurement strategy π_t, collecting measurements o_t, and retrieving the target. It is noted that the measurements depend on the EM manipulation via the programmable metasurface, so the efficient measurement strategy can be achieved by changing the control coding pattern of the metasurface. To build a machine for this problem, we explore the probabilistic generative model and its Bayesian inference solution. For the target-sensor scenario with the generative distribution $P_\varphi(s_{\leq t}, o_{\leq t}|\pi_{\leq t})$, the metasurface sensor has a picture of it, which is characterized with a posterior distribution (i.e., recognition function) $Q_\theta(s_{\leq t}|o_{\leq t}, \pi_{\leq t})$. Here, φ and θ encapsulate all trainable parameters defining P_φ and Q_θ, respectively. Then, the generative network P_φ and inference network Q_θ could be achieved by minimizing the following free energy [19], i.e.,

$$\mathcal{F} = \mathbb{E}_{Q_\theta(s_{\leq t}|o_{\leq t}, \pi_{\leq t})}\left[\ln\left(\frac{Q_\theta(s_{\leq t}|o_{\leq t}, \pi_{\leq t})}{P_\varphi(s_{\leq t}, o_{\leq t}|\pi_{\leq t})}\right)\right] \tag{10.3}$$

Under well-known Markov chain approximation, (10.3) can be expressed as:

$$\mathcal{F} = \sum_{t=1}^{T} J_t \tag{10.4}$$

where

$$J_t = J_{t-1} - \mathbb{E}_{Q_\theta(s_{\le t-1}|o_{\le t-1},\pi_{\le t-1})} \left[\mathbb{E}_{Q_\theta(s_t|o_t,s_{\le t-1},\pi_t)} \underbrace{\ln\left(P_\varphi\left(o_t|s_t,\pi_t\right)\right)}_{\text{Likelihood}} \right]$$

$$- \mathbb{E}_{Q_\theta(s_{\le t-1}|o_{\le t-1},\pi_{\le t-1})} \left[\mathbb{E}_{Q_\theta(s_t|o_t,s_{\le t-1},\pi_t)} \underbrace{\ln\left(\frac{P_\varphi(s_t|s_{\le t-1})}{Q_\theta(s_t|o_t,s_{\le t-1},\pi_t)}\right)}_{\text{KL divergence}} \right].$$

$\mathbb{E}_{Q_\theta(s_t|o_t,s_{\le t-1},\pi_t)} \ln\left(P_\varphi\left(o_t|s_t,\pi_t\right)\right)$ is the likelihood or observation accuracy at time t, while $\mathbb{E}_{Q_\theta(s_t|o_t,s_{\le t-1},\pi_t)} \ln\left(\frac{P_\varphi(s_t|s_{\le t-1})}{Q_\theta(s_t|o_t,s_{\le t-1},\pi_t)}\right)$, Kullback–Leibler (KL) distance or relative entropy, characterizes the recognition complexity of $Q_\theta(s_t|o_t,s_{\le t-1},\pi_t)$ at t. In order to facilitate numerical implementations, several assumptions have made:

(i) $Q_\theta\left(s_t|o_t,s_{\le t-1},\pi_t\right) = \mathcal{N}(s_t|f_\theta\left(o_t,\pi_t\right),\alpha^2 I)\mathcal{N}(s_t|s_{t-1},\beta^2 I)$, where the nonlinear function f_θ is modeled with a U-net artificial neural network [41] in Figure 10.10(b), α^2 and β^2 are two trainable parameters.
(ii) $P_\varphi\left(s_t|s_{\le t-1}\right)$ is modeled with the so-called Brown motion [6].
(iii) $P_\varphi\left(o_t|s_t,\pi_t\right)$ is represented with a physical-model-based neural network, i.e., $P_\varphi\left(o_t|s_t,\pi_t\right) = \mathcal{N}(o_t|\mathcal{A}_{\pi_t}s_t,\gamma^2 I)$, where $\mathcal{A}_{\pi_t}$ is a linear operator defined in [33], and γ^2 is a trainable parameter.

Now, one can determine the measurement strategy π_t, and generative network P_φ and inference network Q_θ by minimizing (10.4) by exploring variational autoencoder method [32].

10.5.3   Results

The metasurface sensor was deployed in a real-world indoor environment, leading to the seriously noisy measurements. In particular, such sensor works at around 2.4 GHz, thus the acquired signals are inevitably disturbed by unwanted but unknown inband wireless signals (like Wi-Fi, Bluetooth, etc.) everywhere. In addition, a plenty of unwanted interferences exist, which arise from surrounding environment such as walls, furniture, and so on. These disturbances were remarkably dominant over the acquired signals carrying the target's information, and more importantly, the in-band inferences from commodity wireless signals were usually statistically non-stationary. Therefore, in order to realize the acceptable sensing tasks, it is urgently demanded to develop denoise methods.

To this end, the denoise model proposed by Wang *et al.* is discussed here. Assuming that a transmitter at r_T gives rise to a frequency-domain signal $s(\omega)$, and a point-like object with reflection coefficient σ (r_o) is situated at r_o, where ω is the angular frequency. Such point-like target model makes sense and following discussions can be readily extended for the case of extended objects in terms of linear supposition principle in context of Born scattering approximation [1,6]. Then, the echo acquired by the receiver at r_R reads:

$$y\,(r_R;\omega,\mathcal{C}) \approx s(\omega)\sigma\,(r_o)\,g\,(r_R,r_o;\omega)\,g\,(r_T,r_o;\omega)$$

$$+ \; s(\omega)\sigma\,(r_o)\,g\,(r_R,r_o;\omega)\left(\sum_n \Gamma_n^{\mathcal{C}}(\omega)g\,(r_T,r_n;\omega)\,g\,(r_o,r_n;\omega)\right) \qquad (10.5)$$

$$+ \; s(\omega)g\,(r_R,r_T;\omega) + s(\omega)\sum_n \Gamma_n^{\mathcal{C}}(\omega)g\,(r_R,r_n;\omega)\,g\,(r_T,r_n;\omega)$$

$$+ \; \varepsilon(\omega)$$

Herein, $g\,(r_R,r_o;\omega)$ denotes the so-called Green's function of considered physical environment that characterizes the system response at r_R given a radio source at r_o. $\Gamma_n^{\mathcal{C}}(\omega)$ represents the reflection coefficient of the nth meta-atom at r_n, when the metasurface is configured with the control coding pattern $\mathcal{C}$. Note that the summation is performed over the meta-atoms. Moreover, $\varepsilon(\omega)$ accounts for disturbances from aforementioned in-band inferences, environment clutters, system noise, and others. In (10.5), other possible multiple-scattering terms have been ignored due to the deployment of the directional transmitting and receiving antennas. Note that the first and second terms in the right hand of (10.5) carry the target's information; while other terms are usually target-independent. It is trivial to remove the third and fourth terms by exploring a simple background removal operation; however, to filter out the last term is challenging due to its statistically non-stationary nature for the real-time application demand, since conventional filter-based methods, such as the time-frequency filtering, principal component filtering, and others, are typically computationally prohibitive. To this end, we designed an end-to-end deep convolutional network, termed as Filter-CNN, to map the noisy signal after background removal to desired denoised signal. The training was performed in a GPU personal computer with Nvidia GTX2080Ti and computation parameters: the optimizer was ADAM, batch-size = 128, initial learning rate = 0.01, and iteration epochs = 100. Such training costs 2 h. Once the Filter-CNN was trained well, the filtering time for a group of $20 \times 1,000$ data was about 0.3 s.

Selected results of the denoised signals are presented in Figure 10.11(b) and (e), where 18 random control coding patterns of reprogrammable metasurface are used, and a human target stands quietly in indoor environment. For comparison, corresponding down-converted signals are also provided in Figure 10.11(a) and (d). It can be clearly observed from Figure 10.11(a)(b) and (d)(e) that the overwhelming unwanted inferences can be well filtered out using our Filter-CNN. Recall (10.5), the first term characterizes the direct arrival from the source to receiver, and thus is out of control of the programmable metasurface. To demonstrate the role of the metasurface on the compressive measurements, the first term in (10.5) is removed by mean-value filter with respect to the slow time (i.e., measurement index), and corresponding results are plotted in Figure 10.11(c) and (f). From these figures, we can see that the programmable metasurface can flexibly manipulate the acquisition of the microwave signals carrying the target's information, implying that the target's

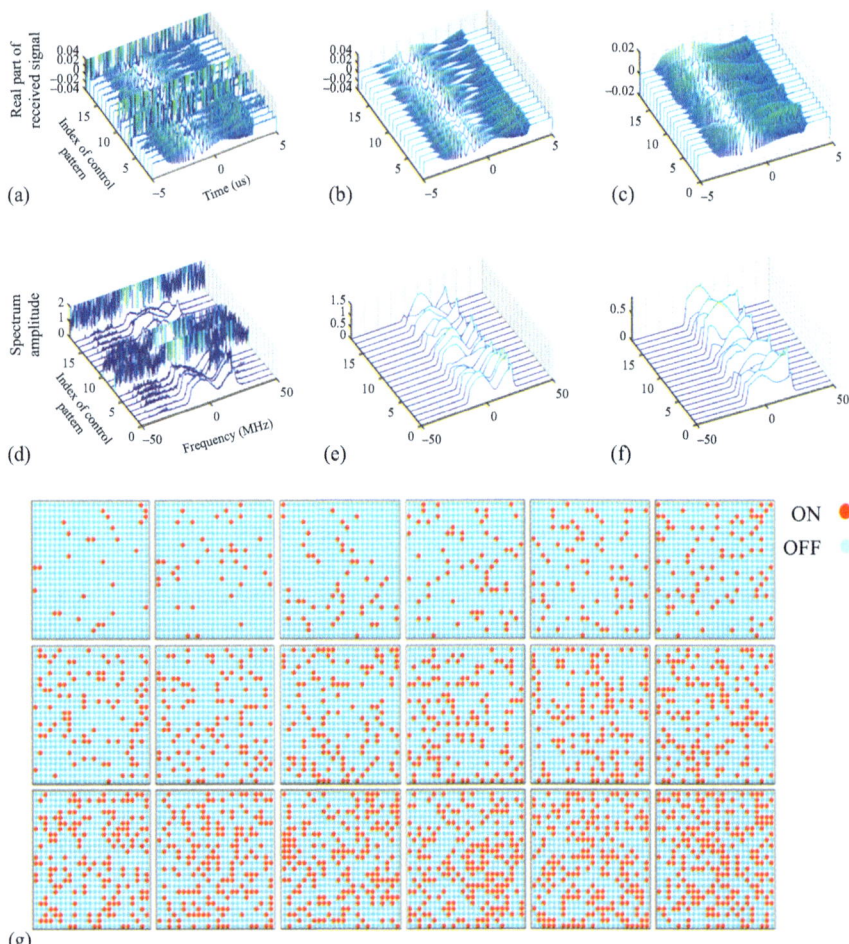

(g)

Figure 10.11   Experimental results of the signal denoise [33]. (a) Real parts of time-domain down-converted signals where 18 random control patterns of metasurface are considered. This figure clearly shows very serious in-band and out-of-band disturbances. (b) Real parts of 18 denoised time-domain signals through the proposed Filter-CNN. (c) Real parts of mean-valued-filtered signals. (d and f) are the spectrum amplitudes. (g) 18 random control patterns of the metasurface used in this set of experiments.

information can be efficiently captured by a fixed receiver in a compressive way. Additionally, 18 random control coding patterns of the metasurface involved in Figure 10.11(a)–(f) have been plotted in Figure 10.11(g). Recall (10.3), one interesting conclusion can be observed, i.e., the good measurements imply the acquired signals with good signal-to-noise (SNR), which can be achieved by controlling the coding patterns of the reprogrammable metasurface. Intuitively, such measurement strategy can be conceived by designing the control pattern of the metasurface such that the resultant radiation beams are focused towards the target, where the prior on the target's location can be estimated from the image obtained at last time. Results with focusing measurement strategy corresponding to Figure 10.11 are presented in Figure 10.12, from which above conclusions can be drawn again. Moreover, we can observe that the SNRs of acquired radio signals can be remarkably improved by using the focusing measurement strategy.

The performance of the developed metasurface sensor for the in situ imaging of human freely moving in our lab was examined experimentally. To train the metasurface sensor in a supervised manner, a commercial optical binocular camera ZED2 from

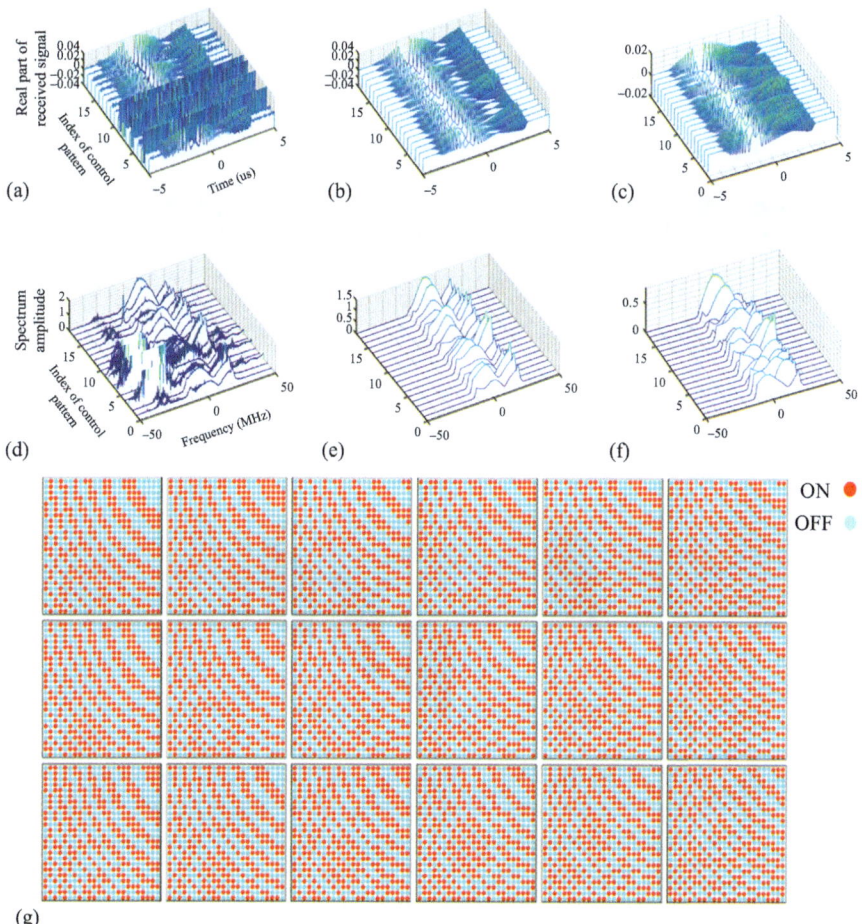

Figure 10.12   Experimental results of the signal denoise [33]. The 18 focusing control patterns of the programmable metasurface shown in Figure 10.12(g) are considered.

Stereolabs [43] was embedded. Then, the optical videos by the ZED2 were utilized as the labeled training samples, where a sequence of processes including background removal, segmentation and IUV-transformation through Densepose [44] are involved. The metasurface sensor was trained by inviting one person acting freely in the lab, and tested by another person. In this work, 8×10^4 pairs of labeled training videos have been collected. The metasurface sensor, once being well trained, can produce high-fidelity videos of the test person with the frame rate of about 20 Hz.

A series of IUV microwave images at several selected moments from a video recorded by the metasurface sensor is shown in Figure 10.13(a), from which one can

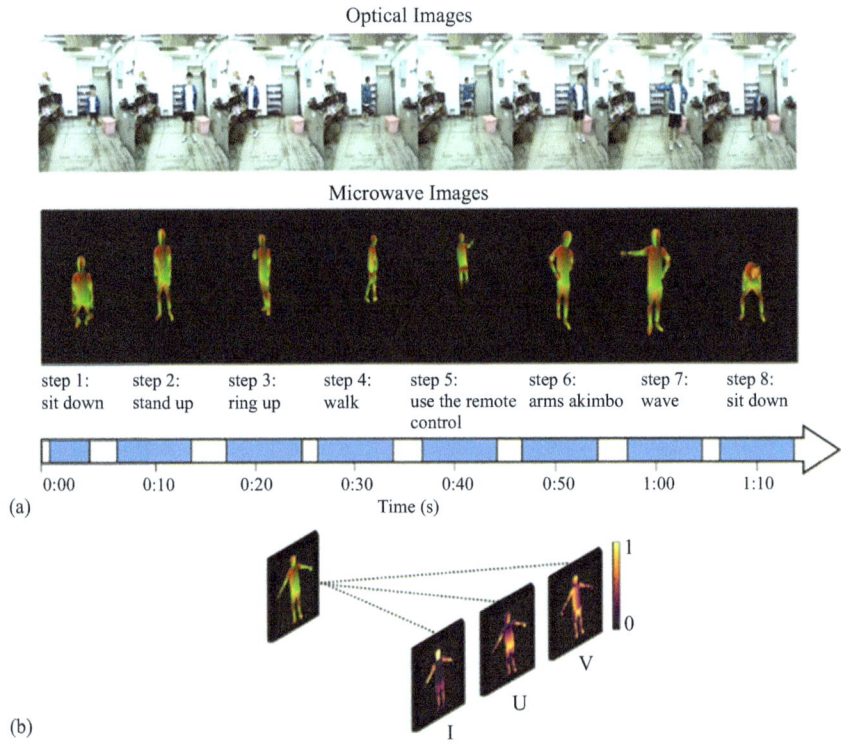

Figure 10.13   (a) *Experimental in situ imaging results of the test person freely acting in indoor environment as shown in Figure 10.9(a). (Top) The optical RGB images at selected moments that are recorded by optical ZED2 camera. (Middle) The IUV images recorded by the intelligent metasurface sensor at selected moments corresponding to those in top row. (Bottom) Time line. (b) An IUV image has three channels: I-channel, U-channel, and V-channel. The UV channels provide the result of mapping all human pixels from the RGB image to the 3D surface of the human body [33].*

readily recognize the actions of the test person in indoor environment, for instance, sitting down, standing up, making phone calls, walking, turning on the air conditioner, waving hands, and so on. More details about the IUV microwave images can be found in Figure 10.13(b), which shows that an IUV image has three channels: I-channel, U-channel, and V-channel. The I-channel image is the classification of pixels belonging to either background or different parts of body, which provides a coarse estimate of surface coordinates. The UV channels indicate the results of mapping all human pixels of an RGB image to the 3D surface of the human body. As discussed in Section 10.5.2, the metasurface sensor has the intelligence enabled by the adaptive data acquisition and processing.

The metasurface sensor has great performance of through-wall sensing. To show this, a set of experimental results are shown here, where the target freely acts in corridor outside the room with a 60-cm thickness load-bearing concrete wall. The training and test procedures are the same as those in indoor case. Figure 10.14 reports through-wall IUV images at selected moments recorded by the intelligent metasurface sensor. It can be observed that the image quality in outdoor environment is comparable to those in indoor environment and that the actions of the test person behind a 60 cm-thickness concrete wall remains to be clearly identified. In a word, the intelligent metasurface sensor developed by Wang *et al.* is capable of enabling us to see clearly

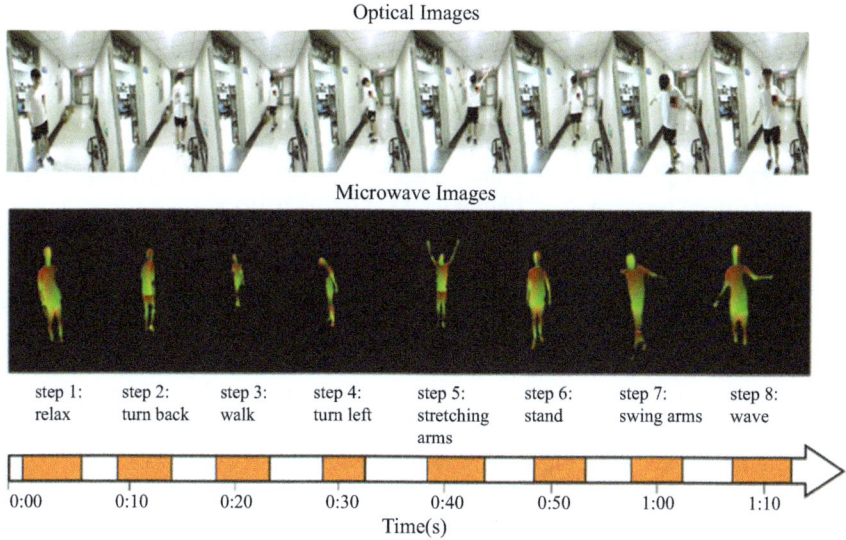

Figure 10.14 *Experimental in situ imaging results of the test person freely acting in a corridor outside the lab with a 60-cm thickness load-bearing concrete wall [33]. (Top) RGB images recorded by optical ZED2 camera at selected moments. (Middle) IUV images by the intelligent metasurface sensor at selected moments corresponding to that in top row. (Bottom) Time line.*

human behaviors behind a 60-cm thickness reinforced concrete wall with high frame rate. Such sensing strategy could open up a promising route toward smart community and beyond, and can be readily transposed to other frequencies and other types of wave phenomena.

References

[1]   Pastorino M. *Microwave Imaging*, New York, NY: John Wiley & Sons, Inc. (2010).

[2]   Brown W. M. Synthetic aperture radar. *IEEE Trans. Aerosp. Electron. Syst.* **3**, 217 (1967).

[3]   Turpin, A., Kapitany, V., Radford, J., *et al.* 3D imaging from multipath temporal echoes. *Phys. Rev. Lett.* **126**, 174301 (2021).

[4]   Hunt, J., Driscoll, T., Mrozack, A., *et al.* Metamaterial apertures for computational imaging. *Science* **339**, 310–313 (2013).

[5]   Picardi G. Radar soundings of the subsurface of Mars. *Science* **310**, 1925–1928 (2005).

[6]   Li, L., Hurtado, M., Xu, F., *et al.* A survey on the low-dimensional-model-based electromagnetic imaging. *Found Trends Signal Process* **12**, 107–199 (2018).

[7]   Duarte, M. F., Davenport, M. A., Takhar, D., *et al.* Single-pixel imaging via compressive sampling. *IEEE Signal Process. Mag.* **25**, 83–91 (2008).

[8]   Patel V. M., Mait, J. N., Prather, D. W. and Hedden, A. S. Computational millimeter wave imaging: problems, progress, and prospects. *IEEE Signal Process. Mag.* **33**, 109–118 (2016).

[9]   Edgar M. P., Gibson G. M. and Padgett, M. J. Principles and prospects for single-pixel imaging. *Nat. Photonics* **13**, 13–20 (2019).

[10] Fenn, A. J., Temme, D. H., Delaney, W. P. and Courtney, W. E. The development of phased-array radar technology. *Lincoln Lab. J.* **12**, 20 (2000).

[11] Xu Q., Chen Y., Wang B. and Liu K. J. R. Radio biometrics: human recognition through a wall. *IEEE Trans. Inf. Forensics Secur.* **12**, 1141–1155 (2017).

[12] Adib F., Hsu C.-Y., Mao H., Katabi D. and Durand F. Capturing the human figure through a wall. *ACM Trans. Graph.* **34**, 1–13 (2015).

[13] Lien, J., Gillian, N., Karagozler, M., *et al.* Soli: ubiquitous gesture sensing with millimeter wave radar. *ACM Trans. Graph.* **35**, 1–19 (2016).

[14] Cui T. J., Qi M. Q., Wan X., Zhao J. and Cheng Q. Coding metamaterials, digital metamaterials and programmable metamaterials. *Light Sci. Appl.* **3**, e218–e218 (2014).

[15] Li L. and Cui T. J. Information metamaterials – from effective media to real-time information processing systems. *Nanophotonics* **8**, 703–724 (2019).

[16] Arbabi, E., Arbabi, A., Kamali, S. M., Horie, Y., Faraji-Dana, M. and Faraon, A. MEMS-tunable dielectric metasurface lens. *Nat. Commun.* **9**, 812 (2018).

[17] Li, L., Cui, T. J., Ji, W., *et al.* Electromagnetic reprogrammable coding-metasurface holograms. *Nat. Commun.* **8**, 197 (2017).

[18] Zhao, H., Shuang, Y., Wei, M., Cui, T. J., Hougne, P. D. and Li, L. Metasurface-assisted massive backscatter wireless communication with commodity Wi-Fi signals. *Nat. Commun.* **11**, 3926 (2020).

[19] Shuang, Y., Wei, M., Ruan, H., Wang, L. and Li, L. Controllable manipulation of Wi-Fi signals using tunable metasurface. *J Radars*, **10**(2), 313–325 (2021).

[20] Zhao, J., Yang, X., Dai, J. Y., *et al.* Programmable time-domain digital-coding metasurface for non-linear harmonic manipulation and new wireless communication systems. *Natl. Sci. Rev.* **6**, 231–238 (2019).

[21] Li, L., Wang, L. G., Teixeira, F. L., Liu, C., Nehorai, A. and Cui, T. J. DeepNIS: Deep Neural Network for Nonlinear Electromagnetic Inverse Scattering. *IEEE Trans. Antennas Propagat.* **67**, 1819–1825 (2019).

[22] Mehta, P., Bukov, M., Wang, C. H., *et al.* A high-bias, low-variance introduction to machine learning for physicists. *Phys Rep* **810**, 1–124 (2019).

[23] Sinha A., Lee J., Li S. and Barbastathis G. Lensless computational imaging through deep learning. *Optica* **4**, 1117 (2017).

[24] Goodfellow I., Bengio Y. and Courville A. *Deep Learning*. New York, NY: MIT Press (2016).

[25] Sleasman T., Imani M. F., Gollub J. N. and Smith D. R. Dynamic metamaterial aperture for microwave imaging. *Appl. Phys. Lett.* **107**, 204104 (2015).

[26] Sleasman T., Imani M. F., Gollub J. N. and Smith, D. R. Microwave imaging using a disordered cavity with a dynamically tunable impedance surface. *Phys. Rev. Appl* **6**, 054019 (2016).

[27] del Hougne, P., Imani M. F., Diebold A. V., Horstmeyer R. and Smith D. R. Learned integrated sensing pipeline: reconfigurable metasurface transceivers as trainable physical layer in an artificial neural network. *Adv. Sci.* **7**, 1901913 (2019).

[28] del Hougne P., Imani M. F., Fink M., Smith D. R. and Lerosey G. Precise Localization of multiple noncooperative objects in a disordered cavity by wave font shaping. *Phys. Rev. Lett.* **121**, 063901 (2018).

[29] Liu C., Ma Q., Luo Z.J., *et al.* A programmable diffractive deep neural network based on a digital-coding metasurface array. *Nat. Electron.* **5**, 113–122 (2022).

[30] Li, L., Ruan, H., Liu, C., *et al.* Machine-learning reprogrammable metasurface imager. *Nat. Commun.* **10**, 1082 (2019).

[31] Li, L., Shuang, Y., Ma, Q., *et al.* Intelligent metasurface imager and recognizer. *Light Sci. Appl.* **8**, 97 (2019).

[32] Li, H. Y., Zhao, H. T., Wei, M. L., *et al.* Intelligent electromagnetic sensing with learnable data acquisition and processing. *Patterns* **1**, 100006 (2020).

[33] Wang, Z., Zhang, H., Zhao, H., Cui, T. J. and Li, L. Intelligent electromagnetic metasurface camera: system design and experimental results, *Nanophotonics*, to appear.

[34] LeCun Y., Bengio Y. and Hinton G. Deep learning *Nature* **521**(28), 436–444 (2015).

[35] Kingma D. P. and Welling, M. Auto-Encoding Variational Bayes. arXiv:1312.6114 (2014).

[36] Doersch, C. Tutorial on Variational Autoencoders. arXiv:1606.05908 (2016).

[37]   Vedula, S., Senouf O., Zurakhov G., *et al.* Learning beamforming in ultrasound imaging. *Proc. Mach. Learn. Res.* **102**, 493–511 (2019).

[38]   Chakrabarti, A. Learning sensor multiplexing design through back-propagation, in: *Advances in Neural Information Processing Systems*, pp. 3081–3089 (2016).

[39]   Li L., Jafarpour B. and Mohammad-Khaninezhad, M. R. A simultaneous perturbation stochastic approximation algorithm for coupled well placement and control optimization under geologic uncertainty. *Comput. Geosci.* **17**, 167–188 (2013).

[40]   Friston K. The free-energy principle: a unified brain theory? *Nat. Rev. Neuro.*, **11**, 127–138 (2010).

[41]   Ronneberger O., Fischer P. and Brox T. U-net: convolutional networks for biomedical image segmentation, in: *International Conference on Medical Image Computing and Computer-Assisted Intervention*. New York, NY: Springer, pp. 234–241 (2015).

[42]   Kingma D. and Ba J. Adam: a method for stochastic optimization. *arXiv preprint arXiv:1412.6980*, 2014 (ICLR 2015).

[43]   https://www.stereolabs.com/zed-2/

[44]   Alp Guler R., Neverova N. and Kokkinos I. DensePose: dense human pose estimation in the wild, in *IEEE Conference on Computer Vision and Pattern Recognition*, pp. 7297–7306 (2018) (Code is available at http://densepose.org/).

Chapter 11

Deep learning techniques for metamaterials and metasurfaces design

Tao Shan[1], Maokun Li[1], Fan Yang[1] and Shenheng Xu[1]

11.1 Introduction

Metamaterials, artificially engineered composite structures, are rationally designed for effective material parameters with exotic properties in a periodic or non-periodic way [1–4]. The intriguing and unprecedented material properties and corresponding applications have been already reported, such as negative refraction [5,6], invisibility cloaking [7–9], perfect lensing [10], chirality [11,12], etc. Metasurfaces, the two-dimensional equivalence of bulk metamaterials, consists of single-layer or few-layer planar engineered structures that are periodically or quasi-periodically distributed on an ultra-thin surface [13,14]. Metasurfaces demonstrate many novel photonic and electromagnetic phenomena by manipulating the reflection or transmission on a surface or interface, such as broadband diffusion [15], anomalous reflection or refraction [16], arbitrary beamforming [17], polarization conversion [18,19], etc.

The principal driving force behind the dramatic developments of metamaterials and metasurfaces lies in the continuous advancement of design techniques [20,21]. The design of a metamaterial or metasurface involves the forward modeling part and the inverse design part, as shown in Figure 11.1. Here, we denote m as the data space that contains the structural parameters to construct metamaterials/metasurfaces and denote d as the model space that collects the user-defined properties of metamaterials/metasurfaces. The forward modeling is the mapping from m to d, and it can be described by:

$$d = F(m), \tag{11.1}$$

where F is the forward operator. Forward modeling is an essential tool to assist human intuition-guided metamaterial designs by simulating iteratively to provide feedback. The fundamental limitation here is the computational efficiency that results in a trade-off between the computational time and the simulation accuracy. In an opposite direction, the inverse design aims to find the optimal m given the desired d:

$$m = I(d), \tag{11.2}$$

[1]Beijing National Research Center for Information Science and Technology (BNRist), Department of Electronic Engineering, Tsinghua University, China

where I denotes the inverse operator. The goal of inverse design is to directly determine the optimal design in m given the desired properties d by approximating the inverse of F^{-1}. The inverse design is usually an ill-posed problem that sets up barriers to efficient design, including non-uniqueness and instability. The design complexity grows sharply due to the exponential growth of candidate designs when the degrees of freedom increase, as shown in Figure 11.2 [22]. The growing design complexity is a long-standing challenge for the efficient design of both metamaterials and metasurfaces.

The rapid development of deep learning (DL) has significantly accelerated the pace in many fields [23], such as computer vision [24], speech recognition [25],

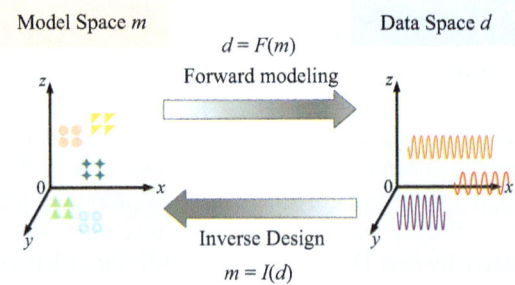

Figure 11.1   Forward modeling and inverse design of metamaterials and metasurfaces

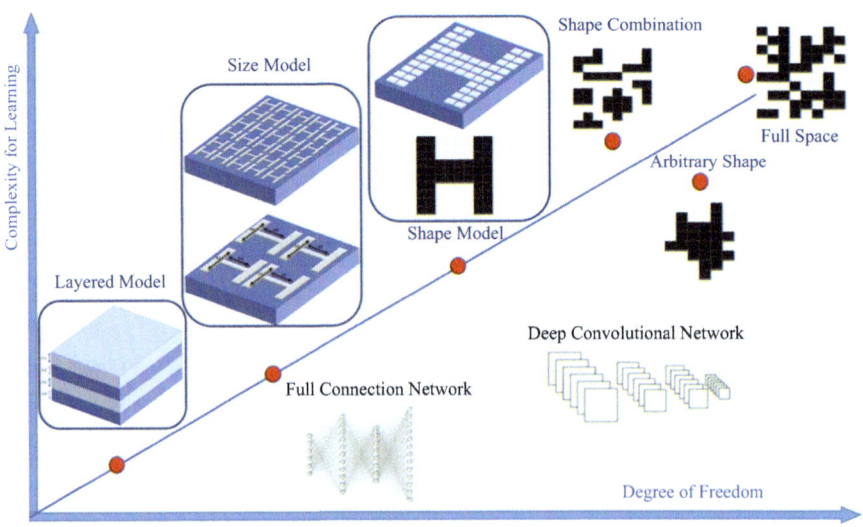

Figure 11.2   Relationship between the design complexity and the degree of freedom (source: [22])

machine translation [26], etc. Recently, many works have been reported to apply DL to accelerate the design of metamaterials and metasurfaces [20,21]. On the one hand, deep learning can abstract the inner law from massive data with powerful learning capacity and approximating ability. This makes DL capable of efficiently approximating a target function, which is advantageous for designing metamaterials and metasurfaces. By building and training deep neural networks (DNNs) to replace the parts with high complexity, DL can help mitigate the computational loads of forward modeling and inverse design. On the other hand, DL is good at compactly extracting features from a large amount of data. This can help build the compact representation of geometrical or structural features of metamaterials and metasurfaces, which further overcomes the "curse of dimensionality."

In this chapter, we offer a brief review of recent advances on the applications of DL into the design of metamaterials and metasurfaces. The design strategies are categorized into four groups: discriminative learning approach, generative learning approach, reinforcement learning approach, deep learning and optimization hybrid approach.

11.2 Discriminative learning approach

The discriminative learning approach builds the DNN models to learn the inverse mapping from the data space to the model space:

$$m = \mathcal{N}(d, \theta), \tag{11.3}$$

where $\mathcal{N}$ denotes the DNN model to approximate the inverse operators I in (11.2) and θ is the parameter set of the DNN model. The DNN architectures, training strategies, and data generating schemes diversify to accommodate different design goals and address the intrinsic difficulties of the inverse design.

The first and straightforward approach for inverse design is to build a direct DNN model to learn the inverse mapping directly, and it is categorized as the direct DNN model in this chapter. By generating labeled pairs (m, d) as the training data set, the built DNN model is trained in a supervised way to perform reliable predictions, as described in (11.3).

Various architectures of DNNs are considered to perform different design tasks. The adaptive artificial neural network (ANN) is presented for the inverse design of the thin film metamaterials consisting of graphene and Si_3N_4 [27]. The adaptive ANN introduces the adaptive batch normalization into a standard ANN structure to reduce the output error. Taking the optical spectrum as input, the adaptive ANN can directly predict the thickness of each layer in the graphene-based metamaterial with high precision and fast computing speed. Inspired by applying ANNs to model the optical spectra of metamaterials, the ANNs are applied to the inverse design of metamaterials by switching the inputs and outputs of the forward modeling ANNs [28]. In [28], two concentric multilayered cylinder metamaterials are designed by training the ANN model to predict each core layer's refractive index and radius regarding a given optical response. The deep learning approach is presented to design the metasurfaces to

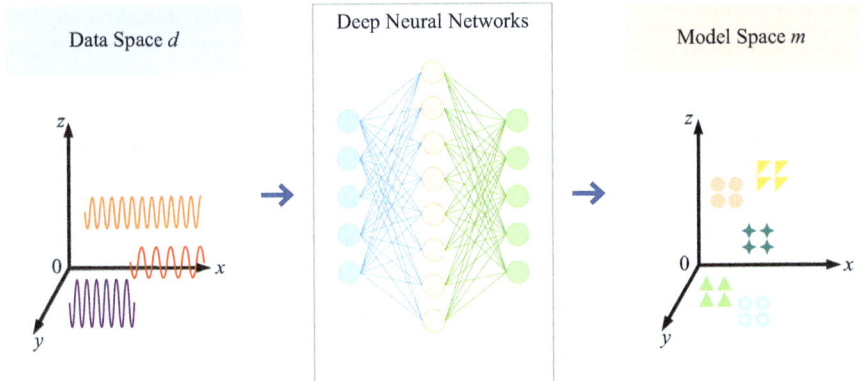

Figure 11.3   Workflow of the inverse design of metamaterial or metasurface based on a direct DNN

achieve single/multiple resonances at the desired working frequency in [29]. Eight annular models are generated and numbered in digital codes as the basic unit cells to construct the metasurface. The DNN models take the S-parameters of TE and TM modes and output the digital codes of the eight annular models. The automatic design approach is also applied to design the metasurface based on the eight predefined ring-shaped patterns with the in [30]. The FCN model is trained to design the bifocal metalens to achieve two independent foci of the orthogonally polarized light [31]. The FCN model predicts two control parameters of unit cells in the metalens by taking the transmission intensities of TE and TM polarized plane waves as input. The optical illusion is customized by intelligently arranging the elements of form-free metasurfaces based on deep learning techniques in [32], as shown in Figure 11.4. Two neural networks are trained to predict a one-hot code representing the metasurface configuration given the near-field or the far-field pattern. A conformal metasurface with curved shapes is taken as a proof of concept to achieve a wide range of in situ applications.

The convolutional neural networks (CNNs) have proven good at extracting hierarchical features from the input geometries, which is suitable for the design task. The weight sharing of CNNs can also reduce the number of parameters in the whole design process. A CNN model is adopted for accurate and fast inverse design of the plasmonic metasurfaces [33]. The input of the CNN model is the absorption spectra, and the output is a vector of parameters to optimize. As the unit cell in the plasmonic metasurface is symmetric, the absorption spectra do not change when mirroring the unit cell along the symmetry axis, leading to the one-to-many problem in the inverse design. A single CNN model cannot learn such one-to-many mapping, and its performance could deteriorate when the training data contains the one-to-many. Therefore, the symmetry of unit cells is restricted when applying the FDTD to generate training data in [33]. The performance of the CNN and ANN models is further compared, and the CNN model demonstrates better computational

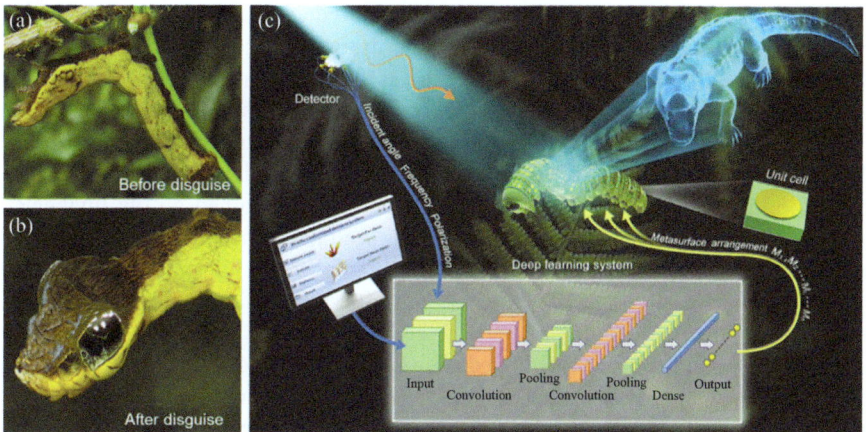

Figure 11.4   *A sketch of the global metasurface design for intelligent optical illusion (source: [32])*

precision and generalization ability. The convolutional autoencoder is connected with the CNN to design the broadband metasurface absorber in [34]. The convolutional autoencoder is trained to encode the spectra into the low-dimensional latent vectors. Then the CNN takes the latent vectors as input and predicts the structural parameters.

Autoencoder helps transform the data and model spaces into the reduced ones to address the many-to-one problem. The dimensionality reduction is performed based on the autoencoder to convert the many-to-one problem to the one-to-one problem in the inverse design of reconfigurable metasurfaces based on the plasmonic phase-change materials [35]. Due to the many-to-one mapping between the original model and data space, the autoencoders are trained to compress the original data and model spaces into the reduced latent ones. The mapping path between the reduced latent data and the model spaces is one-to-one. In the inverse design process, the reduced structural parameters can be inferred regarding the encoded optical responses, and then the original structural parameters can be decoded based on the reduced ones.

The evolutionary algorithms and machine learning techniques are studied and compared to design novel graphene metamaterials with wideband plasmon-induced transparency effect as the design target in [36]. The machine learning techniques applied in the forward modeling and inverse design include the K nearest neighbor, random forest, decision tree, and ANN model with the FDTD and Monte Carlo simulation for generating training data. The genetic algorithm (GA) is applied to tune the hyper-parameters of the ANN model for better performance. Numerical results show that all machine learning techniques are effective and random forest has minimal computing time. Furthermore, the GA, PSO, quantum GA are compared in the design task of broad bandwidth transmission spectrum, and the non-dominated sorting GA is adopted for the optimization of multiple performance metrics.

The transfer training scheme is also applied to reduce the computational load of training a large-scale DNN model. The inverse design of functional metasurfaces is performed based on the transfer learning in [37]. The metasurface is first encoded as an array of meta-atoms where 1 represents the copper, and 0 denotes vacuum. The transfer learning network (TLN) regards the predictions of reflection phases as a classification problem. The TLN is built to predict reflection phases given the encoded meta-atom arrays based on the Inception V3 network pretrained on the ImageNet data set. Only the fully connected layer of the pretrained Inception V3 network is adjusted by freezing the convolutional and pooling layers. A full-phase-span library of meta-atoms is established based on the TLN for metasurfaces' fast and accurate design.

Second, the bidirectional DNN is designed to address the one-to-many problem in the inverse design of metamaterials and metasurfaces, which is also known as the tandem model and first proposed in [38]. The training procedure of the bidirectional DNN can be divided into two phases, as shown in Figure 11.5. Here we give a brief introduction of the commonly-applied bidirectional DNN approach. First, a DNN model is trained to learn the forward modeling from the model space m to the data space d:

$$df = \mathcal{N}_f(m, \theta_f), \tag{11.4}$$

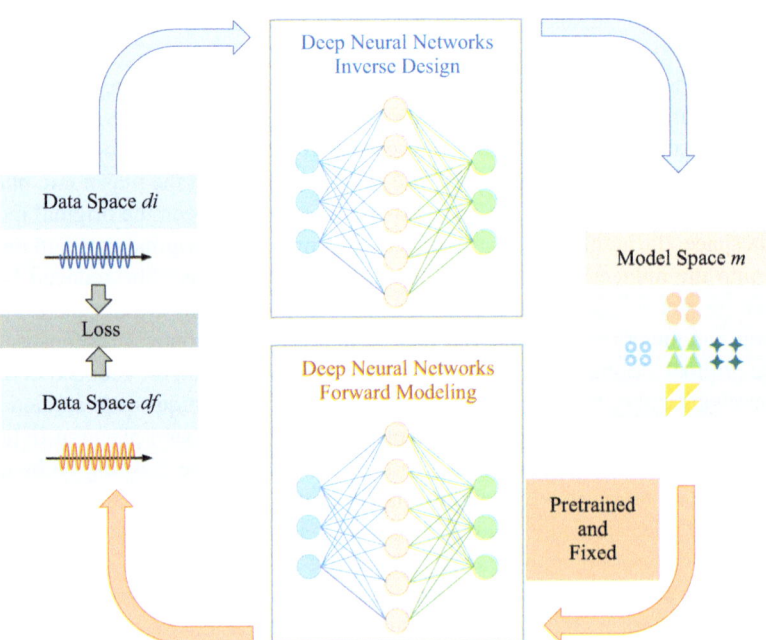

Figure 11.5   Workflow of the inverse design of metamaterial or metasurface based on a bidirectional DNN

Once the training is finished, the forward DNN model serves as the forward simulator with its weights fixed. Then the inverse DNN model is trained to learn the inverse mapping from the data space *d* to the model space *m*:

$$m = \mathcal{N}_i(di, \theta_i), \tag{11.5}$$

The objective function of the inverse DNN model is to measure the discrepancy between the desired spectra and the predicted spectra, and the predicted spectra are generated by the forward DNN model regarding the model parameters produced by the inverse DNN model. The objective function can be written as:

$$obj = \mathcal{M}(di, \mathcal{N}_f(\mathcal{N}_i(di, \theta_i), \theta_f)), \tag{11.6}$$

where $\mathcal{M}$ denotes the metric function to evaluate the error. It is noted that θ_i is to train, and θ_f is fixed. The workflow of the bidirectional DNN model is depicted in Figure 11.5. The bidirectional DNN model enforces the spectra of inverse design to be the same as the desired ones instead of minimizing the difference between the desired and predicted geometry parameters. Therefore, the one-to-many or the non-uniqueness existing in the direct DNN model can be avoided.

The deep-learning-based approach stacks two bidirectional DNN models for the inverse design of chiral metamaterials, including the primary network and the auxiliary network, in [39], as shown in Figure 11.6. The primary network focuses on the mappings between the reflection spectra and the design parameters. The auxiliary network is trained to model the mappings between the chiroptical response and the design parameters. The forward and inverse combiners are created to post-process the outputs of the primary and auxiliary networks. The proposed approach is further verified to design chiral metamaterials regarding the on-demand requirements of chiral performance, and it demonstrates high computational efficiency and accuracy. The design of three-layered spherical core-shell nanoparticles is assisted by a bidirectional DNN model comprising a design network and a spectrum network in [40]. The spectrum and design networks establish the forward and inverse relationship between the thickness and information of materials and the spectra of electric and magnetic dipoles. Compared to (11.6), the objective function additionally introduces the error between the desired and predicted design parameters in [40]:

$$obj = \alpha \mathcal{M}(di, \mathcal{N}_f(\mathcal{N}_i(di, \theta_i), \theta_f)) + (1 - \alpha)\mathcal{M}_m(mi, mt), \tag{11.7}$$

where α is the weight, *mi* and *mt* are the predicted and the desired model parameters, $\mathcal{M}_m$ is the metric function to evaluate the error. The proposed method is applied to inversely design the core-shell nanoparticles to tune the electric dipole resonances at various wavelengths, achieve the isolated magnetic dipole resonances, and produce the electric and magnetic dipole resonances simultaneously. The metamaterials consisting of split ring resonators are inversely designed by building a bidirectional DNN model in [41]. The bidirectional DNN model consists of forward and reverse neural networks trained separately to learn the forward and inverse mappings between the structural parameters and the *x*-/*y*-polarized reflectance. The structural parameters to design include the line width, open angle, and inner ring radius of the resonators. The high-quality factor resonance is designed by building a bidirectional ANN model that

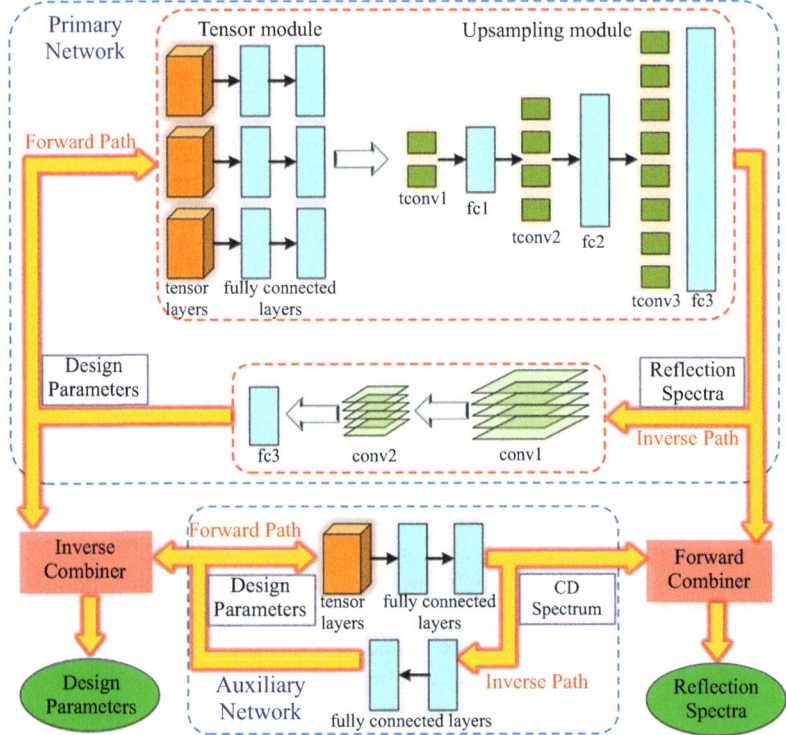

*Figure 11.6   Designing chiral metamaterials based on the deep-learning model
(Reprinted (adapted) with permission from [39]. Copyright 2022
American Chemical Society.)*

consists of two independent ANNs for forward and inverse mappings in [42]. The proposed DL-based approach is validated by designing a single optoacoustic metasurface
with the non-radiating toroidal dipoles as the building blocks. Inspired by [38], the
tandem model is built for the inverse design of mid-infrared graphene-based metamaterials with the desired optical spectra [43], as shown in Figure 11.7. The forward
DNN has six hidden layers, and it is pre-trained to be combined with the inverse DNN
with only two hidden layers. All-dielectric metasurfaces with cylindrical meta-atoms
are designed by a bidirectional neural network model in [44]. Leveraging spectral
scalability, the proposed model can address the challenges of dimension mismatch
and spectral generalizability in the inverse design of metasurfaces. Instead of building one forward predictor, multiple DNN models can be built as different forward
predictors of different design targets. A tandem neural network is utilized to design
a focusing metamirror to achieve high reflectivity and a minimal phase mismatch
in [45]. Two forward predictors are trained to predict the phase and reflectivity given
the input geometries. Then the inverse generator is connected with the two trained
forward predictors and trained to produce designs that meet the requirements of phase
and reflectivity at the same time.

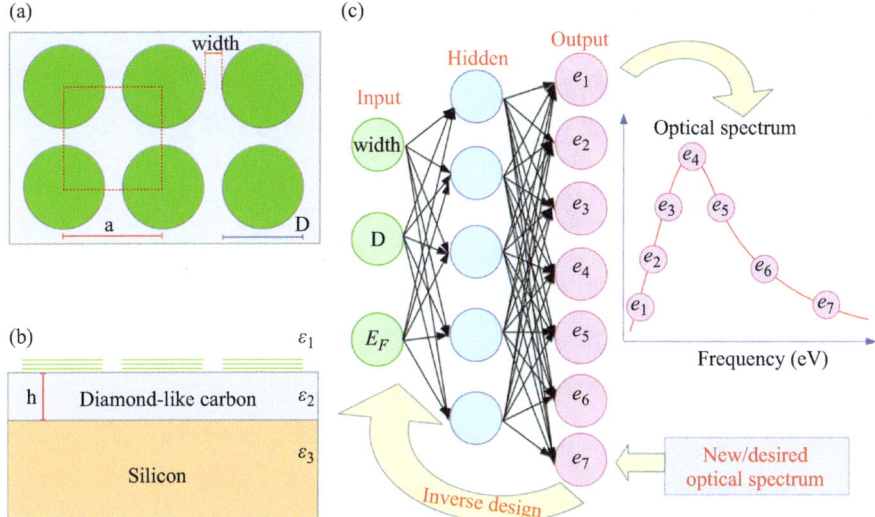

Figure 11.7   Designing the graphene plasmons based on a tandem neural network, (a) top-down view, (b) side view and (c) a forward modeling neural network (source: [43])

The DNN architectures proven effective are also adopted in building the bidirectional DNN model, such as autoencoders and residual networks. The bidirectional autoencoder model is built for the fast design of the gap plasmon-based metasurfaces that performs the polarization conversion of light in [46]. The encoder model compresses the optical responses into the low dimensional structural parameters of metasurfaces, and then the decoder is trained to output the optical responses regarding the structural parameters. The discrepancies between the spectra and structural parameters are combined to constrain the training of the bidirectional autoencoder. The tandem autoencoder is built to overcome the non-uniqueness existing in the inverse design of plasmonic metasurface structural color in [47]. The forward DNN model is first trained to predict color given the geometry. The tandem autoencoder is trained to minimize the discrepancy between the input and predicted color by cascading the pre-trained forward DNN model to the inverse DNN model. A tandem residual neural network is built to design the multiplexed supercells to construct the metal-insulator-metal metasurface in [48], as shown in Figure 11.8. The 1D CNNs are taken as basic modules of the residual neural network. The ability of a tandem network is validated to address the non-uniqueness problem in the inverse design. The alternative training strategy is implemented to improve the performance of specific design task. The bidirectional neural network is trained to achieve both forward characterization and inverse design of metasurfaces at the same time [49]. The geometry-predicting-network (GPN) and spectrum-predicting-network (SPN) are introduced based on the FCNs. Training GPN and SPN simultaneously can improve performance instead of training them alternatively, which is verified by the numerical experiment.

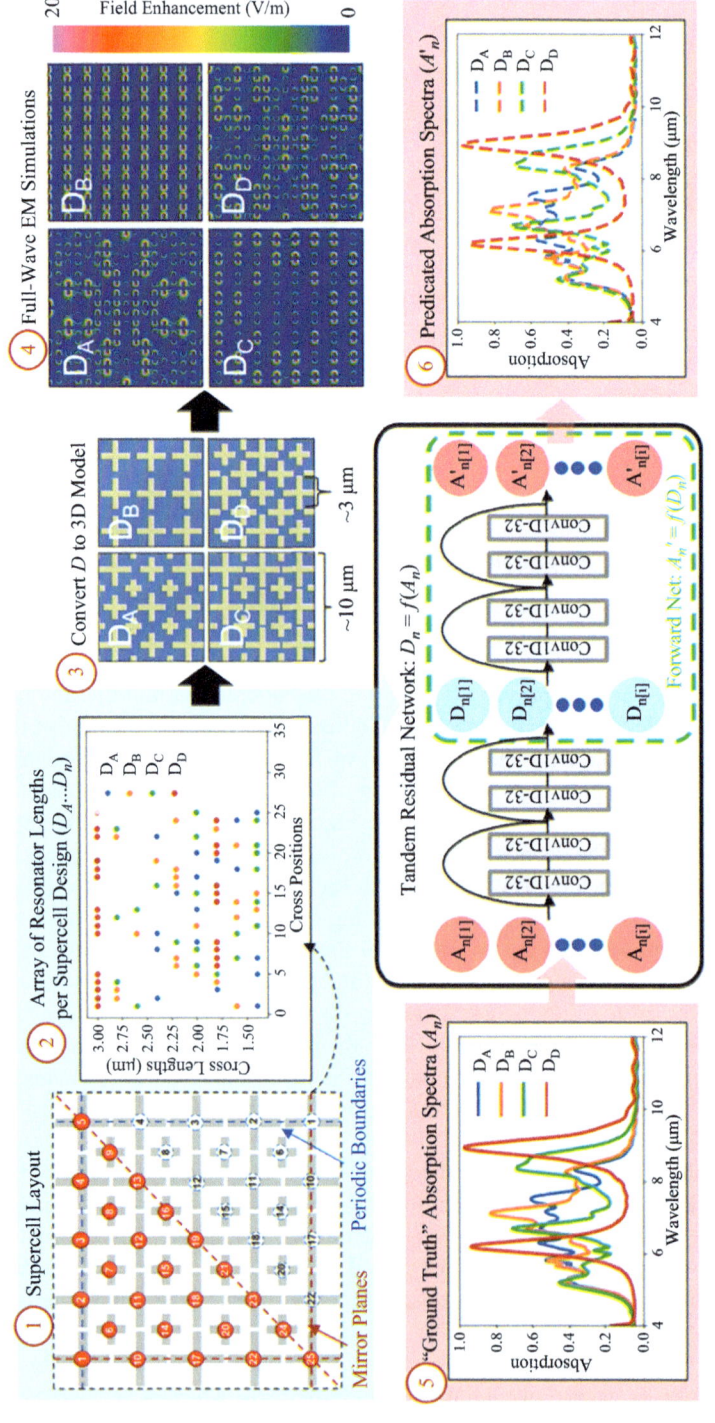

Figure 11.8 The sketch of designing multiplexed supercell metasurface based on the tandem residual neural network (source: [48])

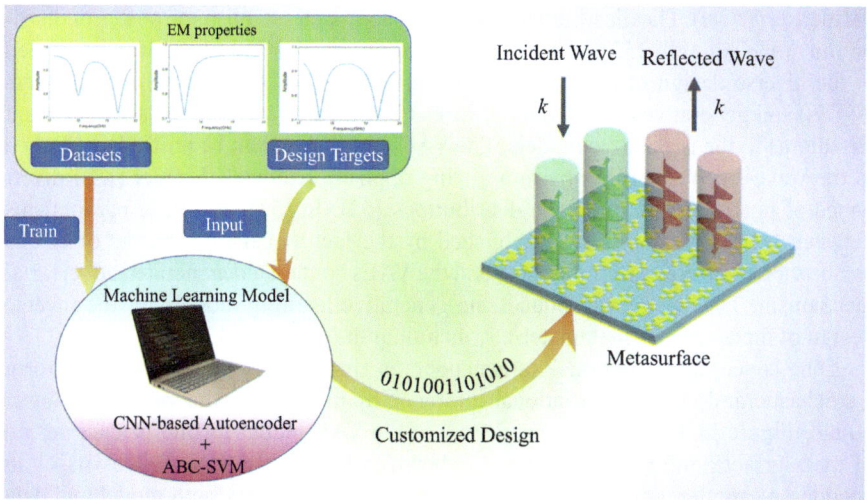

Figure 11.9   The deep learning-based approach for a generic metasurface with the matrix encoding method (source: [51])

Unlike the bidirectional DNN model, two different deep learning models can be first trained separately and then combined to finish the design task. The hybrid approach based on the autoencoder (AE) and the multilayer perceptron (MLP) is proposed in [50] for the automatic design of metasurfaces. Inspired by the similarity between the S-parameters and the acoustic signal, the S-parameters are first smoothed by a Gaussian filter, then transformed by Cestrum analysis. Taking the pre-processed S-parameters as input, the AE functions to compress and extract the compact features with reduced dimensionality. The MLP with sigmoid function is trained with the L-2 regularization to match the extracted features and the metasurface structures. The effectiveness of the proposed approach is validated by designing a triple-band absorber with the desired absorption rate. The convolutional neural network-based autoencoder (CNN-AE), support vector machine (SVM), and the artificial bee colony (ABC) are combined for the inverse design of EM metasurfaces in [51], as shown in Figure 11.9. The CNN-AE is trained to compress the input EM property as the representative features. The ABC optimizes the SVM with the Gaussian kernels to match the metasurface structures to the representative features produced by the CNN-AE. Compared to the MLP applied in [50], the SVM model can overcome the dependence on the nonlinear activation function.

11.3   Generative learning approach

The generative learning approach aims to learn the joint distribution of the input and output [52]. In the context of the inverse design, the joint distribution can be defined as $p(d, m)$. The discriminative learning approach establish the posterior distribution $p(m|d)$ directly [52]. The generative learning models have been widely applied in the inverse design of metasurfaces/metamaterials [20]. Variational autoencoders (VAEs) and generative adversarial networks (GANs) are two different and important paradigms in the generative models. The VAEs and GANs can be unified and linked to the wake-sleep algorithm by minimizing opposite Kullback–Leibler (KL) divergence of posterior and inference distributions [53]. In [53], the close parallelisms between GANs and VAEs are established by the fact that the generators of GANs can be viewed as posterior inference, and the VAEs contain a degenerated adversarial mechanism. This section introduces the generative learning models for the inverse design of metasurfaces/metamaterials, including the GANs and VAEs.

The concept of variational autoencoders is first introduced in [54] to perform a stochastic and efficient variational inference by minimizing the variational lower bound. Figure 11.10 shows the workflow of the VAE model for the inverse design of metasurfaces and metamaterials. As shown in Figure 11.10, VAE consists of an encoder, a sampler, and a decoder. The encoder takes as input both model and data space, then encodes them into the continuous Gaussian distributed latent variable space with a reduced dimension. Conditioned by the data space, the decoder reconstructs the desired model space regarding the latent variable z produced based on the Gaussian distribution by the sampler. Let $q_\theta(z|x)$ and $p_\phi(x|z)$ denote the posterior distribution and likelihood distribution of encoder and decoder respectively, the Evidence Lower Bound (ELBO) can be written as [54]:

$$\mathcal{L}(\theta, \phi, x_i) = -D_{KL}\left(q_\theta\left(z \mid x_i\right) \| p(z)\right) + E_{q_\theta(z|x_i)}\left[\log p_\phi\left(x_i \mid z\right)\right] \tag{11.8}$$

ELBO is the core of VAEs, which is the lower bound of the target that VAEs aims to maximize. It can be observed that the KL divergence regularizes the form of the posterior distribution in the ELBO.

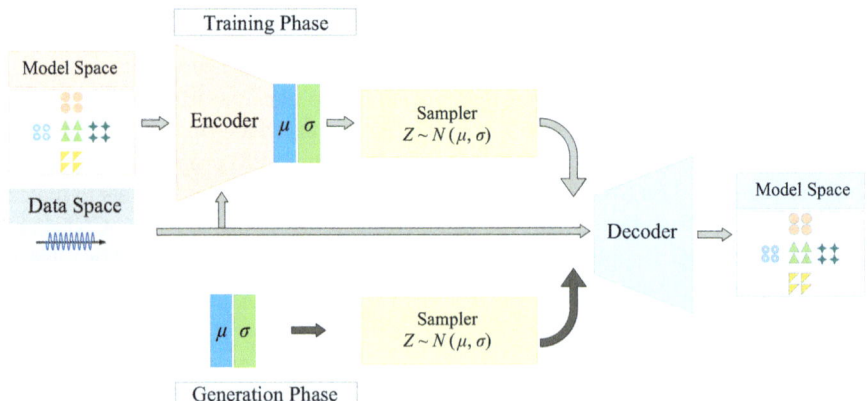

Figure 11.10   Workflow of a variational autoencoder model for the inverse design of metasurfaces and metamaterials

The probabilistic representation of the metamaterial structure is established via a VAE model to accelerate the inverse design of metamaterial in a semisupervised training manner [55]. The VAE model encodes the design geometry and optical response into the latent space with the predefined Gaussian prior distribution. The optimal structure is retrieved by the VAE model based on the sampled latent space and the desired optical response. A self-supervised VAE model is built for the inverse design of nanophotonic structures in the reflective metasurfaces [56]. The encoder of VAE is allowed to simultaneously predict the latent vectors and the reflection spectra. The decoder can reconstruct the desired metasurface layout regarding the sampled latent vectors and the reflection spectra. The objective function of the VAE model is defined as [56]:

$$
\begin{aligned}
L &= L_{latent} + L_{recon} + \alpha L_{spec} \\
&= KL[E_\phi(z|x)||P(z|x)] - \mathcal{E}_{E_\phi(z|x)}[\log(D_\theta(x|y,z))] + \alpha(y - \hat{y})^2,
\end{aligned}
\tag{11.9}
$$

where E_ϕ and D_θ denote the encoder and the decoder parameterized by ϕ and ϕ, z is the latent variable, $P(z|x)$ is the standard Gaussian distribution of z, x, y and $\hat{y}$ are input layout, output spectra and labeled spectra respectively. The adversarial autoencoders (AAEs), consisting of an encoder, a decoder, and a discriminator, are adopted for the topology optimization of metasurface for thermal emitter [57], as shown in Figure 11.11. Similar to VAEs, the encoder tries to compress the given topology into a compact latent space, the decoder is trained to perform a reliable reconstruction. However, the learning of AAEs is adversarial by building a discriminator to differentiate between the latent space and the predefined topology distribution. Compared to the VAEs, the advantage of AAEs lies in that the distribution of latent space can be defined precisely, not limited to the normal distribution.

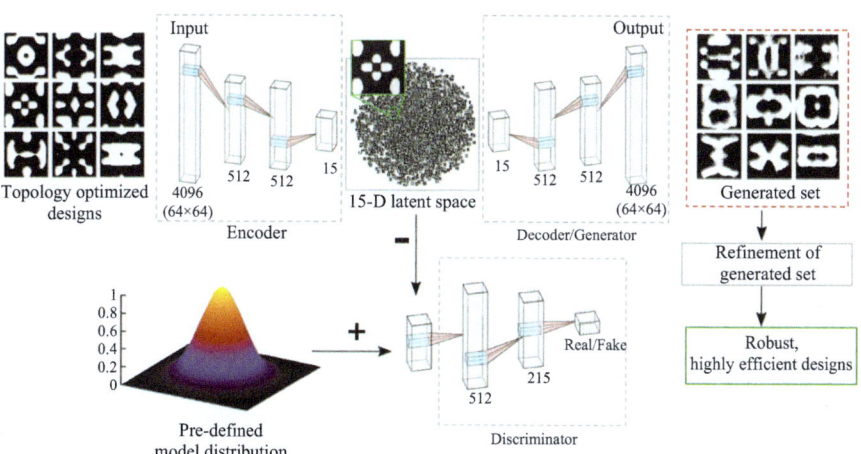

Figure 11.11   Adversarial autoencoder-assisted topology optimization of metasurfaces (source: [57])

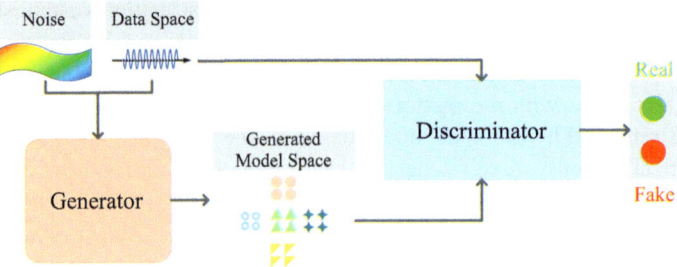

Figure 11.12   Workflow of a generative adversarial network model for the inverse design of metasurfaces and metamaterials

Generative adversarial networks are first proposed in [58]. GANs usually consist of a generative model and a discriminative model. Figure 11.12 depicts the workflow of applying GANs to the inverse design of metasurfaces and metamaterials. The generator produces the structural parameters regarding the input noise or/and property data. The discriminator is trained to differentiate the generated predictions from the desired ground truth. The generator G and discriminator D try to solve the minmax problems [58]:

$$\min_G \max_D V(D, G) = \mathbb{E}_{x \sim p_{\text{data}}(x)}[\log D(x)] + \mathbb{E}_{z \sim p_z(z)}[\log(1 - D(G(z)))] \quad (11.10)$$

where p_{data} and p_z denote the probability distributions of data and noise. The generator reaches optimal when the generator's distribution p_g equals p_{data}. Such an adversarial training process can ultimately lead to mutual improvement of both the generator and discriminator.

The plain GANs are first applied to solve the design task. The deep learning-based approach is proposed as a candidate solution to address the challenge of designing random and complex metasurfaces in [22], as shown in Figure 11.13. The purely reflective metasurface is taken as an example, with the co-polarized reflectance (coPR) as the design target. Drawing on bidirectional neural networks and GANs, the proposed network consists of three parts: an inverse generator to generate the candidate design, a pre-trained forward predictor to predict the coPR, and a discriminator to differentiate the predicted design. The alternate training scheme is adopted to tune the generator and discriminator.

Besides, the conditional generative adversarial networks (CGANs) and the conditional deep convolutional generative adversarial networks (cDCGANs) are adopted by introducing conditions to control the design process. The conditional deep convolutional generative adversarial networks (cDCGANs) are first applied in the data-driven design of nanophotonic antennae given the desired reflection spectra [59]. The cDC-GAN, a GAN variant, can overcome the instability of vanilla GAN and generate a stable Nash equilibrium solution. Conditioned by the reflection spectrum, the generator takes the noise as input and produces the corresponding probability distribution

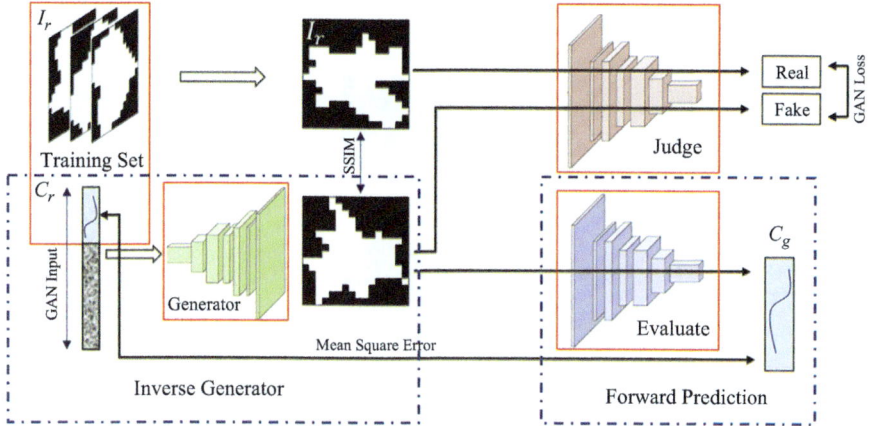

Figure 11.13   Generative adversarial network for the metasurface design
(source: [22])

function (PDF) of the antenna. The discriminator tries to differentiate between the desired and generator-produced designs of metasurfaces. The cDCGAN is trained to design the anisotropic metasurface to achieve full phase properties in ultrawide-band [60]. The generator and discriminator are trained alternatively to minimize two independent pre-defined loss functions. The pre-trained forward predictor constitutes a loop with the cDCGAN to select the candidate designs. The design task can also be viewed as an image processing problem to solve under the framework of image processing techniques. The image-based deep learning framework is presented to achieve the inverse design of materials and structures in the context of nanophoton-ics [61]. The structure of the applied cDCGAN is shown in Figure 11.14. Two classes of absorbing metasurfaces are considered: metal-insulator-metal and hybrid dielectric metasurfaces. The information of metasurfaces is first encoded into an RGB image, including the geometries, material properties, and thicknesses. The spectra are fed into the generator along with a latent vector for generating the corresponding candi-date design image. The discriminator attempts to differentiate the generated image from the ground truth. The 3D metasurface models can be reconstructed based on the generated image. The generative data-driven approach based on GAN is presented for the high-throughput inverse design of photonic crystals (PhC) [62]. In GAN, the generator is trained to produce the PhC unit cells that can deceive the discriminator with the noise as input. The discriminator is tasked with differentiating the PhC unit cells produced by the generator. The CGAN, widely applied in the image-to-image translation, is also employed to the inverse design of PhC in [62]. Compared to the GAN, CGAN adopts another conditional input to guide the inverse design. The condi-tional input is the discretized inclusion outline of the PhC and the CGAN can translate the conditioned outline to a permittivity profile hosting the desired properties. The conditional GAN (CGAN) and Wasserstein GAN (WGAN) are combined to design

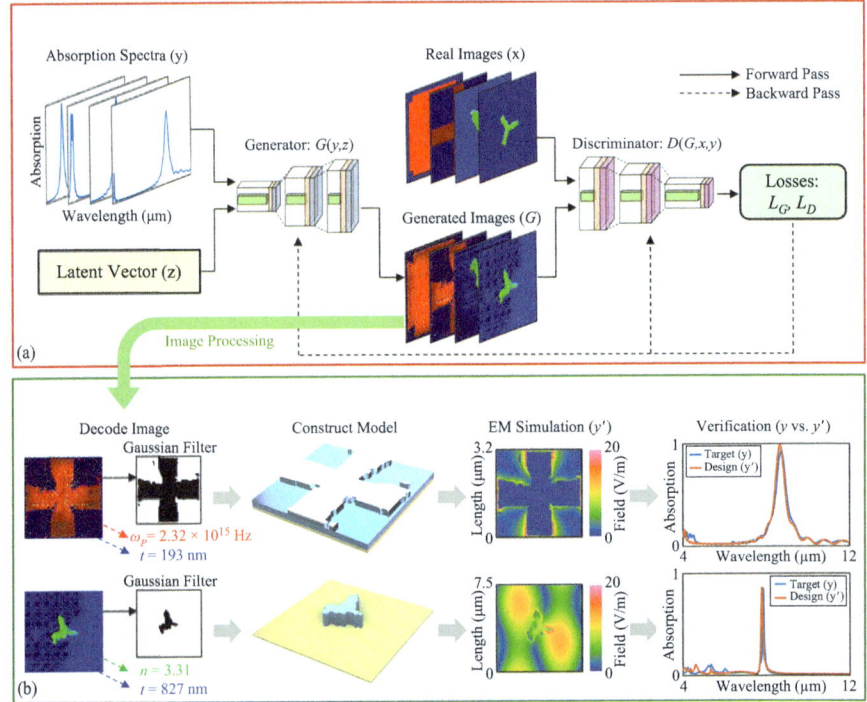

Figure 11.14   Conditional deep convolutional generative adversarial network for the design of photonic structures, (a) conditional deep convolutional generative adversarial network (cDCGAN) training and (b) multi-class metasurface design (source: [61])

multifunctional metasurfaces in [63], as shown in Figure 11.15. The proposed GAN model is trained to produce a target design given the sampled Gaussian noise vector conditioned with the pre-defined auxiliary information. The Wasserstein distance between the target design y and the generated design y' is adopted to stabilize the training [63]:

$$W\left(P_{\text{data}}, P_{\text{G}}\right) \approx \sup_{\|D\|_{L} \leq 1} \left\{ \mathbb{E}_{y \sim P_{\text{dat}}}\left[D(y \mid x)\right] - \mathbb{E}_{y' \sim P_{G}}\left[D\left(y' \mid x\right)\right] \right\}, \qquad (11.11)$$

where P_{data} and P_{G} are the true and generated EM response sets, D is the Wasserstein distance produced by the discriminator.

The predictor or critic is introduced to add another spectra or property constrain for the training of the generator in addition to the geometry constrain from a discriminator. A generative model is presented based on the GAN for the inverse design of unit cell patterns of metasurfaces [64]. The architecture of the generative model is illustrated in Figure 11.16. The proposed generative model consists of a generator, a critic, and a simulator. The simulator is a pre-trained neural network for estimating the optical spectra given the specific metasurface pattern. The simulator can reduce the

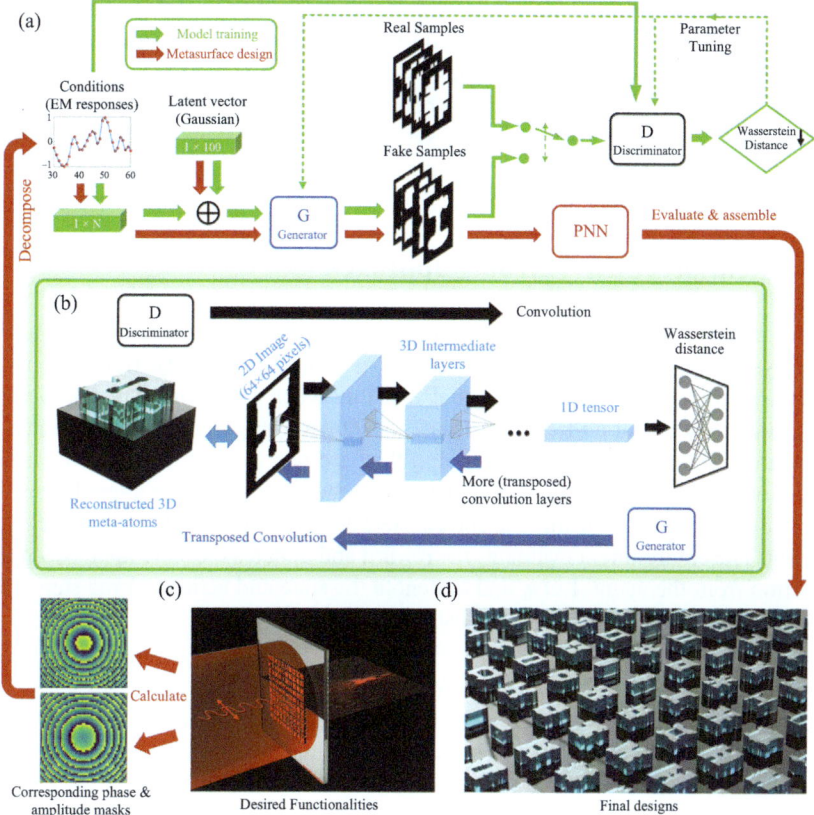

Figure 11.15 The neural network for designing meta-atoms (source: [63])

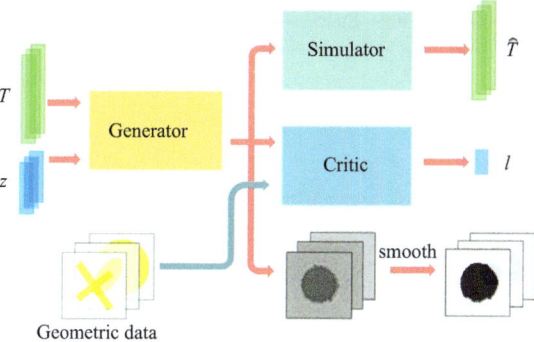

Figure 11.16 Architecture of the generative model for metasurface design. (Reprinted (adapted) with permission from [64]. Copyright 2022 American Chemical Society.)

time-consuming electromagnetic full-wave simulation and control the accuracy of the optical spectra of the generator-produced metasurfaces. The generator produces the metasurface patterns in response to the desired optical spectra along with the noise. The critic is trained to assess the distance between the distributions of the user-defined and generator-produced geometric data sets. Such distance can guide the generator to produce patterns that have similar features to the user-defined ones.

11.4　Reinforcement learning approach

Reinforcement learning (RL), a framework of experience-driven autonomous learning, plays a vital role in the field of artificial intelligence [65]. RL trains agents to learn the optimal actions or make optimal decisions when interacting with the specific environment via a trial-and-error process. As shown in Figure 11.17, the agent and environment interact with each other, and they are two primary components in the perception-action-learning system of the RL. In the interaction, the agent learns the policy to take actions regarding the current state and the rewards returned by the environment. The environment evaluates the reward and updates the state upon accepting the action from the agent. Let s_t and a_t denote the state and action at time step t, the policy π is the posterior probability distribution of a_t given s_t [65]:

$$\pi = p(a_t|s_t) \tag{11.12}$$

The environment produces the reward r_t when receiving the action a_t. Before the terminal of actions at time step T, a series of a_t, s_t and r_t is produced, and the rewards are accumulated as the return R that can be written as [65]:

$$R = \sum_{t=0}^{t=T} \gamma_t r_t, \tag{11.13}$$

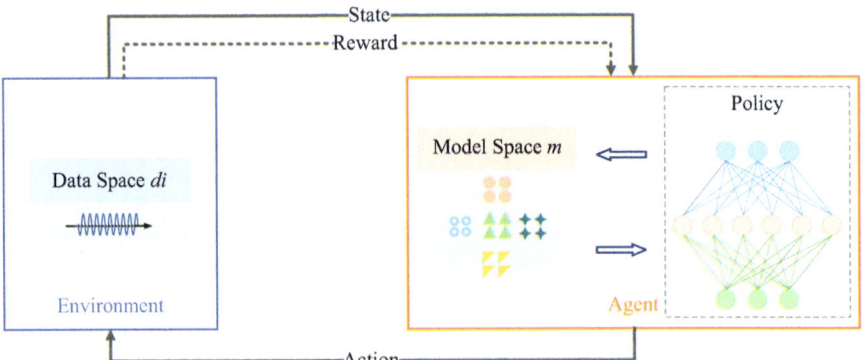

Figure 11.17　Workflow of the reinforcement learning approach for the inverse design of metamaterials or metasurfaces

where γ_t denotes the discount factor to weigh the rewards at different time steps. In RL, the ultimate goal is to find an optimal policy that can produce the maximum expected return R [65]:

$$\pi_{best} = \arg\max_{\pi} E(R|\pi) \tag{11.14}$$

The Markov property is widely applied in RL by assuming that the next state is only dependent on the current state. The traditional RL faces several long-standing challenges, such as lack of scalability and limited dimensionalities of the state and action spaces. The introduction of deep learning enables RL to solve high-dimensional intractable decision-making problems [65]. The deep reinforcement learning (DL) has been applied in a wide range of problems, such as video games [66], robotics [67], go playing [68], and indoor navigation [69]. Various DRL algorithms have been reported, such as the value functions-based method [70], policy search-based method [71], and actor-critic method [72].

The plain DRL can make design process of metamaterials or metasurfaces more efficient by carefully defining the roles of the agent, environment and reward in DRL. DRL is applied to design the optical multilayer films by viewing the task as a sequence generation problem [73]. Reinforcement learning is implemented by applying proximal policy optimization to train the sequence generation network. The sequence generation network is built by combining a gated recurrent unit (GRU) and two multilayer perceptrons (MLPs). The GRU can predict the material and thickness of each layer with the memory of the history sequence. Two MLPs are trained to predict the logit vectors of materials and thicknesses corresponding to the pre-defined sets. With the simulator named as TMM in [74], the cumulative reward of the RL process is defined as:

$$G(\mathscr{S}) = 1 - \frac{1}{K}\sum_{k=0}^{K}\frac{1}{J}\sum_{j=0}^{J-1}\left|T^{\mathscr{S}}\left(\lambda_j,\delta_k\right) - \tilde{T}\left(\lambda_j,\delta_k\right)\right| \tag{11.15}$$

where $T^{\mathscr{S}}\left(\lambda_j,\delta_k\right)$ is the spectrum of the generated structure $\mathscr{S}$ given the wavelength λ_j and incident angle δ_k, $\tilde{T}\left(\lambda_j,\delta_k\right)$ is the ground truth. Let $\pi_\theta(a|s)$ denote the policy of agent parameterized by θ for generating the action a given the state s, the goal of the RL process is to maximize the expected rewards [74]:

$$J(\theta) = \mathbb{E}_{\mathscr{S}\sim\pi_\theta}[G(\mathscr{S})] \tag{11.16}$$

The parameters θ is optimized by the gradient descent method where the proximal policy optimization (PPO) is applied to calculate the gradient with respect to θ [74]:

$$g = \nabla_\theta \mathbb{E}_{\mathscr{S}\sim\pi_\theta}\left[\min\left(r(\theta)A_{\theta_v}(\mathscr{S}), clip\left(r(\theta), 1-\varepsilon, 1+\varepsilon\right)A_{\theta_v}(\mathscr{S})\right)\right], \tag{11.17}$$

where $r(\theta) = \frac{P_\theta(\mathscr{S})}{P_{\theta_{old}}(\mathscr{S})}$ is to weigh the importance, *clip* disincentivizes the large update steps, ε is a hyperparameter to affect the actual update size. The effectiveness of the proposed approach is validated by two optical design tasks: ultra-wideband absorber and incandescent light bulb filter. The structural design task of the 1D freeform

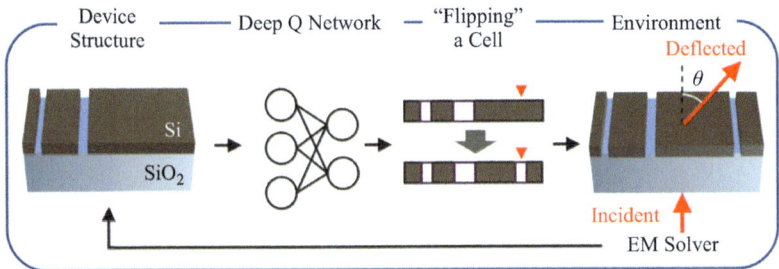

Figure 11.18   Design of 1D metagratings based on reinforcement learning.
(Reprinted (adapted) with permission from [75]. Copyright 2022
American Chemical Society.)

metagrating is viewed as a DRL problem in [75], as summarized in Figure 11.18. The 1D metagrating is represented by a 1D array, of which cells filled with silicon are denoted as $+1$ and ones filled with air denoted as -1. In the DRL, an action switches the $+1$ and -1 of the 1D array. The environment simulates the deflection efficiency and returns the rewards to the agent given the current metagrating structure.

The double deep Q learning (DDQL) is an important variant of DRL for better stability and performance. The DDQL is applied to optimize the color generation by designing the parameters of dielectric nanostructures [76]. The color generation depends on the reflection or transmission spectra that are determined by the structural parameters of nanostructures. In the DDQL, the return formulated in (11.13) is replaced by a value function $Q(s, a)$. The $Q(s, a)$ can be derived from the Bellman equation [76]:

$$Q(s, a) = r(s, a) + \gamma \max_a Q(s', a) \tag{11.18}$$

With $Q(s, a)$ representing the maximum discounted future reward, the policy is learned to choose the action with the highest Q at a specific state. The state in DRL contains the dimensions of silicon-based nanostructures, including the nanodisk diameter and thickness, the distance between nanodisks, Si_3N_4 layer thickness. The action guides the geometrical parameters of the nanostructures. The reward is defined based on the difference between the predicted and desired colors. Two similar fully connected networks are built in DDQL. The first is the main network to map from the state space to Q, and the second is the target network to estimate the Q value of the action. The proposed approach is validated to design the nanostructures with purer red, green and blue colors. The double deep Q-learning network (DDQN) is trained to learn the optimal policy of designing the biomimetic ultra-broadband perfect absorbers [77]. The absorbers are made up of chromium and adopt the moth-eye structures. In DDQN, two neural networks are built: the target network for predicting the Q value of the specific state and the policy network for choosing the action. The loss function of the target network is defined as the Huber loss as [77]:

$$L = \begin{cases} \frac{1}{2}(x_i - y_i)^2 & \text{for } |x_i - y_i| \leq 1 \\ |x_i - y_i| - \frac{1}{2}, & \text{otherwise} \end{cases} \tag{11.19}$$

The state consists of the geometrical parameters of absorbers to optimize, including the periodicity of the unit cell, the diameter, height, curvature radius of the moth-eye structure, the material and thickness of the spacer layer, the material of the metallic substrate. The environment is the moth-eye structure absorber, and the reward is defined by penalizing the low absorption [77]:

$$\text{reward} = \begin{cases} -10 & \text{if absorption} < 85\% \\ \left(\frac{\text{absorption}}{90}\right)^9 - 1 & \text{if absorption} > 85\% \\ 10,000 \text{ (and end the episode)} & \text{if absorption} > 99\% \end{cases} \tag{11.20}$$

The penalization on the low absorption can enforce the agent of RL to reach the terminal state quickly. The absorptions are simulated by the commercially FDTD solver with the perfectly matched layers. The DDQL is also applied to design the metasurface holograms with the efficiency as the target in [78]. The main and auxiliary networks are built, the former for predicting actions and the latter for updating the weights of the main work. The action taken by the agent is defined as the changes of geometries and materials.

The traditional RL algorithm can be coupled with DL to perform efficient search of an optimum. The inverse design strategy for structural color is designed by combining the supervised learning (SL) models and the reinforcement learning algorithm in [79]. The proposed strategy comprises three steps: dataset establishment, SL models, and RL algorithms. The dataset is established by applying the finite element method to simulate the color properties of different geometries, as shown in Figure 11.19(a). Three SL models are trained to learn the mappings between the geometries and colors: the forward kernel ridge regression (KRR) model for forward mapping from color properties to geometries, support vector classification (SVC) model for classifying the color brightness, and the backward KRR model for the inverse mapping from geometries to color properties. Due to the intrinsic non-uniqueness of inverse design, the RL is implemented to find the optimal geometry given the desired color. The backward KRR model is utilized to generate a reliable and reasonable geometry as the initial guess of the RL algorithm regarding the desired color property. Then the greedy algorithm is employed to search the geometry space to find the current optimal value. The SVC model monitors whether the structural color is qualified or not during the iterative process. The forward KRR model helps guide the update of the geometry. The workflow of the RL algorithm is depicted in Figure 11.19(b).

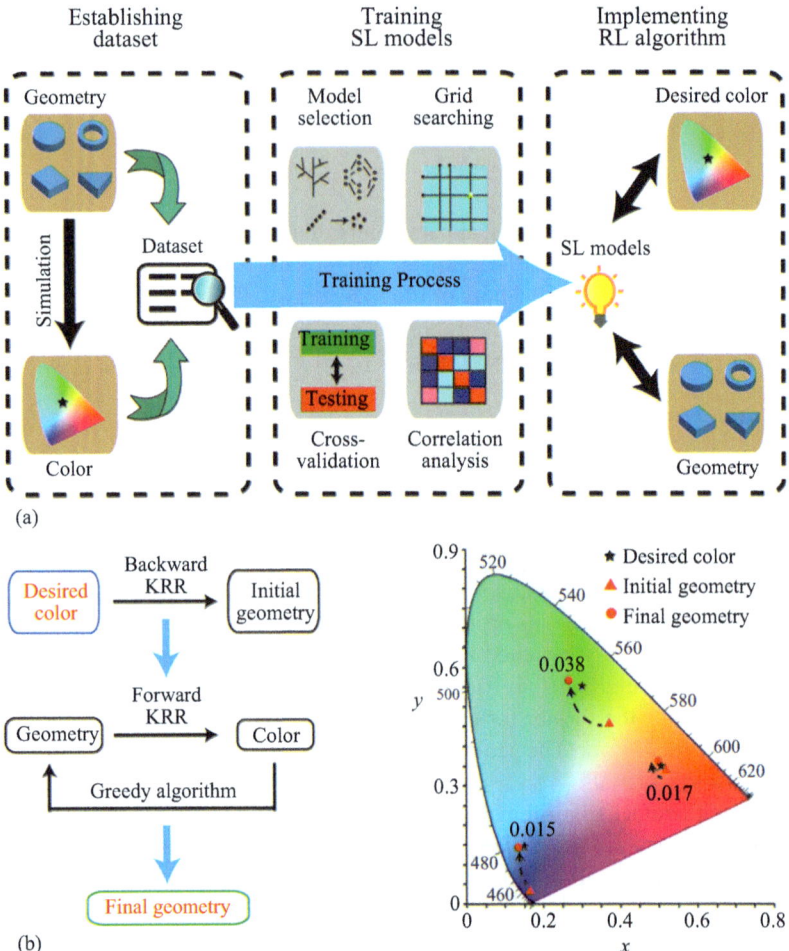

Figure 11.19   The inverse design strategy for structural color, (a) workflow of the inverse design strategy for structural color and (b) workflow of the RL process (source: [79])

11.5   Deep learning and optimization hybrid approach

The hybrid approach combines the advantages of both deep learning techniques and optimization methods. Figure 11.20 shows the workflow of the hybrid approach. The optimization method searches the model space m or the encoded latent space z via an iterative process. The searching strategies of the model space m include the gradient based strategy and the non-gradient based strategy. In the gradient-based strategy, the neural adjoint (NA) method is adopted to calculate the required gradient based on the DNN model. In the non-gradient-based strategy, the DNN is usually trained

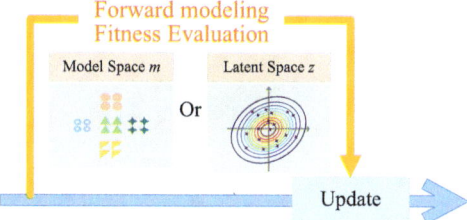

Figure 11.20   *Workflow of the deep learning and optimization hybrid approach*

to predict the corresponding properties with significantly reduced computing time. The latent space *z* is usually generated by the generative learning approach, such as VAEs, GANs, and their variants. The latent space *z* can enable an efficient search by reducing the dimensionality of the model space *m*. The trained generator of VAE or GAN needs to be employed to perform the mapping from *m* to *z* for evaluating the fitness to guide the optimization process.

The first non-gradient-based strategy to search the model space *m* is the direct search method. The inverse design of all-dielectric metasurfaces is accelerated by combining the fast forward dictionary search (FFDS) and deep learning techniques [80]. The DNN model is built by combining the FCN and CNN to perform reliable spectra predictions regarding the geometric data. The entire set of spectra is generated regarding all geometric combinations with a total of 13^8 spectra. The FFDS can search the entire spectra set to find the optimal metasurface structure given the desired spectra. The forward ANN model is built to accelerate the design of subwavelength grating (SWG) couplers in [81]. The FDTD is applied to perform the optical simulation to generate training data. The brute-force parametric sweep is connected with the trained ANN model by taking advantage of the ultrafast speed of the ANN predictions, which is further validated in the design task of the polarization-insensitive SWG couplers. The direct search algorithm is combined with deep learning techniques to design the broadband and wide-field-of-view metalenses consisting of free-form meta-atoms in [82]. The DNN model is trained to produce meta-atom designs by exploring a wide range of geometric degrees of freedom in an efficient way. This is also suited for the direct search algorithm that is good at handling meta-atoms with large varieties. The 3D all-dielectric metasurfaces are designed by the objective-driven deep learning approach in [83]. The design loop is summarized in Figure 11.21. The FCN model is trained for forward modeling to reduce the computing time of full-wave electromagnetic simulation with reliable and accurate predictions of transmission and phase spectra. The closed-loop design framework is proposed by connecting the FCN model with the meta-atom model generator. The FCN model predicts the EM properties of the current design, and the model generator tunes the structural parameters based on the discrepancy between the predicted and desired EM properties.

Another non-gradient-based strategy to search the model space *m* is the evolutionary approach, such as particle swarm (PSO), differential evolution (DE), and genetic algorithm (GA). The deep learning technique and binary particle swarm optimization

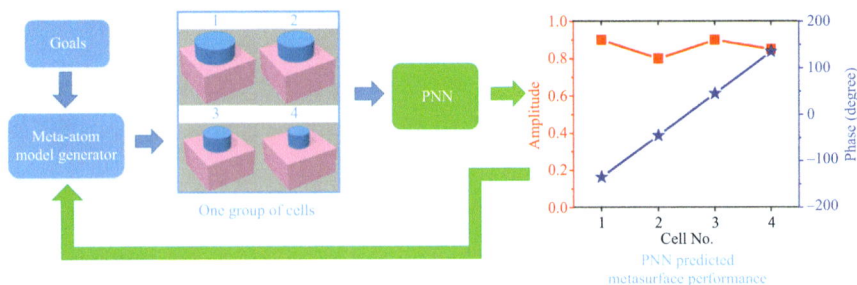

Figure 11.21   Design loop of meta-atom (Reprinted (adapted) with permission from [83]. Copyright 2022 American Chemical Society.)

(BPSO) are coupled for the design of the anisotropic metasurfaces [84]. The metasurfaces are represented by meta-atoms in the digital coding manner. The CNN model is trained to predict the phase responses given the coding patterns, which further accelerates the calculation of fitness in the BPSO. Three metasurfaces are designed and fabricated by the proposed strategy to achieve dual-beam forming, and these two beams host different polarizations. The meta-atom is optimized based on the genetic algorithm with the Inception V3 network as the forward simulator in [85]. The Inception V3 network can significantly improve efficiency by directly mapping the phase and meta-atoms. The GA can search for the optimal meta-atoms with the desired phase responses at orthogonal polarization. A hybrid approach combines the particle swarm optimization and the DNN model for the inverse design of high-contrast-index gratings, which is termed the metamodel-based optimization (MmBO) scheme in [86]. The MmBO scheme is computationally faster than the metaheuristics ones by applying the trained DNN model to evaluate the candidate designs. The MmBO scheme is extended to design the 3D all-dielectric metasurfaces to reproduce the desired colors in [87]. The hybrid approach for designing the metamolecules is presented by consolidating the compositional pattern-producing network (CPPN), and cooperative coevolution (CC) in [88]. The CPPN is trained in a pixel-to-pixel manner by encoding the pixel coordinates into a low-dimensional latent space. Then the CC is combined with the CPPN to optimize an independent meta-atom via an iterative process. The design of the whole metamolecules is finished by assembling all the optimized independent meta-atoms. The surrogate-assisted DE algorithm is presented to design the extended unit cell metagratings in [89], as shown in Figure 11.22. The performances of ResNet, DenseNet-I, and DenseNet-II are compared for the forward modeling between the geometrical parameters and the corresponding spectra. In the iterative process of the DE algorithm, the trained DenseNet-II can provide fast and reliable spectra predictions to accelerate the evaluation of fitness. The PSO is applied to efficiently search the configurations of metamaterial absorbers with the pre-trained DNN models predicting the reflection coefficients [90]. The absorption and diffusion of the meta-atoms are considered in the optimization process of PSO. A deep learning-based inverse strategy is presented for few-layer metasurfaces by combining the CNN

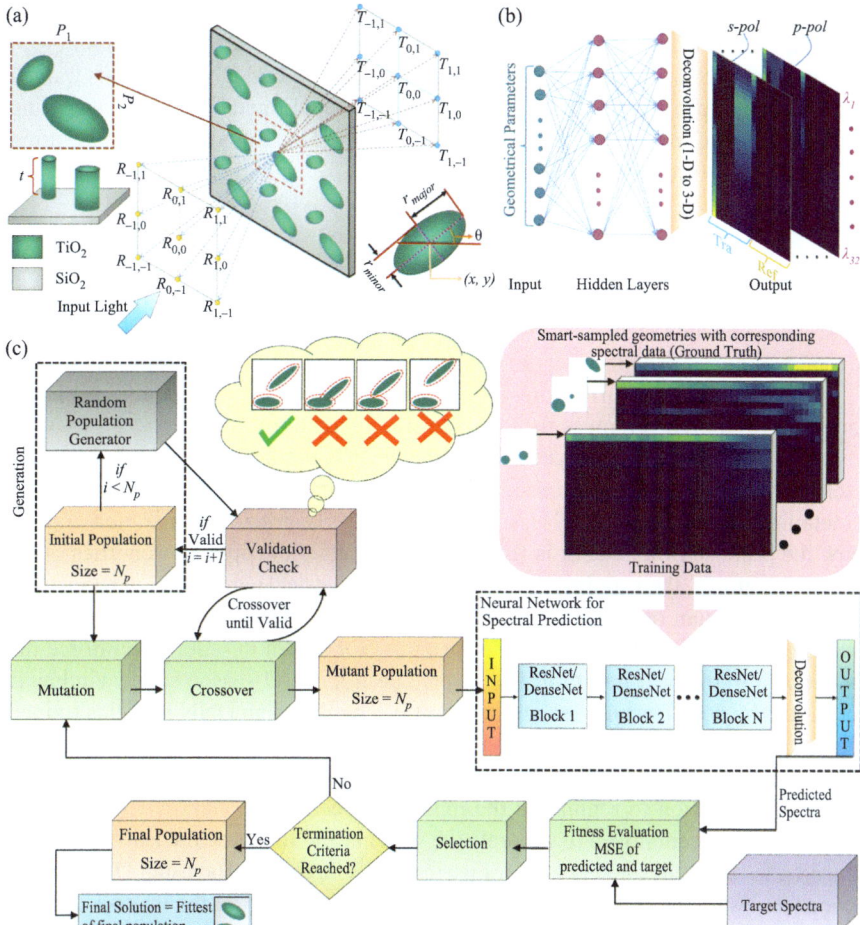

Figure 11.22   Surrogate-assisted differential evolution optimization for metagrating design (source: [89])

model, evolution algorithm, and matrix theory of multiple optics in [91]. The CNN model is trained to predict the scattering matrix given the monolayer metasurface. The scattering matrix of the few-layer metasurface is further calculated to evaluate the fitness score for the evolution algorithm according to the matrix theory.

The DNN model can also help predict good initializations for the optimization algorithm instead of accelerating the property evaluations. The surrogate-assisted approach is presented for the inverse optics design, and a 16-layered thin-film is designed to validate its effectiveness [92]. The forward DNN model is first trained to learn the forward mapping from the geometries to the spectra. Then the forward DNN is coupled with the DE methods to find or select good initial guesses of the layouts, which can further accelerate the computing of the differential evolution algorithm. A similar approach is applied for the inverse design of multilayered thin-films in [93] by pairing the DNN model with the DE method. The forward DNN model is trained to perform good preselections or initializations for the DE methods. Active sampling is also introduced to improve the performance of the DNN model by incorporating an uncertainty estimation in the loss function [93]:

$$
L(y, \hat{y}) = \frac{\sum_{i=1}^{B} \left(\frac{(y-\hat{y})^2}{2\sigma_i^2} + \frac{1}{2} \log \sigma_i^2 \right)}{B} + \lambda \sum_{j=1}^{K} |w_k|^2, \tag{11.21}
$$

where B is the batch size, σ_i is hyperparameters learned by the DNN model, K is the total number of weights, λ controls the strength of the regularizations. In (11.21), $\sum_{j=1}^{K} |w_k|^2$ is the $L2$ regularization of the DNN weights, and $\frac{1}{2} \log \sigma_i^2$ forces the DNN model to assign high variances for the undesired predictions.

The Bayesian framework is coupled with the DNN model to solve the design task. A self-learning framework is presented to model and optimize the optical chirality of metallic nanostructures in [94]. Consolidating the Bayesian optimization and the CNN, the proposed framework is first trained to predict the optical properties given the nanostructures, then applied to optimize the geometrical parameters of nanostructures. The Bayesian optimization, a derivative-free algorithm, is applied to sample a set of optimized inputs based on the current CNN model in order to reinforce the CNN model in the next iteration. The inverse design of VO_2-based smart window is accelerated by deep learning techniques in [95]. The Bayesian DNN model is trained to learn the forward mapping between the structural parameters and the resultant merits. Compared to the plain DNN model, the Bayesian DNN model can incorporate the prior information to regularize the training process. The Markov Chain Monte Carlo (MCMC) is employed to generate the weights and biases of the Bayesian DNN according to the predefined posterior distribution. The trained Bayesian DNN model is connected with the classical trust region algorithm to perform the inverse design.

Many other effective optimization methods are also connected with the DNN models for the efficient design of metamaterials and metasurfaces. The interior-point optimization algorithm is accelerated for the inverse design of metagratings by training the fast forward DNN model [96]. The forward DNN model is trained to establish the reliable mapping between the structural shape r_i and the diffraction efficiency of unit cells in the metagratings. The trained forward DNN model can significantly reduce the computing time of interior-point optimization compared to the full-wave simulation. The truncated Newtonian optimization algorithm is in conjunction with the DNN models to design the integrated Bragg grating devices in [97]. The waveguide neural network is first trained to learn the forward mapping between the effective index and waveguide geometry for the first two transverse electric (TE) and transverse magnetic (TM) modes. Then the Bragg grating neural network is trained for the forward modeling between the Bragg grating geometry and the corresponding responses with the training data generated by the waveguide neural network. Forward and inverse design strategies are proposed with the fast forward modeling parameterized by the neural network. In the inverse design, the truncated Newtonian optimization algorithm applies the Bragg grating neural network to accelerate the cost function calculation. The active learning is applied to replace the EM simulations in the inverse design of photonic metamaterial in [98]. Compared to the plain DNN model, active learning measures the uncertainty of DNN predictions, and the uncertain points are selected and kept aside for another training in an iterative manner. Active learning demonstrates the reduced training time and the improved performance, which further accelerate the inverse design process. Instead of gradient-based methods, the approach based on the conservative convex separable approximations is connected with the active-learning model. The phase-modulating dielectric metasurface is designed by combining the transfer learning and genetic algorithms in [99]. The real FCN model is first trained to establish the forward mapping from the geometries to the real parts of the complex transmission coefficients. The imaginary FCN model shares the same architecture as the real one. Trained in a transfer manner, the imaginary FCN model initializes its weights of the first three layers to be the same as the ones of the real FCN model. The transfer learning scheme can significantly reduce the training time with improved prediction accuracy. The GA is applied to perform the inverse design process with the trained real and imaginary FCN models as the EM simulators. As a proof of concept, two deflectors and metalenses are designed based on the proposed approach.

The gradient-based strategy for searching the model space is implemented based on the commonly-applied NA method. NA method is inspired by the classical adjoint method for inverse design [100]. An adjoint method is an essential approach for inverse design in engineering. The forward modeling from m to d can be described with a non-linear forward operator, as described in (11.1). In the inverse design, the adjoint method can identify the local optimum guided by the descent directions based on the gradient $\frac{\partial F}{\partial m}$ with respect to m. The NA method consists of two steps, as depicted in Figure 11.23. The first step is to train the DNN F_{NN} to learn the mapping from

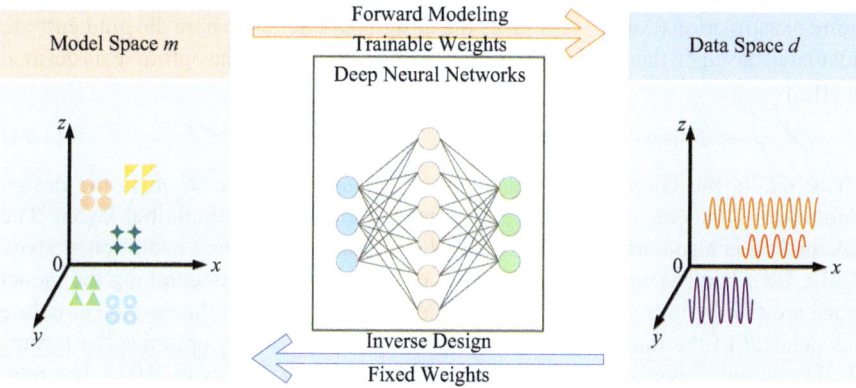

Figure 11.23   Workflow of the inverse design of metamaterial or metasurface based on the neural adjoint method

m to d by approximating the forward operator F. The second step is to optimize the geometrical parameters by using the gradient calculated by the backpropagation [101]:

$$\hat{m}^{i+1} = \hat{m}^i - \alpha \frac{\partial \mathcal{L}\left(F_{NN}\left(\hat{m}^i\right), d\right)}{\partial m}\Bigg|_{m=\hat{m}^i}, \qquad (11.22)$$

where $\hat{m}$ and d denote the geometrical parameters to optimize and the desired property, $\mathcal{L}$ is the loss function, and α is the learning rate. It is noted that the weight of F_{NN} is fixed in the second step. The initial guess $\hat{m}_0$ of the geometry is generated based on the specific distribution.

The NA method can be utilized to solve the design task without the connections with other optimization methods. The NA method is applied to design the metamaterial consisting of four cylinders given the desired electromagnetic spectrum in [101]. The geometrical parameters to optimize are the radii and heights of the cylinders. A highly accurate forward simulator is then yielded based on the DNN model by generating 40,000 samples for training with a significant reduction in computing time. With the trained forward simulator, the inverse design can be implemented according to (11.22). Additional to (11.22), the boundary loss is introduced to improve the performance of the NA method [101]:

$$\mathcal{L}_b = ReLU(|\hat{m} - \mu_x| - \frac{1}{2}R_x), \qquad (11.23)$$

where μ_x and R_x are the mean and the range of the training data. Numerical experiments demonstrate that the NA method needs more computational costs but achieves better performances.

In fact, the NA method is usually combined with other effective optimizers to efficiently calculate the required gradients. The efficient grating couplers on a Si-on-insulator (SOI) platform are designed by the NA method given the specific operating wavelength in [102]. The FCN model is trained to learn the forward mapping from the design parameters to the corresponding coupling efficiency spectrum with the data set generated by the 2D FDTD simulations. Once the FCN model is trained, the adaptive moment estimation (Adam) optimizes the design parameters, where the gradients are calculated based on the backpropagation. The loss function of the optimizer is defined as [102]:

$$\mathcal{L} = -CE + p \cdot ||m - m_{init}||^2, \qquad (11.24)$$

where CE is the coupling efficiency, p is a penalty parameter, m is the design parameters, $||m - m_{init}||^2$ penalizes the large difference from the initial layout. The NA method is also compared to the covariance matrix adaptation evolution strategy (CMA-ES). The NA method can find the optimal structure by searching the model space around the initial structure, while CMA-ES can produce the optimal structure independent of the initial one. The NA method is proposed to optimize the layouts of 2D photonic crystal nanocavities to achieve better Q factors in [103]. The relationship between the Q factors and the positions of air holes is first established by a four-layer CNN model. The trained CNN model can calculate the gradient of Q factor with respect to structural parameters at high speed, which further benefits the gradient-based structural optimization. With the Momentum method as an optimizer, the loss function of the structural optimization is defined as [103]:

$$\mathcal{L} = |\log_{10} Q_{CNN} - \log_{10} Q_{target}|^2 + \frac{1}{2}\lambda \sum_{ij} |\vec{d}_{i,j}|^2, \tag{11.25}$$

where $\vec{d}_{i,j}$ is the displacement vectors, Q_{CNN} and Q_{target} are the predicted and desired Q factors, λ weighs the regularization term, $\sum_{ij} |\vec{d}_{i,j}|^2$ penalizes the large displacement of air holes.

The first commonly used generative model is GANs, especially the CGANs that can including different controlling conditions. The latent spaces generated by the generative models can significantly reduce the computational load of searching during the optimization process. The evolutionary search strategy has been widely employed to find the optimal design within the latent space. The CGAN is combined with iterative optimization for the inverse design of high-performance metagratings with the desired deflection angles and operating wavelengths [104]. The sketch of the CGAN assisted topology optimization is depicted in Figure 11.24. The input of the generator includes the Gaussian random variables, operating wavelength, and the output deflection angle, and the output is the binary image of the metasurface unit cell. The discriminator tries to differentiate between ground truth and the metasurfaces produced by the generator. The iterative topology optimization is applied to further refine the metasurfaces designed by CGAN. The optimized metasurface structures are fed back to train the CGAN for improved performance. Because the metasurface structure designed by CGAN is near the local optimum, only a few iterations are needed for the refinement. A cyclical framework combines the DNN models with the GA to design nanophotonic metasurfaces to achieve the desired optical responses [105]. The simulation neural network is built based on the CNN and FCN to predict the optical responses given the geometries. The cGAN is trained to generate the candidate structural designs by modeling the distribution of the design space. The pseudo GA

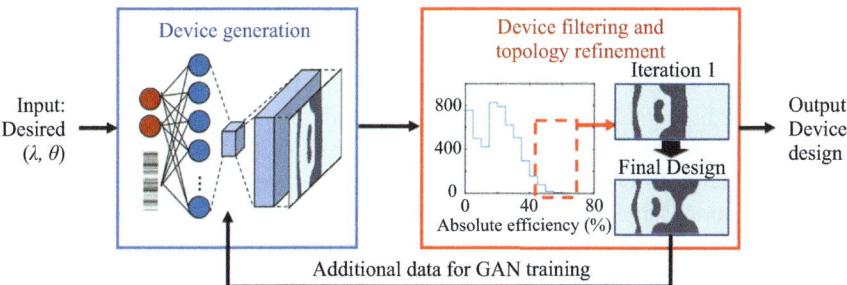

Figure 11.24   CGAN assisted topology optimization (Reprinted (adapted) with permission from [104]. Copyright 2022 American Chemical Society.)

searches the design space to produce the optimal metasurface boasting the desired optical response.

The second generative model is VAEs that can produce continuously distributed latent vectors given different input. A hybrid strategy is proposed for the automated design of the engineered photonic materials by combining the VAE model and the evolution strategy [106]. The VAE is trained to encode all possible metasurface structures into the compact and continuous latent space. With the encoded latent space, the evolution strategy is modified to search the optimal latent vector for the desired metasurface design. The single-layered metasurfaces are designed by the proposed strategy to validate the efficacy by predefining specific structural parameters. The proposed strategy can be generalized to more complicated metasurfaces by including more structural parameters in the latent space encoded by the VAE. The conditional AAE (c-AAE) is coupled with the differential evolution (DE) optimizer to perform the multiparametric global optimization of nanophotonic metadevices [107]. The architecture of c-AAE and the corresponding hybrid strategy are illustrated in Figure 11.25. The conditional AAE is trained to build the compressed latent model space of metadevices design with a physics-driven regularization that contains a predefined model space distribution and a binary vector. The conditional Visual Geometry Groupnet (c-VGGnet) is built for the rapid efficiency estimation regarding the model generated by the trained c-AAE generator. With the trained c-VAE and c-VGGnet, the DE optimizer is applied to search within the compressed latent model space. In the iterative process of DE optimizer, the c-VAE generator produces the model regarding the latent vector, and the c-VGGnet estimates the corresponding efficiency to help the DE optimizer perform evolution. Dual-and triple-layer electromagnetic metasurfaces are designed based on the VAE model to achieve the desired scattering coefficients in [108]. Each layer of metasurfaces is viewed as an image where 0 denotes the non-metallic part, and 1 denotes the metallic part. The VAE model is trained to encode the metasurface structures into the latent vectors with the forward predictor as the regularizer. The forward predictor is based on the FCN model and trained to predict the scattering responses given the latent vectors. The training data of the FCN model is generated by the generalized scattering matrix-based method, and the interlayer coupling is included in this way. The PSO is implemented with the trained VAE to search the latent space for the optimal metasurface design. The metasurface retroreflectors are designed by combining the machine learning and the evolution optimization algorithm for the incident waves of arbitrary directions in [109]. The metasurface is first encoded into 2 bit digital codes representing four different elements. The conditional VAE (cVAE) functions as a design generator to produce the coding matrix conditioned by the target angle range. The FCN model is trained for the fast and accurate prediction of monostatic radar cross section. The GA can search the design space to efficiently locate the optimum with the cVAE generation initializations and the FCN model to evaluate the fitness.

Besides, a single DNN generator can also be trained instead of adopting the paradigm of a GAN or a VAE. A hybrid global optimization strategy is presented in [110] to optimize the structures of metagratings to achieve better efficiencies. Figure 11.26 summarizes the proposed hybrid global optimization strategy. The

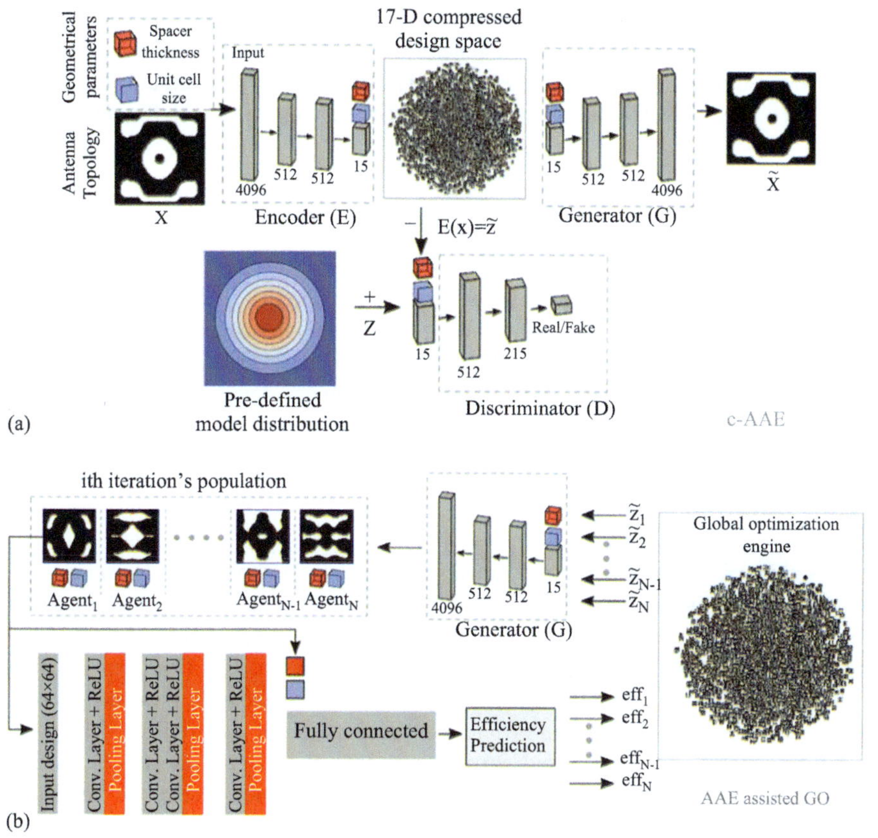

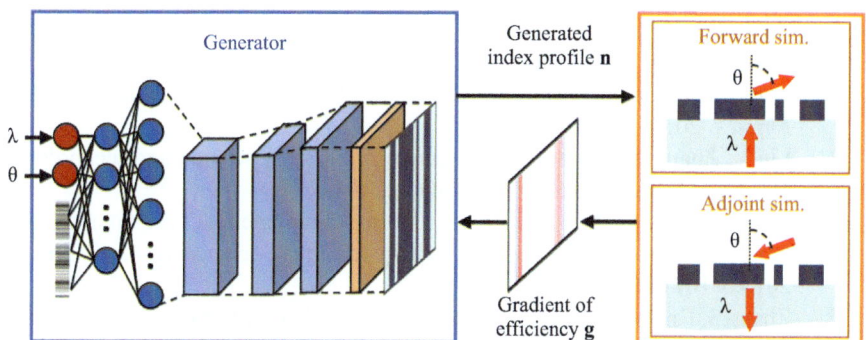

Figure 11.25   *Deep learning-assisted global optimization for photonic devices,*
(a) conditional adversarial autoencoder and (b) c-AAE assisted DE
algorithm for photonic devices (source: [107])

Figure 11.26   *Global optimization of dielectric metasurfaces based on a physics*
driven generator (Reprinted (adapted) with permission from [110].
Copyright 2022 American Chemical Society.)

proposed strategy combines a conditional generator and adjoint electromagnetic simulations. The input of the conditional generator includes the operating wavelength λ, the desired outgoing angle θ, and the sampled noise vector z. The output is the device instances n. The objective function for the conditional generator is defined as [110]:

$$L = -\frac{1}{M} \sum_{m=1}^{M} \exp\left(\frac{\text{Eff}^{(m)} - \text{Eff}_{\max}\left(\lambda^{(m)}, \theta^{(m)}\right)}{\sigma}\right) \mathbf{n}^{(m)} \cdot \mathbf{g}^{(m)} + \beta R, \qquad (11.26)$$

where $m = 1, \ldots, M$ is the index of device, $\text{Eff}_{\max}\left(\lambda^{(m)}, \theta^{(m)}\right)$ denotes the maximum efficiency at the specific pair of wavelength and angle, and $\text{Eff}^{(m)}$ denote the efficiency of the generator-produced device instance, $\mathbf{g}^{(m)}$ is the efficiency gradient with respect to $\mathbf{n}^{(m)}$ and it is calculated by the adjoint electromagnetic simulations, $R = \left|\mathbf{n}^{(m)}\right| \cdot \left(2 - \left|\mathbf{n}^{(m)}\right|\right)$ is the regularization term to force the binary device structures and α is the weight of regularization term.

11.6   Summary

In this chapter, we have reviewed various approaches to apply deep learning techniques into the inverse design of metamaterials and metasurfaces, including the discriminative learning approach, generative learning approach, reinforcement learning approach, deep learning and optimization hybrid approach. The deep learning techniques play an important role in two aspects of inverse designs. First, deep learning can provide an accurate and efficient approximation of a complicated function costly to evaluate. Second, deep learning can extract and generate the high-level features of geometries in a hierarchical and compact manner. These two important characteristics of deep learning poses a great potential for the future design tools of metamaterials and metasurfaces. Deep learning is bound to become a pivotal tool in the inverse design of metamaterials and metasurfaces.

References

[1]   Kadic M, Bückmann T, Schittny R, *et al.* Metamaterials beyond electromagnetism. *Reports on Progress in Physics.* 2013;76(12):126501.

[2]   Kadic M, Milton GW, van Hecke M, *et al.* 3D metamaterials. *Nature Reviews Physics.* 2019;1(3):198–210.

[3]   Liu Y, Zhang X. Metamaterials: a new frontier of science and technology. *Chemical Society Reviews.* 2011;40(5):2494.

[4]   Cui TJ, Qi MQ, Wan X, *et al.* Coding metamaterials, digital metamaterials and programmable metamaterials. *Light: Science & Applications.* 2014;3(10):e218–e218.

[5]   Zhang S, Fan W, Panoiu N, *et al.* Experimental demonstration of near-infrared negative-index metamaterials. *Physical Review Letters.* 2005;95(13):137404.

[6]   García-Meca C, Hurtado J, Martí J, *et al.* Low-loss multilayered metamaterial exhibiting a negative index of refraction at visible wavelengths. *Physical Review Letters.* 2011;106(6):067402.

[7] Schurig D, Mock JJ, Justice B, *et al.* Metamaterial electromagnetic cloak at microwave frequencies. *Science.* 2006;314(5801):977–980.

[8] Zhang S, Xia C, Fang N. Broadband acoustic cloak for ultrasound waves. *Physical Review Letters.* 2011;106(2):024301.

[9] Guenneau S, Amra C, Veynante D. Transformation thermodynamics: cloaking and concentrating heat flux. *Optics Express.* 2012;20(7):8207–8218.

[10] Pendry JB. Negative refraction makes a perfect lens. *Physical Review Letters.* 2000;85(18):3966.

[11] Pendry J. A chiral route to negative refraction. *Science.* 2004;306(5700): 1353–1355.

[12] Frenzel T, Kadic M, Wegener M. Three-dimensional mechanical metamaterials with a twist. *Science.* 2017;358(6366):1072–1074.

[13] Chen HT, Taylor AJ, Yu N. A review of metasurfaces: physics and applications. *Reports on Progress in Physics.* 2016;79(7):076401.

[14] Yang F, Rahmat-Samii Y. *Surface Electromagnetics: With Applications in Antenna, Microwave, and Optical Engineering.* Cambridge: Cambridge University Press; 2019.

[15] Gao LH, Cheng Q, Yang J, *et al.* Broadband diffusion of terahertz waves by multi-bit coding metasurfaces. *Light: Science & Applications.* 2015;4(9):e324–e324.

[16] Yu N, Genevet P, Kats MA, *et al.* Light propagation with phase discontinuities: generalized laws of reflection and refraction. *Science.* 2011;334(6054): 333–337.

[17] Shan T, Pan X, Li M, *et al.* Coding Programmable metasurfaces based on deep learning techniques. *IEEE Journal on Emerging and Selected Topics in Circuits and Systems.* 2020;10(1):114–125.

[18] Cong L, Cao W, Zhang X, *et al.* A perfect metamaterial polarization rotator. *Applied Physics Letters.* 2013;103(17):171107.

[19] Cheng YZ, Withayachumnankul W, Upadhyay A, *et al.* Ultrabroadband reflective polarization convertor for terahertz waves. *Applied Physics Letters.* 2014;105(18):181111.

[20] Khatib O, Ren S, Malof J, *et al.* Deep learning the electromagnetic properties of metamaterials—a comprehensive review. *Advanced Functional Materials.* 2021;31(31):2101748.

[21] So S, Badloe T, Noh J, *et al.* Deep learning enabled inverse design in nanophotonics. *Nanophotonics.* 2020 May;9(5):1041–1057.

[22] Zhang T, Kee CY, Ang YS, *et al.* Deep learning-based design of broadband GHz complex and random metasurfaces. *APL Photonics.* 2021 Oct;6(10):106101.

[23] LeCun Y, Bengio Y, Hinton G. Deep learning. *Nature.* 2015;521(7553): 436–444.

[24] Voulodimos A, Doulamis N, Doulamis A, *et al.* Deep learning for computer vision: a brief review. *Computational Intelligence and Neuroscience.* 2018;2018:7068349.

[25] Zhang Z, Geiger J, Pohjalainen J, *et al.* Deep learning for environmentally robust speech recognition: an overview of recent developments.

ACM *Transactions on Intelligent Systems and Technology (TIST)*. 2018;
9(5):1–28.

[26]   Singh SP, Kumar A, Darbari H, *et al.* Machine translation using deep
learning: An overview. In: *2017 International Conference on Computer,
Communications and Electronics (Comptelix)*. IEEE; 2017. p. 162–167.

[27]   Chen Y, Zhu J, Xie Y, *et al.* Smart inverse design of graphene-based photonic metamaterials by an adaptive artificial neural network. *Nanoscale*.
2019;11(19):9749–9755.

[28]   Akashi N, Toma M, Kajikawa K. Design by neural network of concentric multilayered cylindrical metamaterials. *Applied Physics Express*.
2020;13(4):042003.

[29]   Ghorbani F, Shabanpour J, Beyraghi S, *et al.* A deep learning approach for
inverse design of the metasurface for dual-polarized waves. *Applied Physics
A*. 2021;127(11):869.

[30]   Ghorbani F, Beyraghi S, Shabanpour J, *et al.* Deep neural network-based automatic metasurface design with a wide frequency range. *Scientific Reports*.
2021;11(1):7102.

[31]   Gu Y, Hao R, Li EP. Independent bifocal metalens design based on
deep learning algebra. *IEEE Photonics Technology Letters*. 2021;33(8):
403–406.

[32]   Jia Y, Qian C, Fan Z, *et al.* In situ customized illusion enabled by global
metasurface reconstruction. *Advanced Functional Materials*. 2022;32(19):
2109331.

[33]   Lin R, Zhai Y, Xiong C, *et al.* Inverse design of plasmonic metasurfaces by
convolutional neural network. *Optics Letters*. 2020;45(6):1362–1365.

[34]   Ma J, Huang Y, Pu M, *et al.* Inverse design of broadband metasurface absorber
based on convolutional autoencoder network and inverse design network.
Journal of Physics D: Applied Physics. 2020;53(46):464002.

[35]   Kiarashinejad Y, Abdollahramezani S, Adibi A. Deep learning approach based
on dimensionality reduction for designing electromagnetic nanostructures.
NPJ Computational Materials. 2020 Feb;6(1):1–12.

[36]   Zhang T, Liu Q, Dan Y, *et al.* Machine learning and evolutionary algorithm studies of graphene metamaterials for optimized plasmon-induced
transparency. *Optics Express*. 2020;28(13):18899–18916.

[37]   Zhu R, Qiu T, Wang J, *et al.* Phase-to-pattern inverse design paradigm
for fast realization of functional metasurfaces via transfer learning. *Nature
Communications*. 2021;12(1):2974.

[38]   Liu D, Tan Y, Khoram E, *et al.* Training deep neural networks for the inverse
design of nanophotonic structures. *Acs Photonics*. 2018;5(4):1365–1369.

[39]   Ma W, Cheng F, Liu Y. Deep-learning-enabled on-demand design of chiral
metamaterials. *ACS Nano*. 2018;12(6):6326–6334.

[40]   So S, Mun J, Rho J. Simultaneous inverse design of materials and structures via deep learning: demonstration of dipole resonance engineering
using core–shell nanoparticles. *ACS Applied Materials & Interfaces*.
2019;11(27):24264–24268.

[41]   Hou Z, Tang T, Shen J, *et al.* Prediction network of metamaterial with split ring resonator based on deep learning. *Nanoscale Research Letters.* 2020;15(1):1–8.

[42]   Xu L, Rahmani M, Ma Y, *et al.* Enhanced light–matter interactions in dielectric nanostructures via machine-learning approach. *Advanced Photonics.* 2020;2(2):026003.

[43]   Phan AD, Nguyen CV, Linh PT, *et al.* Deep learning for the inverse design of mid-infrared graphene plasmons. *Crystals.* 2020;10(2):125.

[44]   Tanriover I, Hadibrata W, Aydin K. Physics-based approach for a neural networks enabled design of all-dielectric metasurfaces. *ACS Photonics.* 2020;7(8):1957–1964.

[45]   Shelling Neto L, Dickmann J, Kroker S. Deep learning assisted design of high reflectivity metamirrors. *Optics Express.* 2022;30(2):986.

[46]   Mall A, Patil A, Tamboli D, *et al.* Fast design of plasmonic metasurfaces enabled by deep learning. *Journal of Physics D: Applied Physics.* 2020;53(49):49LT01.

[47]   Roberts NB, Keshavarz Hedayati M. A deep learning approach to the forward prediction and inverse design of plasmonic metasurface structural color. *Applied Physics Letters.* 2021;119(6):061101.

[48]   Yeung C, Tsai JM, King B, *et al.* Multiplexed supercell metasurface design and optimization with tandem residual networks. *Nanophotonics.* 2021;10(3):1133–1143.

[49]   Malkiel I, Mrejen M, Nagler A, *et al.* Plasmonic nanostructure design and characterization via deep learning. *Light: Science & Applications.* 2018;7(1):60.

[50]   Qiu T, Shi X, Wang J, *et al.* Deep learning: a rapid and efficient route to automatic metasurface design. *Advanced Science.* 2019;6(12):1900128.

[51]   Shi X, Qiu T, Wang J, *et al.* Metasurface inverse design using machine learning approaches. *Journal of Physics D: Applied Physics.* 2020;53(27):275105.

[52]   Ng A, Jordan, M. On discriminative vs. generative classifiers: A comparison of logistic regression and naive bayes. *Advances in Neural Information Processing Systems.* 2001;14:1–8.

[53]   Hu Z, Yang Z, Salakhutdinov R, *et al.* On Unifying Deep Generative Models. arXiv:170600550 [cs, stat]. 2018 Jul.

[54]   Kingma DP, Welling M. Auto-Encoding Variational Bayes. arXiv:13126114 [cs, stat]. 2014 May.

[55]   Ma W, Cheng F, Xu Y, *et al.* Probabilistic representation and inverse design of metamaterials based on a deep generative model with semi-supervised learning strategy. *Advanced Materials.* 2019;31(35):1901111.

[56]   Ma W, Liu Y. A data-efficient self-supervised deep learning model for design and characterization of nanophotonic structures. *Science China Physics, Mechanics & Astronomy.* 2020;63(8):284212.

[57]   Kudyshev ZA, Kildishev AV, Shalaev VM, *et al.* Machine-learning-assisted metasurface design for high-efficiency thermal emitter optimization. *Applied Physics Reviews.* 2020;7(2):021407.

[58] Goodfellow IJ, Pouget-Abadie J, Mirza M, *et al.* Generative Adversarial Networks. arXiv:14062661 [cs, stat]. 2014 Jun.

[59] So S, Rho J. Designing nanophotonic structures using conditional deep convolutional generative adversarial networks. *Nanophotonics*. 2019; 8(7):1255–1261.

[60] Wang HP, Li YB, Li H, *et al.* Deep learning designs of anisotropic metasurfaces in ultra wideband based on generative adversarial networks. *Advanced Intelligent Systems*. 2020;2(9):2000068.

[61] Yeung C, Tsai R, Pham B, *et al.* Global inverse design across multiple photonic structure classes using generative deep learning. *Advanced Optical Materials*. 2021;9(20):2100548.

[62] Christensen T, Loh C, Picek S, *et al.* Predictive and generative machine learning models for photonic crystals. *Nanophotonics*. 2020;9(13):4183–4192.

[63] An S, Zheng B, Tang H, *et al.* Multifunctional metasurface design with a generative adversarial network. *Advanced Optical Materials*. 2021;9(5):2001433.

[64] Liu Z, Zhu D, Rodrigues SP, *et al.* Generative model for the inverse design of metasurfaces. *Nano Letters*. 2018;18(10):6570–6576.

[65] Arulkumaran K, Deisenroth MP, Brundage M, *et al.* A brief survey of deep reinforcement learning. *IEEE Signal Processing Magazine*. 2017;34(6):26–38.

[66] Mnih V, Kavukcuoglu K, Silver D, *et al.* Human-level control through deep reinforcement learning. *Nature*. 2015;518(7540):529–533.

[67] Levine S, Finn C, Darrell T, *et al.* End-to-end training of deep visuomotor policies. *The Journal of Machine Learning Research*. 2016;17(1):1334–1373.

[68] Silver D, Huang A, Maddison CJ, *et al.* Mastering the game of Go with deep neural networks and tree search. *Nature*. 2016;529(7587):484–489.

[69] Zhu Y, Mottaghi R, Kolve E, *et al.* Target-driven visual navigation in indoor scenes using deep reinforcement learning. In: *2017 IEEE International Conference on Robotics and Automation (ICRA)*. IEEE; 2017. p. 3357–3364.

[70] Watkins CJ, Dayan P. Q-learning. *Machine Learning*. 1992;8(3):279–292.

[71] Salimans T, Ho J, Chen X, *et al.* Evolution strategies as a scalable alternative to reinforcement learning. arXiv preprint arXiv:170303864. 2017.

[72] Schulman J, Moritz P, Levine S, *et al.* High-Dimensional Continuous Control Using Generalized Advantage Estimation. arXiv preprint arXiv:150602438. 2015.

[73] Wang H, Zheng Z, Ji C, *et al.* Automated multi-layer optical design via deep reinforcement learning. *Machine Learning: Science and Technology*. 2021;2(2):025013.

[74] Byrnes SJ. Multilayer Optical Calculations. arXiv preprint arXiv:160302720. 2016.

[75] Seo D, Nam DW, Park J, *et al.* Structural optimization of a one-dimensional freeform metagrating deflector via deep reinforcement learning. *ACS Photonics*. 2022;9(2):452–458.

[76] Sajedian I, Badloe T, Rho J. Optimisation of colour generation from dielectric nanostructures using reinforcement learning. *Optics Express*. 2019;27(4):5874–5883.

[77]   Badloe T, Kim I, Rho J. Biomimetic ultra-broadband perfect absorbers optimised with reinforcement learning. *Physical Chemistry Chemical Physics*. 2020;22(4):2337–2342.

[78]   Sajedian I, Lee H, Rho J. Double-deep Q-learning to increase the efficiency of metasurface holograms. *Scientific Reports*. 2019;9(1):10899.

[79]   Huang Z, Liu X, Zang J. The inverse design of structural color using machine learning. *Nanoscale*. 2019;11(45):21748–21758.

[80]   Nadell CC, Huang B, Malof JM, *et al*. Deep learning for accelerated all-dielectric metasurface design. *Optics Express*. 2019;27(20):27523.

[81]   Gostimirovic D, Ye WN. An open-source artificial neural network model for polarization-insensitive silicon-on-insulator subwavelength grating couplers. *IEEE Journal of Selected Topics in Quantum Electronics*. 2019; 25(3):1–5.

[82]   Yang F, An S, Shalaginov MY, *et al*. Design of broadband and wide-field-of-view metalenses. *Optics Letters*. 2021;46(22):5735.

[83]   An S, Fowler C, Zheng B, *et al*. A deep learning approach for objective-driven all-dielectric metasurface design. *ACS Photonics*. 2019;6(12):3196–3207.

[84]   Zhang Q, Liu C, Wan X, *et al*. Machine-learning designs of anisotropic digital coding metasurfaces. *Advanced Theory and Simulations*. 2019;2(2): 1800132.

[85]   Zhu R, Qiu T, Wang J, *et al*. Multiplexing the aperture of a metasurface: inverse design via deep-learning-forward genetic algorithm. *Journal of Physics D: Applied Physics*. 2020;53(45):455002.

[86]   Kalt V, González-Alcalde AK, Es-Saidi S, *et al*. Metamodeling of high-contrast-index gratings for color reproduction. *Journal of the Optical Society of America A*. 2019;36(1):79.

[87]   González-Alcalde AK, Salas-Montiel R, Kalt V, *et al*. Engineering colors in all-dielectric metasurfaces: metamodeling approach. *Optics Letters*. 2020;45(1):89.

[88]   Liu Z, Zhu D, Lee KT, *et al*. Compounding meta-atoms into metamolecules with hybrid artificial intelligence techniques. *Advanced Materials*. 2020;32(6):1904790.

[89]   Panda SS, Hegde RS. A learning based approach for designing extended unit cell metagratings. *Nanophotonics*. 2022;11(2):345–358.

[90]   Chen J, Ding W, Li XM, *et al*. Absorption and diffusion enabled ultrathin broadband metamaterial absorber designed by deep neural network and PSO. *IEEE Antennas and Wireless Propagation Letters*. 2021;20(10):1993–1997.

[91]   Li Z, Liu W, Ma D, *et al*. Inverse design of few-layer metasurfaces empowered by the matrix theory of multilayer optics. *Physical Review Applied*. 2022;17(2):024008.

[92]   Hegde RS. Accelerating optics design optimizations with deep learning. *Optical Engineering*. 2019;58(6):065103.

[93]   Hegde RS. Photonics inverse design: pairing deep neural networks with evolutionary algorithms. *IEEE Journal of Selected Topics in Quantum Electronics*. 2020;26(1):1–8.

[94]   Li Y, Xu Y, Jiang M, *et al*. Self-learning perfect optical chirality via a deep neural network. *Physical Review Letters*. 2019;123(21):213902.

[95]   Balin I, Garmider V, Long Y, *et al*. Training artificial neural network for optimization of nanostructured VO_2-based smart window performance. *Optics Express*. 2019;27(16):A1030.

[96]   Inampudi S, Mosallaei H. Neural network based design of metagratings. *Applied Physics Letters*. 2018;112(24):241102.

[97]   Hammond AM, Camacho RM. Designing integrated photonic devices using artificial neural networks. *Optics Express*. 2019;27(21):29620–29638.

[98]   Pestourie R, Mroueh Y, Nguyen TV, *et al*. Active learning of deep surrogates for PDEs: application to metasurface design. *NPJ Computational Materials*. 2020;6(1):164.

[99]   Xu D, Luo Y, Luo J, *et al*. Efficient design of a dielectric metasurface with transfer learning and genetic algorithm. *Optical Materials Express*. 2021;11(7):1852.

[100]   Bendsøe MP, Kikuchi N. Generating optimal topologies in structural design using a homogenization method. *Computer Methods in Applied Mechanics and Engineering*. 1988;71(2):197–224.

[101]   Ren S, Padilla W, Malof J. Benchmarking deep inverse models over time, and the neural-adjoint method. *Advances in Neural Information Processing Systems*. 2020;33:38–48.

[102]   Miyatake Y, Sekine N, Toprasertpong K, *et al*. Computational design of efficient grating couplers using artificial intelligence. *Japanese Journal of Applied Physics*. 2020;59(SG):SGGE09.

[103]   Asano T, Noda S. Optimization of photonic crystal nanocavities based on deep learning. *Optics Express*. 2018;26(25):32704–32717.

[104]   Jiang J, Sell D, Hoyer S, *et al*. Free-form diffractive metagrating design based on generative adversarial networks. *ACS Nano*. 2019;13(8):8872–8878.

[105]   Mall A, Patil A, Sethi A, *et al*. A cyclical deep learning based framework for simultaneous inverse and forward design of nanophotonic metasurfaces. *Scientific Reports*. 2020;10(1):19427.

[106]   Liu Z, Raju L, Zhu D, *et al*. A hybrid strategy for the discovery and design of photonic structures. *IEEE Journal on Emerging and Selected Topics in Circuits and Systems*. 2020;10(1):126–135.

[107]   Kudyshev ZA, Kildishev AV, Shalaev VM, *et al*. Machine learning–assisted global optimization of photonic devices. *Nanophotonics*. 2021; 10(1):371–383.

[108]   Naseri P, Hum SV. A generative machine learning-based approach for inverse design of multilayer metasurfaces. *IEEE Transactions on Antennas and Propagation*. 2021;69(9):5725–5739.

[109]   Lin H, Hou J, jin J, *et al*. Machine-learning-assisted inverse design of scattering enhanced metasurface. *Optics Express*. 2022;30(2):3076.

[110]   Jiang J, Fan JA. Global optimization of dielectric metasurfaces using a physics-driven neural network. *Nano Letters*. 2019;19(8):5366–5372.

Chapter 12

Deep learning techniques for microwave circuit modeling

Jing Jin[1], Sayed Alireza Sadrossadat[2], Feng Feng[3], Weicong Na[4] and Qi-Jun Zhang[5]

This chapter provides a description of deep learning as applied to microwave circuit modeling. Microwave circuit modeling is an important area of computer-aided design for fast and accurate microwave design and optimization. In recent years, rapid development of modern electronic devices/systems and wireless communications requires various customized microwave circuits. Subsequently, the modeling of microwave circuits becomes more complex and more challenging due to the demand for higher functionality, better reliability, and shorter design cycle. As a result, there is a need for more accurate, more effective, and more efficient modeling techniques for microwave circuits. To address this issue, deep learning has been introduced into the area of microwave circuit modeling. Deep learning is a class of machine learning that utilizes artificial neural networks with many layers to learn the complex input–output relationships. It has been highly successful in solving complex and challenging problems such as pattern recognition and classification. The powerful learning ability also makes it a suitable choice for modeling the complex input–output relationship of microwave circuits. Researchers have investigated a variety of important applications utilizing the ability of deep learning to perform microwave circuit modeling.

12.1 Introduction

Microwave circuit modeling plays an important role in the area of computer-aided design for fast and accurate microwave optimization and design. The developed microwave circuit models allow fast simulation and optimization and subsequently can be implemented in high-level circuit designs or computer-aided tuning of microwave circuits. Different modeling techniques, such as artificial neural network technique

[1]College of Physical Science and Technology, Central China Normal University, China
[2]Department of Computer Engineering, Yazd University, Iran
[3]School of Microelectronics, Tianjin University, China
[4]Faculty of Information Technology, Beijing University of Technology, China
[5]Department of Electronics, Carleton University, Canada

[1,2], Kriging [3,4], support vector machine (SVM) [5], and polynomial-based surrogate modeling [6], have been reported for modeling of microwave circuits. Kriging, SVM, and polynomial models have good generalization capability when training data are limited [6], while artificial neural network is well suited to the case when the amount of training data is large.

Artificial neural network (ANN) has been recognized as a powerful technique in the area of microwave modeling and design [7–15]. ANNs can be trained from the simulated or measured data to learn the nonlinear input–output relationships of microwave components/circuits. These trained ANN models can then be used to provide fast answers to the tasks they have learned [16]. This makes ANN an efficient alternative to empirical model or electromagnetic simulation for microwave modeling. Applications of ANNs have been reported in microwave filter modeling and design [17–19], power amplifier modeling [20–22], nonlinear microwave device modeling [23–27], parametric modeling of microwave components [28,29], multiphysics parametric modeling and optimization [30–32], coplanar waveguide (CPW) circuit modeling [33], and microwave component design [34]. These applications in the area of microwave modeling and design are achieved mostly using shallow neural networks, i.e., ANNs with one or two hidden layers.

In microwave modeling field, there are situations where the model input–output relationship is highly nonlinear, which makes the modeling problem harder. To address this kind of complicated microwave modeling problem using neural networks, there are usually two possible solutions. One is to add more hidden neurons to the shallow neural network, and the other is to add more hidden layers. It has been proved that the neural network with many hidden layers can perform significantly better than the shallow neural network when both neural networks have the same number of training parameters [35,36].

In recent years, there has been growing interest in the neural network community in neural networks with many hidden layers, known as deep neural network [37–41]. Typically, the numbers of network layers in a deep neural network range from five to more than a thousand [42]. Deep neural networks have been recognized to be very powerful at modeling intricate relationships in large data sets [37]. Outstanding results have been produced by deep neural networks in a variety of challenging fields, such as image recognition [43], speech recognition [44], language processing [45], machine translation [46], and sentiment analysis [47]. Due to its powerful learning ability, deep neural networks have also been reported in the field of microwave circuit modeling to learn the highly nonlinear input–output relationships that are beyond the capability of shallow neural networks [48–61].

As the rapid development of computing technology, researchers have developed various deep neural network structures, including feedforward deep neural network [42], RNN [62], convolutional neural network (CNN) [63], and so on. A basic type of deep neural network structure is the feedforward deep neural network such as MLP with many hidden layers, where the number of hidden layers should be three or more. The feedforward deep neural network can be used to learn the relationship between a fixed-size input and a fixed-size output [37]. Recently, the feedforward deep neural network technique has been reported to address modeling problems with high-dimensional inputs in the area of microwave modeling [48,49]. Batch normalization has been incorporated into the feedforward deep neural network for automated modeling of microwave components [52]. The feedforward neural networks such as MLP are mostly used for modeling of linear components. Due to their architecture, feedforward neural networks can not capture time-dependency very well. Therefore, exploring new techniques for modeling time-dependent signals is crucial in nonlinear modeling area. In order to model the dynamic behavior of a nonlinear microwave device, time-dependent training signals should be generated and passed to the network structure. Due to this time-dependency of training signals, RNN structures (a specific type of deep neural network that contains feedback loops involving delay units) are exploited for nonlinear circuit modeling [55–57].

In this chapter, an overview of deep learning as applied to microwave circuit modeling is provided. The feedforward deep neural network and the vanishing gradient problem during its training process are introduced. A hybrid feedforward deep neural network that can be trained without the vanishing gradient problem is then presented [48]. Also, the RNNs for nonlinear circuit modeling are presented, including the global-feedback RNN [70], the adjoint recurrent neural network (ARNN) [71], the global-feedback deep RNN (DRNN) [55], the local-feedback deep RNN (LFDRNN) [56], and long short-term memory (LSTM) [57]. Following the overview of different deep neural network methods for microwave circuit modeling, several application examples are presented to demonstrate the deep neural network modeling techniques. Subsequently, the proper usage of different methods in different practical situations is discussed. Finally, a conclusion for the chapter is provided.

12.2 Feedforward deep neural network for microwave circuit modeling

In this section, an overview of recent feedforward deep neural network-based methods for microwave circuit modeling is presented. Feedforward artificial neural networks have been reported for microwave circuit modeling for years [10–18]. These applications are achieved mostly using shallow neural networks. In recent years, as the development of the microwave components and communication systems, the microwave modeling problem becomes more and more complicated. To address this kind of microwave circuit modeling problems that are beyond the capability of feedforward shallow neural networks, the feedforward deep neural networks were presented for microwave modeling [48,52]. The feedforward deep neural network, such as MLP with many hidden layers, is a basic type of deep neural network structure. The training of the feedforward deep neural network is not easy because of the vanishing gradient problem [64,65]. This section provides an introduction to the vanishing gradient problem in the feedforward deep neural network. A hybrid feedforward deep neural network that can be trained without the vanishing gradient problem is also reviewed [48].

12.2.1   *Feedforward deep neural network and the vanishing gradient problem*

12.2.1.1   Feedforward deep neural network structure

A neural network is an information processing system that is composed of two types of basic components, namely, neurons and links. The neurons are the processing elements, and the links are the interconnections between neurons [7]. In a neural network, each neuron (except for the neurons at the input layer) receives stimuli (inputs) from the neighboring neurons connected to it, processes the information, and produces an output. Neurons can process information in different ways, and the connections from the neurons to one another can be different. Different information processing elements and different connection manners between the neurons can construct different neural network structures.

A basic type of deep neural network structure is the feedforward deep neural network such as MLP with many hidden layers. The structure of MLP with many hidden layers is the same as the MLP structure shown in Figure 12.1, where the number of hidden layers should be three or more [66]. Let x be a vector of size $N_x \times 1$ containing the external inputs to the neural network model (e.g., design parameters of a given microwave component). Let y be a vector of size $N_y \times 1$ containing the outputs from the neural network model (e.g., EM responses of the given microwave component). The input–output relationship of the neural network model is given by [67]

$$y = y(x, w). \tag{12.1}$$

where w is a vector containing all the weight parameters representing various interconnections in the neural network model.

12.2.1.2   Vanishing gradient problem in training of feedforward deep neural network

In order to represent the input–output relationship of a microwave circuit, the feedforward deep neural network needs to be trained through a set of training data generated from EM/physics simulation or measurement. Let sample pairs (x_n, d_n), $n = 1, 2, ..., N_{tr}$, represent the training data, where d_n is simulated/measured output data (i.e., the desired outputs of the feedforward deep neural network) for the input x_n, and N_{tr} is the number of training samples. The objective of the training process is to determine the weight vector w such that the difference between neural network outputs and the desired outputs is minimized. The difference, also known as the training error, is formulated as [67]

$$E(w) = \frac{1}{2} \sum_{n=1}^{N_{tr}} \|y(x_n, w) - d_n\|^2. \tag{12.2}$$

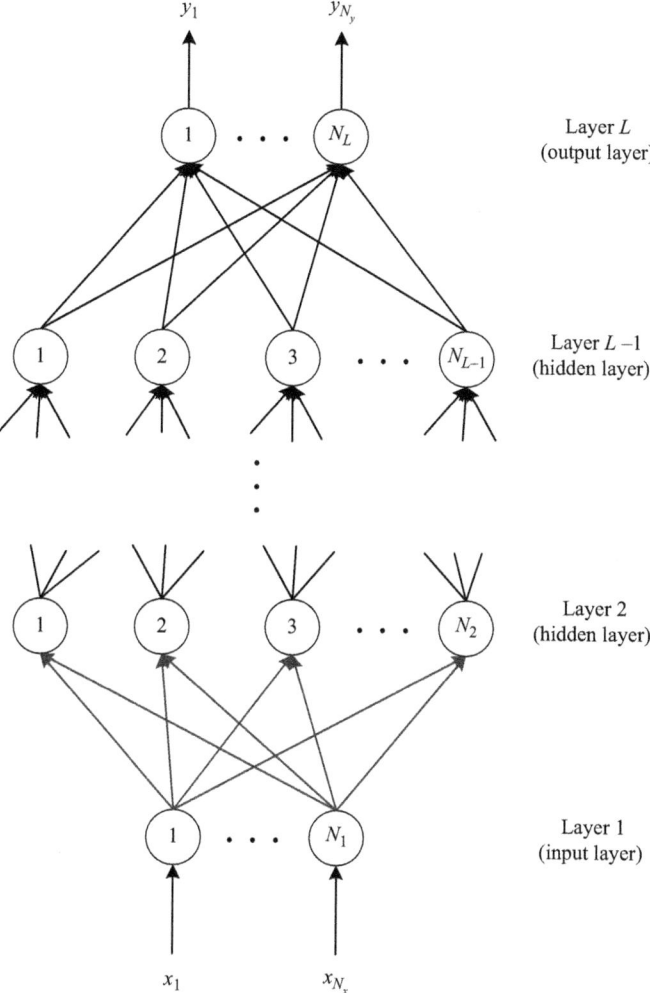

Figure 12.1   Multilayer perceptrons (MLP) neural network structure containing one input layer, one output layer, and one or more hidden layers [67]

The training of deep neural networks is affected by the selection of nonlinear activation functions. In conventional (shallow) ANNs, the most commonly used activation functions are smooth switch functions such as the sigmoid function, formulated as [67].

$$\sigma(\gamma) = \frac{1}{(1 + e^{-\gamma})} \tag{12.3}$$

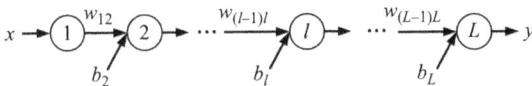

Figure 12.2   A simple deep neural network with one input and one output to illustrate the vanishing gradient problem

The symbol γ represents the total input to the hidden neuron. Shallow neural networks with sigmoid functions usually can be trained successfully with gradient-based learning methods, while deep neural networks with sigmoid functions cannot be trained effectively using gradient-based method because of the vanishing gradient problem [68]. A simple deep neural network with one input and one output, as shown in Figure 12.2, is used to illustrate the vanishing gradient problem. Each hidden layer in this simple deep neural network has one hidden neuron. All hidden neurons employ the sigmoid function as the activation function. L is defined as the total number of layers of the simple deep neural network. Let l be defined as the index of layers in the neural network. Let x represent the input and y represent the output. Let $E(y)$ be defined as the error function representing the difference between deep neural network outputs and the desired outputs from the training data. In the training process, the derivatives of the error function $E(y)$ with respect to the weight between the neuron in the $(l-1)$th and the lth layer can be calculated by [68]

$$\frac{\partial E(y)}{\partial w_{(l-1)l}} = \frac{\partial E(y)}{\partial y}\sigma(\gamma_{l-1})\prod_{h=l}^{L-1}\left(w_{h(h+1)}\sigma'(\gamma_h)\right) \tag{12.4a}$$

$$\sigma'(\gamma) = \sigma(\gamma)(1-\sigma(\gamma)) \tag{12.4b}$$

where $w_{(l-1)l}$ is the weight between the neuron in the $(l-1)$th and the lth layer, γ_l represents the input to the neuron in the lth layer. $\sigma'(\,\cdot\,)$ represents the gradient of the sigmoid function formulated in (12.3).

According to (12.3) and (12.4b), $\sigma'(\,\cdot\,)$ is always in the range of

$$0 < \sigma'(\gamma) \le 0.25 \tag{12.5}$$

Thus we can derive

$$\left|\frac{\partial E(y)}{\partial w_{(l-1)l}}\right| \le \left|\frac{\partial E(y)}{\partial y}\right| 0.25^{(L-l)} \tag{12.6}$$

The absolute value of the derivative $\frac{\partial E(y)}{\partial y}$ is a constant for each training data. Equation (12.6) shows that derivatives for the weights close to the input layer will be much smaller than that close to the output layer. The gradient will decrease exponentially and approach 0 gradually with the increase of L and the decrease of l, which means that weights close to the input layer cannot be trained effectively. This is known as the vanishing gradient problem for training deep neural networks [68].

12.2.2 A hybrid feedforward deep neural network

Recently, researchers have presented a hybrid feedforward deep neural network in the area of microwave circuit modeling to address the challenges due to high-dimensional model inputs [48]. The hybrid feedforward deep neural network can be trained without the vanishing gradient problem.

12.2.2.1 Structure of the hybrid feedforward deep neural network

The structure of the feedforward deep neural network model is shown in Figure 12.3 [48]. It is a fully connected neural network with many hidden layers. In order to reduce the number of hidden neurons as well as avoid the vanishing gradient problem, both sigmoid functions and smooth ReLUs are utilized as activation functions for the hybrid feedforward deep neural network model. The sigmoid function is expressed in (12.3). The smooth ReLU function is formulated as [48]

$$f_s(\gamma) = \begin{cases} \gamma & \text{if } \gamma > \alpha \\ \frac{1}{4\alpha}\gamma^2 + \frac{1}{2}\gamma + \frac{1}{4}\alpha & \text{if } -\alpha \leq \gamma \leq \alpha \\ 0 & \text{if } \gamma < -\alpha \end{cases} \tag{12.7}$$

The hidden layers close to the input layer employ sigmoid functions as activation functions, while the rest of hidden layers utilize the smooth ReLUs as activation functions. The external inputs to the hybrid feedforward deep neural network model are defined as $x = [x_1 \ x_2 \ x_3 \ ... \ x_n]^T$, where n represents the number of the model inputs. The outputs of the feedforward deep neural network model are defined as $y = [y_1 \ y_2 \ y_3 \ ... \ y_m]^T$, where m represents the number of the model outputs. Let p represent the number of hidden layers with sigmoid functions, and q represent the number of hidden layers with smooth ReLUs. The total number of layers of the hybrid deep neural network will be $L = p + q + 2$, including the input, output, and hidden layers.

12.2.2.2 Training of the hybrid feedforward deep neural network

The hybrid feedforward deep neural network model needs to be trained with the simulated or measured circuit data before it can be used in microwave design. Let (x_k, d_k), $k = 1, 2, ..., T_r$, represent the training data, where d_k represents the desired outputs of the deep neural network model for inputs x_k, and T_r represents the total number of training samples. The training process is a process to optimize the weight vector w for the deep neural network so that the error function can be minimized. The standard error function [67] is expressed as

$$E(w) = \sum_{k=1}^{T_r} \left(\frac{1}{2} \sum_{j=1}^{m} (y_j(x_k, w) - d_{jk})^2 \right) \tag{12.8}$$

where $y_j(x_k, w)$ represents the jth output of the feedforward deep neural network model for x_k, and d_{jk} is the jth element of d_k.

To train the hybrid deep neural network, the derivatives of the error function formulated in (12.8) with respect to all the weight parameters in the deep neural

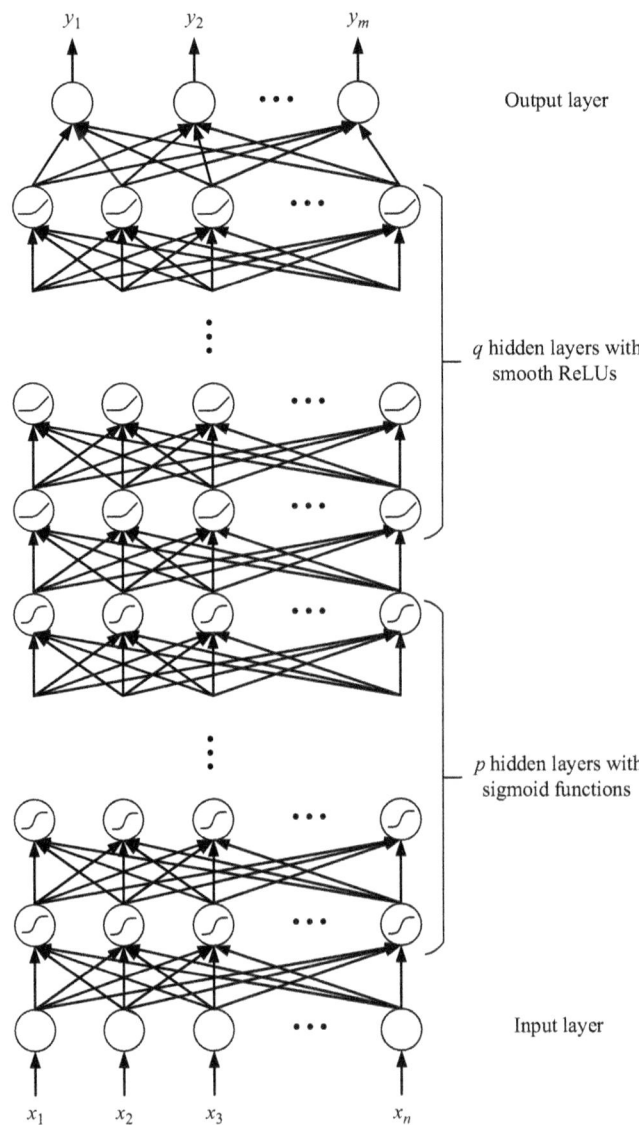

Figure 12.3   *Structure of the hybrid deep neural network model for microwave components. The hidden layers in the hybrid deep neural network are divided into two parts. p hidden layers close to the input layer employ sigmoid functions as neuron activation functions, while the rest q hidden layers utilize the smooth ReLUs as neuron activation functions [48].*

network model need to be calculated. The details of deriving these derivatives can be found in [48]. With the derivatives, the hybrid feedforward deep neural network can be trained using a three-stage deep learning algorithm. The flow diagram of the three-stage training for the feedforward hybrid deep neural network model is shown in Figure 12.4.

During the three-stage training process, the same number of neurons are used in each hidden layer. In Stage I, the deep neural network with only sigmoid hidden layers is trained so that the number of sigmoid layers, i.e., the value of p, can be determined. The training algorithm starts with $p = 3$. There usually be no vanishing gradient problem in a deep neural network with three sigmoid hidden layers. The deep neural network with p sigmoid hidden layers is trained to make the training error as small as possible. After training, if the accuracy requirement is satisfied, the deep neural network training is finished and there is no need for Stages II and III. Otherwise, a new hidden layer with sigmoid functions is added again and again until the accuracy requirement of the model is satisfied or until the vanishing gradient problem begins to appear. Upon encountering the vanishing gradient problem, the last added sigmoid

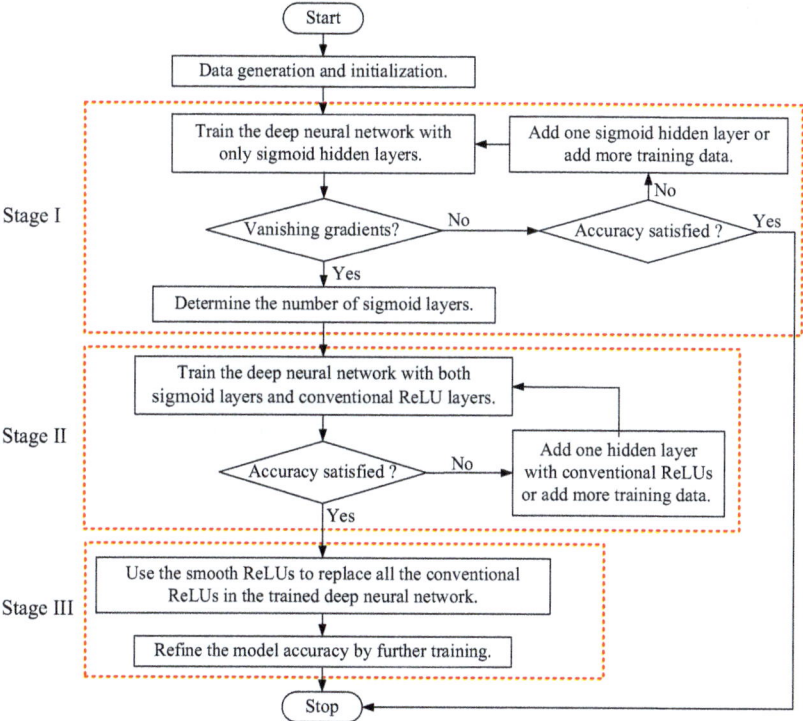

Figure 12.4   The flow diagram of the process for training the feedforward hybrid deep neural network model [48]

hidden layer needs to be deleted, the total number of sigmoid layers p can be fixed, and the training is proceeded to Stage II.

In Stage II, instead of adding hidden layers with smooth ReLUs directly, hidden layers with conventional ReLUs (which will not lead to the vanishing gradient problem) [37] are temporarily added above the trained sigmoid hidden layers. A new hidden layer with conventional ReLUs is added again and again until the accuracy requirement of the deep neural network model is satisfied. In this way, the number of ReLU hidden layers, i.e., the value of q, can be determined. Then, the training is proceeded to Stage III.

In Stage III, the conventional non-smooth ReLUs in the deep neural network model trained from Stages I and II are replaced by the smooth ReLUs. The replacement may slightly affect the output values of the deep neural network model. The model accuracy is refined by further training the deep neural network with p sigomid hidden layers and q smooth ReLU hidden layers. The deep neural network obtained from Stage III will provide smooth input–output relationships required for microwave modeling.

12.3   Recurrent neural networks for microwave circuit modeling

In this section, an overview of recent recurrent neural network (RNN)-based methods for nonlinear device macromodeling is presented. RNN with global-feedback was the early neural network-based method introduced for nonlinear device modeling [70]. After that, few works have been done to improve the performance of RNN in computer-aided design area. Recently, some more advanced recurrent neural networks such as adjoint recurrent neural network (ARNN) [71], global-feedback deep recurrent neural network (DRNN) [55], local-feedback deep recurrent neural network (LFDRNN) [56], and LSTM [57] have been introduced in nonlinear macromodeling area. Deep structures gained lots of attention recently in machine learning area due to their superior capability to catch the complex input–output relationship but their structures suffer from modeling long-term dependencies and difficult training. Therefore, efforts have been done to develop recurrent structures with higher performance such as LSTM approach. LSTM can combat the vanishing gradient problem. It achieves this goal with specific structure that can capture long-term dependencies in training signals and will be explained later in this section.

12.3.1   Global-feedback recurrent neural network

The neural-network learning capability can be used to learn the input–output behavior directly from measured or simulated input–output data of the original circuit, avoiding otherwise manual effort of developing equivalent-circuit topology. The universal approximation property of full RNN confirms that the model has a theoretical base of representing the full analogue behavior of the circuit with good accuracy. Also, the evaluation of the RNN from input to output is very fast, which makes it a great candidate to be used as a strong tool for modeling of nonlinear devices. There are two major types of RNNs. One with global-feedback which connects output layer to the input layer and has been used for nonlinear circuit macromodeling [70], and the other one with local-feedback which connects each hidden neuron's output to its input internally and has been mostly used in machine learning area but recently introduced in computer-aided design area [56]. RNNs are often used for data where time is of the essence and the order of the data is important. In global-feedback type of RNN, the output at any time depends on the history of inputs and outputs at previous time steps. In local-feedback type of RNN, the output at any time depends on the input and the state of the network at that moment as well as the previous moments.

12.3.1.1 Global-feedback RNN structure

The global-feedback RNN was first used for modeling of nonlinear circuits [70]. This structure is similar to feedforward multi-layer perceptron except there are feedbacks from the outputs to the inputs. At any time of the network, in addition to the input and some delays from the input, the outputs of the previous moments are involved to calculate the outputs at current time according to the following formula [70]:

$$y(t) = f_{ANN}(u(t), ..., u(t - m\tau), y(t - \tau), ..., y(t - n\tau), W) \qquad (12.9)$$

where W is the weight matrix connecting neurons of the RNN structure, $u(t)$ is the external input of the original circuit (could be voltages of nodes or currents of branches), $y(t)$ is the final output of the original circuit (could also be voltages or currents), m is the number of delays for the input signal and n is the number of delays for the output signal. The structure of a conventional 3-layer RNN is shown in Figure 12.5. As it can be seen from the figure, there are global feedbacks which

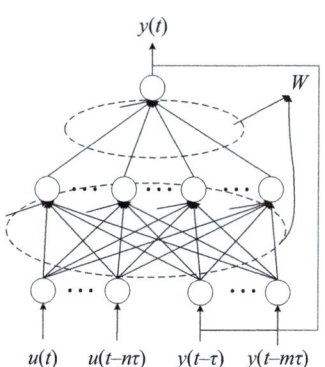

Figure 12.5   The structure of the global-feedback RNN [70]

connects the output layer directly to the input layer. The error function for the global-feedback RNN is defined as [70],

$$E^m = \frac{1}{2} \int_0^t \left\| y^m(t) - y_d^m(t) \right\|^2 dt \tag{12.10}$$

where $y_d^m(t)$ and $y^m(t)$ are the desired training output signal obtained from simulation or measurement and corresponding RNN output signal, respectively, and E^m is the error between these two signals. The total error E is defined as follows where S is the number of training signals,

$$E = \sum_{m=1}^{S} E^m \tag{12.11}$$

12.3.1.2   Global-feedback RNN training

The training of neural networks is an optimization process where the objective function is the error function and the design variables are the weights connecting neurons inside the structure. The goal of this optimization problem is to adjust the weights such that the error is minimized and the neural network can truly represent the behavior of the original circuits. In order to solve this optimization problem using gradient-based methods, gradients of the error function w.r.t. weights should be obtained. As RNN structure computation is a recursive procedure through time, therefore the gradients should also be calculated recursively through time which is much harder than feedforward neural networks. It was shown in [70] that the gradients of global-feedback RNN can be calculated as follows,

$$\frac{dy^m(t)}{dW} = \frac{\partial y^m(t)}{\partial W} + \sum_{D=1}^{n} \frac{\partial y^m(t)}{\partial y^m(t - D * \tau)} \frac{\partial y^m(t - D * \tau)}{\partial W} \tag{12.12}$$

where the first part, $\frac{\partial y^m(t)}{\partial W}$, is the backpropagation procedure through layers, and the second part is the backpropagation procedure through time which is a recursive formula.

12.3.2   Adjoint recurrent neural network

An advanced version of global-feedback RNN is developed recently [71]. This method is called ARNN. Similar to RNN, ARNN can be trained by input–output training waveforms obtained from the original nonlinear circuit without requiring knowledge about internal details of the device. In ARNN method, the time derivative of the original circuit outputs will be provided as additional training data to train the model. In this way, the training process will be more efficient requiring less training data (smaller number of time steps). This sensitivity-based training concept originally was developed for feedforward neural networks in [72] and has also been applied to stated-space dynamic neural network [73]. This concept has been extended to global-feedback RNNs and will be explained here.

As mentioned earlier, the use of sensitivity information in training results in better performance and more accurate model. In another way, using time sensitivity information leads to similar accurate model with a smaller number of training time steps. The structure of the ARNN is demonstrated in Figure 12.6. In this figure, the original weight matrix divided into several sub-matrices, where W_u is the weight matrix connecting external input and its delays to the hidden layer, W_y is the weight matrix connecting output delays to the hidden layer, W_o is the weight matrix between hidden layer and RNN output, and W_{ub}, W_{yb}, and W_{ob} are their corresponding bias matrices.

The total error for mth training waveform of the ARNN structure is defined as [71],

$$E_t^m = \frac{1}{2}k \int_0^T \left\| y^m - y_d^m \right\|^2 dt + \frac{1}{2}k' \int_0^T \left\| \dot{y}^m - \dot{y}_d^m \right\|^2 dt \qquad (12.13)$$

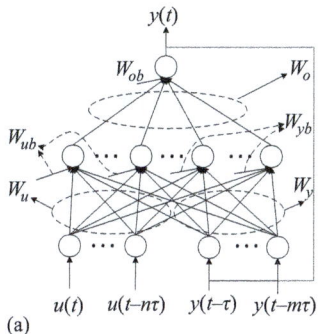

(a)

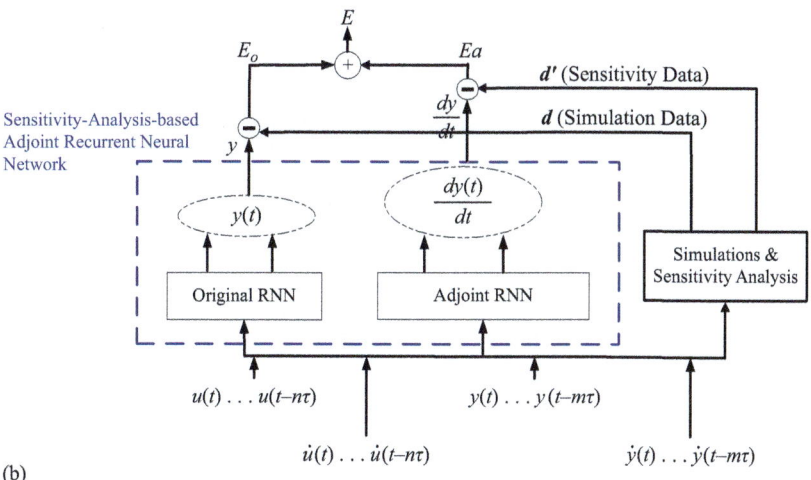

(b)

Figure 12.6   (a) Structure of the original global-feedback recurrent neural network with internal weights. (b) Adjoint global-feedback recurrent neural network with sensitivity data [71].

where the first part is the original error function (similar to the one already calculated in (12.10)), and the second part is the adjoint error function. k and k' are scaling factors. In order to train the ARNN model, the gradients should be computed. The formulations to calculate the gradients are provided in [71]. The training process is to adjust the weights in order to minimized the error which is the difference between desired outputs and ARNN outputs.

12.3.3   Global-feedback deep recurrent neural network

12.3.3.1   DRNN structure and training

RNNs are naturally deep in time because the current state of the network is a function of the state histories in all previous time steps due to the existing feedback connections in RNN architectures. Global-feedback deep RNN (DRNN) structure was developed in [55] for nonlinear macromodeling which benefits from depth in both space and time dimensions. The structure of DRNN is shown in Figure 12.7.

The input–output relationship of DRNN follows the same equation as (12.9) except that there are much more weights in several hidden layers involved in the mathematical representation. Training of the DRNN is also similar to the conventional RNN (12.12) but a longer gradient computation is required. This long derivation makes this structure more exposed to vanishing or exploding gradient problem.

12.3.3.2   The advantage of deep structures

The growing complexity of training data motivated researches in speech recognition, image processing, and many other areas to move toward deep neural networks [37]. As in deep structure, training data will be passed through both multiple time and space layers and nonlinearities, deep structures are more powerful to capture complex nonlinear relationships between inputs and outputs while shallow structures lack

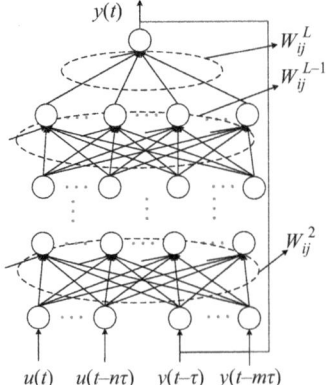

Figure 12.7   Global-feedback deep recurrent neural network (DRNN) structure [55]

hierarchical processing through space layers. In fact, each stacked space layer adds a level of nonlinearity to the model that cannot be represented by single layer structures. Also, for representing some functions, the deep model is exponentially more efficient than the shallow counterpart [66]. Therefore, the deep networks can model nonlinear relationships with considerably more accuracy and better generalization using similar amount of training data.

12.3.4 Local-feedback deep recurrent neural network

The local-feedback deep recurrent neural network (LFDRNN) was first introduced for modeling nonlinear circuits in [56]. The structure is similar to global-feedback RNN but feedback exists on neurons themselves. It can remove the redundant inputs generated by global-feedback RNN and capture more complex time-dependencies compared to global-feedback RNN. Noteworthy to mention that, it can have more weight parameters than original RNN used in microwave computer-aided design area. The structure of the LFDRNN is shown in Figure 12.8. In this figure, $u(k)$ and $y(k)$ are the input and output signals at time k, $\hat{y}(k)$ is the desired output at time k, N_u, N_y, and k_p are the number of inputs, outputs and circuit parameters, respectively. V^l represents the weight vector connecting neurons of two adjacent vertical layers, and

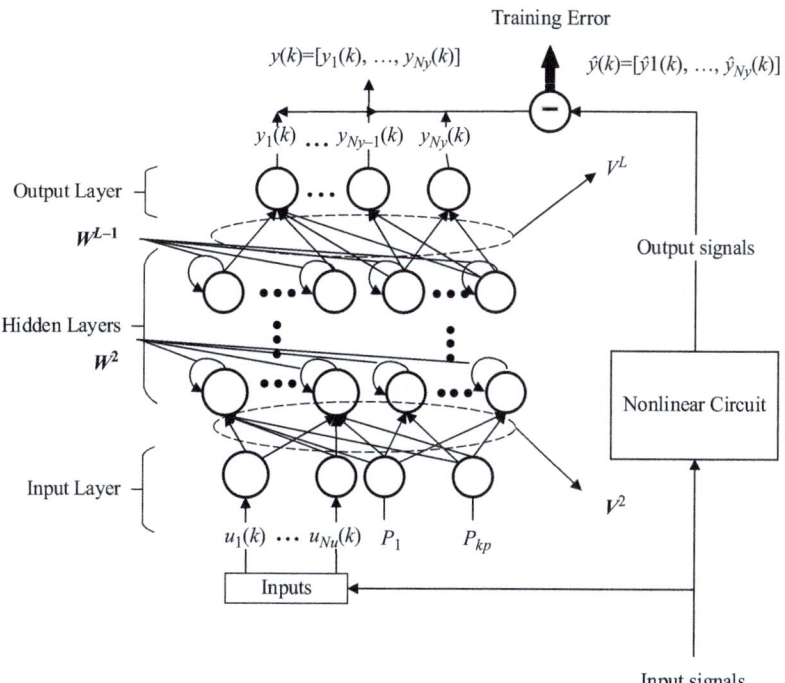

Figure 12.8 The LFDRNN-based macromodel structure according to [56]

W^l represents the weight vector connecting each neuron state at current time to its previous state at previous time. Circuit parameters are represented by P_i, and L is the number layers (including input and output layers).

The LFDRNN outputs can be formulated as [56],

$$y(k) = f_{LFDRNN}(u(k), P) \tag{12.14}$$

where f_{LFDRNN} represents the nonlinear relationship between the inputs and outputs of LFDRNN. The error function for mth training waveform can be defined as [56],

$$E_m(W, V) = \frac{1}{2} \sum_{j=1}^{N_y} \sum_{k=1}^{N_t} (y_{jm}(k, W, V) - \hat{y}_{jm}(k))^2 \tag{12.15}$$

where N_t is the number of time steps, and N_y is the number of model outputs. Subsequently, the total error for all training waveforms can be obtained by,

$$E = \sum_{m=1}^{S} E_m(W, V) \tag{12.16}$$

where S is the total number of training signals obtained from circuit simulation or measurement. Unlike the global-feedback RNN where the output of each neuron at each vertical layer is a function of the neurons outputs at bottom layer, the output of each neuron in LFDRNN will be computed as follows [56],

$$y_t^l = \Psi(W^l y_{t-1}^l + V^l y_t^{l-1}) \quad 2 \leq l \leq L \tag{12.17}$$

where y_{t-1}^l is the layer lth output at $(t-1)$th time step, y_t^{l-1} is the layer $(l-1)$th output at tth time step, and Ψ is the neuron's activation function (sigmoid or a hyperbolic tangent function). Similar to the conventional RNN with global feedback, the gradients w.r.t. weights should be calculated recursively through time but for both through time and through layer weight parameters [56].

12.3.5 Long short-term memory neural network

12.3.5.1 LSTM and vanishing gradient problem

Conventional recurrent neural networks suffer from vanishing gradient problem when the structure becomes deep through time or layers. When the depth of the layers or number of time steps increases, according to the chain rule, the gradients values must be multiplied to reach the initial layer or time step. As maximum value of the gradient of the sigmoid function is limited to 0.25, multiplying small numbers causes the gradient value to tend to zero. Therefore, it makes training very difficult and the weight updates become very slow. As a result, it leads to large training time. LSTM by its special structure tries to alleviate this problem in an efficient way. The LSTM is a recurrent type of neural network that was first introduced by Hochreiter and Schmidhuber [74]. It can combat the gradient vanishing problem with gating techniques and internal interactions. For the first time, it has been applied in computer-aided design area in [57] for modeling nonlinear components. The LSTM neural network uses the sequence of input–output and histories of input data as well as the network state in previous times to obtain each output in the current time frame. It works like a memory cell with its internal architecture which can forget, remember or pass the data to the next time step.

12.3.5.2   LSTM structure

The LSTM structure is shown in Figure 12.9 where N_1 is the number of LSTM blocks, h_t^k and c_t^k are the output and state of the kth LSTM block at the kth time step, respectively. Each LSTM block contains d memory cells as shown in Figure 12.10. The delays are passed to the block and the final cell output will be the main output of the block. Each LSTM cell (as shown in Figure 12.11 according to [57]) includes three gates: Input, Forget, and Output Gates, and another component called New Memory Cell. These three modules determine what information to remember or to forget. In each cell, hidden state (h_t) and cell state (c_t) will be passed to the next cell.

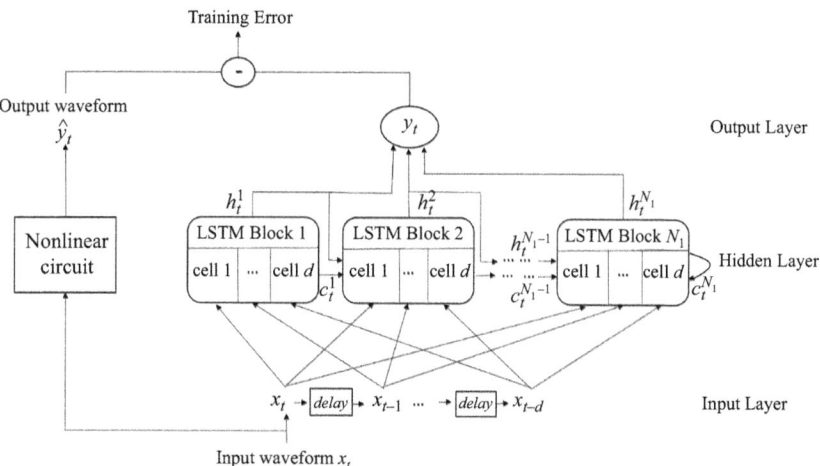

Figure 12.9   The LSTM structure according to [57]

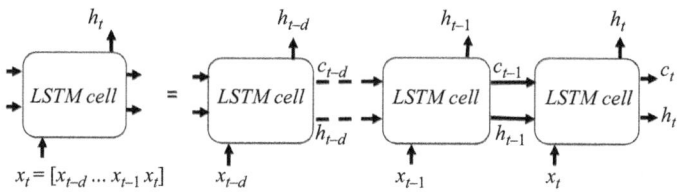

Figure 12.10   The LSTM block with d memory cells unrolled through time [57]

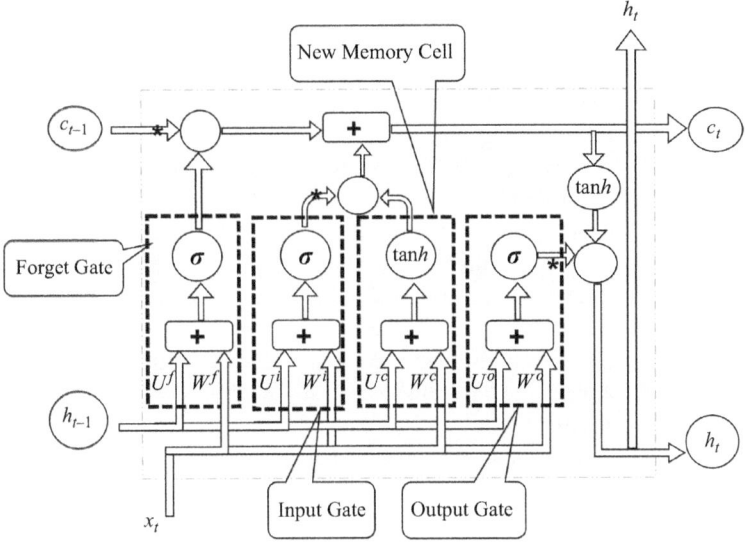

Figure 12.11   The LSTM cell structure [57]

The cell state is responsible for keeping long-term dependencies, and the hidden state is for short-term dependencies.

12.3.5.3   LSTM training

Similar to other recurrent neural network training algorithms, the total error function which is defined as follows should be minimized [57],

$$E_T = \frac{1}{2} \sum_{m=1}^{S} \sum_{t=1}^{N_t} E_t^m \tag{12.18}$$

where E_t^m is the error for the mth training waveform at time step t, which is formulated as

$$E_t^m = (y_t^m - \hat{y}_t^m)^2 \tag{12.19}$$

In order to proceed with training, the multivariate chain rule to calculate the gradients is as follows [57],

$$\begin{aligned}
\frac{\partial E_t}{\partial W} &= \frac{\partial E_t}{\partial \boldsymbol{h}_t} \frac{\partial \boldsymbol{h}_t}{\partial \boldsymbol{c}_t} \frac{\partial \boldsymbol{c}_t}{\partial \boldsymbol{c}_{t-1}} \cdots \frac{\partial \boldsymbol{c}_{t-(d-1)}}{\partial \boldsymbol{c}_{t-d}} \frac{\partial \boldsymbol{c}_{t-d}}{\partial W} \\
&= \frac{\partial E_t}{\partial \boldsymbol{h}_t} \frac{\partial \boldsymbol{h}_t}{\partial \boldsymbol{c}_t} \left(\prod_{k=1}^{d} \frac{\partial \boldsymbol{c}_{t-(k-1)}}{\partial \boldsymbol{c}_{t-k}} \right) \frac{\partial \boldsymbol{c}_{t-d}}{\partial W}
\end{aligned} \tag{12.20}$$

In (12.20), the recursive term $\frac{\partial c_{t-(k-1)}}{\partial c_{t-k}}$ can cause the gradients to be vanished. The cell state gradient was computed as,

$$\frac{\partial c_{t-(k-1)}}{\partial c_{t-k}} = A_k + B_k + D_k + E_k \tag{12.21}$$

where the formulations of A_k, B_k, D_k, and E_k can be found in [57]. This summation is almost equal to the forget gate output according to [57]. Therefore, in the back-propagation through time process of LSTM training, the multiplication of several terms of gradients can much better be prevented to be vanished and it makes LSTM structure a much better candidate for deep recurrent network-based macromodeling of microwave components.

12.4 Application examples of deep neural network for microwave modeling

12.4.1 High-dimensional parameter-extraction modeling using the hybrid feedforward deep neural network

In this application example, the hybrid feedforward deep neural network technique is used to develop a parameter-extraction model for a fourth-order multicoupled cavity filter with a 4-GHz center frequency and a 40-MHz bandwidth [69]. The objective of the parameter-extraction model in this application example is to estimate the coupling parameters from the given S-parameters. The ideal coupling matrix for this filter is

$$M_{ideal} = \begin{bmatrix} -0.0157 & 0.8950 & 0 & -0.2346 \\ 0.8950 & -0.010 & 0.8080 & 0 \\ 0 & 0.8080 & -0.010 & 0.8950 \\ -0.2346 & 0 & 0.8950 & -0.0157 \end{bmatrix} \tag{12.22}$$

The outputs of the deep neural network model are the eight nonzero coupling parameters in the M_{ideal} matrix, i.e., $y = [M_{11} \ M_{22} \ M_{33} \ M_{44} \ M_{12} \ M_{23} \ M_{34} \ M_{14}]^T$. The inputs to the deep neural network model are S_{11} in dB at 35 frequency samples, i.e., $x = [dB(S_{11}(f_1)) \ dB(S_{11}(f_2)) \ \ldots \ dB(S_{11}(f_{34})) \ dB(S_{11}(f_{35}))]^T$, where $f_1, f_2, \ldots, f_{34}, f_{35}$ are 35 frequency samples in the frequency range of 3,930–4,070 MHz. A tolerance of ± 0.3 for every nonzero coupling parameter is used to generate the training and test data.

The hybrid feedforward deep neural network technique is utilized to develop the parameter-extraction model for the fourth-order filter. The test error threshold for this modeling example is set as 2%. The number of hidden neurons used in each hidden layer is 100 for this example. After training with the three-stage training algorithm, the final deep neural network obtained has eight hidden layers with sigmoid functions and three hidden layers with smooth ReLUs. The training and test error of the trained deep neural network model are 1.22% and 1.88%, respectively [48]. The corresponding error between S-parameters from the test data and those calculated from extracted coupling parameters is 0.87%.

In order to examine the model accuracy, the parameter-extraction model developed using the hybrid feedforward deep neural network technique is used to extract the coupling parameters for a slightly and a highly detuned filter. The S-parameters of a slightly and a highly detuned filter are fed into the trained hybrid deep neural network model, which will extract the coupling parameters for these two cases. The desired and the extracted coupling parameters for both test cases are shown in Table 12.1. Figure 12.12 shows the comparison between the S-parameters from the desired coupling matrix and that from the extracted coupling matrix for both the slightly and the highly detuned filters. According to the results in Table 12.1 and Figure 12.12, a good match between the desired and extracted coupling parameters along with an excellent match between the responses from desired and extracted coupling matrices have been achieved for both the slightly and the highly detuned filters. The test results reveal that the hybrid feedforward deep neural network technique is suitable for the parameter-extraction modeling of microwave filters. Unlike the conventional optimization method, the hybrid feedforward deep neural network model does not need to simulate the filter circuit iteratively for each detuned case. Once the model is developed, it can be used to provide parameter extraction solutions for both the slightly and highly detuned filters instantly as long as the detuned filters are in the range of the training data.

The hybrid feedforward deep neural network model is compared with the shallow neural network for this example. A shallow neural network with two hidden layers and 253 hidden neurons in each hidden layer is trained with the training data. This shallow neural network has the similar number of training parameters (i.e., weight parameters and biases) as that of the deep neural network with eight hidden layers and 100 hidden neurons in each layer from Stage I of the three-stage training. The comparison results of the shallow and deep neural networks are summarized in Table 12.2 [48]. The test error of the shallow neural network is 3.85% while that of the deep neural network is 2.62%. It shows that deep neural network is more effective for improving modeling accuracy than shallow neural network when both neural networks have same type of activation functions and similar number of training parameters.

Table 12.1   *Comparison between the desired and the extracted coupling parameters for the slightly and highly detuned filter [48]*

Coupling parameters	Desired (slightly)	Extracted (slightly)	Desired (highly)	Extracted highly
M_{11}	−0.0607	−0.0677	−0.2657	−0.2703
M_{22}	0.060	0.0588	−0.240	−0.2443
M_{33}	−0.040	−0.0369	0.260	0.2545
M_{44}	0.0343	0.0370	0.2343	0.2355
M_{12}	0.7950	0.7912	1.1450	1.1348
M_{23}	0.9080	0.9035	1.0580	1.0658
M_{34}	0.7950	0.8025	0.6450	0.6421
M_{14}	−0.1346	−0.1344	−0.4846	−0.4917

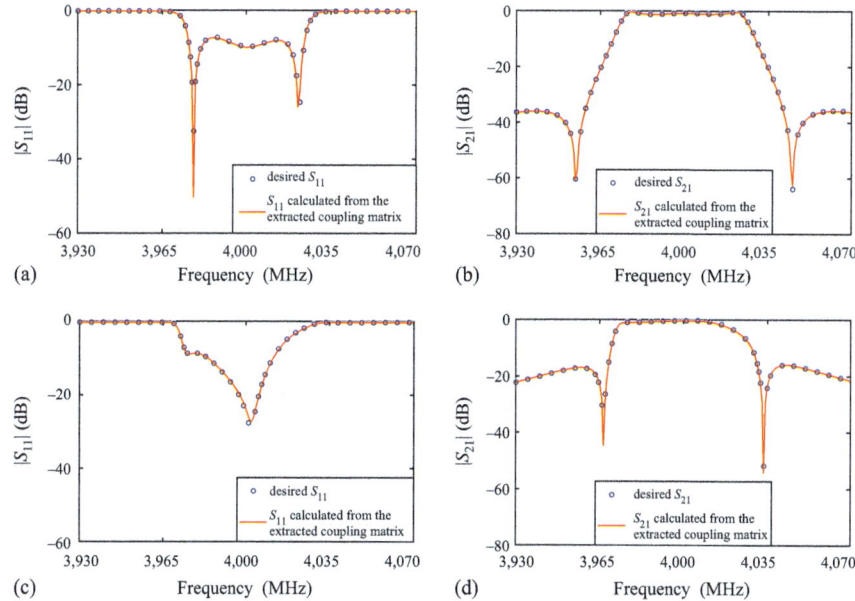

Figure 12.12   *The comparison of S-parameters from the desired (o) and the extracted (–) coupling matrix for the slightly and the highly detuned fourth-order filter [48]. (a) Return loss S_{11} for the slightly detuned case. (b) Insertion loss S_{21} for the slightly detuned case. (c) Return loss S_{11} for the highly detuned case. (d) Insertion loss S_{21} for the highly detuned case.*

Table 12.2   *Comparison of shallow and deep neural networks with similar number of training parameters for parameter-extraction modeling of the fourth-order filter example [48]*

Neural network structure	Hidden neurons per layer	Number of training parameters	Training error	Test error
Shallow neural network (2 sigmoid hidden layers)	253	75k	3.29%	3.85%
Deep neural network (8 sigmoid hidden layers)	100	75k	2.04%	2.62%

In order to illustrate the benefit of combining sigmoid function with ReLU, the hybrid deep neural network is compared with deep neural networks with only conventional ReLUs. In this chapter, the deep neural network with only conventional ReLUs is named as pure ReLU deep neural network. Pure ReLU deep neural networks

Table 12.3   Comparison of the hybrid feedforward deep neural network and pure ReLU deep neural network for parameter-extraction modeling of the fourth-order filter example [48]

Structure of deep neural networks	Training error	Test error
13-layer hybrid deep neural network[a]	1.22%	1.88%
13-layer pure ReLU network	2.73%	3.38%
16-layer pure ReLU network	2.23%	3.27%

[a]The 13-layer hybrid deep neural network has eight hidden layers with sigmoid functions and three hidden layers with smooth ReLUs

with same number of layers as or more number of layers than that of the final hybrid deep neural network model are utilized to develop the parameter-extraction model for the fourth-order filter example. The number of hidden neurons in each hidden layer is 100 for both the hybrid deep neural network and the pure ReLU deep neural networks. The comparison of results is shown in Table 12.3 [48]. The hybrid feedforward deep neural network can achieve a better test error than pure ReLU deep neural network with same number of hidden neurons. Adding more hidden neurons in pure ReLU deep neural network can reduce the test error. In other words, pure ReLU deep neural network will need much more hidden neurons to achieve similar accuracy as the hybrid deep neural network [48].

12.4.2   Macromodeling of audio amplifier using long short-term memory neural network

In this example, the LSTM neural network is used for macromodeling of an audio amplifier as shown in Figure 12.13. Training waveforms were generated with different amplitudes (50 mV, 70 mV, 90 V) and different frequencies (1–2 kHz with step size of 0.2 kHz). Also, three signals were generated as test waveforms that were not used in training as follows,

–   (Amplitude, frequency) = (90 mV, 1.5 kHz)
–   (Amplitude, frequency) = (60 mV, 2.0 kHz)
–   (Amplitude, frequency) = (80 mV, 1.3 kHz)

Table 12.4 shows the comparison between the LSTM neural network and conventional global-feedback RNN methods. As it can be seen from the table, the LSTM method outperforms RNN in both speed and accuracy. Also, Table 12.5 demonstrates the speedup comparison between the transistor-level model and the LSTM-based model. As it can be seen from this table, the obtained model from LSTM technique is much faster than the existing transistor-level model of this device. Figure 12.14 demonstrates the comparison between outputs of the transistor-level model, RNN-based model, and LSTM-based model. As shown in this figure, LSTM-based model

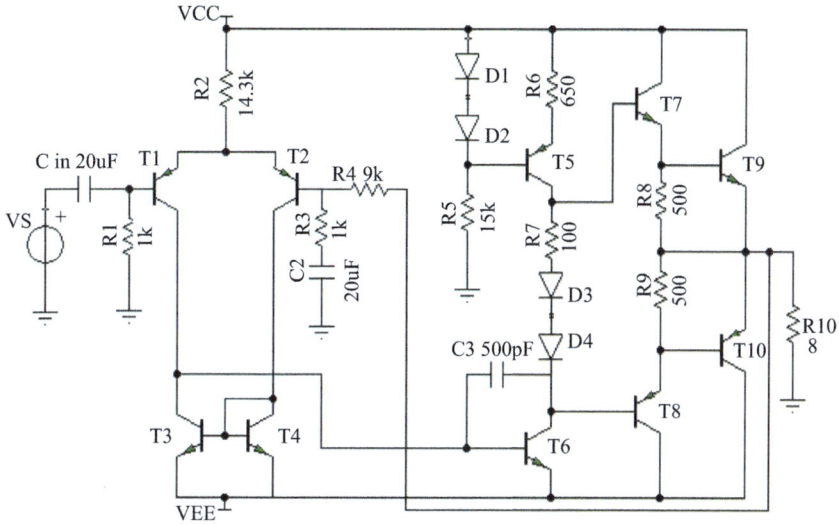

Figure 12.13   The schematic of audio amplifier [57]

Table 12.4   Comparison of training and test errors and times of LSTM-based and RNN-based models [57]

Model type	Structure	Training error	Test error	Train time	Test time
LSTM	$N_1 = 3, d = 2$	$5.9 * 10^{-5}$	$14 * 10^{-5}$	300.104 s	0.009 s
RNN	Hidden neurons $= 50$ No. of delays $= 2$	$18 * 10^{-5}$	$15 * 10^{-4}$	1957.430 s	0.013 s

Table 12.5   Speedup comparison between the transistor-level model and LSTM-based model of the audio amplifier [57]

Model	Structure	Speedup
Transistor-level	–	1 (reference for comparison)
LSTM-based	$N_1 = 3, d = 2$	11.1

output matches the transistor-level model very well and better than RNN-based model output.

The aforementioned RNN techniques for macromodeling of components and circuits can be applied to any time scale and are not limited to the frequency range shown in this example. As long as the time-dependent training waveforms are ready

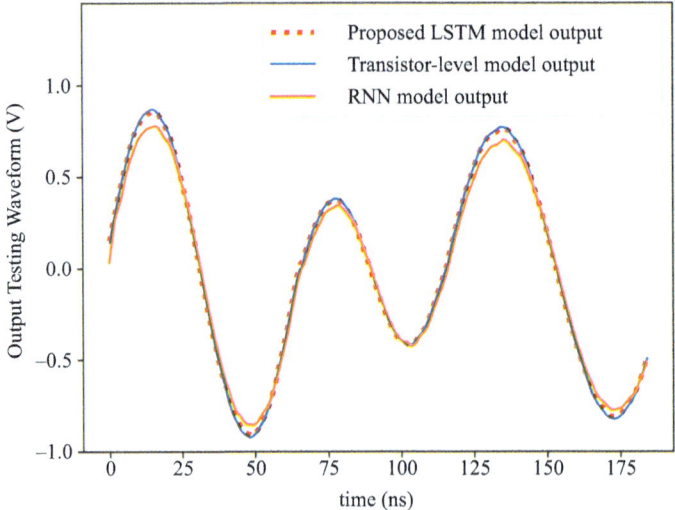

Figure 12.14   Comparison of the outputs of transistor-level, RNN-based and LSTM-based models for audio amplifier [57]

or can be generated, all the RNN techniques reviewed in this chapter can be used for nonlinear modeling of microwave circuits as well.

12.5   Discussion

This chapter has presented some recent advances in deep neural networks for microwave circuit modeling. Here, we discuss the proper usage of different methods in different practical situations. The powerful learning of deep neural networks makes it a suitable tool for modeling the complex input–output relationship of microwave circuits. For passive linear device modeling, the feedforward neural network is a suitable choice. If the nonlinearity of the input-output relationship of the passive linear device is not very high, the feedforward shallow neural network such as three or four-layer MLP can be used. If the nonlinearity of the input–output relationship of the passive linear device is very high and consequently beyond the capability of shallow neural networks, the feedforward deep neural network becomes a useful method.

When modeling time-domain dynamic behavior of the nonlinear devices, the recurrent neural network is more suitable. If the dynamic behavior has relatively simple time-dependencies and input–output relationship, the global-feedback RNN and the ARNN can be used to model the circuit behavior. With the exploration of the time sensitivity information, the ARNN method could obtain similar accurate model with a smaller number of training time steps. If the dynamic behavior of nonlinear circuit has more complex input–output relationship, the global-feedback DRNN are more powerful to capture complex nonlinear behaviors. The local-feedback DRNN is a further advance of the global-feedback DRNN, where the redundant inputs generated by global-feedback RNN are removed. Compared to global-feedback RNN, the local-feedback RNN can capture more complex time-dependencies. When the depth of the layers or the number of time steps increases, the LSTM neural network can be used to overcome the vanishing gradient problem.

12.6 Conclusion

An overview of recent advances in deep neural network techniques for microwave circuit modeling has been provided in this chapter. We have first introduced the feedforward deep neural network and the vanishing gradient problem in the training process. Then, we have described the hybrid deep neural network technique which can be trained without the vanishing gradient problem and can address the challenges due to high-dimensional inputs. Following that, we have reviewed various recurrent neural networks for nonlinear circuit modeling. Lastly, we have demonstrated two application examples using the recently developed deep neural network techniques. The deep neural networks can be used to learn the complex input–output relationship for microwave circuit modeling, which could be beyond the capability of shallow neural networks. The trained deep neural network models can provide fast solutions to the tasks they have learned and can be subsequently used in the high-level circuit and system design.

References

[1] F. Feng, C. Zhang, J. Ma, and Q.-J. Zhang, "Parametric modeling of EM behavior of microwave components using combined neural networks and pole-residue-based transfer functions," *IEEE Trans. Microw. Theory Techn.*, vol. 64, no. 1, pp. 60–77, 2016.

[2] H. Kabir, Y. Wang, M. Yu, and Q.-J. Zhang, "Neural network inverse modeling and applications to microwave filter design," *IEEE Trans. Microw. Theory Techn.*, vol. 56, no. 4, pp. 867–879, 2008.

[3] S. Koziel, L. Leifsson, I. Couckuyt, and T. Dhaene, "Robust variable-fidelity optimization of microwave filters using co-kriging and trust regions," *Microw. Opt Technol. Lett.*, vol. 55, no. 4, pp. 765–769, 2013.

[4] N. Leszczynska, I. Couckuyt, T. Dhaene, and M. Mrozowski, "Low-cost surrogate models for microwave filters," *IEEE Microw. Wireless Compon. Lett.*, vol. 26, no. 12, pp. 969–971, 2016.

[5] G. Angiulli, M. Cacciola, and M. Versaci, "Microwave devices and antennas modelling by support vector regression machines," *IEEE Trans. Magnetics*, vol. 43, no. 4, pp. 1589–1592, 2007.

[6]   J. L. Chávez-Hurtado and J. E. Rayas-Sánchez, "Polynomial-based surrogate modeling of RF and microwave circuits in frequency domain exploiting the multinomial theorem," *IEEE Trans. Microw. Theory Techn.,* vol. 64, no. 12, pp. 4371–4381, 2016.

[7]   Q.-J. Zhang, K. C. Gupta, and V. K. Devabhaktuni, "Artificial neural networks for RF and microwave design-From theory to practice," *IEEE Trans. Microw. Theory Techn.,* vol. 51, no. 4, pp. 1339–1350, 2003.

[8]   J. E. Rayas-Sánchez, "EM-based optimization of microwave circuits using artificial neural networks: the state-of-the-art," *IEEE Trans. Microw. Theory Techn.,* vol. 52, no. 1, pp. 420–435, 2004.

[9]   M. B. Steer, J. W. Bandler, and C. M. Snowden, "Computer-aided design of RF and microwave circuits and systems," *IEEE Trans. Microw. Theory Techn.,* vol. 50, no. 3, pp. 996–1005, 2002.

[10]   H. Kabir, L. Zhang, M. Yu, P. H. Aaen, J. Wood, and Q.-J. Zhang, "Smart modeling of microwave devices," *IEEE Microw. Mag.,* vol. 11, no. 3, pp. 105–118, 2010.

[11]   F. Mkadem and S. Boumaiza, "Physically inspired neural network model for RF power amplifier behavioral modeling and digital predistortion," *IEEE Trans. Microw. Theory Techn.,* vol. 59, no. 4, pp. 913–923, 2011.

[12]   S. Liao, H. Kabir, Y. Cao, J. Xu, Q. J. Zhang, and J. Ma, "Neural-network modeling for 3-D substructures based on spatial EM-field coupling in finite-element method," *IEEE Trans. Microw. Theory Techn.,* vol. 59, no. 1, pp. 21–38, 2011.

[13]   S. A. Sadrossadat, Y. Cao, and Q. J. Zhang, "Parametric modeling of microwave passive components using sensitivity-analysis-based adjoint neural-network technique," *IEEE Trans. Microw. Theory Techn.,* vol. 61, no. 5, pp. 1733–1747, 2013.

[14]   W. C. Na and Q. J. Zhang, "Automated parametric modeling of microwave components using combined neural network and interpolation techniques," in *IEEE MTT-S Int. Dig.,* Seattle, WA, USA, June 2013.

[15]   W. Na and Q. J. Zhang, "Automated knowledge-based neural network modeling for microwave applications," *IEEE Microw. Wireless Compon. Lett.,* vol. 24, no. 7, pp. 499–501, 2014.

[16]   V. K. Devabhaktuni, M. Yagoub, Y. Fang, J. Xu, and Q.-J. Zhang, "Neural networks for microwave modeling: model development issues and nonlinear modeling techniques," *Int. J. RF Microw. Comput.-Aided Eng.,* vol. 11, pp. 4–21, 2001.

[17]   C. Zhang, J. Jin, W. Na, Q. J. Zhang, and M. Yu, "Multivalued neural network inverse modeling and applications to microwave filters," *IEEE Trans. Microw. Theory Techn.,* vol. 66, no. 8, pp. 3781–3797, 2018.

[18]   S. Li, Y. Wang, M. Yu, and A. Panariello, "Efficient modeling of Ku-band high power dielectric resonator filter with applications of neural networks," *IEEE Trans. Microw. Theory Techn.,* vol. 67, no. 8, pp. 3427–3435, 2019.

[19]   P. Zhao and K. Wu, "Homotopy optimization of microwave and millimeter-wave filters based on neural network model," *IEEE Trans. Microw. Theory Techn.,* vol. 68, no. 4, pp. 1390–1400, 2020.

[20] M. Isaksson, D. Wisell, and D. Ronnow, "Wide-band dynamic modeling of power amplifiers using radial-basis function neural networks," *IEEE Trans. Microw. Theory Techn.*, vol. 53, no. 11, pp. 3422–3428, 2005.

[21] D. Schreurs, M. O'Droma, A. A. Goacher, and M. Gadringer, *RF Power Amplifier Behavioral Modeling.* New York, NY: Cambridge University Press, 2008.

[22] S. Wang, M. Roger, J. Sarrazin, and C. Lelandais-Perrault, "Hyperparameter optimization of two-hidden-layer neural networks for power amplifiers behavioral modeling using genetic algorithms," *IEEE Microw. Wireless Compon. Lett.*, vol. 29, no. 12, pp. 802–805, 2019.

[23] W. Liu, W. Na, L. Zhu, J. Ma, and Q. J. Zhang, "A Wiener-type dynamic neural network approach to the modeling of nonlinear microwave devices," *IEEE Trans. Microw. Theory Techn.*, vol. 65, no. 6, pp. 2043–2062, 2017.

[24] L. Zhu, Q.-J. Zhang, K. Liu, Y. Ma, B. Peng, and S. Yan, "A novel dynamic neuro-space mapping approach for nonlinear microwave device modeling," *IEEE Microw. Wireless Compon. Lett.*, vol. 26, no. 2, pp. 131–133, 2016.

[25] Y. Long, Z. Zhong, and Y.-X. Guo, "A novel 4-D artificial-neural-network-based hybrid large-signal model of GaAs pHEMTs," *IEEE Trans. Microw. Theory Techn.*, vol. 64, no. 6, pp. 1752–1762, 2016.

[26] A.-D. Huang, Z. Zhong, W. Wu, and Y.-X. Guo, "An artificial neural network-based electrothermal model for GaN HEMTs with dynamic trapping effects consideration," *IEEE Trans. Microw. Theory Techn.*, vol. 64, no. 8, pp. 2519–2528, 2016.

[27] Z. Zhao, L. Zhang, F. Feng, W. Zhang, and Q. J. Zhang, "Space mapping technique using decomposed mappings for GaN HEMT modeling," *IEEE Trans. Microw. Theory Techn.*, vol. 68, no. 8, pp. 3318–3341, 2020.

[28] W. Na, F. Feng, C. Zhang, and Q. J. Zhang, "A unified automated parametric modeling algorithm using knowledge-based neural network and l_1 optimization," *IEEE Trans. Microw. Theory Techn.*, vol. 65, no. 3, pp. 729–745, 2017.

[29] F. Feng, V. Gongal-Reddy, C. Zhang, J. Ma, and Q. J. Zhang, "Parametric modeling of microwave components using adjoint neural networks and pole-residue transfer functions with EM sensitivity analysis," *IEEE Trans. Microw. Theory Techn.*, vol. 65, no. 6, pp. 1955–1975, 2017.

[30] W. Zhang, F. Feng, V. Gongal-Reddy, *et al.*, "Space mapping approach to electromagnetic centric multiphysics parametric modeling of microwave components," *IEEE Trans. Microw. Theory Techn.*, vol. 66, no. 7, pp. 3169–3185, 2018.

[31] W. Na, W. Zhang, S. Yan, F. Feng, W. Zhang, and Y. Zhang, "Automated neural network-based multiphysics parametric modeling of microwave components," *IEEE Access,* vol. 7, pp. 141153–141160, 2019.

[32] W. Zhang, F. Feng, W. Liu, S. *et al.*, "Advanced parallel space-mapping-based multiphysics optimization for high-power microwave filters," *IEEE Trans. Microw. Theory Techn.*, vol. 69, no. 5, pp. 2470–2484, 2021.

[33] P. M. Watson and K. C. Gupta, "Design and optimization of CPW circuits using EM-ANN models for CPW components," *IEEE Trans. Microw. Theory Techn.*, vol. 45, no. 12, pp. 2515–2523, 1997.

[34]   J. Monzó-Cabrera, J. L. Pedreno-Molina, A. Lozano-Gurrrero, and A. Toledo-Moreo, "A novel design of a robust ten-port microwave reflectometer with autonomous calibration by using neural networks," *IEEE Trans. Microw. Theory Tech.*, vol. 56, no. 12, pp. 2972–2978, 2008.

[35]   F. Seide, G. Li, and D. Yu, "Conversational speech transcription using context-dependent deep neural networks," in *Proc. Conf. Int. Speech Comm. Assoc.*, Florence, Italy, August 2011, pp. 437–440.

[36]   M. Telgarsky, "Benefits of depth in neural networks," in *Proc. Conf. Learn. Theory*, New York, NY, June 2016, pp. 1517–1539.

[37]   Y. LeCun, Y. Bengio, and G. Hinton, "Deep learning," *Nature*, vol. 521, pp. 436–444, 2015.

[38]   G. E. Hinton, S. Osindero, and Y.-W. Teh, "A fast learning algorithm for deep belief nets," *Neural Computat.*, vol. 18, no. 7, pp. 1527–1554, 2006.

[39]   A. Krizhevsky, I. Sutskever, and G. Hinton, "ImageNet classification with deep convolutional neural networks," in *Advances Neural Inf. Process. Syst.*, pp. 1097–1105, 2012.

[40]   A. Graves, A.-R. Mohamed, and G. Hinton, "Speech recognition with deep recurrent neural networks," in *Proc. Int. Conf. Acoust. Speech Signal Process.*, Vancouver, BC, Canada, May 2013, pp. 6645–6649.

[41]   G. W. Cottrell, "New life for neural networks," *Science*, vol. 313, no. 5786, pp. 454–455, 2006.

[42]   V. Sze, Y.-H. Chen, T.-J. Yang, and J. S. Emer, "Efficient processing of deep neural networks: a tutorial and survey," *Proc. IEEE*, vol. 105, no. 12, pp. 2295–2329, 2017.

[43]   C. Farabet, C. Couprie, L. Najman, and Y. Lecun, "Learning hierarchical features for scene labeling," *IEEE Trans. Pattern Anal. Mach. Intell.*, vol. 35, no. 8, pp. 1915–1929, 2013.

[44]   D. Chen and B. K.-W. Mak, "Multitask learning of deep neural networks for low-resource speech recognition," *IEEE Trans. Audio, Speech, Language Process.*, vol. 23, no. 7, pp. 1172–1183, 2015.

[45]   R. Collobert and J. Weston, "A unified architecture for natural language processing: deep neural networks with multitask learning," in *Proc. Int. Conf. Mach. Learn.*, Helsinki, Finland, July 2008, pp. 160–167.

[46]   K. Cho, B. V. Merrienboer, C. Gulcehre, *et al.*, "Learning phrase representations using RNN encoder–decoder for statistical machine translation," in *Proc. Conf. Empirical Methods Natural Language Process.*, Doha, Qatar, October 2014, pp. 1724–1734.

[47]   X. Glorot, A. Bordes, and Y. Bengio, "Domain adaptation for large-scale sentiment classification: a deep learning approach," in *Proc. Int. Conf. Mach. Learn.*, Bellevue, Washington, USA, July 2011, pp. 513–520.

[48]   J. Jin, C. Zhang, F. Feng, W. Na, J. Ma, and Q. J. Zhang, "Deep neural network technique for high-dimensional microwave modeling and applications to parameter extraction of microwave filters," *IEEE Trans. Microw. Theory Techn.*, vol. 67, no. 10, pp. 4140–4155, 2019.

[49] S. Zhang, X. Hu, Z. Liu, *et al.*, "Deep neural network behavioral modeling based on transfer learning for broadband wireless power amplifier," *IEEE Microw. Wireless Compon. Lett.*, vol. 31, no. 7, pp. 917–920, 2021.

[50] E. Zhu, Z. Wei, X. X. Xu, and W.-Y. Yin, "Fourier subspace-based deep learning method for inverse design of frequency selective surface," *IEEE Trans. Antennas Propag.*, Early Access, July 2021.

[51] X. Yu, X. Hu, Z. Liu, C. Wang, W. Wang, and F. M. Ghannouchi, "A method to select optimal deep neural network model for power amplifiers," *IEEE Microw. Wireless Compon. Lett.*, vol. 31, no. 2, pp. 145–148, 2021.

[52] W. Na, K. Liu, W. Zhang, H. Xie, and D. Jin, "Deep neural network with batch normalization for automated modeling of microwave components," in *2020 IEEE MTT-S International Conference on Numerical Electromagnetic and Multiphysics Modeling and Optimization (NEMO)*, 2020, pp. 1–3.

[53] J. Jin, F. Feng, J. Zhang, S. Yan, W. Na, and Q. J. Zhang, "A novel deep neural network topology for parametric modeling of passive microwave components," *IEEE Access*, vol. 8, pp. 82273–82285, 2020.

[54] Z. Lin, R. Guo, M. Li, *et al.*, "Low-frequency data prediction with iterative learning for highly nonlinear inverse scattering problems," *IEEE Trans. Microw. Theory Tech.*, vol. 69, no. 10, pp. 4366–4376, 2021.

[55] Z. Naghibi, S. A. Sadrossadat, and S. Safari, "Time-domain modeling of nonlinear circuits using deep recurrent neural network technique," *AEU – Int. J. Electron. Commun.*, vol. 100, pp. 66–74, 2019.

[56] M. Noohi, A. Mirvakili, Sadrossadat, SA. "Modeling and implementation of nonlinear boost converter using local feedback deep recurrent neural network for voltage balancing in energy harvesting applications," *Int. J. Circ. Theor. Appl.* 2021, pp. 1–17, doi:10.1002/cta.3143

[57] M. Moradi A., S. A. Sadrossadat, and V. Derhami, "Long short-term memory neural networks for modeling nonlinear electronic components," *IEEE Trans. Compon. Pack. Manufact. Technol.*, vol. 11, no. 5, pp. 840–847, 2021.

[58] H. Li, Y. Zhang, G. Li, and F. Liu, "Vector decomposed long short-term memory model for behavioral modeling and digital predistortion for wideband RF power amplifiers," *IEEE Access*, vol. 8, pp. 63780–63789, 2020.

[59] G. Pan, Y. Wu, M. Yu, L. Fu, and H. Li, "Inverse modeling for filters using a regularized deep neural network approach," *IEEE Microw. Wireless Compon. Lett.*, vol. 30, no. 5, pp. 457–460, 2020.

[60] Y. Hu, Y. Jin, X. Wu, and J. Chen, "A theory-guided deep neural network for time domain electromagnetic simulation and inversion using a differentiable programming platform," *IEEE Trans. Antennas Propag.*, vol. 70, no. 1, pp. 767–772, 2022.

[61] Y. Mao, Q. Zhan, R. Zhang, D. Wang, W.-F. Huang, and Q. H. Liu, "Fast simulation of electromagnetic fields in doubly periodic structures with a deep fully convolutional network," *IEEE Trans. Antennas Propag.*, vol. 69, no. 5, pp. 2921–2928, 2021.

[62]   H. Mayer, F. Gomez, D. Wierstra, I. Nagy, A. Knoll, and J. Schmidhuber, "A system for robotic heart surgery that learns to tie knots using recurrent neural networks," in *IEEE/RSJ Int. Conf. Intell. Robots Syst.,* Beijing, China, October 2006.

[63]   Y. LeCun, K. Kavukcuoglu, and C. Farabet, "Convolutional neural networks and applications in vision," in *Proc. IEEE Int. Symp. Circuits Syst.,* Paris, France, June 2010.

[64]   Y. Bengio, P. Simard, and P. Frasconi, "Learning long-term dependencies with gradient descent is difficult," *IEEE Trans. Neural Netw.,* vol. 5, no. 2, pp. 157–166, 1994.

[65]   S. Hochreiter, Y. Bengio, P. Frasconi, and J. Schmidhuber, "Gradient flow in recurrent nets: the difficulty of learning long-term dependencies," in S. C. Kremer and J. F. Kolen, Eds. *A Field Guide to Dynamical Recurrent Neural Networks*, Piscataway, NJ: IEEE Press, 2001.

[66]   Y. Bengio, Learning deep architectures for AI, *Found. Trends Mach. Learn.,* vol. 2, no. 1, pp. 1–127, 2009.

[67]   Q. J. Zhang and K. C. Gupta, *Neural Networks for RF and Microwave Design.* Norwood, MA: Artech House, 2000.

[68]   M. A. Nielsen, *Neural Networks and Deep Learning*, Determination Press, 2015.

[69]   H.-T. Hsu, Z. Zhang, K. A. Zaki, and A. E. Atia, "Parameter extraction for symmetric coupled-resonator filters," *IEEE Trans. Microw. Theory Techn.,* vol. 50, no. 12, pp. 2971–2978, 2002.

[70]   Y. Fang, M. C. E. Yagoub, F. Wang, and Q.-J. Zhang, "A new macromodeling approach for nonlinear microwave circuits based on recurrent neural networks," *IEEE Trans. Microw. Theory Techn.,* vol. 48, no. 12, pp. 2335–2344, 2000.

[71]   Z. Naghibi, S. A. Sadrossadat, and S. Safari, "Adjoint recurrent neural network technique for nonlinear electronic component modeling," *Int. J. Circuit Theory Appl.,* vol. 50, no. 4, pp. 1119–1129, 2022.

[72]   S. A. Sadrossadat, Y. Cao Y, and Q. J. Zhang, "Parametric modeling of microwave passive components using sensitivity-analysis based adjoint neural-network technique," *IEEE Trans. Microw. Theory Techn.,* vol. 61, no. 5, pp. 1733–1747, 2013.

[73]   S. A. Sadrossadat, P. Gunupudi, and Q. J. Zhang, "Nonlinear electronic/ photonic component modeling using adjoint state-space dynamic neural network technique," *IEEE Trans. on Compon., Packag., Manuf. Technol.,* vol. 5, no. 11, pp. 1679–1693, 2015.

[74]   S. Hochreiter and J. Schmidhuber, "Long short-term memory," *Neural Comput.,* vol. 9, no. 8, pp. 1735–1780, 1997.

Chapter 13

Concluding remarks, open challenges, and future trends

Marco Salucci[1,2] and Maokun Li[3,4]

13.1 Introduction

Throughout this book, we have seen how deep learning (*DL*) has recently become a very active research field in many electromagnetic (*EM*) engineering applications [1–4].

DL algorithms are sophisticated machine learning (*ML*) techniques that try to mimic human brains to learn how to accurately solve a given task with extraordinary efficiency, robustness, and reliability [5–8]. However, their application to *EM* problems is nowadays in its very early stages and their development is significantly less mature than in other fields (e.g., speech, image, and text recognition).

In this concluding chapter, we try to summarize the main pros and cons of *DL*, together with the main challenges that still need to be solved in such an emerging research field. Finally, some future trends are briefly discussed as well, hopefully fostering future research in using this very powerful paradigm to address paramount challenges in many *EM*-related areas.

13.2 Pros and cons of *DL*

13.2.1 High computational efficiency and accuracy

After being trained with properly-built training datasets, *DL* methods can perform thousands of repetitive tasks with very high prediction accuracy and time saving with respect to traditional algorithms (e.g., full-wave *EM* forward solvers). The overall "quality" of the outputted results never decreases, subject to the availability of training data that correctly represents the problem to be solved [5].

[1]ELEDIA Research Center (ELEDIA@UniTN – University of Trento), DICAM – Department of Civil, Environmental, and Mechanical Engineering, Italy
[2]CNIT – "University of Trento" ELEDIA Research Unit, Italy
[3]Institute of Microwave and Antenna, Department of Electronic Engineering, China
[4]ELEDIA Research Center (ELEDIA@TSINGHUA – Tsinghua University), China

13.2.2   *Bypassing feature engineering*

Feature engineering (*FE*) is often considered a fundamental task in *ML* and it refers to the extraction of highly informative features from raw training data on the underlying input–output relationship to learn. One of the main advantages of *DL* techniques is their capability of automatically performing *FE* without the introduction of domain/physical knowledge from the user. As a matter of fact, *DL* algorithms are capable of analyzing the data in order to find features that humans could miss but that enable a faster learning of the assigned task without being explicitly programmed [5].

13.2.3   *Large amounts of training data*

"In a world with infinite data, and infinite computational resources, there might be little need for any other technique" [9]. Unfortunately, in the real-world current *DL* algorithms need large amount of training data (much larger than in traditional *ML* methods) to tune their parameters and learn a specific task starting from scratch. The more powerful abstraction and generalization capabilities are needed, the more parameters need to be trained. However, the amount of necessary training samples exponentially increases with the number of parameters/the complexity of the network.

13.2.4   *High computational burden*

As already said, building a high-fidelity *DL* model requires a lot of training data. This directly translates into the need for adequate processing power. As a matter of fact, multi-core high-performance graphics processing units (*GPU*s) are often employed to reduce the time consumption of the learning stage. However, such processing units are generally quite expensive and energy-hungry.

13.2.5   *Deep architectures, not learning*

Another issue with current *DL* algorithms is that they perform well at predicting a given input–output relationship but not at "understanding" the context and meaning of the handled data. As a matter of fact, the term "deep" is more referred to the underlying architecture (i.e., the large amount of hidden layers/neurons) rather than the level of understanding of the performed tasks [5].

13.2.6   *Lack of transparency*

One of the most critical limitations of *DL* relies in the fact that we cannot clearly understand how artificial neural networks (*ANN*s) find a particular solution to the assigned problem. Indeed, their output is the result of the "reasoning" coming from the interaction of thousands of artificial neurons arranged into a large number of hidden layers. As a result, *DL* algorithms are often regarded as "black-boxes," and such a lack of transparency may cause (i) a hard detection/prediction of failures as well as (ii) the impossibility for human users to understand or verify how a given model made a particular decision [9].

13.3   Open challenges

13.3.1   The need for less data and higher efficiency

As previously highlighted, one main drawback of *DL* is the need for large amounts of training data to make accurate predictions. However, in many practical scenarios, the available data may be insufficient as compared to the amount of parameters to train. When data is synthetically generated (e.g., by means of full-wave simulations), the time cost of generating many training samples could easily become unaffordable/unfeasible. Transfer learning [10] is nowadays considered an effective recipe to (at least partially) cope with this paramount issue. Differently, many "physics-driven" approaches [11,12] have been recently explored to introduce some physical knowledge about the underlying *EM* mechanisms within the network and therefore significantly reduce the amount of necessary examples. However, paramount efforts are still needed to make the *DL* training process more efficient and less time consuming, achieving high prediction performance with significantly less data.

13.3.2   Handling data outside the training distribution

As *ML* techniques, also *DL* ones are trained using a specific set of training samples. Therefore, they perform well as long as the new test samples to predict belong to the same distribution that was learnt during the off-line phase. An important challenge in *DL* is to make a trained model capable of making accurate predictions also when data exhibiting non-negligible variations from the training data is fed to it.

13.3.3   Improving flexibility and enabling multi-tasking

Generally, a *DL* algorithm need to be trained with data properly describing the specific task at hand, being ineffective to solve any other problems. In other words, existing algorithms are highly specialized and even solving a very similar problem often requires a re-training to guarantee accurate solutions. One big challenge currently unsolved is the development of "multi-task" *DL* methodologies capable of performing several problems with the same architecture and training. Pioneer attempts in this directions have demonstrated that it is possible to build an *ANN* capable of simultaneously solving a number of problems including image/speech recognition and translation [13].

13.3.4   Counteracting over-fitting

Another big challenge in *DL* (inherited from *ML*) is to avoid over-fitting, which refers to the inability of a trained model to predict well on previously unseen data since it has been too much customized to the training data. Often, such an issue is observed in complex models where the number of parameters to tune is relatively larger than the available training data.

13.4 Future trends

13.4.1 Few shot, one shot, and zero shot learning

A very attractive recent trend in *DL* is aimed at solving one of its major drawbacks, that is the need for extremely large training databases. Few shot (*FSL*), one shot (*OSL*), and zero shot (*ZSL*) learning refer to a class of new techniques that enable building effective and accurate predictors from very limited or even no training data. Among them, *ZSL* is clearly the most exciting concept that probably will attract more and more attention by the scientific community in the near future. In classification problems, it is inspired by the ability of humans in recognising unseen objects by exploiting the knowledge distilled from seen classes [14]. Guan *et al.* in [14] provide this clear example of *ZSL*: "*if a child has seen a horse before and learned from a textbook that a zebra looks very similar to a horse but has black and white stripes, s/he would then have no problem in recognising a zebra when seeing one.*"

13.4.2 Foundation models

Another emerging and exciting research trend in *DL* is currently represented by the so-called foundation models. A foundation model can be defined as "*any model that is trained on broad data (generally using self-supervision at scale) that can be adapted (e.g., fine-tuned) to a wide range of downstream tasks*" [15]. Such models, based on existing *DL* techniques but comprising billions of parameters, exhibit surprising capabilities such as generating novel text or even images/video from text, and are attracting a lot of researchers in many scientific fields.

13.4.3 Attention schemes and transformers

Attention mechanisms and transformer architectures are rapidly emerging in many research areas. They enable the generation of *DL* models that dynamically choose how much significance/weight should be given to each portion of the input data, revolutionizing the way machines cope with and understand text, speech, and images [16,17]. Pioneer applications of attention schemes can be found in the field of microwave inverse scattering to improve the resolution in the retrieved images by highlighting the unknown scatterers and inhibiting undesired artifacts within the background [18]. However, the application of such *DL* techniques to *EM* problems is currently at the beginning.

13.4.4 Deep symbolic reinforcement learning

Current deep reinforcement learning (*DRL*) models lack the ability to "reason" on an abstract level, and consequently it is hard to implement high-level cognitive functions including hypothesis-based reasoning, transfer learning, and analogical reasoning [19]. Current trends in this field are aimed at developing innovative *DRL* architectures that are capable of learning "symbolic rules" that are easily comprehensible to humans, differently from traditional techniques whose decision process is totally opaque and almost impossible to verify.

13.5   Conclusions

Nowadays, *DL* must be regarded as a highly efficient paradigm for solving many prediction problems in *EM*, provided that the exploited model has been trained with enough data and the test set closely resembles the training database. *DL* is not a "magic wand" and it should not be merely employed as a black-box. Moreover, recalling the *no-free-lunch* theorems [20,21], there is no *ML/DL* algorithm universally performing better than others in any type of prediction problem.

A deeper understanding of the human brain could provide us with some suggestions on how to develop better and better *DL* algorithms. However, the human mind should be just an inspiration and not necessarily something we must copy.

References

[1]   Campbell S.D., Jenkins R.P., O'Connor P.J., and Werner D. 'The explosion of Artificial Intelligence in antennas and propagation: how deep learning is advancing our state of the art'. *IEEE Antennas Propag. Mag.* 2021;**63**(3), pp. 16–27.

[2]   Massa A., Marcantonio D., Chen X., Li M., and Salucci M. 'DNNs as applied to electromagnetics, antennas, and propagation – a review'. *IEEE Antennas Wireless Propag. Lett.* 2019;**18**(11), pp. 2225–2229.

[3]   Salucci M., Arrebola M., Shan T., and Li M. 'Artificial Intelligence: new frontiers in real-time inverse scattering and electromagnetic imaging'. *IEEE Trans. Antennas Propag.*, 2022;**70**(8), pp. 6349–6364, doi: 10.1109/TAP.2022.317755.

[4]   Li M., Guo R., Zhang K., *et al.* 'Machine learning in electromagnetics with applications to biomedical imaging: a review'. *IEEE Antennas Propag. Mag.* 2021;**63**(3), pp. 39–51.

[5]   Goodfellow I., Bengio Y., and Courville A. *Deep Learning*. Cambridge, MA: MIT Press; 2016.

[6]   LeCun Y., Bengio Y., and Hinton G. 'Deep learning'. *Nature*. 2015;**521**(7553), pp. 436–444.

[7]   Schmidhuber J. 'Deep learning in neural networks: an overview'. *Neural Netw.* 2015;**61**, pp. 85–117.

[8]   Dong S., Wang P., and Abbas K. 'A survey on deep learning and its applications'. *Comput. Sci. Rev.* 2021;**40**, p. 100379.

[9]   Marcus G. 'Deep learning: a critical appraisal'. arXiv:1801.00631.

[10]   Zhuang F., Qi Z., Duan K., *et al.* 'A comprehensive survey on transfer learning'. *Proc. IEEE.* 2021;**109**(1), pp. 43–76.

[11]   Guo R., Shan T., Song X., *et al.* 'Physics embedded deep neural network for solving volume integral equation: 2D case'. *IEEE Trans. Antennas Propag.* 2022;**70**(8), pp. 6135–6147, doi: 10.1109/TAP.2021. 3070152.

[12] Guo R., Lin Z., Shan T., *et al.* 'Physics embedded deep neural network for solving full-wave inverse scattering problems'. *IEEE Trans. Antennas Propag.* 2022;**70**(8), pp. 6148–6159, doi: 10.1109/TAP.2021.3102135.

[13] Kaiser L., Gomez A.N., Shazeer N., *et al.* 'One model to learn them all'. arXiv:1706.05137

[14] Guan J., Lu Z., Xiang T., Li A., Zhao A., and Wen J.-R. 'Zero and few shot learning with semantic feature synthesis and competitive learning'. *IEEE Trans. Pattern Anal. Mach. Intell.* 2021;**43**(7), pp. 2510–2523.

[15] Bommasani R., Hudson D.A., Adeli E., *et al.* 'On the opportunities and risks of foundation models'. arXiv:2108.07258.

[16] Lin T., Wang Y., Liu X., and Qiu X. 'A survey of transformers'. arXiv:2106.04554.

[17] Vaswani A., Shazeer N., Parmar N., *et al.* 'Attention is all you need'. arXiv:1706.03762

[18] Ye X., Bai Y., Song R., Xu R., and An J. 'An inhomogeneous background imaging method based on generative adversarial network'. *IEEE Trans. Microw. Theory Techn.* 2020;**68**(11), pp. 4684–4693.

[19] Garnelo M., Arulkumaran K., and Shanahan M. 'Towards deep symbolic reinforcement learning'. arXiv:1609.05518

[20] Wolpert D.H. and Macready W.G. 'No free lunch theorems for optimization'. *IEEE Trans. Evol. Comput.* 1997;**1**(1), pp. 67–82.

[21] Wolpert D.H. 'The supervised learning No-Free-Lunch theorems'. in Roy R., Koppen M., Ovaska S., Furuhashi T., Hoffmann F. (eds.). *Soft Computing and Industry*, London: Springer, 2022, pp. 25–42.

Index

Printed and bound by CPI Group (UK) Ltd, Croydon, CR0 4YY

21/01/2025

01823630-0007